Tropical Forest Ecosystems in Africa and South America: A Comparative Review

Tropical Forest Ecosystems in Africa and South America: A Comparative Review

EDITED BY
BETTY J. MEGGERS, EDWARD S. AYENSU,
AND W. DONALD DUCKWORTH

SMITHSONIAN INSTITUTION PRESS • CITY OF WASHINGTON • 1973

Published by the Smithsonian Institution Press.
Distributed in the United States and Canada by George Braziller, Inc.
Distributed throughout the rest of the world by Feffer & Simons, Inc.

Designed by Crimilda Pontes. Printed in the United States of America.

Library of Congress Cataloging in Publication Data

Main entry under title:
Tropical forest ecosystems in Africa and South America.

"A symposium planned by the Association for Tropical Biology to be held in Ghana, West Africa, during February of 1971."
Includes bibliographies.
1. Forest ecology—Tropics—Addresses, essays, lectures. 2. Forest ecology—Africa—Addresses, essays, lectures. 3. Forest ecology—South America—Addresses, essays, lectures. I. Meggers, Betty J., ed.
II. Ayensu, Edward S., ed. III. Duckworth, W. Donald, ed. IV. Association for Tropical Biology.
QH541.5.F6T76 574.5′264 72-8342 ISBN 0-87474-125-4

Preface

This volume is the outgrowth of a symposium planned by the Association for Tropical Biology to be held in Ghana, West Africa, during February of 1971. Cancellation of the meeting was forced by lack of funding, but the preliminary drafts of the papers, provided in advance to the Organizing Committee, were of such high caliber and great interest that we felt their publication would accomplish some of the goals of the symposium; namely to draw attention to the remarkable nature of the lowland tropical forest ecosystems of Africa and South America and to assess the current status of our knowledge—or more accurately, our ignorance—about them.

An exhaustive comparison of these two ecosystems, or even of the forms of life that are treated in the individual papers, is obviously beyond our capacity. Rather, the aim has been to highlight some of the unique features of these tropical forest biomes, to call attention to certain remarkable convergences that have developed, and to identify some of the adaptive problems facing plants and animals, including man. The authors were instructed to write for the non-specialist, although not for a popular audience. While some papers contain more technical detail than others, none is beyond the comprehension of a serious student of the tropics, whether his field is botany, zoology, geography, anthropology, or development planning. Indeed, it is our hope that individuals, regardless of specialty, will profit by the opportunity to see how their findings relate to those of their colleagues in other disciplines, will note any disagreements that appear to exist, and will be stimulated to attempt to resolve them as well as to fill the gaps that still loom large even in the better studied aspects of the two geographical areas. The "development" programs underway, particularly in Amazonia, add urgency to the task, not only because the ecosystems are in danger of destruction with incalculable loss of information to mankind, but because even now we know enough to avoid calamitous mistakes. We feel sure that we speak for all of the contributors in expressing the hope that this volume will have some impact on those with power to influence the character of future human intervention in the lowland tropical forest ecosystems of Africa and South America.

Planning of the Fourth International Symposium of the Association for Tropical Biology was authorized during the annual meeting held in Mayaguez, Puerto

Rico, in June 1969. The Organizing Committee, consisting of Betty J. Meggers, Chairman, Edward S. Ayensu, and W. Donald Duckworth, planned the program and selected the participants. Three of those scheduled to take part in the symposium were unable to furnish manuscripts and substitutes have been invited to replace them. The volume thus represents the programmed coverage, but lacks some of the cohesion that might have resulted had the contributors been able to consult with one another in advance of publication, as was originally planned. They all have been prompt in responding to the demands placed upon them and extremely cooperative in providing well organized and informative papers. It has been a great pleasure to work with them all, and we welcome this opportunity to offer publicly our thanks. Finally, we wish to extend to J. P. M. Brenan, who was President of the Association for Tropical Biology in 1971, our special appreciation for preparing the synthesis of the papers that forms the Introduction.

Betty J. Meggers
Edward S. Ayensu
W. Donald Duckworth
Editors

March 1, 1972

Contents

CONTENTS

Introduction

J. P. M. BRENAN

In terms of the area of their basins, the Amazon and the Congo are the largest and second largest rivers in the world. As such, they have long fascinated not only geographers and naturalists, but the general public. Only during the past century has the veil been sufficiently lifted to reveal their general features, and even today something of the romance of the unknown clings to them. Although a general geographical picture of these two great river drainages can now be drawn with some accuracy, a lack of information is the principal obstacle to more detailed ecological analysis. This situation is clearly evident in the various papers included in this volume.

There are obvious parallels and resemblances, both geographical and physiognomic, between Amazonia and the Congo Basin. Both their rivers lie within the tropics, in regions of plentiful rainfall, and both drain vast tracts clothed with evergreen rain forest. These apparent similarities, however, raise many questions: Do the animals and plants that prevail in these regions in such profusion and luxuriance show more than a superficial resemblance? Do their systematic relationships indicate a relatively recent common origin? What are the causes of such resemblances or differences? Can we make fruitful comparisons between these two areas in terms of their potential for human utilization? To answer this last question means studying not only the aboriginal peoples of Africa and South America, but also the effects on the rain forest ecosystem of agricultural and industrial development backed by the technical knowledge of the 20th century. These problems are certainly not solved in this volume, but the papers that follow make worthwhile contributions towards their understanding and indicate some of the most useful directions that future investigations should take. I will attempt to summarize what seem to me some of the more important points made in the various contributions ranging over different fields of natural science. In performing this function, it is difficult for a botanist not to let his personal interest run away with him. Although I cannot hope to have wholly avoided this, I have kept the danger well in mind! On the whole, I have been greatly struck by the consistency of the picture jointly formed by the botanical and zoological evidence.

J. P. M. BRENAN, Royal Botanic Gardens, Kew, Richmond, Surrey, England.

To begin with, how similar are the two regions ecologically? Although incidental comparisons and contrasts are made in several of the contributions, a special comparative study of the limnology of the Amazon and Congo river systems is made by Marlier, who shows that there are considerable resemblances but also marked differences, and that the latter outweigh the former. The lower reaches of the Congo are impeded by rapids not far from its mouth and the river is also much shallower than the Amazon. By contrast, tides extend up the Amazon for 850 kilometers, allowing ocean-going vessels to sail far upriver. Roberts, in discussing fish fauna, points out that the black, white, and clear water streams of Amazonia with their important ecological differences are duplicated (although in different proportions) in the Congo Basin. The striking general resemblances in appearance of the terrestrial vegetation in the two areas are discussed by Richards.

Considerable areas of tropical Africa and smaller portions of South America are covered by savanna rather than rain forest, although forest and savanna often coexist in a complex ecological mosaic. The biological relationship of these two very different formations has been much discussed, for it is a vital factor in assessing the evolutionary history of faunas and floras. Jago considers African grasshoppers in this context and draws some very interesting conclusions that would seem potentially more widely applicable. The tropical forest is "not a great generator of new species," although it favors oddity and specialization, and the forest species seem less adaptable and thus more prone to extinction through changing conditions than the savanna species. Conversely, the savanna is to be regarded as an area favorable to active variation and speciation. It is suggested that the equator, with its differing climatic conditions to the north and south, is of importance as a "generator of species" among grasshoppers. Whatever its validity for other animal groups, this does not seem to have been equally significant a factor with plants.

In their paper on the growth habits of tropical trees, Tomlinson and Gill show that existing information and deductions reached therefrom have been too much governed by observations on trees growing in the North Temperate Zone, which are unrepresentative of trees in general, and consequently offer little guidance for analyzing the much greater variety encountered in the tropics. Ideas of what is "normal" and what is "abnormal" have thus been inverted. Detailed growth analysis of tropical trees has barely begun, but offers immense scope for future research. Botanists in the tropics, please take note!

"Afro-American complexes" of plants and animals have naturally attracted attention. They are inevitably each split into two separate parts by the intervention of the Atlantic Ocean; why, in spite of this, do they exist at all? Predictably, answers (reliable ones, at least) are hard to find, but the fascination of the search persists. Two examples of such complexes from the plant kingdom are carefully discussed: the resin-producing members of the tribes Cynometreae-Amherstieae in the Leguminosae by Langenheim and the striking family Velloziaceae by Ayensu. In both cases, the amphiatlantic connection is beyond dispute.

Four leguminous genera are involved (*Hymenaea, Trachylobium, Copaifera,* and *Guibourtia*) and the transoceanic links between them are based on both morphological and chemical evidence, the latter from resin analyses. "Genetic interchange" is postulated between the ancestral African and American representatives, followed by evolution leading mainly to distinct genera. Long-distance ocean dispersal is concluded to be the most likely mechanism, in preference to convergent evolution or land connection between the continents.

The strange looking Velloziaceae have been hitherto taxonomically confused and Ayensu and L. B. Smith provide a new key to the genera. A wide range of morphological, biological, and anatomical evidence leads to the conclusion that the family originated in Africa, where *Xerophyta* is the most primitive genus. Evolutionary development in the family was dependent on a specialized geological substrate: bare rocks of granite, gneiss, or quartzite. In spite of an African origin, species are much more numerous in America than in Africa, another warning of the danger of equating the place of present-day maximum concentration of species with the place of origin of a genus or group!

In order to assess the relationship between the fauna and flora of Amazonia and Equatorial Africa, it is essential to compare them, to see how much there is in common, and to weigh the differences. Important comparative studies of the fishes of the Congo and Amazon basins are made by Roberts, of the amphibians and reptiles by Laurent, of the ants by Brown, and of the diptera parasitic on vertebrates by Zumpt. The implications of the last-named group in the spread of human and animal diseases need no emphasis here. A similar analysis is made by Thorne in his discussion of floristic relationships between tropical Africa and tropical America. At the specific level, the two areas have very little in common, either in flora or fauna, except for obvious transoceanic introductions, deliberate or accidental, and those occasional instances (for example, plants whose seeds or fruits are widely dispersed by sea currents) where the Atlantic Ocean is not the formidable barrier to dispersal that it obviously is to the vast majority of plants and animals.

Heiser's account of variation in the bottle gourd (*Lagenaria siceraria*) illustrates the history of one pantropical plant that achieved its wide range in prehistoric times, partly through the activities of man and partly because of the capacity of gourds to float for long periods in the sea without loss of viability of their cargo of seeds. One subspecies (*siceraria*) is common to Africa and America, the other (*asiatica*) to Asia and the Pacific. An African origin for the species is suggested, with subsequent introduction by man into Asia followed by mutation and selection giving rise to subspecies *asiatica*. Subspecies *siceraria* is believed to have spread from Africa to America by sea.

No doubt a number of species now established on both sides of the Atlantic are attributable to such activities as the slave trade. There are also the "tramps"—those plants and animals that somehow make their way, often by devious means, from one country to another in the wake of man's activities.

At the generic level, the two areas have much more in common. Brown, for example, mentions 96 genera of ants as being endemic to the forest regions of Africa or South America, while 29 are pantropical. Other groups may show much higher proportions of endemic genera, but nevertheless generally significantly large numbers of pantropical ones. The physiognomic resemblances between the vegetation of the two rivers is thus to be seen as convergence, and analogous ecological niches are generally filled by quite different species in the two areas.

Amadon makes a general comparison of the avifaunas of the Congo and Amazon forests, which is particularly useful as these two faunas are considered relatively well known down to the species level. The general picture is one of great differences between the two regions, some points in common notwithstanding. Nineteen families occur in the Congo but not in the Amazon, 30 in the Amazon but not in the Congo. The greater specific diversity in the Amazon is again emphasized (roughly twice as many species as in Africa). Only a handful of genera, but no species, are common to both areas.

Bourlière explores the comparative ecology in South America and Africa of rain forest mammals which, interestingly, are nearly equal in numbers of genera and species. The stratification of trees and shrubs in the rain forest is correlated with stratification of animal species. Different ecological niches are provided by different food materials (fruits, leaves, etc.) and the availability of food controls the size and distribution of populations of individual species. Bourlière's paper illustrates the real value of comparative ecological studies covering different continents. Observations and data collected from one continent may be important and relevant to another, notwithstanding the fact that the faunas may be distinct taxonomically and their ecological similarities due to parallelism and convergence.

What then is the history of the faunas and floras of these two areas? Before pursuing this point, it is important to note one significant fact particularly emphasized in the papers by Richards and Vanzolini. This is that Africa is not only biologically less diversified than America, but that its flora and fauna show a striking poverty in comparison not only with America but with rain forest areas in Asia. The reasons for this are not fully settled, but are probably (as briefly indicated in Moore's paper on palms) related to major climatic changes in continental Africa resulting in the large scale extermination of past faunas and floras.

Richards produces striking figures for the family of palms, normally so conspicuous and abundant a feature of tropical vegetation: in West Africa, there are 13 genera and 24 species; in Africa as a whole, 15 genera and 50 species (Moore gives 16 genera and 117 species). In the Americas, there are 92 genera and 1100 species, while Asia and Australasia possess 107 genera and 1150 species. There are more genera and species of palms on Madagascar than in the whole of continental Africa, but the palms are only one example among the many groups that attest to the relative poverty of African flora.

The role of palms in the tropical ecosystems of Africa and South America is explored in greater depth by Moore, who expands his view to encompass other parts of the world. The paucity of palms in continental Africa is contrasted with the relative richness of the Indian Ocean islands, though both of these areas are far outnumbered by South America and in particular the eastern tropics. Moore sees the key to understanding of the distribution of palms in continental drift, "the reality of which now can scarcely be denied." In spite of the wealth of palms, tropical Asia shows no particular concentration of primitive forms, unlike South America. Accordingly, an austral origin of palms in western Gondwanaland is suggested.

Vanzolini explains the multiplication of species in Amazonia in terms of drastic climatic fluctuations coupled with rapid rates of differentiation in multiple peripheral topographical refuge areas. He postulates a greater topographical uniformity in Africa and hence a reduction in the number of refuges. Amadon's comparison of the avifaunas, already mentioned, makes the point that in spite of a much greater total avifauna, the Amazon is only "modestly richer" in species than the Congo when single localities are compared. He does not consider speciation in the tropics to be inherently greater than in temperate regions, but believes that there has been less "turnover and extinction" in the tropics.

What is the significance of continental drift? Has it a role in explaining the resemblances as well as the differences between the modern floras and faunas of the two continents? The paper by Delevoryas does not answer this question directly, but by surveying the history of continental drift provides some clues. The first rift in Pangea, as the primitive landmass is called, occurred during the Triassic and separated a northern Laurasia from a southern Gondwana. Additional rifts occurred in the later Triassic and Jurassic. Floras were then "essentially similar" over vast distances; by contrast "at present, floras are infinitely more diverse over smaller areas than they were in the Jurassic." In other words, continental drift is by no means as significant a factor in explaining biological differences between continents as some have believed.

Botanists, for example A. C. Smith and Thorne (although Moore takes an opposite view for palms), and zoologists, for example Laurent, seem to agree with Delevoryas that although continental drift may have played a part in the original separation of Africa and America, it cannot be invoked as a relevant feature in comparing the present-day floras and faunas of the two continents. Smith considers that the separation of Africa and America dates from the Jurassic or very early Cretaceous, when there were probably no flowering plants in their flora, with the possible exception of the "annonalean complex."

Brown lists a large number of ant genera (30) shared by Africa and Asia or Australia. No genera are restricted to Africa and America, but 12 are found in America, India, and Australia. This disparity in geographical distribution is discussed more fully for plants in Smith's important paper, in which he develops the view that the origin of phanerogams is to be sought in the Asia-Australia region, whence they spread to Africa by one route and to America by an entirely different one. On that basis, the present dissimilarities between the two floras are easily explicable. There are, however, a handful of groups with striking amphiatlantic distributions (usually predominantly American with outliers in Africa), which are considered either to have arrived by human intervention or by occasional long-distance dispersal, perhaps as far back as the Tertiary. Laurent suggests that some amphibia may have crossed the Atlantic on "flotsam and jetsam," and Brown considers that ants may occasionally have traveled in the shelter of floating trees. Johnson and Bowden view the Atlantic as an impassable barrier to insects at present, at least in the zone between 25 degrees N and 25 degrees S latitude, but sound a timely warning about the possible role of aircraft in the accidental transmittal of noxious insects (or fungus spores) in both directions.

Last, but far from least, Amazonia and the Congo Basin remain to be examined as habitats for man. To do this, it is necessary to consider not only the relationship of aboriginal man to his environment, but also how these regions can come to terms with 20th century cultural development.

The paper by Meggers on Amazonia kills, one hopes once and for all, the idea that this area has immense potential natural wealth, awaiting only the technological resources of western civilization for fruitful exploitation. She explains why most of the land away from the main river (the *terra firme*) is an inhospitable area, in which a luxuriant vegetation disguises a nutritionally deficient ecosystem, and shows that the aboriginal cultures, too casually dismissed or, worse, destroyed, were in delicate balance with their environment in subtle ways, which modern attempts at exploitation of the terra firme have failed, at their cost, to understand.

Miracle considers the Congo Basin as a habitat for man, describing its physical structure, its population fluctuation through time, and the manner in which agricultural methods have changed in recent centuries. In general, the picture is that of a rural area with a sparse population mainly dependent on field crops. The latter may be numerous and varied, reflecting a large influx of new cultigens over the past three centuries; one tribe is said to cultivate more than 50 different crops. Few

cattle are kept (perhaps the tsetse fly is one cause of this) and nonvegetal protein comes mainly from fish, game, and (in third place!) insects. Hunger is uncommon except in childhood (see Lowenstein, p. 297) and for short periods before harvest. Although the population of the Congo Basin today is probably larger than at any time in its history, there is an unexplained paradox: Though evidence suggests that a maximum has been reached in some areas, parts of the Nigerian rain forest support at least 25 times as many people per square kilometer as live in the Congo Basin.

The concept of a balance between primitive man and his environment is probed more thoroughly by Lowenstein in his analysis of human adaptability in tropical Africa and America. A direct comparison is difficult because the aboriginal African counterparts of the South American Indians have been largely displaced by the comparatively recent Bantu influx, leaving only a few remnants, such as the pygmies in the eastern Congo and the Bushmen, now surviving only beyond the limits of the forest. There is thus a marked difference between Africa and South America. The Amazonian Indian has become closely and well adapted to his environment, but is very vulnerable to change and to alien disease. In Africa, the Bantu has become quite well adapted to a changing environment and is comparatively resistant to disease after a vulnerable childhood period.

Except in a few exceptionally favored areas, the humid tropics in general and the rain forest areas in particular are characterized by impoverished soils, low population densities and, paradoxically (in the forest areas), a vegetation of remarkable profusion and luxuriance. But the latter is a façade, an outward magnificence cloaking the poverty beneath. Limited stores of nutrients rapidly recycled in a very efficient manner, a process apparently evolved over a very long period of time, form the flimsy basis of a delicately balanced ecosystem which, as Fosberg points out, seems well-nigh defenceless against the menace of temperate agricultural practices heedlessly imported into the tropics.

The tropical rain forest is disappearing at an alarming rate and it is only too easy to find vast tracts of grassland where the forest that once covered the land is gone—for good. It is, however, at least possible that some of the causes of this destruction, such as burning and uncontrolled overgrazing, antedate the introduction of temperate agricultural techniques in the African tropics. The picture is a gloomy one. Although Fosberg's forecast seems a jeremiad, and preachers and prophets have an uncomfortable record of being proven right by events, some counteracting gleams of light are to be found in the important paper by Sioli, to which I now wish to turn.

The poverty in nutrients of the soil of Amazonia away from the rivers is again emphasized by Sioli and he describes schemes designed to bring agricultural or forestry prosperity that have failed because those who planned them did not understand the limitations of the environment. The picture is, however, not altogether dark, for he shows how certain other approaches succeeded because they were carefully and thoughtfully designed. Justified tribute is paid to the policies of Felisberto C. de Camargo, formerly Director of the Instituto Agronômico do Norte in Brazil. What Sioli terms the "mentality of extractiveness" must be restrained and the basis for every intervention by man in the vulnerable ecosystems of Amazonia and tropical Africa must be an understanding of their ecology.

It is inevitable that the constant growth of the world's population and the increasing desire of poor people in the tropics to share in the prosperity and amenities of life will intensify the impetus towards change and development in Amazonia in the immediate future. Agriculture and forestry no doubt are the most promising activities, but they must be carried out with solicitude and comprehension of the unique characteristics of the natural environment. It is to be hoped that the papers assembled here will stimulate efforts to increase our knowledge of the plants and animals inhabiting the lowland tropical forest ecosystems of Africa and South America. With this knowledge, it may be possible to learn before it is too late how to utilize and live with these fragile communities without destroying them—perhaps also to appreciate that they possess an innate value not to be measured in terms of money or economic return.

Postdrifting Mesozoic Floral Evolution

T. DELEVORYAS

There seems to have been a considerable shift in opinion concerning continental drifting in recent years, especially among workers on the western side of the Atlantic Ocean. Americans had been reluctant, on the whole, to accept the concept of drifting continents and regarded some of their colleagues in the eastern hemisphere as suspect for insisting that drifting had occurred. At the present time, however, there seems to be little or no discussion centered on whether drifting did or did not occur; currently the principal subjects concern the timing of the drifting and the mechanisms that might have been responsible. There seems to have been a change in the thinking of biologists as well. Earlier, those who tended to believe in drifting of continents assembled all the possible biological data available to demonstrate that there had been a single land mass involving present-day southern hemisphere areas as well as the Indian subcontinent. Now that belief in drifting is more fashionable, the trend in biological thinking seems to be to try to interpret biotas, past and present, as having been influenced by separation of continental masses.

This paper is intended to serve as a kind of background for others in this series. It is an attempt to trace, in a general way, floral changes throughout the world in past ages, with emphasis on Africa and South America. It is in no way a complete survey of all the known kinds of fossil plants in these regions. The emphasis is, rather, on some of the more familiar and significant types of plants with a fairly well-known history.

T. Delevoryas, Department of Biology, Peabody Museum of Natural History, Yale University, New Haven, Connecticut 06520. This work has been supported by National Science Foundation Grant No. GB20999X.

Brief Survey of Continental Positions

Perhaps it would be useful, at this time, to trace some of the historical aspects of continental position. Dietz and Holden (1970) have recently summarized current ideas of the events involved. The concept of drifting continents calls for a landmass, called Pangea, that incorporated all of the continents as we know them today. The northern part of North America was contiguous with Europe; southern North America and South America were contiguous with the west coast of Africa, while Antarctica, India, and Australia were grouped together in the vicinity of what is now southeastern Africa. This is the configuration that was supposed to have held true through the Permian Period, or until about 225 million years ago. Separation within this continental mass is believed to have been initiated by rifts into which lava poured in what was to become ocean floor. The first separation occurred in the Triassic Period, with a rift dividing Pangea into the northern Laurasia, incorporating what is now North America, Europe, and Asia, and a southern Gondwana that included the present-day South America, Africa, India, Australia, and the Antarctic continent. Later in the Triassic there began a separation of some of the Gondwana land masses, with South America and Africa maintaining close connection and India separating in one direction and Australia and Antarctica, still contiguous, separating in another. During the Jurassic Period, a rift was initiated between Africa and South America that became more pronounced during the Cretaceous Period. Finally, later in the Cretaceous, India became attached to the Asian continent and Australia and Antarctica separated. Continued drifting during the Cenozoic Era eventually brought about the continental relationships that exist at the present time.

This history of continental juxtaposition and subsequent separation provides the principal explanation for the biotic affinity between the various continents. The Gondwana continents, furthermore, remained contiguous for a much longer period of time than the landmasses in what is now the northern hemisphere. Although continental configurations certainly provide a physical basis for biological homogeneity, interactions between taxa are generally assumed to play a significant role in the production of diversity. The conspicuous uniformity of floras over great areas seems to indicate, however, that this process was unimportant until post-Jurassic times.

History of Floras

I would like to continue the discussion by considering floras represented by fossil plants in rocks that existed before any drifting is thought to have taken place. In fact, the components of these extinct floras provided a very important piece of evidence to biologists in the past, who felt that there must have been a single landmass, Pangea, with the later Gondwanaland as the southern component. Among some of the best known fossil floras are those of the Carboniferous Period, a time when extensive coal deposits were being formed in many parts of the world from the lush plant life that existed then. When one attempts to survey the floras of the Carboniferous Period, he is immediately struck with the pronounced homogeneity of the plant assemblages in all parts of the world. Although "floral zones" are recognizable, the *kinds* of plants involved are extremely similar in all parts of the world. While this discussion is centered on tropical America and Africa, it is of interest and importance that the same kinds of plants that existed in these southern hemisphere regions occurred in practically all other parts of the world where fossil plants of that age are preserved. Furthermore, the discussion cannot be confined solely to Africa and South America because the fossil record is incomplete in these areas and knowledge of the fossil floras in other regions at a given time makes it easier to reconstruct what the plants might have been at a given place where there is no fossil record.

The most common Coal Age plants such as the lepidodendrons, calamites, cordaitaleans, seed ferns, and an abundance of ferns (hence the name "Age of Ferns" often applied to the Carboniferous Period) have been found in deposits of similar ages in North and South America, in Antarctica, Europe, Asia, Africa, and Australia. For some reason, however, India does not fully conform to this distribution, although the apparent difference may be more one of interpretation than biology. India's Permo-Carboniferous floras on the whole differ from those of other places, although certain elements are held in common. In spite of the nonconformity of India's Permo-Carboniferous floras, one is struck with the relative uniformity of plant types throughout the world. So striking is it that we are hard pressed to suggest an explanation. Even if continents had been closer than they are now, or actually contiguous, it would still be difficult to understand why there is not more variation among plant types. Over landmasses of considerably less area at the present time there is infinitely more variation. The most probable explanation would be the fact that since Permo-Carboniferous time (in fact, since Jurassic time), evolution of plant types apparently proceeded at an accelerated rate and, as a result, there are simply more kinds of plants in existence. With a greater abundance of species there would obviously result a greater competition for niches on a very fine level. Niches filled in diverse ways would create more diverse kinds of niches, and so on, with the result that still more kinds of organisms would evolve. Even conceding that the numbers of kinds of plants during the Carboniferous Period were relatively small, we find it difficult to understand the mechanisms of distribution; they must have been extremely efficient, whatever they may have been.

Maheshwari (1968) raises an interesting point concerning the equating of "northern" plants with "southern" forms during the Carboniferous and Permian periods. One example he used is *Sphenophyllum* (Figure 1),[1] generally regarded to have been a universally distributed genus in the late Paleozoic. A "southern" form was recognized in India; in addition this "southern" form was reported from Queensland (Ball 1912, Walkom

[1] Figures have been selected to show some of the representative plant types referred to in the text. An attempt has been made to utilize material from Gondwana continents where possible, although there are some exceptions.

1922), Southern Rhodesia (Walton 1929, Lacey 1961, Lacey and Huard-Moine 1966), Nyasaland (Lacey 1961), and Argentina (Archangelsky 1958). Maheshwari maintains, however, that the "southern" form is not really *Sphenophyllum* but belongs, instead, to a different genus, *Trizygia* (Feistmantel 1880). This change seems to me to be a subjective evaluation of the situation which is certainly permissible, but not necessarily imperative to acceptance. Furthermore, the basic idea in comparing these floras is to determine biological affinity, and there seems no doubt on the basis of what is known now that even if *Trizygia* is in reality a genus distinct from *Sphenophyllum*, it is very closely allied and represents evidence that the same type of plant group existed in India and elsewhere in the southern hemisphere as existed farther north. The same interpretation holds true for *Annularia* (Figure 2) (leaves assignable to plants of the Calamitales) and *Stellotheca* (Figure 4) for which Maheshwari (1968) presents the same kind of argument. Surange and Prakash (1960), as well as Maheshwari, suggest that because there is no evidence for the Calamitales in the southern hemisphere, these two genera must remain distinct. I, on the other hand, would argue that because there are leaves such as *Stellotheca* in the southern hemisphere, there is enough evidence to demonstrate that there were calamitaleans there. But even if the plants in question turn out not to be calamitaleans as we know them from the northern hemisphere floras, they represent the same group of plants (sphenophytes) with obvious natural affinity.

The transition from the Carboniferous to the Permian was a gradual one, with a large number of plant groups persisting relatively unchanged in the Permian. There were a number of changes, however, and in the case of persistent forms, the relative abundance was often different. Lycophytes, for example, while still important in the Permian, were less abundant than they were in the Carboniferous.

In addition to the lycophytes and sphenophytes of the types known from the Carboniferous were many ferns, seed ferns, and certain conifers. Of the last, especially important were a group in the order Voltziales, abundant in the late Paleozoic and early Mesozoic. These trees had a superficial resemblance to some of the extant araucarias and played an important part in the subsequent evolution of conifers.

Perhaps more important for our consideration at the moment is the occurrence of a few plant types that were not quite universal in dispersal but were confined, rather, to parts of the earth now in the southern hemisphere and in India. The first plant group that immediately comes to mind is the assemblage referred to as the glossopterids. *Glossopteris* (Figure 3) is a genus defined by Brongniart (1828) to include more or less strap-shaped or spatulate leaves possessing a midvein constituted largely by the converging and descending lateral veins. These laterals are characteristically anastomosed. A similar genus, *Gangamopteris* McCoy (1875) has leaves with a similar shape, but the midvein is absent. The usual distribution of *Glossopteris* is in rocks of Permian age in South America, Africa, India, Australia, and the Antarctic continent. In fact, the distribution of fossil glossopterids was for a long time used as an important piece of evidence to demonstrate that at one time these Gondwana continents were continuous. Certainly a distribution of these lands on the globe such as is found at the present time would make it extremely difficult to understand the distribution of *Glossopteris*. In the same deposits are found unusual axes, often in the form of compressions or casts, that show a peculiar type of segmentation; for this reason they have been called *Vertebraria*. It is generally assumed (e.g., Schopf 1965) that these axes were really parts of the plant that bore *Glossopteris* leaves, and at least one author (Schopf 1965) has suggested that they were probably roots on the basis of a study of axes with anatomical details preserved. In fact, the size of some of the axes would suggest that *Glossopteris* was a tree. It should be stressed at this point that *Glossopteris* alone did not tie together floras of the Gondwana continents. Rather, there were plant assemblages, often referred to as the *Glossopteris* flora, that shared many elements. In fact, it is becoming more apparent that *Glossopteris* does not represent one kind of plant but instead is a generalized leaf form associated with plants that are not necessarily biologically related. In recent years descriptions of fruiting structures belonging to plants with glossopterid foliage have been appearing in the paleobotanical literature, and there are a number of different kinds involved. Furthermore, I have re-

FIGURES 1–3. 1, *Sphenophyllum emarginatum,* fragment of plant (× 1). 2, *Annularia stellata,* several nodes of a twig (× 1). 3, *Glossopteris* sp. (× 1).

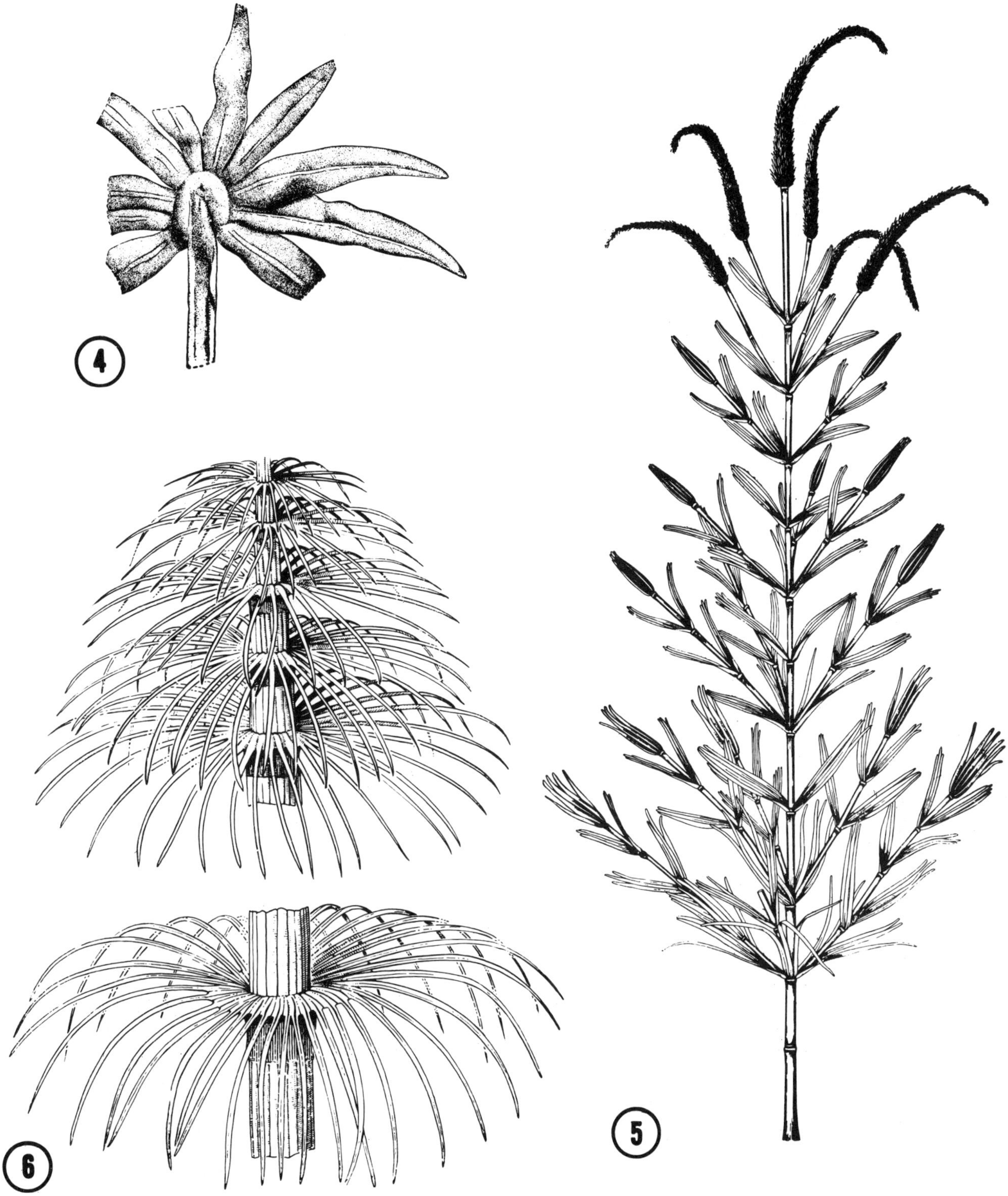

FIGURES 4–6. 4, *Stellotheca robusta,* node (× 2) (From Surange and Prakash 1960). 5, *Schizoneura paradoxa,* reconstruction (× 1/6) (From Mägdefrau 1968). 6, *Phyllotheca equisetitoides,* reconstruction of portion of plants (× 1/2) (After Rasskazova, in Boureau, 1964.)

cently described leaves indistinguishable from those of *Glossopteris* from rocks of Middle Jurassic age in Mexico (Delevoryas 1969). Thus *Glossopteris,* itself, is not the unifying member among southern hemisphere Permian floras.

Ferns such as *Pecopteris, Asterotheca,* and *Sphenopteris,* familiar Carboniferous genera, were also conspicuous members of the Gondwana Permian. Another plant with fernlike leaves (although not necessarily a fern), *Gondwanidium,* like *Glossopteris,* is found in the Permian of Gondwana regions. An additional "southern" genus, *Schizoneura* (Figure 5), was a plant that supposedly had large paired leaves with many veins that eventually split to produce the effect of whorls of linear leaves. It was thought to have reached the size of a small tree. *Phyllotheca* (Figure 6), also a sphenophyte, is another characteristic component of *Glossopteris* floras.

There have been attempts to reconstruct the landscape during "Lower Gondwana" time (equivalent to Upper Carboniferous and Permian periods), as well as to infer climatic conditions. It is generally assumed that there had been glacial activity in Gondwana lands at the end of the Paleozoic, and various authors interpret the effects of glaciation in different ways. Seward (1941) pictured a rather severe landscape with glaciers and icebergs in the immediate background. Plumstead (1966) felt that the plant assemblages as represented by fossils were indicative of a more lush growth; but she, too, felt that glaciers, especially in valleys, were in near proximity. Recently Rigby (1969) postulated that there was, indeed, a lush vegetation in Lower Gondwana times, but that it occurred in an area with gentle relief some time after the glaciers had retreated. The climate was presumed to have been temperate, with seasonal fluctuation of temperature and/or moisture. Wood from gymnospermous plants shows well-defined growth rings; this phenomenon is recorded from the Lower Gondwana of both South America and Africa. Another piece of evidence that there were climatic rhythmic cycles are lycophyte stems with leaf scars preserved on the surface. These cycles are reflected in the spacing of the scars on various parts of a stem.

Rigby (1969) makes an interesting point in remarking that in a given locality representing a given time zone, there are a large number of relatively few kinds of plants. In other words, the diversity that we are accustomed to in modern floras or even in fossil floras of later ages, is not represented in Lower Gondwana fossil localities. This reinforces a remark made above that in the Paleozoic there were fewer kinds of plants spread over greater land areas than there are at present.

Although the climate probably was not especially mild (in the sense of tropical) during Lower Gondwana time, it most likely did not get too severe because some of the trees, especially gymnospermous plants, attained considerable size. Furthermore, many ferns were present, and most likely the herbaceous ferns constituted the ground cover at that time.

Middle Gondwana formations are generally equivalent to the Triassic Period, during which time considerable evolution occurred among plant groups. Again, there were certain types of plants that were comparable to those in the northern hemisphere, with additional, quite striking, strictly "southern" types. Among remnants of older kinds of vascular plant groups that were rather widespread is the genus *Neocalamites* (Figure 7), in many ways comparable to the Late Paleozoic arborescent *Calamites. Neocalamites* has been reported from the Triassic of both South America and Africa, and was a common form in many other places as well. Similarly, the related *Equisetites,* which resembles so closely the present-day *Equisetum,* occurred in the same regions. Another widespread, almost ubiquitous, genus, *Cladophlebis* (Figure 8), had representatives in both South America and Africa during the Triassic. The significance of *Cladophlebis* and its distribution is unclear, however, because it is obviously an "artificial" genus, with more than one natural family represented.

Among the characteristically "southern" types that occurred in both Africa and South America is the enigmatic *Dicroidium* (Figures 9, 10), a presumed seed plant with fernlike, coriaceous leaves. There is considerable variation in leaf form, ranging from pinnately divided (Figure 10) to practically entire (Figure 9). It occurs nowhere except in regions that were once part of the ancient Gondwanaland. *Thinnfeldia* is also a fernlike plant (probably pteridospermous) that had representatives on both sides of the Atlantic as well as in other Gondwana countries. *Xylopteris* (or *Stenopteris* according to some authors) (Figure 11) is another plant with fernlike leaves that were finely

FIGURES 7–10. 7, *Neocalamites carreri* (× 1). 8, *Cladophlebis australis,* fragment of frond (× 1). 9, *Dicroidium coriaceum,* parts of three fronds (× 1). 10, *Dicroidium feistmanteli,* frond fragment (× 1). (Jain and Delevoryas 1967.)

FIGURES 11–14. 11, *Xylopteris rigida,* frond (× 1). (From Jain and Delevoryas 1967.) 12, *Sagenopteris nilssoniana,* leaf (× ¾). (After Schenk, in Mägdefrau, 1968). 13, *Noeggerathiopsis* sp., leaf (× 1). (From Jain and Delevoryas 1967.) 14, *Ginkgo huttoni,* leaves (× 1).

divided with linear pinnules. A curious plant, *Rhexoxylon* (Figure 15), has been found in South America and South Africa. It is recognizable from stem sections on the basis of multiple conducting regions separated by living tissue; it is not too unlike certain lianas in general internal structure. It seems to have occurred nowhere else and serves to link the floras of Africa and South America more closely during the Triassic Period.

Other plants, not exclusively "southern" but with representatives in both South America and Africa during the Triassic, had affinities with the maidenhair tree, *Ginkgo biloba*. One of these is *Sphenobaiera*, with a leaf having segments finer than those in the modern *Ginkgo* leaf. Other ginkgophytes are *Ginkgoites* and *Baiera*. Another seed plant, *Phoenicopsis*, with elongated, strap-shaped leaves, had Triassic representatives in both Africa and South America. Actually, plants with strap-shaped leaves and probable gymnospermous affinity were widespread throughout the Triassic, and quite likely more than one natural group is involved. Until fruiting structures are known, however, their biological significance will remain uncertain.

The well-known *Sagenopteris* (Figure 12), with palmately compound leaves, was widespread in the Triassic, with remains in both Africa and South America. Their leaf once led to their identification as primitive angiosperms but they are now generally considered related to the seed ferns. *Dictyophyllum* (Figure 16), a genus based on foliage with pinnatifid divisions with net venation, resembles the extant *Dipteris*. It occurs in both South America and South Africa (as well as in more northerly latitudes). *Chiropteris*, with palmately divided leaves and net venation, is thought to have been a fern and has similar distribution.

Certain Paleozoic plant groups persisted into the Triassic Period and retained their same general distribution. The noteworthy *Glossopteris* is one such example. Some leaves resembling those of *Glossopteris* from the Triassic, however, are from plants considered to have affinities with the cycadophytes, which are not biologically related to the Paleozoic *Glossopteris*. Another group that persisted into the Mesozoic is the group of coniferlike cordaitaleans. Leaf genera such as *Pelourdea* and *Noeggerathiopsis* (Figure 13) appear to be the morphological equivalent of cordaitean leaves. (The leaf

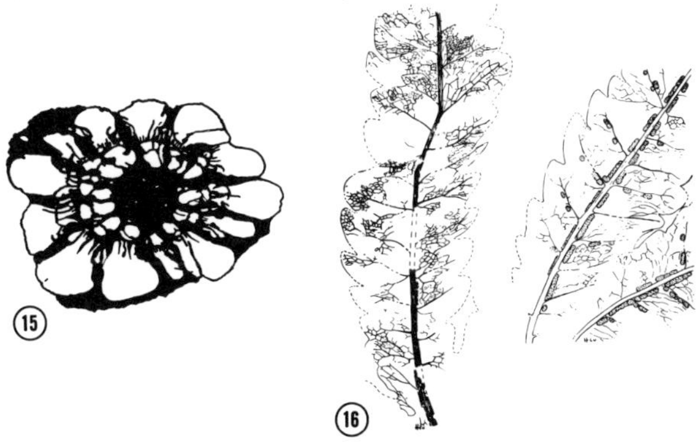

FIGURES 15–16. 15, *Rhexoxylon tetrapteridoides*, diagrammatic transverse section of stem (× 1) (From Walton 1923). 16, *Dictyophyllum ellenbergi* (× 1) (From Fabre and Greber 1960).

of *Noeggerathiopsis* figured bears some similarity to the leaf of a ginkgophyte, *Ginkgoidium* [Frenguelli 1946], that has a bifid leaf. The sinus in the leaf in Figure 13 appears to be a preservational feature, however.) Nothing is known yet of the reproductive structures of plants with leaves such as *Pelourdea* and *Noeggerathiopsis*, but epidermal features of these leaves are quite close to those in leaves of *Cordaites*. *Schizoneura*, a common plant in the Lower Gondwana, persisted into the Mesozoic relatively unchanged and, as mentioned above, the Triassic *Neocalamites* differs insignificantly from the Paleozoic *Calamites* and may be considered an extension of the latter into the Mesozoic. In addition, there was an abundance of fernlike plants, placed in a range of form genera, that seemed to have been universally present throughout the late Paleozoic and early part of the Mesozoic. Some of these ferns undoubtedly belonged to the same or related families. Others must simply have represented a convergence in leaf form. Some conifers, especially members of the important order Voltziales, had a distribution that ranged from the late Paleozoic to the early Mesozoic.

In summary, during the Triassic Period floral assemblages in Africa and South America had much in common, being composed of certain widespread genera as well as several characteristically "southern" forms. Before the end of the Triassic, the separation of Africa from Laurasia had occurred, yet there was still a pronounced similarity between floras of Africa and more northern regions. The separation of Africa and South America supposedly had just begun, and the relative uniformity between the floras of these continents would logically be assumed to be a continuation of the uniformity that had existed during the Paleozoic.

Jurassic plant distributions in South America and Africa are much harder to compare because the fossil record in Africa is less complete than in the Triassic, although South American Jurassic plants occur with satisfactory abundance (Menéndez 1969). There are many other Jurassic floras known, however, in widely separated regions of the world, and because of the impressive uniformity of the floral types in these areas, it can be assumed with almost complete assurance that the same general kinds of plants were in evidence in Africa and South America as well.

Equisetaleans were represented by *Neocalamites* and *Equisetum* (the latter is often referred to as *Equisetites* in the paleobotanical literature, but it is virtually indistinguishable from the modern *Equisetum*). Ferns are extremely abundant in the Jurassic, represented by many genera, some of which are referable to modern families (e.g., Osmundaceae, Marattiaceae, Gleicheniaceae, Cyatheaceae, Dipteridaceae, and others). *Thinnfeldia*, a possible seed fern, also persisted into the Jurassic. The group with the most important distribution is the Cycadophyta. In fact, the Jurassic Period most likely was the time of maximum development of the group. In the Cycadophyta are two principal orders, the Cycadales (true cycads) and the Cycadeoidales (or Bennettitales). Cycadophyte foliage is extremely abundant in Jurassic rocks, but without careful examination of epidermal features, it is impossible to distinguish leaves of the two orders. When cuticle is preserved on compressions, however, there are obvious means of differentiating cycads from cycadeoids. Cycadophytes occurred all over the world in the Jurassic, including Alaska, Europe, Africa, India, Asia, Australia, and South America. Many genera were spread over these regions and were extremely similar. Not only were the leaves much alike, but reproductive structures in the form of microsporophylls and ovulate cones were universally similar. Even on contiguous continents, it would be difficult to comprehend such uniformity over vast areas. The rift that developed between South America and Africa during the Jurassic Period served to provide an additional barrier between floras on both sides of the Atlantic, but the barrier is not reflected in floral differences.

Ginkgophytes were also worldwide, and some differed only little from the extant *Ginkgo biloba*. In fact, some workers refer to certain Jurassic forms as *Ginkgo* (Figure 14), insisting that there was no significant difference between the plants in the Jurassic and the genus that survives today. There do seem to be some climatic effects on the distribution of *Ginkgo*, however, since its presence is more noticeable in cooler, temperate regions than in more tropical regions. For example, I have found no ginkgophytes in Middle Jurassic deposits in Oaxaca, Mexico, but they are abundant in floras of the same age to the north and to the south.

Conifers had a similar distribution, with more frequent occurrence in Jurassic floras reflecting a temperate climatic situation. Some of the present-day families seemed to have originated during or just before the Jurassic Period (e.g., Taxodiaceae, Podocarpaceae).

It was not until the Cretaceous Period, especially after the invasion of large areas by angiosperms, that significant changes became noticeable in floras of continents that were once part of Gondwanaland. There is no need to enter, at this point, into a prolonged discussion concerning the time and place of origin of angiosperms. The important factor here is not the actual time of origin, but the subsequent evolution and disperal during the Cretaceous Period. If angiosperms had, indeed, been on earth for any significant period of time before the Cretaceous, their influence on the known floral components was negligible. With their infinitely more varied reproductive methods, perennating systems, and modifications to a wide range of niches, one would expect a more profound effect on the persistent cryptogamic and gymnospermous floras of the Triassic and Jurassic. Furthermore, fossil floras do not show a replacement in the early Cretaceous of Jurassic kinds of plants. Rather, there had evolved distinct types of early Cretaceous floras which, in turn, were superseded by invading angiosperms (Delevoryas 1971).

Other papers in this series are concerned with floral similarities and differences that occur at the present time between Africa and South America. The principal point I wish to make is that the differences between floral assemblages in various parts of the world are relatively recent occurrences. While it is possible to identify floral provinces in the geologic past, the differences in floral types were not so profound even as late as the Jurassic Period as they are now. Continental separation certainly is a major cause of differentiation, but probably not the most important one; it is notable, in fact, how long separation had existed before biological differences became significant. The effect of environment on the evolution of organisms is obviously more than simply an influence of the physical surroundings, since past ages witnessed just as diverse topographic and climatological situations as exist at the present time, but their impact was less than might have been anticipated. Interbiotic relationships apparently had increasingly profound effects on the evolution of floras in the Cretaceous and later. With more kinds of organisms involved in a variety of climatological, topographical, and biological combinations, conditions became ripe for the proliferation of evolutionary series culminating in the diversity manifested in the world today.

References

Archangelsky, S. 1958. Estudio geológico y paleontológico del Bajo de la Leona (Santa Cruz). Acta Geol. Lilloana 2:5-133.

Ball, L. C. 1912. Mount Mulligan coalfield. Queensland Geol. Surv. Publ. 237:1-39.

Boureau, E. 1964. Traité de Paléobotanique. Vol. 3: Sphenophyta, Noeggerathiophyta. Masson et Cie. Paris.

Brongniart, A. 1828. Histoire des végétaux fossiles ou recherches botaniques et géologiques sur les végétaux renfermés dans les diverses couches du globe. G. Dufour et Ed. D'Ocagne. Paris.

Delevoryas, T. 1969. Glossopterid leaves from the Middle Jurassic of Oaxaca, Mexico. Science 165:895-896.

———. 1971. Biotic provinces and the Jurassic-Cretaceous floral transition. Proc. North Amer. Paleont. Conv. Sept. 1969. Part L: 1660-1674.

Dietz, R. S., and J. C. Holden. 1970. Reconstruction of Pangea: Breakup and dispersion of continents. Jour. Geophys. Res. 75:4939-4956.

Fabre, J., and C. Greber. 1960. Présence d'un *Dictyophyllum* dans la flore Molteno du Basutoland (Afrique australe). Bull. Soc. Geol. France, ser. 7, 2:178-182.

Feistmantel, O. 1880. The fossil flora of the Gondwana system, II: The flora of the Damuda and Panchet divisions. Mem. Geol. Surv. India, Palaeont. Ind., 12th ser., 3:77-149.

Frenguelli, J. 1946. Contribuciones al conocimiento de la flora del Gondwana superior en la Argentina, XXXIII; Ginkgoales de los estratos de Potrerillos en la Precordillera de Mendoza. Not. Mus. La Plata 11. Pal. 87:101-127.

Jain, R. K., and T. Delevoryas. 1967. A Middle Triassic flora from the Cacheuta formation, Minas de Petroleo, Argentina. Palaeontology 10:564-589.

Lacey, W. S. 1961. Studies in the Karroo floras of Rhodesia and Nyasaland, Part 1: A geological account of the plant-bearing deposits. Proc. Trans. Rhodesia Sci. Assoc. 49:26-53.

Lacey, W. S., and D. Huard-Moine. 1966. Karroo floras of Rhodesia and Malawi, Part 2: The *Glossopteris* flora in the Wankie District of Southern Rhodesia. Symp. Floristics Stratigr. Gondwanaland. Birbal Sahni Inst. Palaeobot. Lucknow. Pp. 13-125.

Mägdefrau, K. 1968. Paläobiologie der Pflanzen. 4th edition. Gustav Fischer Verlag. Jena.

Maheshwari, H. K. 1968. Studies in the *Glossopteris* flora of India, 38: Remarks on *Trizygia speciosa* Royle with reference to the genus *Sphenophyllum* Koenig. Palaeobotanist 16:283-287.

McCoy, F. 1875. Prodromus of the palaeontology of Victoria, or figures and descriptions of Victorian organic remains. Victoria Geological Survey; Decade II:1-37.

Menéndez, C. A. 1969. Die fossilen Floren Südamerikas. Pp. 519-561, *in* Fittkau, E. J., et al. (Editors). Biogeography and ecology in South America. Dr. W. Junk Publ. The Hague.

Plumstead, E. P. 1966. The story of South Africa's coal. Optima (Dec. 1966): 186-202.

Rigby, J. F. 1969. The Lower Gondwana scene. Bol. Paran. Geociencias 27:3-13.

Schopf, J. M. 1965. Anatomy of the axis of Vertebraria. *In* Hadley, J. B. (Editor). Geology and paleontology of the Antarctic. Amer. Geophys. Union, Antarctic Res. Ser. 6: 217-228.

Seward, A. C. 1941. Plant life through the ages. Cambridge Univ. Press. Cambridge.

Surange, K. R., and G. Prakash. 1960. Studies in the *Glossopteris* flora of India, 12: *Stellotheca robusta* nov. comb.: A new equisetaceous plant from the Lower Gondwanas of India. Palaeobotanist 9:49-52.

Walkom, A. B. 1922. Palaeozoic floras of Queensland, Part 1: The flora of the lower and upper Bowen series. Queensland Geol. Surv. Publ. 270:1-45.

Walton, J. 1923. On *Rhexoxylon*, Bancroft—a Triassic genus of plants exhibiting a liane-type of vascular organisation. Phil. Trans. Roy. Soc. London 212B:79-109.

———. 1929. The fossil flora of the Karroo system in the Wankie District, Southern Rhodesia. Bull. Geol. Surv. S. Rhodesia 15:62-75.

Africa, the "Odd Man Out"

PAUL W. RICHARDS

It is one of the fundamental generalizations of biogeography that there are far-reaching similarities between ecosystems in similar environments in widely separated parts of the world. This holds for both the temperate and tropical zones and is well illustrated by the lowland vegetation of tropical America, tropical Africa, and the fragmented area extending from southeast Asia to tropical Australia and the western Pacific which I shall call Indo-Malaysia. In each of these three main tropical sectors there is a range of climax vegetation from permanently humid rain forest through various types of semideciduous and deciduous forest to savanna, semidesert and desert. In addition, in all three areas there are edaphically determined ecosystems, such as freshwater swamp forests and mangrove communities, which are found in similar physiographic situations in a wide range of tropical climates. The corresponding types of vegetation in each sector show close resemblance in the structure and physiognomy of their component species to their counterparts in the other two; for instance the lowland rain forest of South America is like that of Malaya and the Congo, not only in its broader features but in many details such as the prevalence of certain shapes and sizes of leaves, the buttressing of the trees, and in the occurrence of cauliflorous flowers in many lianes and small to medium-sized trees.

It is generally agreed that these similarities represent adaptations to similar climates and they become even more striking when comparisons are made between habitats in different continents in which the topography and soil as well as the climate are alike. A remarkable example of this is the forest and scrub vegetation of the bleached podsolic soils ("white sands," Lowland Tropical Podzols of Richards 1941, 1966) which are widespread in South America and parts of southeast Asia (though not in Africa). In western and southern Borneo, as well as in Thailand, Cambodia, Sumatra, and other neighboring countries, the climax vegetation of these "white sands" is a type of tropical rain forest known as Heath forest or "kerangas." This is strikingly different in general appearance as well as in its species composition from the Mixed Dipterocarp forest, which is the climax on the red and yellow latosols that are the prevailing soil types in the humid tropics of the Indo-Malaysian region as a whole. In South America, in the Guianas and Amazonia, especially in the basin of the Rio Negro, even larger areas of a very similar type of bleached sandy soil are associated with distinctive types of rain forest, quite different from those encountered on other soils in the same area: examples of these are the Wallaba [*Eperua*] forest of the Guianas, the forest associations variously known as pseudocaatinga, campina, carrasco alta, etc., in the neighborhood of Manaus and elsewhere in Amazonia. As pointed out long ago (Richards 1936), no species and very few genera of trees are common to the Heath forest of Borneo and the Guiana Wallaba forest, but there is a very far-reaching resemblance in physiognomy and structural features between the two associations; both differ from the climax forest on latosols in their own area and in exactly similar ways. In these instances we are dealing with what Emberger called "homoécies"—equivalent environments. The floristic make-up of the two communities is completely different but each is adapted to similar conditions in which not only the climate but almost all the environmental factors

PAUL W. RICHARDS, School of Plant Biology, University College of North Wales, Bangor, United Kingdom.

are similar. This is true not only of the evergreen forest zones, but of deciduous forests and savannas, though as yet detailed comparisons of the latter formations in America, Africa, and Indo-Malaysia have hardly been attempted.

The convergent resemblances between corresponding vegetation types in distant tropical areas are undoubtedly the result of selection. Owing to barriers and long-continued isolation, the floras themselves are very unlike; for example in Borneo and tropical America no plant species (excepting some weeds and perhaps a few seashore and aquatic plants) are the same, and only a proportion of the genera and families. Selection pressures, acting on very diverse genetical material, have produced plant communities with a large number of features in common.

But it is with the differences rather than the resemblances that I wish to deal here, in particular certain important differences between the tropical forests of Africa and those of the other two tropical sectors. It is these which have led me to conclude that whenever differences are found between the biota of Africa, America, and Indo-Malaysia, it is Africa that is almost always the "odd man out." Africa always seems to differ from the other two more than they do from each other. I would like briefly to review some of the facts which appear to bear out this contention—which is not an entirely new one—and then to discuss some possible explanations.

Three differences that seem to me significant are (1) the relative poverty of the African flora, (2) the wide area of distribution of African species, and (3) the poverty of Africa in certain plant groups.

The relative poverty of the African flora. Compared to most temperate countries the flora of tropical Africa is very rich in species and it may seem strange to describe it as relatively poor. Yet anyone with wide plant-collecting experience in the tropics of both hemispheres would agree that *tropical* Africa (excluding southern Africa) is much less rich that the other tropical regions. Reliable data to support this are hard to obtain and meaningful comparisons are difficult to make, but a few figures will give some indication of the difference. According to F. N. Hepper, the present editor of the second edition of Hutchinson and Dalziel's *Flora of West Tropical Africa* (1954-1972), it will include about 7000 species of flowering plants. This flora deals with the portion of tropical Africa from west (the former British) Cameroons westwards: there are probably considerably more species in the rest of tropical Africa and the exact number can only be guessed, but there can be little doubt that the total for the whole of tropical Africa is much smaller than for the American tropics or the Indo-Malaysian region. The area discussed in the *Flora Malesiana* alone is estimated to contain about 20,000 species of angiosperms (van Steenis 1938) and even the Malay Peninsula, an area smaller than Ghana and one with a much smaller range of climates, is said to have not less than 9000 species (Corner, in Richards 1966:229). The New World tropics are probably at least as rich as Indo-Malaysia. The Panama Canal Zone, a tiny area, has about 2000 species (Standley 1928) and I would guess that the flora of Costa Rica, which is smaller than most states of the United States and about one-fifth the size of Ghana (though admittedly much more diverse in topography) may equal or exceed that of the whole of West Africa in the number of plant species.

The relatively small number of species recorded from tropical Africa is certainly not due to insufficient knowledge. West Africa must rank as one of the botanically best explored areas of the tropics and the *Flora of West Tropical Africa* is generally agreed to be one of the most complete floras of a large tropical region that has yet been published.

The wide area of distribution of African species. In the forests of tropical Africa, as noted long ago by Chevalier (1917) and Mildbraed (1922), a very large proportion of the species have wide, though sometimes discontinuous, ranges. Species of endemic or regionally localized distribution are relatively few. This is in strong contrast to the situation in South and Central America and in Malaya, Borneo, New Guinea, and other parts of Indo-Malaysia where local species are very numerous. In Guyana (i.e., the former British Guiana) several of the commonest large rain forest tree species are scarcely found beyond the national boundaries. In parts of Amazonia, especially the Rio Negro region, there are many species which appear to be endemics of the area (Ducke and Black 1953, Pires 1966:11).

In contrast to this, most of the tree species found in the evergreen forest of Nigeria and the Cameroons, for example, extend far to the south and east or far to the west and many a long way in both directions. It is true that the "Dahomey Gap" is an important boundary for plants as well as animals and a good many species are known which do not cross it; but as knowledge of the African flora increases a considerable number of species previously thought to occur only west or east of Dahomey have proved to have a wider distribution. A large proportion of African trees, lianes, and smaller forest plants extend all the way from Guinea and Sierra Leone through Nigeria and the Cameroons to the Congo and some even further. It is only in a few small areas, e.g., Spanish Guinea and an area in West Africa centering on Liberia (which may have been refugia in some past period of arid climate), that considerable numbers of species with very local distribution seem to exist.

The wide distribution of many African species is probably connected with the relative poverty of the total flora. In Africa, as elsewhere in the tropics, small areas may be very species-rich. For example in the rain forest of the Cameroons sample plots 400 × 400 feet (c. 1.5 ha) may have as many as 109 species of trees reaching 1-foot (30 cm) girth and over (Richards 1963), a large number compared with many samples of rain forest (though smaller than in some forests of Malaya and Borneo, cf. Ashton 1964). Since most of the species are wide-ranging the species-richness of such a plot does not indicate that the whole region has an exceptionally rich flora.

Poverty of Africa in certain plant groups. Perhaps related in some way with the general poverty of the flora is the curious fact that Africa is remarkably poor in certain families of plants which are generally regarded as characteristically tropical. The most conspicuous examples are the palms and the orchids, but the same is true of other families such as the Lauraceae, Myrtaceae, and Myristicaceae. Oaks (*Quercus* and *Lithocarpus* spp.) are abundant in both Central America and Indo-Malaysia, where they occur in both lowland and montane forests, but are absent in tropical Africa.

In West Africa (as defined in the *Flora of West Tropical Africa*) there are only 13 genera and about 24 species of palms and these figures include the almost certainly non-native coconut and *Nypa fruticans*, a native of Indo-Malaysia which was introduced into Nigeria early in this century and is now established in the Cross River estuary and the Niger delta. Even in the Golfo Dulce, a small area on the Pacific coast of Costa Rica, there are more genera of palms than in the whole of West Africa and nearly as many species (Allen 1956). In the whole of Africa there are only 15 genera of palms and about 50 species, compared with 92 genera and 1140 species in the Americas and 107 genera and 1150 species in tropical Asia and Australasia (Corner 1966).

Though there can be no doubt that Africa has many fewer orchids than tropical America or Indo-Malaysia exact figures are hard to obtain. Many African epiphytic orchids have comparatively insignificant flowers not very attractive to the horticulturist and probably for this reason the family until recently has been somewhat neglected in Africa. In Summerhayes' revision of the Orchidaceae in the second edition of the *Flora of West Tropical Africa* (1968) the number of genera and species has risen to 58 and 403, respectively, compared with 56 and 313 in the first edition (1931). Yet, if every allowance is made for incomplete information the African orchid flora is obviously much poorer than that of other tropical regions, especially Central America and parts of Malaysia. In the Malay Peninsula there are perhaps 107 genera and 800 species of orchids and the Orchidaceae are the largest family of flowering plants (R.E. Holttum, pers. comm.). New Guinea is believed to be considerably richer. Most of the New World tropics are rich in orchids; even the Panama Canal Zone has almost as many genera (57, according to Standley 1928) as the whole of West Africa. The small island of Jamaica has nearly as many genera (57) and half the number of species (W.T. Stearn of the British Museum (Natural History)). Some American countries such as Costa Rica must be even richer, but exact figures are not available.

The Lauraceae are a family of trees which are predominantly tropical, and in the rain forests of both the New World and of Indo-Malaysia they are very numerous in both species and individuals. In the *Flora of West Tropical Africa*, (2nd edition, 1954) only 3 genera and 11 species are enumerated; these include neither forest giants, like the "angelims" and "silverballis" (*Nectandra* and *Ocotea* species) which are so common in South America, nor very abundant small to medium-sized trees like the "medangs" (species of *Litsea* and various other genera) of Malaya which, according to Foxworthy (1927:101) "make up an appreciable part of the composition of all our forests."

Africa is not only relatively poor in certain plant families characteristic of the tropics, but also in certain tropical life forms, notably lianes (see Mildbraed 1922:126) and epiphytes. The relative scarcity of epiphytes in African lowland forests immediately strikes anyone familiar with South and Central American forests and is in part due to the absence of Bromeliaceae, a family represented in Africa only by the single species *Pitcairnia feliciana* which (like many American Bromeliads) is not an epiphyte. It is known only from rocky outcrops in western Guinea. The Bromeliaceae are absent in the eastern tropics as well as in Africa, but there they are replaced to some extent by *Hoya, Dischidia, Myrmecodia,* and other epiphytic genera which are not found in Africa.

The three characteristics mentioned in which tropical Africa seems to differ on the one hand from tropical America and on the other from Indo-Malaysia are particularly marked when lowland areas with a humid climate and a not very severe dry season are compared, that is in areas where tropical rain forest is the potential climax vegetation. If suitable data were available, however, I believe very similar statements could be made about the deciduous forest and savanna regions. Certainly the cerrado of Brazil is much richer in species than the African savanna areas with which I am familiar although this may be less true of African savannas south of the equator. There is little doubt that the deciduous forests of South America and of southeast Asia are richer and more varied botanically than their nearest equivalents in Africa. It is even likely that a similar situation exists in the littoral flora of the three tropical sectors. According to G.W. Lawson of the University of Nairobi, the marine algae of West Africa are poorer in species and less diverse than those of other comparable tropical coasts. If further work establishes the importance of the three characteristics of Africa that I have mentioned and if, as I suspect, they are in some way connected, some theory to account for them is required which can be tested against acceptable evidence.

In seeking an explanation it is natural to think first that these differences must be due to accidents of the geological past—to the making and breaking of land connections and migration routes by continental drift or other means, to past changes of climate and the like. In my opinion, causes of this kind may have to be invoked. It is, or should be, however, a principle of biogeography that before appealing to relatively little known past events we should make quite sure that causes still operating at the present time cannot account for what we are trying to explain. In this instance are there any such causes which might explain, at least in part, the "odd-man-out" characteristics of Africa?

Two factors seem worth consideration, climate and the human impact. The climates (i.e., the range or distribution of different types of climate) of Africa and the other tropical sectors may be less similar than they appear to be. This problem is really one for the climatologists, but even a superficial study of the subject shows that almost none of tropical Africa, however high the annual rainfall, is without a distinct dry season. Debundja in the Cameroons, with an annual average of 10,170 millimeters has two consecutive months each with less than 100 millimeters. Most localities even in the "Cuvette Centrale" of the Congo, which is often cited as an example of an African "ever-wet" climate, have at least one month with less than 100 millimeters and rainless periods of up to 10 consecutive days occur frequently; at Eala there were 17 rainless periods of more than 20 days during 1930-1952 (Evrard 1968). Climates with the uniformly high rainfall, high humidity, uniform temperatures and lack of a regularly recurrent dry season such as exist over a large area in southern Malaya, Borneo and some of the neighboring islands, as well as in the upper Amazon and the Choco of Colombia, are found in Africa only in extremely limited areas, if they exist at all.

The human impact on the natural vegetation of Africa has for several reasons been more severe over a longer period in Africa than in either Amazonia or in large areas of the Indo-Malaysian rain forest. Tropical West Africa (from the Cameroons westwards) has had for a long time a relatively large population of agricultural people who have made their influence felt even in the most remote and inaccessible areas. This has profoundly affected the vegetation. Even in the depths of the so-called primary forest there is often evidence of former human occupation in the form of pottery and charcoal fragments in the soil.

Further evidence suggesting that most supposedly primary African forests are in fact secondary is provided by the frequent reports that many of the larger tree species, especially the "emergent" trees, regenerate poorly, or apparently not at all, by seed; Aubréville (1938) says that the natives of the Ivory Coast believe that certain tree species "ne font jamais des petits." Various conclusions have been drawn from this state of affairs, e.g., that the present forest is not in equilibrium with existing climatic conditions, or that the forest is a mosaic in space and time in which small groups of trees are continually being replaced by groups of different species (Aubréville's "Mosaic Theory," see Richards 1966:49-53). Though the latter explanation may have much truth, the basic reason for the lack of regeneration of many species seems to be that they are light-demanding species with wind- or animal-dispersed fruits biologically similar to those which in the forests of America and Indo-Malaysia are characteristic of middle seral stages; they occur in climax forest only in relatively small numbers in gaps due to windthrows, etc. This lack of regeneration of dominant tree species which is common in Africa is certainly *not* common in presumably primitive forests in the other tropical sectors, where most of the tree species common as large individuals are well represented in the seedling and sapling populations.

Nearer the equator, such as in the Congo Basin, the situation may be somewhat different. I do not know what evidence there is for early human occupation, but agricultural peoples, not to mention primitive nomadic food gatherers, such as the pygmies, seem to have penetrated everywhere and affected almost every ecosystem. In my opinion the "virgin tropical forest" of Africa is a myth. All we can find today are areas of old mature forest here and there which may have been undisturbed for some years but are certainly not unmodified relics of the original pre-human forest.

In tropical America and Indo-Malaysia the situation seems to have been different. I agree with Sioli (see p. 322 herein) that in Amazonia there may still be large areas of forest which up to now have been very little affected by man, and that only in relatively very recent times. An interesting picture of the impact of man on the Amazonian rain forest is given in the recent book by Meggers (1971) in which she makes it clear that until the last two hundred years human populations of sufficient size to have lasting effects on the vegetation were confined to the narrow strips of relatively fertile várzea forest along the main river and some of its larger tributaries. The human population of the far more extensive (and much less productive) terra firme forests was very sparse and its effects correspondingly small.

In the Malay Peninsula also there were very few people until the immigration of the present dominant races in quite recent times. Large areas of the rain forest were probably quite uninhabited or inhabited only by small bands of wandering Sakai food gatherers with no more influence on the vegetation than some of its other animal inhabitants. Most of the interior of Borneo and Sumatra was probably equally uninhabited until recently except for small settlements on the coast and along the larger rivers.

I think it is unlikely that either the lack of large areas with a truly ever-wet climate or the wide extent, high intensity and long duration of human disturbance could entirely account for the special characteristics of the African tropical forest flora. I believe, however, that both factors are important and deserve more serious consideration than they have yet been given. The climatic factor could well be contributory to the relative poverty of the flora and perhaps also to its deficiency in orchids, palms, and other characteristically tropical groups. It is noticeable that the forest flora is richest, and orchids (though perhaps not palms) most numerous, in the areas with the highest rainfall and least seasonal drought, such as Liberia and adjoining areas, eastern Nigeria and the western Cameroons, Spanish Guinea and the "Secteur Forestier Centrale" of the Congo. These are also the areas richest in endemics and species of regionally restricted distribution. In the last named area, according to Evrard (1968), 38 percent of the species present are believed to be endemic.

The long continued human modification of the plant cover might explain the relative poverty of the flora of some regions which have been relatively densely populated for a long time, such as southwestern Nigeria, but could hardly account by itself for the relative poverty of the tropical African flora as a whole. This it seems impossible to explain without some recourse to historical factors: the past history of Africa seems to have been different in some important way from that of America or Southeast Asia. The most plausible explanation may lie in past changes of climate. As a botanist with no specialized knowledge of climatology I get the impression, perhaps wrongly, that during and since the Tertiary, alternations between wet and dry periods have been greater in magnitude or more frequent (or both) in Africa than in Indo-Malaysia or South America. J. Muller's (1970) studies of pollen stratigraphy in northwestern Borneo suggest that at least in the central part of the Indo-Malaysian region plant evolution and the succession of vegetation types has continued since the Cretaceous undisturbed by any major climatic revolutions. In South America the evidence for past changes of climate is as yet rather scanty although there is much evidence to show that large parts of the Amazon forest must have been occupied at some time by drier types of vegetation perhaps similar to the present-day cerrado. There is, however, not much reason for thinking that in tropical South America there have been changes of climate comparable to the drastic and frequently occurring arid and pluvial periods which are so clearly recorded in Africa in lake sediments, by fossil termites' nests, and in the pollen and archeological record.

If it is indeed a fact that tropical Africa has suffered greater vicissitudes of climate than America or Asia, it would be interesting to know why. Even if this problem is at present beyond us, it is at least important to find out whether these climatic changes can account for some of the "odd-man-out" features of Africa. For this it is probably mainly to pollen studies and other paleontological techniques that we must look for an answer.

References

Allen, P. H. 1956. The rain forests of Golfo Dulce. Gainesville, Florida.

Ashton, P. S. 1964. Ecological studies in the Mixed Dipterocarp forests of Brunei State. Oxford For. Mem., No. 25.

Aubréville, A. 1938. La forêt coloniale: les forêts de l'Afrique occidentale française. Ann. Acad. Sci. Colon., Paris, 9:1-245.

Chevalier, A. 1917. La forêt et les bois du Gabon. Les végétaux utiles de l'Afrique tropicale française, fasc. 9, Paris.

Corner, E. J. H. 1966. The natural history of palms. London.

Ducke, A., and G. A. Black. 1953. Phytogeographical notes on the Brazilian Amazon. Anais da Academia Brasileira de Ciências 25 (1) :1-46.

Foxworthy, F. W. 1927. Commercial timber trees of the Malay Peninsula. Malay. For. Rec. 3.

Evrard, C. 1968. Recherches écologiques sur le peuplement forestier des sols hydromorphes de la Cuvette centrale congolaise. Publ. de l'Inst. Nat. Pour l'Étude Agron. du Congo (I.N.É.A.C.), ser. sci. no. 110.

Hutchinson, J., and J. M. Dalziel. 1954-1968. Flora of West Tropical Africa. Second edition. ed. R. W. J. Keay and F. N. Hepper. 3 vols. London.

Mildbraed, J. 1922. Wissenschaftliche Ergebnisse der Zweiten Deutschen Zentral-Afrika-Expedition 1910-1911 unter Führung Adolf Friedrichs, Herzogs zu Mecklenburg. Leipzig.

Meggers, B. J. 1971. Amazonia: Man and culture in a counterfeit paradise. Aldine-Atherton. Chicago and New York

Muller, J. 1970. Palynological evidence on early differentiation of angiosperms. In Evolutionary Events and the Geological Record of Plants (Biological Reviews Symposium). Cambridge, England.

Pires, J. M. 1966. Tipos de vegetação que ocorrem na Amazônia. Area de pesquisas ecológicas do Guamá, 1 ° relatório trimestrial. Inst. de Pesquisas e Experimentação Agropecuárias do Norte, Belém.

Richards, P. W. 1936. Ecological observations on the rain forest of Mount Dulit, Sarawak. Parts 1 and 2. J. Ecol. 24:1-37; 340-360.

————. 1941. Lowland tropical podsols and their vegetation. Nature (London) 148:129-131.

————. 1963. Ecological notes on West African vegetation. II: Lowland forest of the Southern Bakundu Forest Reserve. J. Ecol. 51:123-149.

————. 1966. The tropical rain forest. 4th impression. Cambridge, England. [Originally published 1952.]

Standley, P. C. 1928. Flora of the Panama Canal Zone. Contr. U.S. Nat. Herb. 27.

van Steenis, C. G. G. J. 1938. Recent progress and prospects in the study of the Malaysian flora. Chronica Botanica 4: 392-397.

Floristic Relationships Between Tropical Africa and Tropical America

ROBERT F. THORNE

Introduction

Selected seed-plant taxa are often cited as examples of the spectacular disjunction in distributional area across the Atlantic Ocean between tropical West Africa and tropical South America. Usually the disjunct taxa are presented as biological testimony for the relatively recent occurrence of continental displacement, the conveying apart of fragmented continental masses to their present positions, where they are more or less isolated from one another by the world's oceans. Recently the rather general geological acceptance of the sea-floor spreading concept with ascending convection currents has made the continental drift hypothesis considerably more palatable and even scientifically respectable. Many competent scholars representing various earth and life sciences have presented evidence favoring the hypothesis. It is, therefore, an opportune time to examine the floristic relationships between Africa and South America to determine just how strong they are and whether or not continental displacement offers a rational explanation for them.

Floristic Links

The floristic links between tropical Africa and tropical America are undeniable. Taxonomic groups limited primarily to these areas include every taxonomic rank from family to species. Twelve seed-plant families are restricted largely to Africa (with Madagascar) and tropical America, although several families have representatives in temperate America, and the African representatives of the Loasaceae and Velloziaceae reach Arabia and the Cactaceae (Figure 1) reach the Mascarenes and Ceylon. In fact, no family is confined entirely to tropical mainland Africa and tropical South America, for they all extend into the tropical areas of Central America, Mexico, the Antilles, or Madagascar, or into the temperate regions of North or South America or South Africa. Examination of the *Flora of West Tropical Africa* (Hutchinson and Dalziel, 1927-1936, 1954-1968) revealed 74 genera that are limited essentially to Africa (with Madagascar) and tropical America. At least 111 genera of seed plants (Table 1) are believed to be restricted largely to tropical and warmer temperate America and to Africa, including Madagascar, the Mascarenes, Seychelles, Macaronesia, and other African islands. Of approximately 350 species of seed plants listed as indigenous in the *Flora of West Tropical Africa* that are also reported from America, 108 are apparently known only from Africa and America, these mostly aquatics, maritime species, or herbs with weedy tendencies.

The recognition of the clear floristic relationships between Africa and America is no novelty. The evaluation of the relative importance of these floristic links has, however, seldom been attempted. It is imperative, therefore, to examine thoroughly at this time the strength of these links as compared to the number of taxa that show no African-South American relationships and to the number of taxa that show relationships in other geographic directions.

ROBERT F. THORNE, Rancho Santa Ana Botanic Garden, Claremont, California 91711.

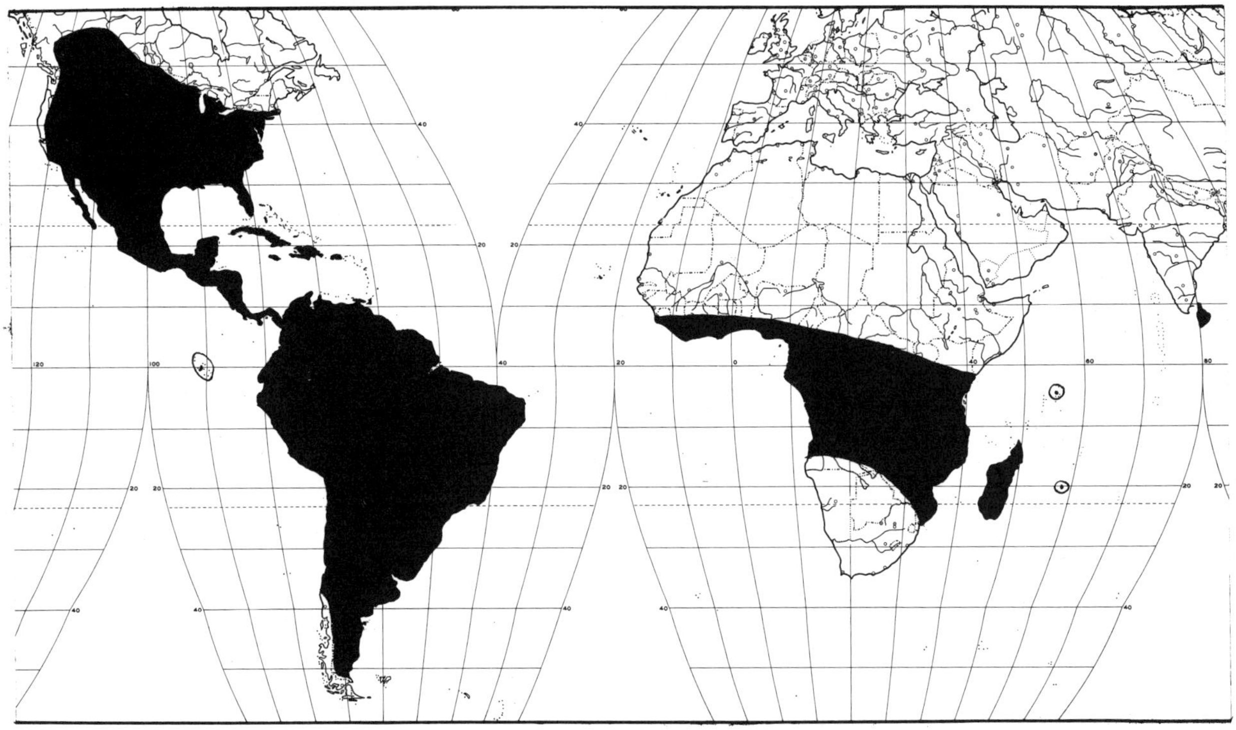

FIGURE 1. Distribution of the Cactaceae, with all the Old World portions of the range attributable to the single cactus *Rhipsalis baccifera*, also widespread in the New World. Map modified from Castellanos and LeLong (1943); base map from Goode Homolosine equal-area projection, prepared by H. M. Leppard. Goode Base Map No. 201HC copyright by the University of Chicago, used with permission from the Department of Geography.

TABLE 1. Genera Essentially Restricted to America and to Africa and/or Madagascar

Genera	*Family*	*Total Species**	*American Species*	*African Species*	*Madagascan Species*
Achyrocline	Asteraceae	20	16	3	1
Acioa	Chrysobalanaceae	33	3	30	0
Amanoa	Euphorbiaceae	10	7	2	1
Andira	Fabaceae	25–35	25–35	1	0
Annona	Annonaceae	117	108	10	1
Aptandra	Olacaceae	4	3	1	0
Ascolepis	Cyperaceae	15	1	15	1
Aspilia	Asteraceae	125	100	21	4
Barbacenia	Velloziaceae	75	57	18	0
Bertiera	Rubiaceae	28	2	16	10
Buforrestia	Commelinaceae	7	1	6	0
Calycobolus	Convolvulaceae	18	6	11	1
Caperonia	Euphorbiaceae	40	30	9	1
Carapa	Meliaceae	15	5	9	0
Carpodiptera	Tiliaceae	8	5	2	1
Cayaponia	Cucurbitaceae	45	43	2	1
Chlorophora	Moraceae	12	5	6	1
Christiana	Tiliaceae	2	1	1	1
Chrysobalanus	Chrysobalanaceae	7	4	3	0
Chrysocoma	Asteraceae	50	38	12	0

TABLE 1. Genera Essentially Restricted to America and to Africa and/or Madagascar—*Continued*

Genera	*Family*	*Total Species**	*American Species*	*African Species*	*Madagascan Species*
Cneorum	Cneoraceae	2	1	1	0
Conocarpus	Combretaceae	2	1	2	0
Copaifera (s.s.)	Fabaceae	30	25	5	0
Coreopsis	Asteraceae	115	76	39	0
Ctenium	Poaceae	20	10	10	1
Desmanthus	Fabaceae	22	15	1	6
Drepanocarpus	Fabaceae	12	11	2	0
Duvernoya	Acanthaceae	38	3	35	0
Ecclinusa	Sapotaceae	21	20	1	0
Echinolaena	Poaceae	7	6	0	1
Eichhornia	Pontederiaceae	7	7	1	1
Elaeis	Arecaceae	2	1	1	1
Eriochrysis	Poaceae	8	4	4	0
Euclasta	Poaceae	1	1	1	0
Gambeya	Sapotaceae	14	1	11	2
Genlisea	Lentibulariaceae	15	11	3	1
Guarea	Meliaceae	170	162	8	0
Guibourtia	Fabaceae	17	3	14	0
Haematoxylum	Fabaceae	3	2	1	0
Heberdenia	Mysinaceae	2	1	1	0
Heisteria	Olacaceae	43	40	3	0
Heteranthera	Pontederiaceae	10	7	3	0
Heteropterys	Malpighiaceae	100	99	1	0
Hirtella	Chrysobalanaceae	50	40	10	1
Hoffmanseggia	Fabaceae	26	20	6	0
Hyptis	Lamiaceae	400	400	2	0
Jaumea	Asteraceae	20	4	16	0
Laguncularia	Combretaceae	1	1	1	0
Legendrea	Convolvulaceae	1	1	1	0
Lindackeria	Flacourtiaceae	18	4	16	0
Lippia	Verbenaceae	220	205	15	0
Loudetia	Poaceae	41	1	38	2
Malouetia	Apocynaceae	25	23	2	0
Maprounea	Euphorbiaceae	4	2	2	0
Mayaca	Mayacaceae	9	8	1	0
Mendoncia	Acanthaceae	68	60	5	3
Menodora	Oleaceae	16	15	2	0
Mitracarpum	Rubiaceae	40	38	2	0
Mostuea	Loganiaceae	8	1	6	1
Neurotheca	Gentianaceae	10	1	9	0
Newtonia	Fabaceae	15	3	12	0
Ocotea	Lauraceae	300–400	280–380	3	18
Olyra	Poaceae	25	25	1	0
Orthoclada	Poaceae	2	1	1	0
Pacouria	Apocynaceae	20	2	17	1
Paepalanthus	Eriocaulaceae	485	484	2	1
Paratheria	Poaceae	2	2	1	1
Parkinsonia	Fabaceae	7	4	3	0
Paullinia	Sapindaceae	180	180	1	0
Pentaclethra	Fabaceae	4	1	3	0
Pentodon	Rubiaceae	2	1	1	1
Phenax	Urticaceae	28	25	0	3
Piriqueta	Turneraceae	29	25	1	3
Pitcairnia	Bromeliaceae	250	249	1	0
Pogonophora	Euphorbiaceae	3–4	2–3	1	0

TABLE 1. Genera Essentially Restricted to America and to Africa and/or Madagascar—*Continued*

Genera	*Family*	*Total Species**	*American Species*	*African Species*	*Madagascan Species*
Pteroglossaspis	Orchidaceae	5	4	1	0
Ptychomeria	Burmanniaceae	22	18	3	1
Ptchopetalum	Olacaceae	7	2	5	0
Raphia	Arecaceae	31	1	30	1
Renealmia	Zingiberaceae	75	53	22	0
Rheedia	Hypericaceae	45	32	0	13
Rogeria	Pedaliaceae	6	1	5	0
Sabicea	Rubiaceae	130	40	82	8
Sacoglottis	Humiriaceae	8	7	1	0
Sauvagesia	Ochnaceae	25	25	1	1
Savia	Euphorbiaceae	25	15	1	9
Schaueria	Acanthaceae	9	8	1	0
Schultesia	Gentianaceae	20	20	1	0
Schwenckia	Solanaceae	26	25	1	0
Sorghastrum	Poaceae	12	7	5	0
Sphaeralcea	Malvaceae	60	56	4	0
Stenocline	Asteraceae	13	9	2	2
Stigmaphyllon	Malpighiaceae	60–70	60–70	1	0
Swartzia	Fabaceae	77	75	2	0
Symmeria	Polygonaceae	1	1	1	0
Symphonia	Hypericaceae	17	1	1	16
Syngonanthus	Eriocaulaceae	196	195	1	1
Tapura	Dichapetalaceae	13	10	3	0
Tarchonanthus	Asteraceae	6	2	4	0
Tetrorchidium	Euphorbiaceae	15	10	5	0
Thalia	Marantaceae	11	10	1	0
Thamnosma	Rutaceae	9	5	4	0
Thyrsodium	Anacardiaceae	8	6	2	0
Trachypogon	Poaceae	10	6	4	1
Trichoneura	Poaceae	9	3	6	0
Trymatococcus	Moraceae	11	3	8	0
Vellozia	Velloziaceae	110	93	17	0
Vismia	Hypericaceae	52	45	7	0
Voyria	Gentianaceae	15	14	1	0
Willkommia	Poaceae	5	2	3	0
Wolffiella	Lemnaceae	8	7	2	0

* Species numbers determined as accurately as possible, but sometimes only approximately, from recent revisions, floras, Willis (1966), Engler (1964), and Index Kewensis (1893-196–).

FAMILIES OF SEED PLANTS

The number of seed-plant families believed to be indigenous to Africa and satellite islands is 234; that for South America with satellite islands, including the Antilles, is 228. Of the 276 different families represented on the two continents and adjacent islands, 186 are found on both continents. (The family concept used here is conservative, c.f., Thorne 1968). Forty-eight of the African and Madagascan families, 22 of them endemic, are not present in America; and 42 of the American families, 23 of them endemic, have no species in Africa or Madagascar. Thus, members of nearly one-third of the 276 families indigenous in Africa and South America have failed to migrate across the Atlantic Ocean or have failed to survive continental displacement. It is rather likely that representatives of 79 families, which have taxa from subfamily to species restricted to Africa and America, have successfully migrated across the Atlantic or have survived continental disruption. Possibly 23 other families with nearly cosmopolitan or pantropical species have also made the crossing successfully, some repeatedly. The remaining 82 wide-ranging families

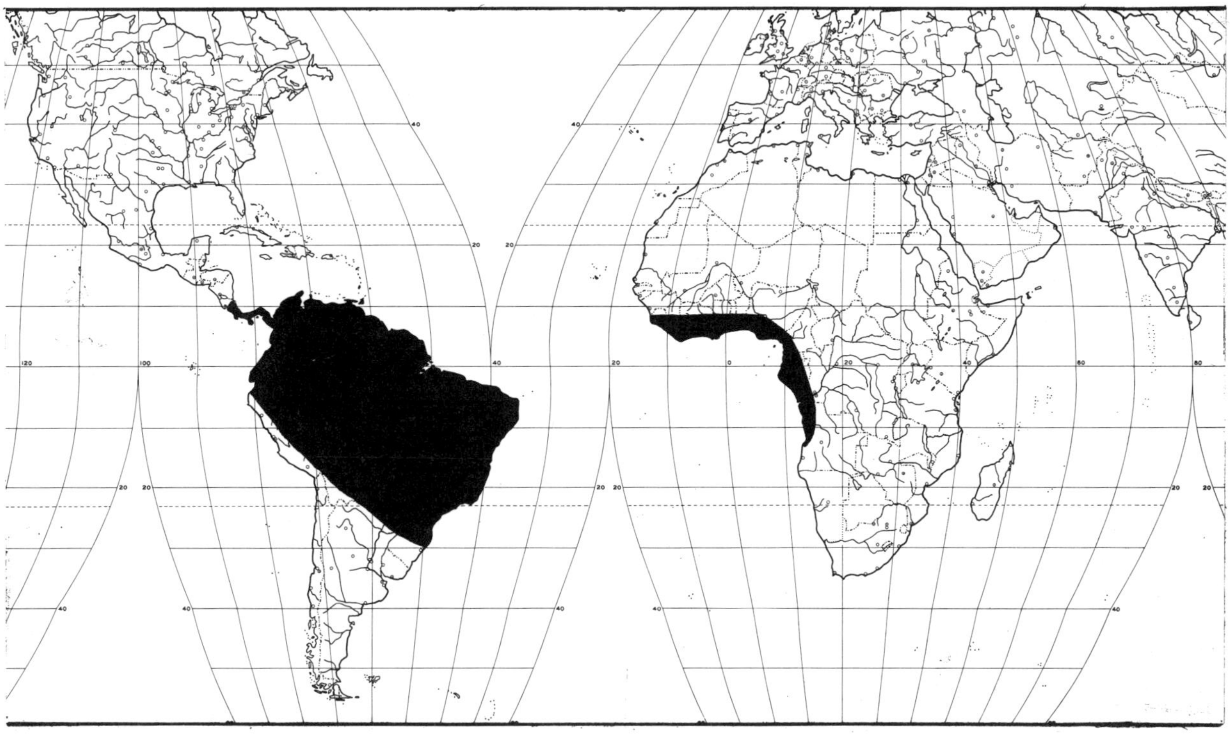

FIGURE 2. Distribution of the Humiriaceae, with just one species, *Sacoglottis gabonensis*, in tropical West Africa. [Map after Cuatrecasas (1961); base map as in Figure 1.]

found on both continents have probably entered the two continents from other continents, mostly from Eurasia or North America. Some of these families have extremely tenuous footholds on one or both continents. The Aceraceae, Betulaceae, Cistaceae, Coriariaceae, Fagaceae, and Pinaceae reach only the Mediterranean coast of Africa and do not penetrate beyond the Sahara Desert. The Clethraceae, Cneoraceae, and Empetraceae are found only on the Macaronesian islands, where they are represented by one species each. Also failing to reach the African mainland are the Chloranthaceae, Trigoniaceae, and Winteraceae, with one species each on Madagascar, and the Elaeocarpaceae, with 18 or more species of *Elaeocarpus* L. on Madagascar and the Mascarene Islands. Similarly in South America the Empetraceae and Restionaceae are represented by one species each in southern Chile and the Cneoraceae and Pinaceae only by Antillean species that are unreported from the South American mainland.

The 12 families restricted primarily to Africa and South America might be expected to provide the strongest evidence for dispersal across an asundered continental mass or across a narrow body of water before continental displacement had gone very far. Therefore, it should be most instructive to examine briefly these few families. Eight of the families, the Bromeliaceae, Cactaceae (Figure 1), Caricaceae, Humiriaceae (Figure 2). Loasaceae, Mayacaceae, Rapateaceae, and Vochysiaceae, with at least 3500 recognized species, are essentially American families with no more than 10 distinct species reported in Africa.

It is astonishing, if one accepts recent continental displacement as the explanation, that only one cactus should be present in tropical Africa considering the 1500 to perhaps 2000 species in 50-100 genera widely scattered throughout the American continents in habitats as varied as coastal to high, cold Andean deserts, cloud forests, lowland rain forests, and maritime thorn scrub. The single cactus is a highly specialized epiphyte, *Rhipsalis baccifera* (J. Miller) W. T. Stearn, whose enormous range from Mexico and Florida to Argentina in the New World and from West Africa to Madagascar, the

Mascarenes, the Seychelles, and Ceylon in the Old World testifies to the efficacy of bird dispersal of the seeds in the succulent, white berries with viscous pulp. A similar extraordinary distribution pattern is shown by the aquatic *Tristichia trifaria* (Bory and Willdenow) Sprengel of the Podostemaceae, which ranges from Mexico to Argentina, Africa, Madagascar, and India.

Likewise widely distributed throughout tropical America is the large monocot family Bromeliaceae with perhaps 1400 species in 60 genera. One species of the large genus *Pitcairnia* L'Heritier, with perhaps 250 species, *P. feliciana* (A. Chevalier) Harms and Mildbraed, is known from Guinea in tropical West Africa. Smaller families, but also with wide ranges in tropical America, are the Loasaceae, with about 200 species in 14 genera (Ernst and Thompson 1963), and the Vochysiaceae, with 200 species in 6 genera. *Fissenia* R. Brown ex Endlicher, the only loasaceous genus outside the New World, has one or two species widely disjunct in Africa between Southwest Africa and Somalia-Arabia. Similarly, the only Old World member of the Vochysiaceae is the genus *Erismadelphus* Mildbraed, with one species in tropical West Africa. Stafleu (1954) has pointed out that one species of the very closely related *Erisma* Rudge of tropical South America is a tree of the Amazonian várzea (floodplain rain forest inundated each year) with fruits well adapted to water transport. With one species each in Africa are the predominantly American families Humiriaceae (Figure 2), Rapateaceae, and Mayacaceae. *Sacoglottis gabonensis* (Baillon) Urban in Martius, of coastal tropical West Africa, belongs to the Humiriaceae, with 8 genera and 49 species especially of Amazonian rain forests (Cuatrecasas 1961). It is closely related to *S. amazonica* Martius. Cuatrecasas believes the family existed in America long before the Tertiary, when it became widely distributed, and that *S. gabonensis* probably originated from drift fruit brought to Africa from the Amazon region by ocean currents. *Sacoglottis* drift endocarps of Amazon and Orinoco origin have been found in the West Indies and even the British Isles. The Rapateaceae, with 16 genera and 80 species mostly of moist, sandy savannas of the Guayana Highlands, have one specialized genus and species, *Maschalocephalus dinklagei* Gilg & K. Schumann in tropical West Africa. Distant cousins of the rapateads, but more fully aquatic, are the Mayacaceae with the sole genus *Mayaca* Aublet, which has four to ten species ranging from the southeastern United States and Mexico to Paraguay and across to Benguela in southwestern Africa. Many aquatic species like *Mayaca baumii* Gürke and many semi-aquatics like the rapateads have close relatives on the other side of the Atlantic or are identical on both sides of the ocean. Presumably they are carried across the Atlantic by ducks (Iltis 1967), other water birds, or shore birds.

Having a slightly stronger foothold in Africa are the Caricaceae, a small family with perhaps 45 species in four genera. They are represented in tropical Africa by the endemic genus *Cylicomorpha* Urban with two species.

Somewhat more evenly divided between Africa and South America are the four remaining families, Canellaceae, Hydnoraceae, Turneraceae, and Velloziaceae. The small family Canellaceae, with about 18 species in 6 genera, has in tropical East Africa three species of the endemic genus *Warburgia* Engler and in Madagascar three species of the endemic *Cinnamosma* Baillon. The other genera and species are tropical American. The chlorophyll-less root-parasites of the Hydnoraceae, like their equally fascinating cousins in the Rafflesiaceae, display disjunctions that are hard to explain, for we are quite ignorant of their means of dispersal. The two genera of the Hydnoraceae are *Hydnora* Thunberg, with a dozen species scattered from Ethiopia to Madagascar and the Cape, and *Prosopanche* de Bary, with six species on the steppes and pampas of Paraguay and Argentina. The Turneraceae, with perhaps 110 species in 5 genera, have the larger number of species in America, 87 in the genera *Turnera* L. and *Piriqueta* Aublet, but greater variability, with 22 species in four genera, in tropical and South Africa, Madagascar, and the Mascarenes. The larger number of genera in the Africa-Madagascar regions may indicate a possible origin there for the family. The other families, except possibly the Hydnoraceae and Velloziaceae, would seem to have probable American origins. The Velloziaceae, the largest of the rather equally distributed families with perhaps 200 species in 3 genera, are discussed adequately elsewhere in this symposium by one of the specialists of the family (see Ayensu, pp. 105 ff. herein).

The patterns of distribution in these 12 families, which are so frequently called upon as evidence for continental drift, can really afford very little comfort to the proponents of recent continental splitting and displacement. Some of them have apparently had wide ranges in South America since early Tertiary time. One must conclude that their meager representation in Africa and Madagascar, perhaps 150 species in 20 genera, of possibly 4300 species in 230 genera, would seem to indicate occasional long-distance dispersal over a wide ocean rather than retention of ancient wide ranges split by continental separation.

SUBFAMILIES AND TRIBES OF SEED PLANTS

Below the family but above the genus are groups belonging to several other larger categories that are restricted to Africa and America. Among them are seven subfamilies: Cyphioideae (Campanulaceae) with 4 genera American, 1 African; Herrerioideae (Liliaceae), 2 South American, 1 Madagascan; Mendoncioideae (Acanthaceae), 1 American, 1 African, and 1 Madagascan; Microteoideae (Phytolaccaceae), 1 American, 1 South African; Napoleonoideae (Lecythidaceae), 1 Amazon region, 2 West African; Siparunoideae (Monimiaceae), 2 American, 1 West African; and Strelitzioideae (Musaceae), 1 tropical American, 1 South African, 1 Madagascan.

Among similarly restricted tribes and subtribes are: Arthropogoneae (Poaceae), 3 American genera, 1 African; Conanthereae (Liliaceae), 4 American, 2 African; Coussareae (Rubiaceae), 2 tropical American, 1 African; Cytineae (Rafflesiaceae), 1 Mexican, 1 Mediterranean, African, and Madagascan; Erismeae (Vochysiaceae), 1 South American, 1 African; Hemimerideae (Scrophulariaceae), 2 American, 2 South African and Madagascan; Lagenocarpeae (Cyperaceae), 6 South American, 3 African and Madagascan; Ravenaleae (Musaceae), 1 northern South American, 1 Madagascan; Swartzieae (Fabaceae), 7 American, 3 African; and Tigridieae (Iridaceae), 3 American, 3 South African.

Just above the generic level are a number of closely related generic pairs not discussed under larger categories. They are listed here, rather incompletely, with the American genus first: *Asclepias* L. (120 species), *Gomphocarpus* R. Brown (50-100), Asclepiadoideae, Apocynaceae; *Langsdorffia* Martius (10)-*Thonningia* Vahl (1-5), Balanophoraceae; *Macrolobium* Schreber (c. 75)-*Gilbertiodendron* J. Leonard (25), Fabaceae; *Myrocarpus* Allemão (4)-*Amphimas* Pierre ex Harms (4), Fabaceae; *Ochthocosmus* Bentham (6)-*Phyllocosmus* Klotzsch (8), Linaceae; *Ophiomenes* Miers (10)-*Oxygone* Schlechter (1), Burmanniaceae; *Piptadenia* Bentham (11)-*Piptadeniastrum* Brenan (1), Fabaceae; *Potalia* Aublet (1)-*Anthocleista* Afzelius (14), Loganiaceae; and *Trilepis* Nees (5)-*Afrotrilepis* (Gilly) Raynal (2), Cyperaceae.

GENERA OF SEED PLANTS

As mentioned above, 74 genera of phanerogams of tropical West Africa are confined primarily to Africa and tropical America and at least 111 genera (Table 1) from all of Africa and Madagascar are limited to Africa and America. These genera, however, are a very small fraction of those found in the tropical regions of the two continents. Analysis of the *Flora of West Tropical Africa* (Hutchinson and Dalziel 1927-1936, 1954-1968) supplied a list of 500 genera endemic to tropical Africa and a total of 684 restricted essentially to Africa and its adjacent islands. Probably no fewer than 1000 genera are endemic to tropical Africa and Madagascar and perhaps another 500 are endemic to South Africa (Good 1964). Good estimates that 3000 seed-plant genera are found only in tropical America, with 500 endemic in Brazil alone. Thus the genera common only to Africa and America are fewer than 2.5 percent of those 4500 limited to one of the two continents. Approximately 450 genera in the *Flora of West Tropical Africa* are pantropical or subcosmopolitan, and about 340 are shared between Africa and Eurasia (many reaching also to the Pacific Ocean area). Probably at least 6500 to 7000 genera form the combined phanerogamic floras of Africa and South America. It is likely that the genera found in both continents, confined to them or much wider ranging, are fewer than 10 percent of the total generic flora of Africa and South America.

Many of the larger genera restricted to Africa and America, as listed in Table 1, are represented by only one or a very few species on one continent. Especially noteworthy here are those American genera with one or a few species in Africa and Madagascar: *Andira* (with 25-35 species), *Ecclinusa* (21), *Heisteria* (43), *Hyptis* (400), *Olyra* (25), *Paepalanthus* (485), *Paullinia* (180), *Phenax* (28), *Pitcairnia* (250), *Sauvagesia* (25), *Schultesia* (20), *Schwenckia* (26), *Stigmaphyllon* (60-70), *Swartzia* (77), and *Syngonanthus* (196). Much fewer and

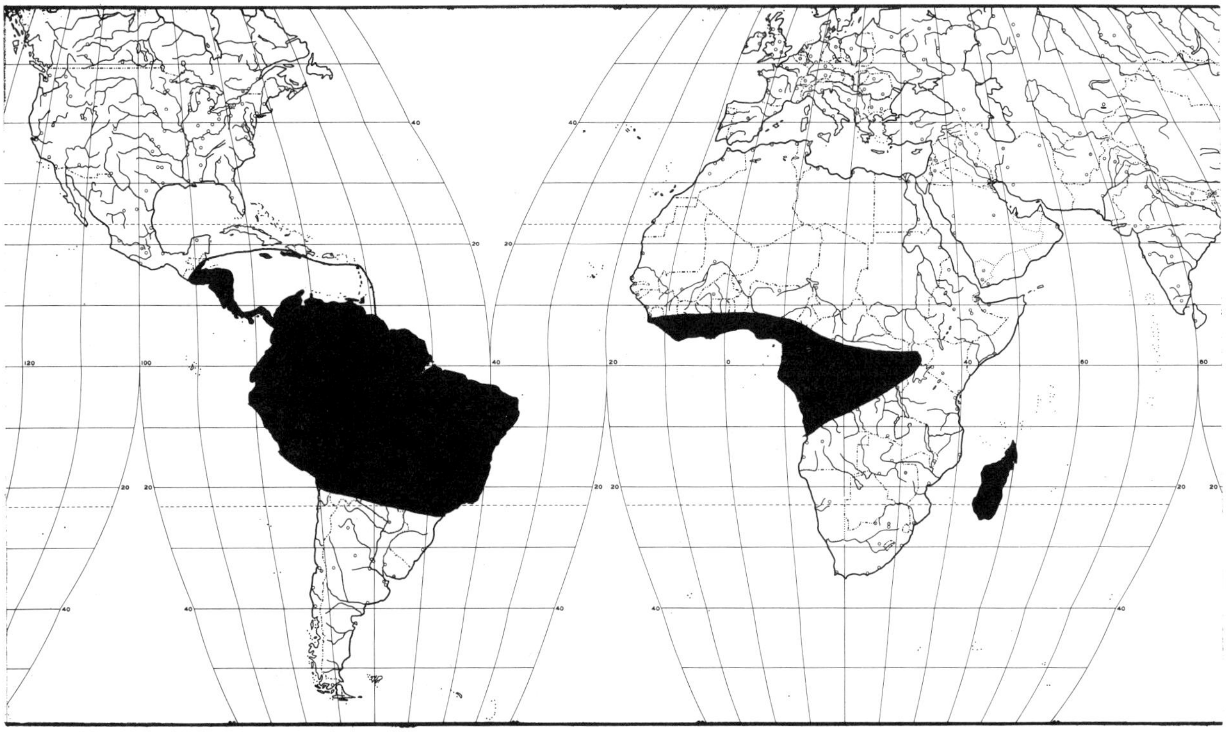

FIGURE 3. Distribution of *Symphonia* L. f., Hypericaceae, mainly of Madagascar but with one species, *S. globulifera*, wide-ranging over tropical Africa and tropical America. Like *Adansonia* in Figure 10 and perhaps the Velloziaceae, *Symphonia* is a good example of a primarily Madagascan group that has spread far from home. [Map modified from Goode (1964); base map as in Figure 1.]

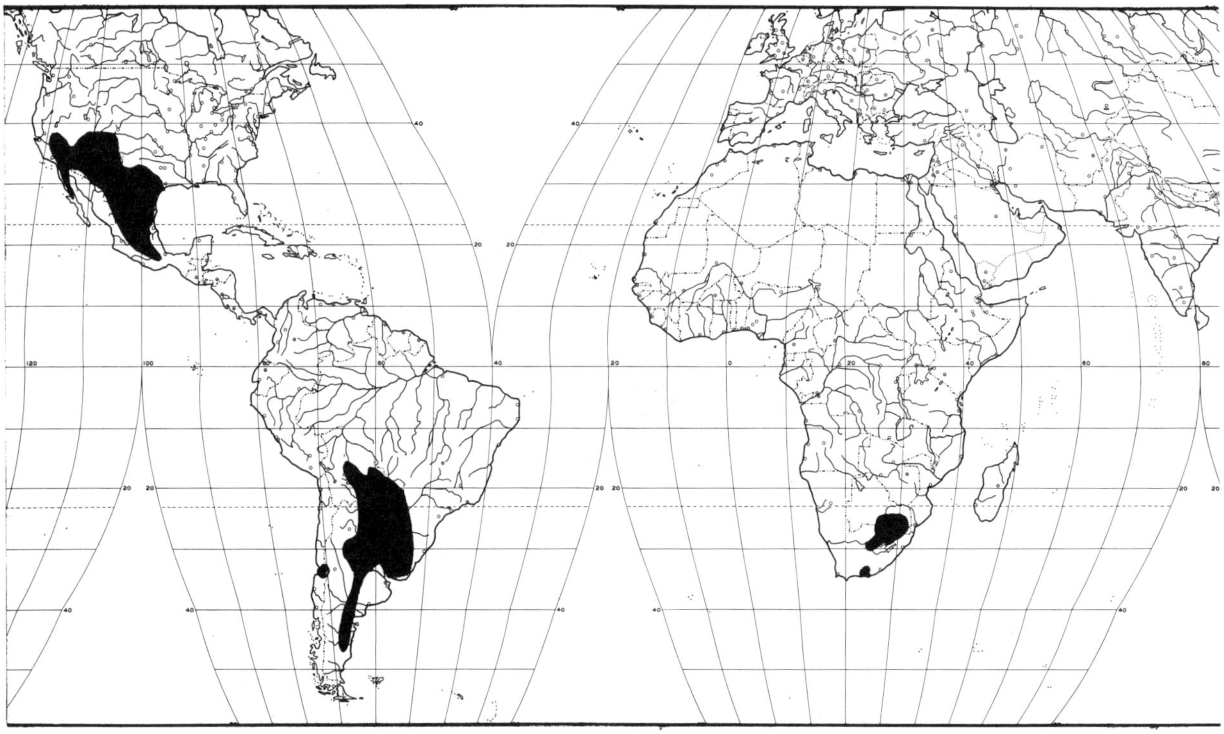

FIGURE 4. Distribution of *Menodora* Humboldt and Bompland, Oleaceae, widely disjunct between North and South America and between America and Africa. It exemplifies the American taxa that reached South Africa but not tropical Africa. [Map modified from Steyermark (1932); base map as in Figure 1.]

smaller are the primarily African and Madagascan genera with one or a very few species in America: *Acioa* (33), *Gambeya* (14), *Jaumea* (20), *Loudetia* (41), *Mostuea* (8), *Raphia* (31), and *Symphonia* (17). Relatively few of the larger restricted genera have more nearly equal representation on both continents. Among them are *Aspilia* (125), *Barbacenia* (75), *Caperonia* (40), *Chrysocoma* (50), *Coreopsis* (115), *Guarea* (170), *Ocotea* (300-400), *Renealmia* (75), *Rheedia* (45), *Sabicea* (130), and *Vellozia* (110). Species of these genera very likely reached the other continent much earlier and had time to radiate in their new environment. The solitary or few African species of *Andira, Eichhornia, Hyptis, Olyra, Paullinia, Sauvagesia,* and *Schultesia,* not to mention the monospecific genera, are, on the other hand, generally considered to be conspecific with American species. Presumably, they are recent immigrants to Africa. A few, in fact, are so weedy throughout their vast range that they may well be early accidental introductions by man.

Tables 2 and 3 list some of the larger endemic or nearly endemic genera of tropical America and of Africa and Madagascar. The total absence of these large, characteristic, and widespread genera from the other continent surely must be considered in interpreting the floristic relationships of Africa and America.

TABLE 2. Largest Genera Restricted Mostly to Tropical America*

Genus	*Family*	*Number of Species***
Pleurothallis	Orchidaceae	1000
Miconia	Melastomataceae	900–700
Epidendrum	Orchidaceae	800–400
Oncidium	Orchidaceae	750–350
Anthurium	Araceae	550
Myrcia	Myrtaceae	500
Tillandsia	Bromeliaceae	500–350
Stelis	Orchidaceae	500–270
Calceolaria	Scrophulariaceae	500–300
Baccharis	Asteraceae	400
Maxillaria	Orchidaceae	300
Masdevallia	Orchidaceae	300–275
Mammillaria	Cactaceae	300–200
Odontoglossum	Orchidaceae	300–200
Phoradendron	Viscaceae	300–190
Philodendron	Araceae	275–250
Dalea	Fabaceae	250
Guatteria	Annonaceae	250
Opuntia	Cactaceae	250
Cuphea	Lythraceae	250–200
Inga	Fabaceae	250–200
Tibouchina	Melastomataceae	250–200
Bactris	Arecaceae	250–180
Cestrum	Solanaceae	250–150
Geonoma	Arecaceae	240
Centropogon	Campanulaceae	230
Serjania	Sapindaceae	215
Siphocampylus	Campanulaceae	215
Leandra	Melastomataceae	200
Palicourea	Rubiaceae	200–150
Vriesia	Bromeliaceae	200–190
Columnea	Gesneriaceae	200–160
Aphelandra	Acanthaceae	200–150
Aegiphila	Verbenaceae	160
Clidemia	Melastomataceae	160–145
Manihot	Euphorbiaceae	150
Psidium	Myrtaceae	150
Adesmia	Fabaceae	150
Bomarea	Liliaceae	150
Besleria	Gesneriaceae	150
Calathea	Marantaceae	150
Coccoloba	Polygonaceae	150–125
Cecropia	Urticaceae	120
Puya	Bromeliaceae	120
Mandevilla	Apocynaceae	114

* Some genera, like *Baccharis, Opuntia,* and *Mammillaria,* also well represented in temperate areas.
** Estimates mostly from Willis (1966) and Engler (1964).

TABLE 3. Largest Genera Restricted Largely to Africa and/or Madagascar*

Genus	*Family*	*Number of Species***
Ruschia	Aizoaceae	350
Aloe	Liliaceae	330–250
Dombeya	Sterculiaceae	350–200
Conophytum	Aizoaceae	300–270
Aspalanthus	Fabaceae	245
Disa	Orchidaceae	200
Lampranthus	Aizoaceae	200–100
Agathosma	Rutaceae	180–170
Selago	Scrophulariaceae	180–150
Phylica	Rhamnaceae	150
Othonna	Asteraceae	150–100
Sutera	Scrophulariaceae	140–130
Protea	Proteaceae	130–100
Cola	Sterculiaceae	125–100
Delosperma	Aizoaceae	120
Cliffortia	Rosaceae	108–80
Aridaria	Aizoaceae	100
Cheiridopsis	Aizoaceae	100
Gravesia	Melastomataceae	100
Moraea	Iridaceae	100
Restio	Restionaceae	100–75
Stapelia	Asclepiadaceae	100–75
Gomphocarpus	Asclepiadaceae	100–50

* Some with representatives as far north as Arabia.
** Estimates mostly from Willis (1966) and Engler (1964).

SPECIES OF SEED PLANTS

The number of species of phanerogams indigenous in both Africa and America is expectedly small. It is especially difficult to determine for widespread species in the tropics which are native and which are introduced in a given area, but surely no more than a few hundred species are truly indigenous to both Africa and America. The number given above of about 350 tropical West African species that are also reported from America is almost certainly greatly inflated by species introduced on one or both continents, for many of the species listed are pantropical ruderals. Even the 108 species of those believed restricted to Africa and America include 20 or 25 that have weedy propensities or are often cultivated by man.

The small number of species common to Africa and America must be viewed in relation to the probable number of species forming the floras of the two continents. DeWolf (1964) has estimated 30,000 species for the 8,000,000 square miles of South and Central America and 40,000 for the 11,500,000 square miles of Africa. DeWolf's estimates are open to question for his figure for Africa seems to be slightly inflated; whereas, his figures for the Americas, southern Asia, and Australia are surely much too low. Good (1964) suggests a more moderate 25,000 to 30,000 species for Africa, and I expect an ultimate figure of approximately 35,000. For temperate South America, Good estimates 12,500 species in 1500 genera. The monumental *Flora Brasiliensis,* started by Martius in 1840 and completed in 1906, contains nearly 23,000 species in 2253 genera but includes cryptogams. Sixty-five years of further botanical exploration in that country must have added many more species, so that an ultimate total of 30,000 might be conservative. Considering the very rich floras of the Andean areas, the tropical Pacific lowlands, Central America, and the West Indies, all with a considerable degree of endemism, I would think a rational estimate for South and Central America and the Antilles would be 50,000 to 60,000 phanerogamic species, nearly double DeWolf's estimate. If we allow a conservative total of 80,000 species for the two continental masses and a generous 500 species native to both continents, the phanerogamic species common to Africa and America would be a mere 0.63 percent of the total seed-plant flora.

A brief consideration of the 108 species native in tropical West Africa and tropical America may be of interest. Approximately 45 species are aquatic or semi-aquatic plants of lakes, marshes, swamps, or wet, sandy open places. A few examples of these are *Bacopa egensis* (Poeppig & Endlicher) Pennell, *Eichhornia natans* (P. Beauvois) Solms-Laubach, *Eleocharis minima* Kunth, *Hydrocotyle bonariensis* Lamarck, *Paepalanthus lamarckii* Kunth, *Utricularia foliosa* L., *Wolffiella welwitschii* (Hegelmaier) Monod, and *Xyris anceps* Lamarck. Aquatic plants are, of course, generally wide-ranging and readily transported by water and shore birds.

Many of the species common to Africa and America are maritime plants of shallow salt or brackish water, salt marshes, salinas, mangrove swamps, sandy strands and dunes, or shore thickets. Since these sea-current-borne plants are often pantropical, relatively few (21) of them are restricted to just the American and West African shores. Of these, a few examples are *Annona glabra* L., *Avicennia germinans* (L.) Stearn, *Chrysobalanus icaco* L., *Conocarpus erectus* L., *Dalbergia ecastophyllum* (L.) Taubert, *Laguncularia racemosa* Gaertner f., *Cyperus ligularis* L., *Philoxerus vermicularis* (L.) R. Brown, *Rhizophora harrisonii* Leechman, *R. racemosa* C. F. W. Meyer, and *Sophora occidentalis* L. Although not necessarily coastal, those species that grow along streams in the interior are similar to the maritime plants in having water-borne seeds or fruits. Some of the riparian or várzea species are *Andira inermis* (Wright) de Candolle, *Christiana africana* de Candolle, *Entada gigas* (L.) Fawcett and Rendle, *Paullinia pinnata* L., *Symmeria paniculata* Bentham, and *Symphonia globulifera* L. f.

About 20 species from fields, roadsides, and other disturbed ground have long been associated with man and very likely owe their wide range to man's peripetetic nature. To be regarded with some suspicion as being relatively recent introductions and not truly indigenous are *Caperonia palustris* (L.) St.-Hilaire, *Cephalostigma perrottetii* A. de Candolle, *Melochia melissifolia* Bentham in Hooker, *Panicum trichoides* Swartz, *Sauvagesia erecta* L., and *Wissadula amplissima* (L.) R. E. Fries, among others. Similarly, often cultivated plants like *Ceiba pentandra* (L.) Gaertner may represent very early and intentional introductions by man.

A small residue of African-American species occurs in rain forest or other noncoastal, nonriparian, nonaquatic, and generally nonruderal habitats. The wide ranges of these plants are difficult to explain because of our ignorance of their methods of dispersal. Some have very small fruits and seeds, some fruits are attractive to birds, and some may have fruits or seeds resistant to immersion in seawater. Among these plants of deep- or open-forested habitats are *Byttneria catalpifolia* Jacquin, *Carapa procera* de Candolle, *Cardiospermum grandiflorum* Swartz, *Eulophia alta* (L.) Fawcett and Rendle, *Eulophidium maculatum* (Lindley) Pfitzer, *Olyra latifolia* L., *Panicum paniculatum* L., *Parinari excelsa* Sabine, and *Peperomia rotundifolia* (L.) Humboldt, Bonpland, and Kunth.

It is, thus, readily evident that the species indigenous in both Africa and America are mostly highly vagile plants whose disseminules can successfully, and probably repeatedly, be transferred by birds or water over large distances.

Another important conclusion from the above analysis of African-American taxa is that migration between the continents has been taking place during a very long time and has continued up to the present. The degree of differentiation between groups reflects the amount of independent evolution that has taken place in each group. Roughly, with many probable exceptions due to varying rates of evolution in different plants, the degree of evolution that has taken place reflects the time that has been available for that evolution. Because the taxa common to the two continents range in differentiation from identical varieties and subspecies to distinct tribes and subfamilies within the same family, we must seek an explanation that permits widely spaced events of immigration across the Atlantic Ocean that have continued throughout the history of the southern hemisphere floras. Continental displacement by itself is inadequate to account for this continuing immigration. Long-distance dispersal, if it is operative today as we are forced to believe it is, surely operated just as successfully in the past.

Intercontinental Relationships

AMPHI-PACIFIC RELATIONSHIPS

According to the usual hypothesis of continental

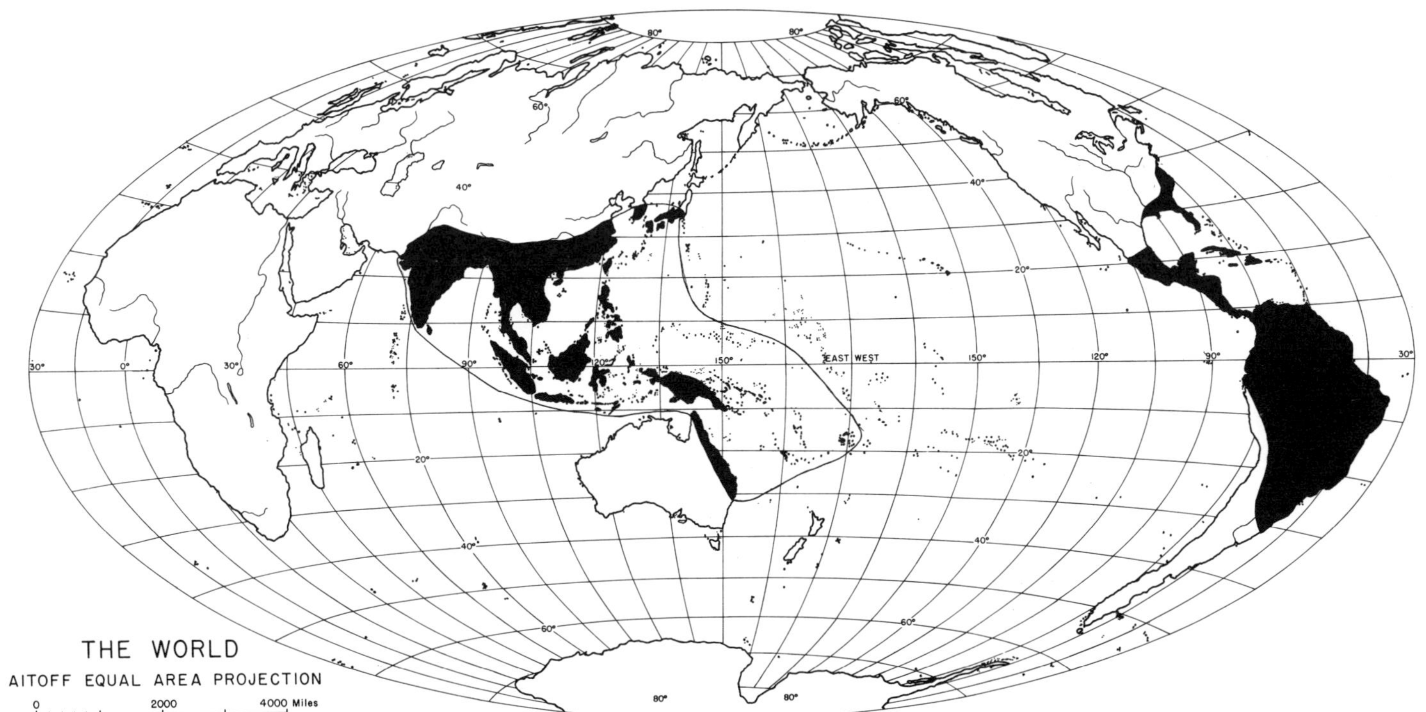

FIGURE 5. Distribution of *Symplocos* Jacq., sole genus of the Symplocaceae, which displays the characteristic amphi-Pacific disjunction that also excludes most of temperate America and Eurasia and all of Africa. [Map heavily modified from Vester (1940); base map is an Aitoff equal-area projection, prepared by Dr. John N. Belkin, Dept. of Zoology, U.C.L.A., and used with his permission.]

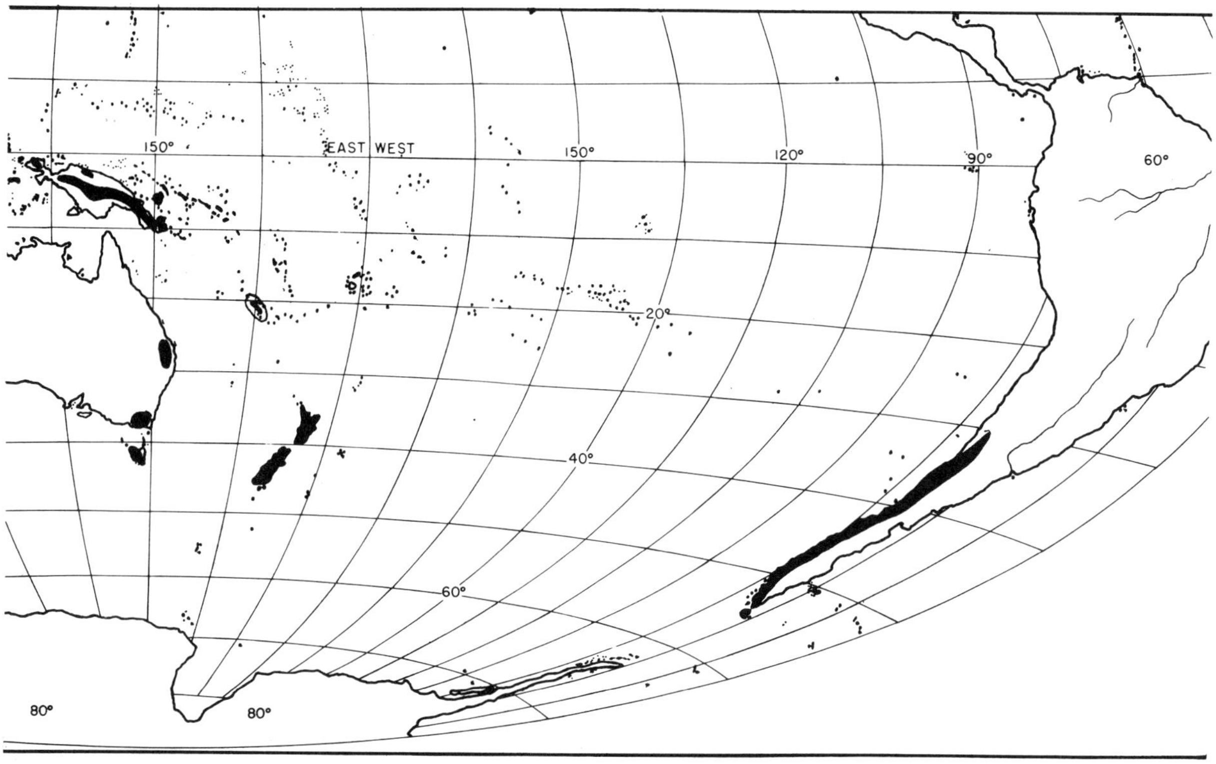

FIGURE 6. Distribution of *Nothofagus* Blume, Sub-Antarctic Beech, characteristic of many genera displaying a probable Antarctic migration route. Two of three subsections in South America have species also in Tasmania. [Map modified from van Steenis and van Balgooy (1966); base map from Aitoff equal-area projection.]

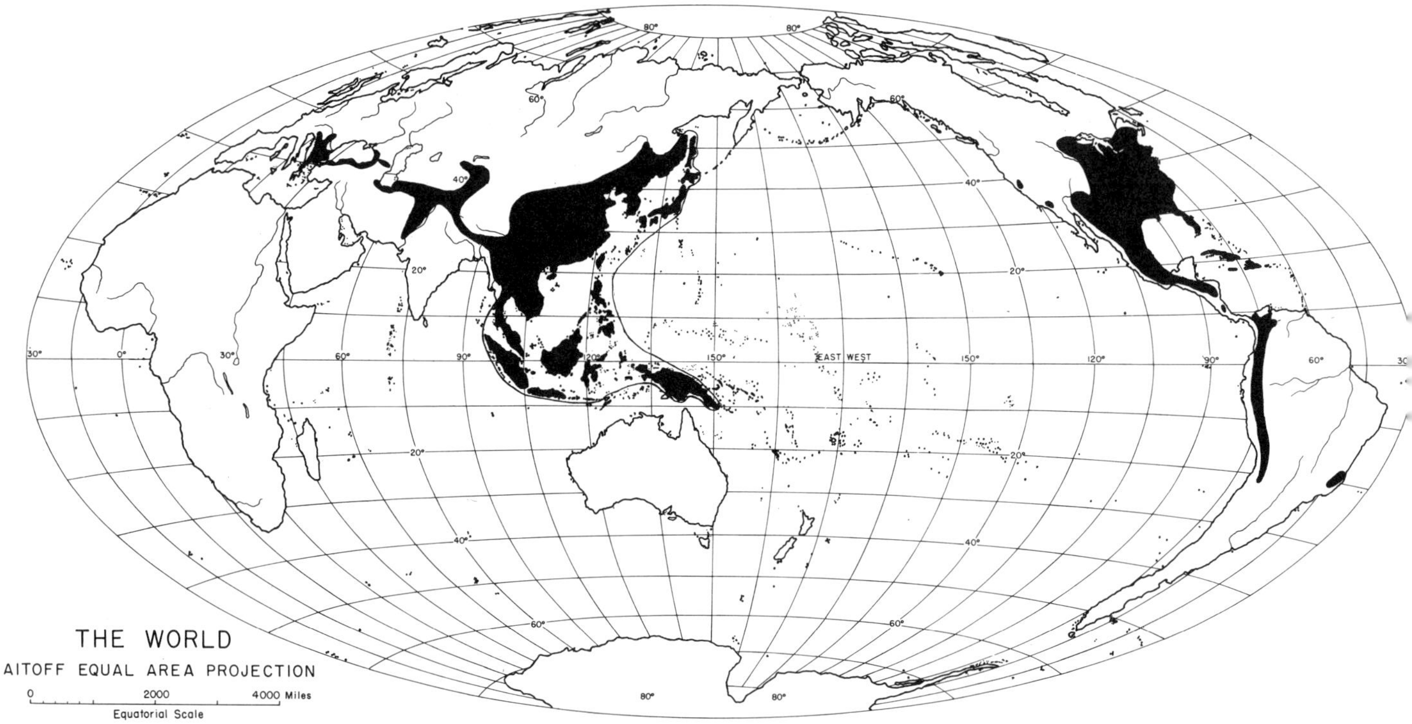

FIGURE 7. Distribution of Juglandaceae, with all the South American and Californian and most of the mainland Eurasian, American, and Mexican range representing the walnuts, *Juglans* L. Like many other genera in Andean South America, as *Alnus* Mill., *Berberis* L., and *Ribes* L., it has almost certainly arrived there via the Beringian-western American cordilleran route. [Map modified from Meusel (1965); base map as in Figure 5.]

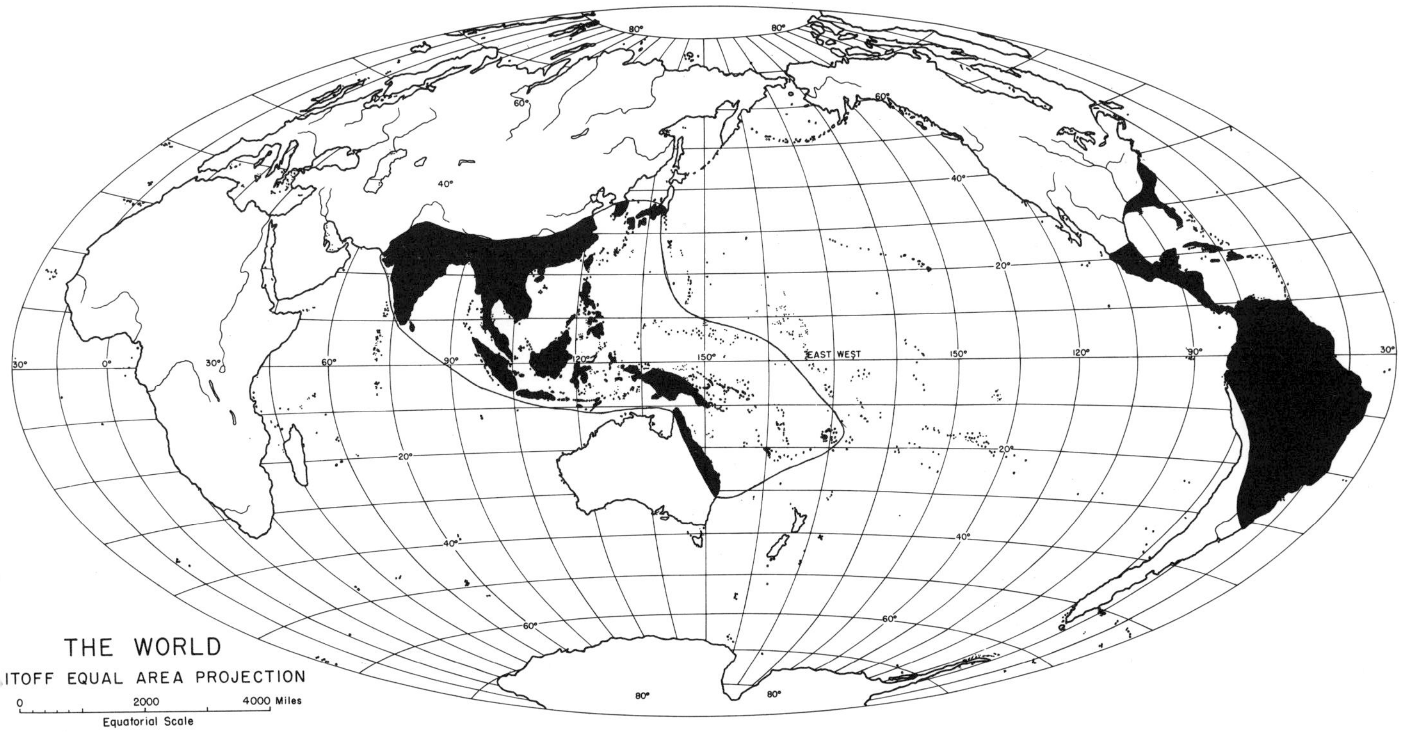

drift, America and Indo-Australasia-Antarctica broke off from opposite sides of the Euro-African landmass and floated toward each other across a once vaster but gradually diminishing Pacific Ocean. We should expect consequently much weaker amphi-Pacific than amphi-Atlantic relationships in the tropics. This, however, is not so. Of 228 phanerogamic families indigenous in South America and satellite islands, 191 are shared with the Asiatic-Australasian borders of the Pacific Ocean and 186 are shared with Africa. The figures are more disparate if the families found only in eastern tropical South America are considered. Of 203 families native in Brazil, the Guianas, and Venezuela, 156 are shared with tropical West Africa and 174 with the much more distant Asia-Australasia. That tropical South America has stronger links with the tropical western Pacific borderlands than with tropical Africa can be explained in part by the efficacy of the present or past, complete or partial, land bridges supplied by Beringia, Central America, and Antarctica and by the highly effective migration route permitted by the western cordillera running with few breaks from Alaska to Patagonia. These migration routes are in addition to the same types of long-distance dispersal across the Pacific as operated across the Atlantic.

On the generic level 152 phanerogamic genera link South America with the western Pacific borderlands compared to the 111 tying South America to Africa. Whereas 74 genera bind tropical West Africa and tropical America, 94 genera link tropical South America (including the Antilles but not Central America and Mexico) with tropical Asia-Australasia (Figure 5). Fifty-eight additional genera of Andean and southern South America bind the continent with temperate parts of Asia and Australasia, 40 apparently via Antarctica (Figure 6) or other southern routes and 18 via Beringia and the western American cordillera (Figure 7). An additional 11 genera belong with the 94 amphi-Pacific tropical genera but were excluded arbitrarily because they have reached the Mascarenes or Madagascar to the southwest. Many of these amphi-Pacific genera are discussed by van Steenis (1962).

The powerful floristic linkage of South America with North America is just what one would expect from the existence of the Central American land bridge. Had a knowledgeable botanist planned it, the existing isthmus could hardly have been made more effective. It is a land bridge-builder's dream with adequate width, variable topography, climate, and substrate, and a most active history of submergence and emergence. Anyone who has collected in the highlands of southern Mexico or Central America can attest to the marvelous mixture there of North and South American taxa of plants and animals. Of the 228 phanerogamic families in South America, 201 are also represented in Mexico and North America, and 7 more occur in Panama.

AFRICAN-ASIAN RELATIONSHIPS

The floristic relationship between tropical West Africa and Asia (including Malesia) is likewise stronger than that between western Africa and tropical America. Western Africa shares 164 of its 177 families with tropical Asia and Malesia as compared with the 156 shared with tropical America. Of 234 families from the whole continent and Madagascar 200 are shared with tropical Asia (and Malesia) but only 186 with America.

A quick analysis of Willis' *A Dictionary of the Flowering Plants and Ferns* (1966) produced a list of nearly 500 genera of seed plants distributed from Africa or Madagascar or adjacent islands to Asia and its adjacent islands (often extending to Australasia or to the Pacific Basin) (Figure 8). The *Flora of West Tropical Africa* alone lists 340 genera occupying areas from West Africa to Asia (and often beyond to the Pacific islands). Most of these genera have major discontinuities in their known ranges, usually skipping the arid lands of North Africa, Arabia, Iran, and Pakistan. Other genera show even greater gaps in their ranges. As extreme examples, *Airyantha* Brummitt (Fabaceae), *Combretodendron* A. Chevalier (Lecythidaceae), *Ctenolophon* Oliver (Figure 9), and *Aeginetia* L. (Orobanchaceae) each have one species in tropical West Africa. *Airyantha* has a second species in Borneo, *Combretodendron* a second in the Philippines, *Ctenolophon* two others in Malesia, and *Aeginetia* about ten more from Ceylon and India to China and New Guinea. With even larger discontinuities are the 32 genera, not included in the 500 mentioned above, that apparently skip from Africa, Madagascar, or the Mascarenes to Australia, New Zealand, New Caledonia, Fiji, or adjacent islands. Some of the more noteworthy of these are *Adansonia* L. (Bombacaceae) (Figure 10) with 9 species in Madagascar and Africa and one in northwestern Australia; *Bulbinella* Kunth (Liliaceae) with 15 in South Africa and 2 in New Zealand; *Cossinia* Commerson ex Lamarck (Sapindaceae) with 2 in the Mascarenes, one in New Caledonia, and one in Fiji; *Cunonia* L. with one in South Africa and 16 in New Caledonia; *Hibbertia* Andrews (Dilleniaceae) with one in Madagascar and about 100 in Australia, New Guinea, New Caledonia, and Fiji; *Keraudrenia* J. Gay and *Rulinga* R. Brown (Sterculiaceae), the former with one in Madagascar and 7 in Australia and the latter with one in Madagascar and 22 in Australia; and *Villarsia* Ventenat (Menyanthaceae) with one in South Africa and 9 in Australia.

Long-distance dispersal is undoubtedly responsible for some of these disjunct ranges. The more likely explanation for others, however, is that they represent the remnants of vast distributional areas that covered a formerly more completely humid East African-Arabian-Iranian-Indian-Malesian-Australasian arc of continental masses and islands. Some may have been dispersed from Madagascar via the Comoros and Seychelles to Ceylon or vice versa; others via Malesia to Fiji; and still others from Papua to New Zealand via New Caledonia. Many genera, and even some species, still display these more complete distribution patterns today. Some genera have xerophytic species or ecotypes that occupy the presently arid lands of North Africa, Arabia, Iran, and Pakistan, thus serving as links between the more mesophytic species in the still humid areas of Africa and southeastern Asia-Malesia-Australasia. Among these taxa especially good examples, many listed by Hepper (1965) and Wild (1965), are afforded by *Acacia nilotica* (L.) Willdenow ex Delile, *Balanites aegyptiaca* (L.) Delile, *Capparis sepiaria* L., *Ceropegia* L. and other Stapelieae, *Commiphora* Jacquin, *Cocculus hirsutus* (L.) Diels, *Cyphostemma setosum* (Roxburgh) Alston, *Flacourtia indica* (Burman f.) Merrill, *Grewia villosa* Willdenow, *Monsonia senegalensis*

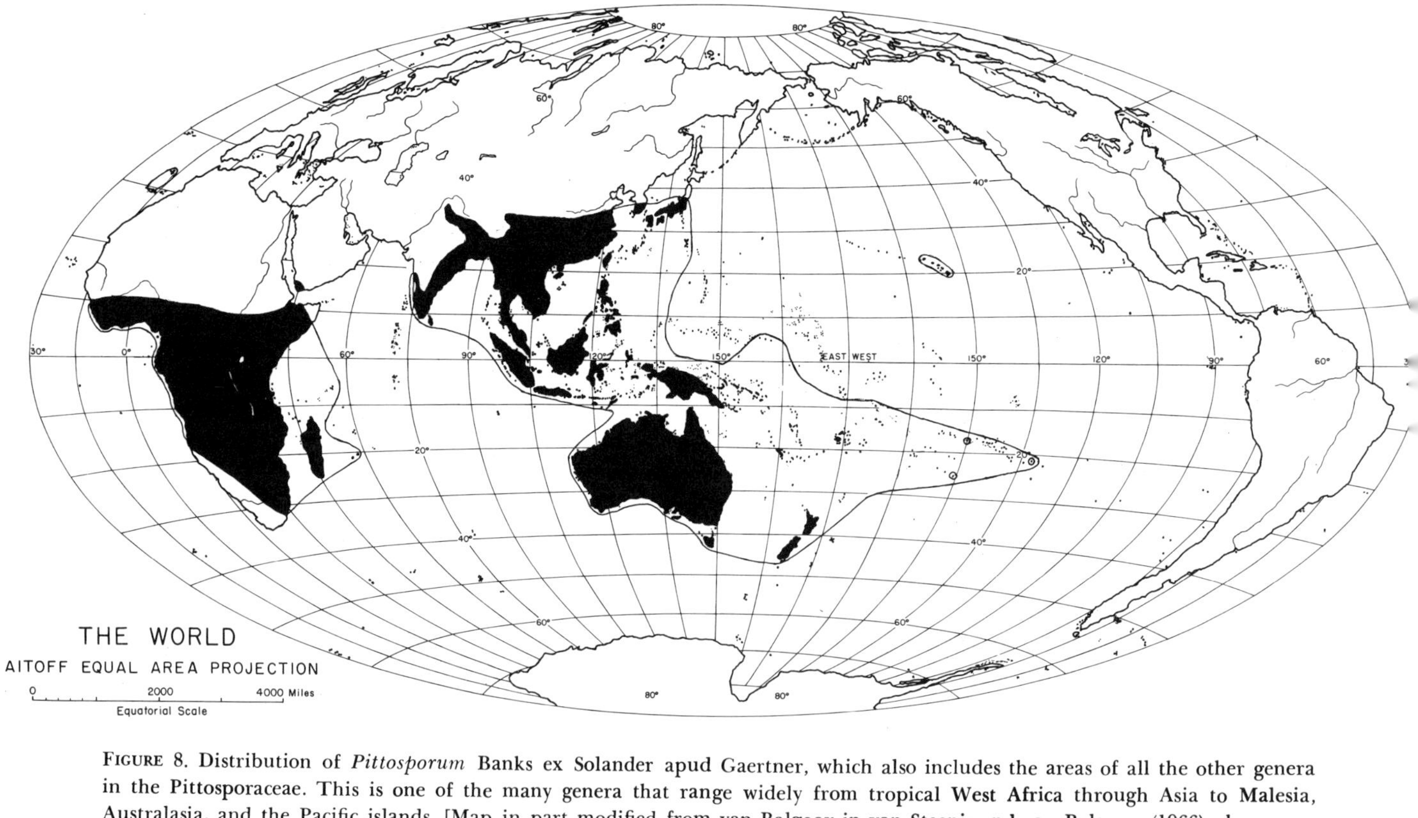

FIGURE 8. Distribution of *Pittosporum* Banks ex Solander apud Gaertner, which also includes the areas of all the other genera in the Pittosporaceae. This is one of the many genera that range widely from tropical West Africa through Asia to Malesia, Australasia, and the Pacific islands. [Map in part modified from van Balgooy in van Steenis and van Balgooy (1966); base map as in Figure 5.]

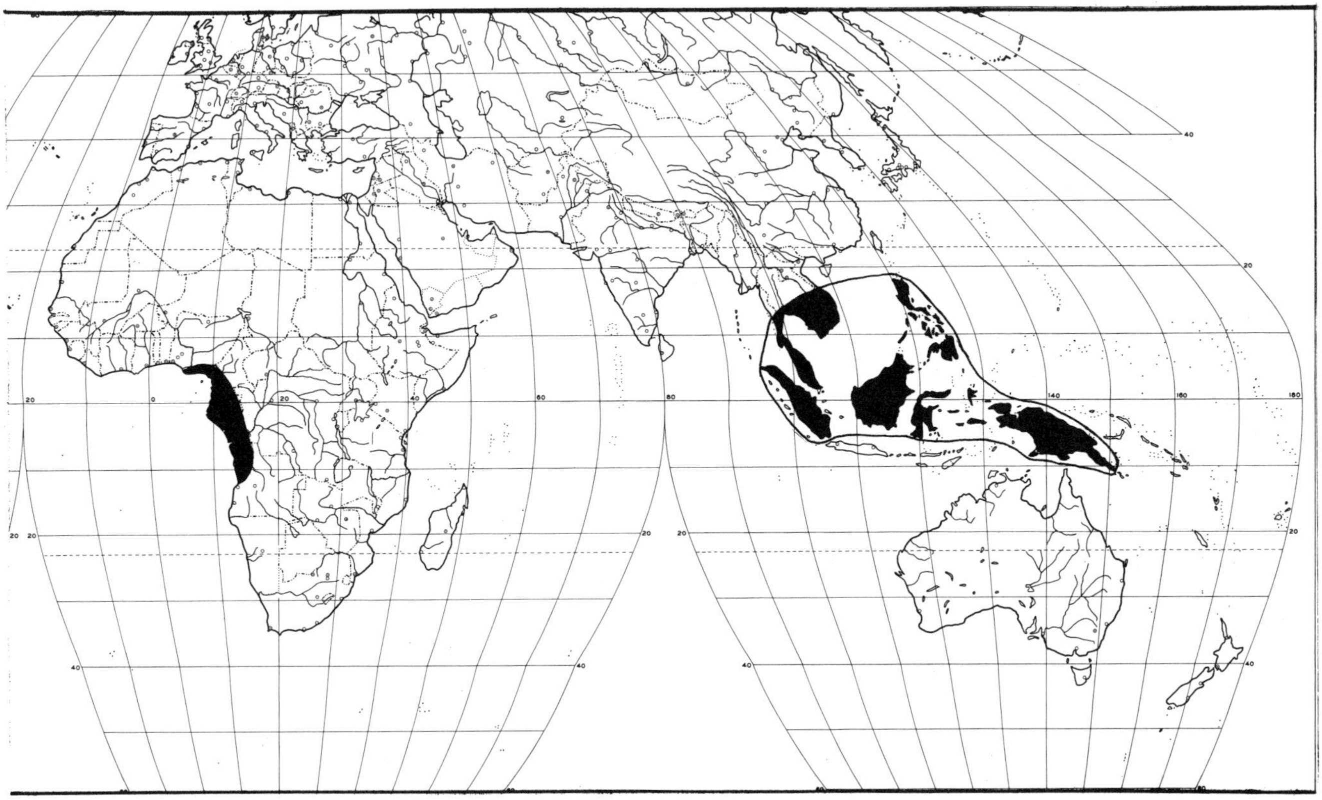

FIGURE 9. Distribution of *Ctenolophon*, single genus of the Ctenolophonaceae, which exemplifies the often widely disjunct ranges of genera of tropical West Africa and Malesia. [Map in part modified from Aubréville (1969) and Hutchinson (1959); base map as in Figure 1.]

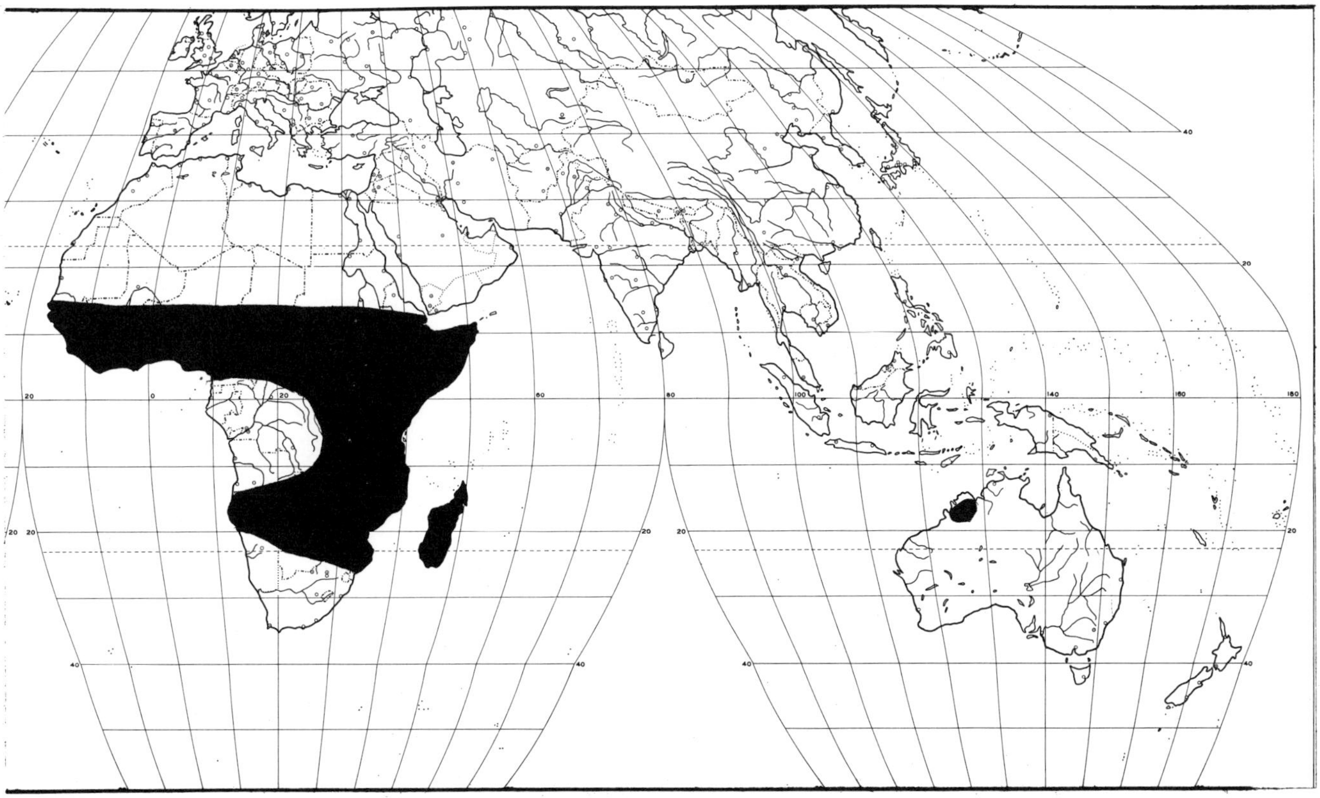

FIGURE 10. Distribution of *Adansonia*, the Baobabs of the Bombacaceae, like *Symphonia* of Figure 3, a primarily Madagascan genus with one wide-ranging species in tropical Africa and another narrowly restricted to the Kimberly District of Western Australia. [Map modified from Hutchinson (1959); base map as in Figure 1.]

Guillemin and Perrottet, *Neurada procumbens* L., *Solanum albicaule* Kotschy ex Dunal in de Candolle, and *Zygophyllum simplex* L.

ANCIENT SEED PLANTS

The oldest generally accepted remains of the flowering plants, the class Angiospermae, are of Early Cretaceous age, no more than 135 million years old. None of the oldest angiosperm fossils is satisfactorily identifiable with extant families; however, some families have been traced back to later Cretaceous age and many to the earliest Tertiary time. Living gymnosperms of several classes had representation from Permian time in the Paleozoic (Florin 1963). If continental displacement were a factor in bringing about the present distribution of seed plants in Africa and South America, an examination of the distributional areas of members of the most ancient living phanerogam families should be expected to reveal evidence of the effects of such drift. Because angiosperm evolution in relation to continental drift is discussed elsewhere in this symposium (Smith, pp. 49 ff. herein), I shall concentrate here on gymnosperm distribution patterns in the Southern Hemisphere.

The Cycadae and Coniferae are the most ancient extant gymnosperm classes represented in Africa and South America. The single order Cycadales of the Cycadae consists of three families, as defined by Johnson (1961), all represented in the Southern Hemisphere. The sole genus *Cycas* L. of the Cycadaceae is mainly western Pacific Ocean, eastern Asiatic, and Indian Ocean in distribution. It reaches Madagascar and the coast of tropical East Africa. The Stangeriaceae has only the monospecific genus *Stangeria* T. Moore of southeastern Africa. The rest of the cycads belong to the Zamiaceae with eight genera in Australia, Africa, and America. No genus is common to two of these continents. *Bowenia* Hooker, *Lepidozamia* Regel, and *Macrozamia* Miquel are restricted to Australia; *Encephalartos* Lehmann to tropical and South Africa; *Ceratozamia* Brongniart and *Dioon* Lindley corr. Miquel to Mexico; *Microcycas* (Miquel) A. de Candolle to Cuba; and *Zamia* L. to tropical America. The main centers of survival of this relict family are tropical America and tropical Australia.

Perhaps the most primitive and ancient of the extant Coniferae are the Araucariaceae. There is an extensive Mesozoic fossil record of araucarians (Florin 1963), some records as early as Upper Triassic, especially in Australasia, India, Europe, North America, and southern South America. There is one early Cretaceous record at the southern tip of Africa. Today the 18 species of *Araucaria* Jussieu are limited to New Guinea, New Caledonia, Norfolk Island, eastern Australia, and South America (Figure 11). The Paraná-pine, *A. angustifolia* (Bertoloni) Kuntze of Brazil, and the Chilean Monkey-puzzle, *A. araucana* (Molina) K. Koch belong to the section *Columbea*, and are related to the Queensland Bunya Bunya-pine, *A. bidwillii* Hooker of the section *Bunya*. The other genus *Agathis* Salisbury ranges only from New Zealand and Fiji to Malaya. Africa has no extant Araucariaceae.

The Pinaceae do not reach the South American mainland and only *Pinus* L. and *Cedrus* Trew have footholds in Africa north of the Sahara Desert. *Cedrus atlantica* (Endlicher) Arnott grows on the Atlas Mountains of North Africa. *Pinus* subgenus *Haploxylon* Koehne, primarily of Asia and North America, is present in the mountains of Europe and reaches Honduras in the highlands of Central America. *Pinus* subgenus *Diploxylon* Koehne is well represented on the Mediterranean shores of North Africa and across northwestern Africa to the Canary Islands. In tropical America the subgenus reaches the Greater Antilles and as far south in Central America as Nicaragua. Surely in both Africa and America, the species of *Pinus* have come from the north. Similarly the Taxaceae, with *Taxus* L. on the western Mediterranean shores of North Africa and in Mexico and Guatemala in Middle America, have migrated from the north.

In the Cupressaceae *Juniperus* L. also has migrated south only to Guatemala and the West Indies in tropical America and in Africa to Macaronesia, the Mediterranean shores of North Africa, and the highlands of East Africa. *Cupressus* L. ranges from Oregon into Mexico and Guatemala and occurs as a small outlier in the central Sahara Desert. The monotypic *Tetraclinis* Masters is restricted to the western Mediterranean region of North Africa, Malta, and Spain. *Austrocedrus* Florin and Boutelje and *Pilgerodendron* Florin of Chile, on the other hand, are closely related to *Libocedrus* Endlicher (sensu stricto) of New Zealand and New Caledonia, *Papuacedrus* Li of New Guinea and the Moluccas, and *Calocedrus* Kurz of southeastern Asia and Pacific North America, all

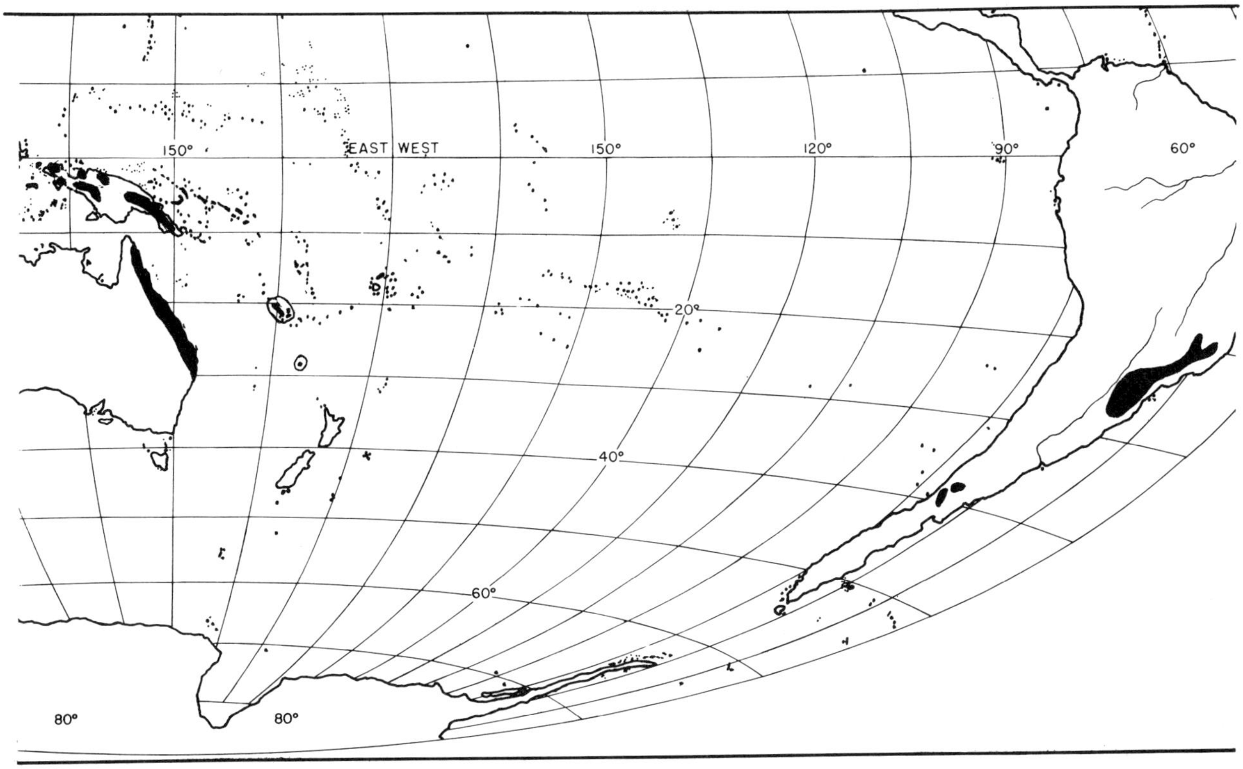

FIGURE 11. Distribution of extant species of *Araucaria*, characteristic of all the conifers of southern South America in its Australasian-Malesian relationships. Like *Nothofagus* of Figure 6 it appears to have arrived in South America via the Antarctic route. [Map modified from Florin (1963) and van Balgooy in van Steenis and van Balgooy (1966); base map as in Figure 5.]

having formerly been included in the circum-Pacific *Libocedrus* (sensu lato). The third Chilean genus, *Fitz-Roya* Hooker f., of the Cupressaceae is related to *Diselma* Hooker f. of Tasmania. *Widdringtonia* Endlicher of tropical and South Africa is believed to be most closely related to members of the *Callitris* Ventenat and *Libocedrus* group of genera of the Australasian-Asiatic region.

Most abundantly represented in all the southern continents are the Podocarpaceae. The large genus *Podocarpus* L'Heritier ex Persoon, including several recent segregate genera recognized by de Laubenfels (1969), centers in Australasia and Malesia and reaches north in America from Patagonia to Mexico and the West Indies, in Australia-Asia from New Zealand, Tasmania, and southwestern Australia to Japan and India, and in Africa from Madagascar and the Cape to Ethiopia, the Cameroons, and São Tomé. The relationships of the South American species in sections *Podocarpus, Stachycarpus*, and *Polypodiopsis* Bertrand (*Decussocarpus* de Laubenfels section *Decussocarpus*), are to species of these sections in New Zealand, New Caledonia, Tasmania, Australia, and Fiji to Malesia. The African species of sections *Podocarpus* and *Afrocarpus* have their closest relationships with the podocarps of southeastern Asia and Malesia. The extensive fossil record of *Podocarpus* in southern South America, western Antarctica, New Zealand, Australia, and India ranges in age from Middle Jurassic or earlier to Pliocene. The few fossils from South Africa are indeterminate Tertiary.

Also present in South America are species, extant or fossil, of four other podocarp genera, *Dacrydium* Solander, *Saxe-Gothaea* Lindley, *Acmopyle* Pilger, and *Dacrycarpus* (Endlicher) de Laubenfels. The other *Dacrydium* species are found in Tasmania, New Zealand, New Caledonia, Fiji, west through Malesia to southeastern Asia. The monotypic *Saxe-Gothaea* is related to *Microcachrys* Hooker f. of Tasmania. Fossils resembling *Acmopyle* have been reported from Argentina, western Antarctica, and

India; whereas, the two living species are relict in New Caledonia and Fiji. Similarly fossils like *Dacrycarpus* have been found in southern South America, western Antarctica, New Zealand, Tasmania, eastern Australia, and India. The extant species, however, are limited to New Zealand and Fiji through Melanesia and Malesia to southeastern Asia. None of these latter genera is represented in Africa.

The data for the Coniferae show that the relatives of the members of the Pinaceae, Taxaceae, and northern genera of Cupressaceae in tropical America and in Africa all represented only in the northern continents. The species of the Araucariaceae, Podocarpaceae and southern Cupressaceae, fossil or extant, in South America or Africa are also not related closely to members of these families in the other continent but most nearly resemble those now relict in the Australasian-Malesian regions. The data led Florin (1963) to the conclusion that continental drift could have had nothing to do with the distribution of the Coniferae. I can only agree with him.

The final class of gymnosperms with members in the southern continents is the Gnetae, known in the fossil record only back to the Eocene epoch. Of the three orders and genera of the class, the extraordinary *Welwitschia bainesii* (Welwitsch) Carrière is peculiar to Southwest Africa and Angola, and *Ephedra* L. with 40 species is absent from Africa but present in South America as well as in North America and Eurasia. Only the circumtropical *Gnetum* L. is found in both Africa and South America, as well as from Indo-Malesia to Fiji. The species of *Gnetum* of western tropical Africa and northern tropical South America and Panama belong to separate subsections of a pantropical section, but in some respects the subsections are more closely related to each other than to the Indomalesian subsection. The lack of a fossil record for the genus and the presence of fleshy seeds suggest that the African and American species may have reached these continents relatively recently from their Indo-Malesian center either independently by separate routes or across the Atlantic by long-distance dispersal.

Similar distribution patterns are displayed by the oldest, most primitive angiosperms. Two diverse and paleontologically ancient families, one aquatic and one terrestrial, can be analyzed here briefly to illustrate the two prevailing patterns among the oldest known flowering plants.

The aquatic Nymphaeales have an extensive fossil record back to the Early Cretaceous despite their obvious specializations for the aquatic way of life. *Nelumbo* Adanson, very different from the other water-lilies and perhaps best treated in its own family or even order (Simon 1970), consists of two closely related species, *N. nucifera* Gaertner of eastern Asia to northern Australia, India, and Iran, and *N. lutea* (Willdenow) Persoon of eastern North America south to Cuba, Jamaica, and Colombia. There is fossil evidence that the genus occurred in northeastern Africa, Europe, Greenland, and western North America (Good 1964). Restricted to the New World are *Cabomba* Aublet (7 species) and *Victoria* Schomburgk (2-3). *Nuphar luteum* (L.) Smith, polymorphic and circumtemperate in the Northern Hemisphere, reaches through the southeastern United States to Cuba. *Brasenia schreberi* Gmelin, like many other aquatics, has a vast but sporadic distribution which includes Africa, eastern Asia, Australia, and North America south to Central America and the West Indies. That of *Ceratophyllum demersum* L. is even wider since its range includes, in addition, Europe and South America south to Argentina. The genus *Nymphaea* L. (35 species) is almost cosmopolitan, absent only from most of the Pacific slope of North America and the Pacific islands, and is well represented by closely related species in all tropical areas, though no species spans the Atlantic Ocean.

The Proteaceae (Figure 12) is an isolated and ancient group which Johnson and Briggs (1963) believe was well defined by Late Cretaceous time. Because of the family's great age and considerable development in both Africa and South America, its distributional pattern in the southern continents is most important. Furthermore, the pattern epitomizes that displayed by so many of the oldest seed-plant taxa in Africa and America. All seven of the South American genera belong to the subfamily Grevilleoideae and four of them, *Gevuina* Molina, *Lomatia* R. Brown, *Oreocallis* R. Brown, and *Orites* R. Brown, with combined range extending along the Andes from southern Chile to Ecuador, also have species in Tasmania, eastern Australia, or New Guinea. These distributional data indicate an Antarctic migration route. The grevilleoids are strongly developed in Australasia and Malesia to eastern Asia, and have traces in Madagascar (a species identified as a *Macadamia* F. Mueller) and South Africa (the monotypic *Brabejum* L.). All the other African proteads, many species in 11 genera, belong to the subfamily Proteoideae, which like the Grevilleoideae, had its probable original center of development in tropical Australia.

Conclusions

The geological evidence for continental displacement has piled up to an impressive extent, and many of the suggested explanations sound most convincing to the nongeologist. Certainly the preceding evidence from the distribution, past and present, of the seed plants cannot be used to deny that continental displacement has taken place. However, I think the botanical evidence just presented argues against continental drift as a significant factor in explaining the distribution of seed plants between Africa and South America. If continental displacement has occurred, the separation of Africa and South America must have attained its present state, or largely so, before the development of the present seed-plant floras of the world, certainly previous to Tertiary and Late Cretaceous time and possibly even previous to Jurassic time.

The relatively tenuous floristic links between tropical Africa and tropical America and the stronger floristic relationships of tropical Africa and tropical South America to other continents appear to rule out continental drift as a valid explanation for wide disjunctions in the ranges of seed plants. Land bridges, complete or partial, like those that exist or have existed in the Central American, Antarctic, Beringian, Malesian, Arabian, and Comoro-Seychellean areas certainly have been most effective in the migration of seed plants. However, for passage across great oceanic gaps between one continent and another and between continents and distant oceanic islands, only one rational explanation is left—long-distance dispersal. Many botanists in the past and a few in the present

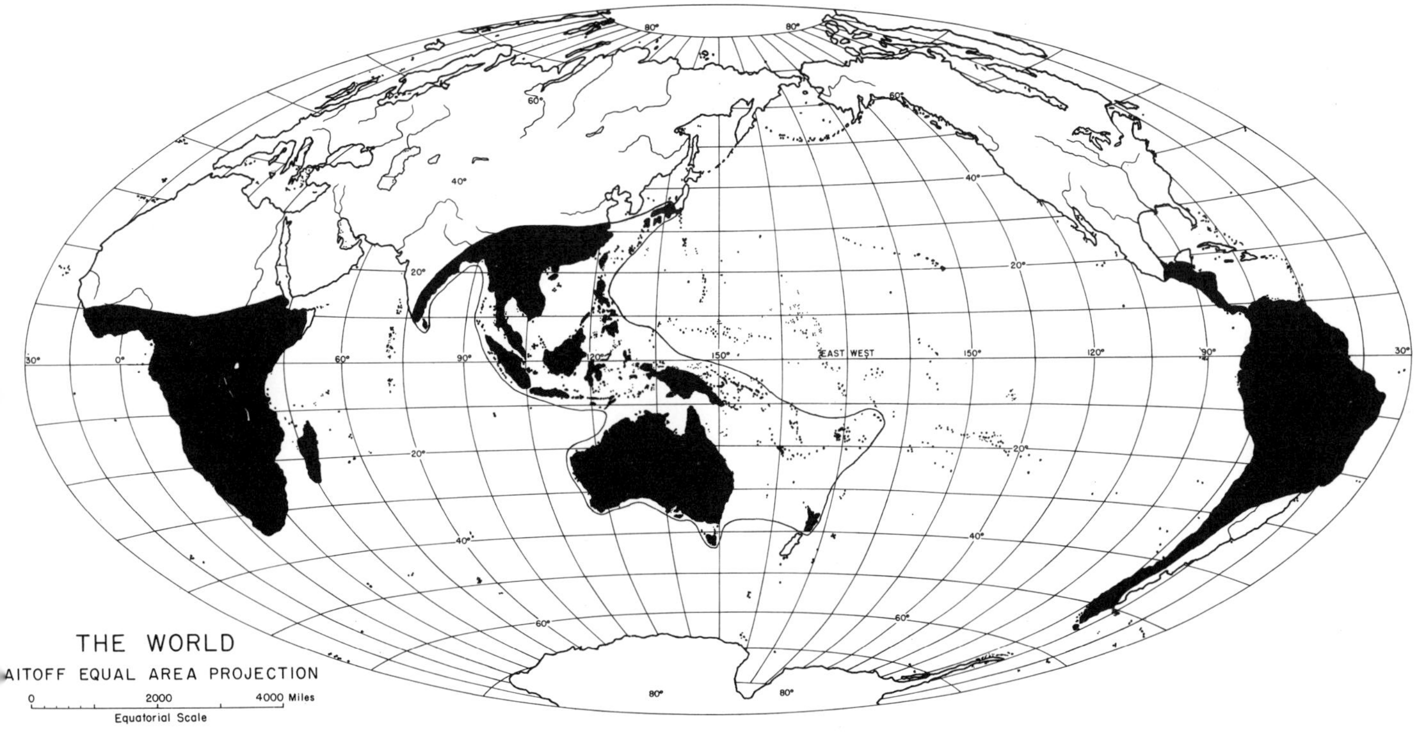

FIGURE 12. Distribution of the Proteaceae, one of many ancient angiosperm and gymnosperm families in which the African and South American taxa are more closely related to the Australasian-Asiatic taxa than to each other. [Map modified from Good (1964); base map as in Figure 5.]

refuse to accept the possibility of occasional, indeed extremely rare, successful long-distance passage and establishment of selected plant disseminules. Indeed, I have long been mystified by the strange aberration afflicting some botanists under which they are unable to visualize long-distance dispersal of tiny fruits and seeds, but can accept without blinking the shifting on command of enormous continental blocks over the surface of the earth like checkers over a board, or the hypothetical elevation from the ocean depths of massive, highly selective, cleverly zigzagging, transoceanic land bridges that soon vanish without a trace. I am not sure there is any cure for this extraordinary phytogeographical affliction, though I would like to suggest as a possible course of treatment a careful reading of Ridley's classic *The Dispersal of Plants throughout the World* (1930), a few of H. B. Guppy's books (1906, 1917), Carlquist's series of papers on "The Biota of Long-distance Dispersal" (1966a, b, c, d, 1967), and the reports of the recent investigations on air-borne organisms by Gressitt, Leech, and Wise (1963), Gressitt and Nakata (1958) and Gressitt and Yoshimoto (1963).

One can hardly expect to witness the accomplishment of any event of long-distance dispersal. Much extrapolation is required from the observational and experimental data that have been recorded. Much more experimental work is needed, especially with birds and water and air currents as vectors. In the meantime, we must develop our case largely on the basis of circumstantial evidence. Toward this end greatly isolated oceanic islands built up from the deep-sea bed by volcanic action are ideal.

Few biologists, and certainly no geologists, would attempt today to explain the Hawaiian Islands as having had other than an isolated, oceanic origin. Like most islands they have been ignored by the continental-drifters though they have not entirely escaped the attention of trans-Pacific land-bridge builders. Hawaii lies approximately 2400 statute miles west-southwest of San Francisco. This is close to the minimal distance of the Hawaiian Islands from any point in America. The eastern Pacific is deep ocean without atolls, guyots, or other vanished island stepping stones. As volcanic outpourings, the islands are recent and relatively ephemeral. The oldest western islands are estimated to be of Miocene age, at most 25 million years old. Yet the archipelago has a relatively rich, though disharmonic, flora with 216 genera of angiosperms (Fosberg 1948). Carlquist (1967) has estimated that 20 percent of the seed plants are of American origin. Thus, there have been at least 52 arrivals from America among 256 angiospermous immigrants from all sources needed to account for the present indigenous phanerogamic flora. Ninety-five of the 256 successful immigrations were recent enough or repeated often enough that the immigrating species have retained their identity. In his analysis of probable modes of arrival from all directions Carlquist estimated that nearly 75 percent of the immigrants were transported by birds, 38.9 percent internally, 23.1 percent adhering externally by barbs, bristles, viscidity, etc., and 12.8 percent in mud on feet or feathers. Ocean currents apparently accounted for 22.8 percent of the immigrants, 14.3 percent by direct flotation and 8.5 percent by rafting on logs or other floating debris. An insignificant 1.4 percent may have arrived by flotation on air currents.

The preceding data have great bearing upon the African-American floristic disjunctions. The distance from Natal on the eastern bulge of Brazil to the nearest point in West Africa is 1800 statute miles, as compared to the 2400 miles from San Francisco to Hawaii. Africa and South America are considerably larger target areas for vectors than the Hawaiian Islands. The interchange can be two-way between two continents rather than one-way between a source continent and a receptive archipelago. The two continents present numerous similar ecological niches that surely facilitate establishment of disseminules from the opposite continent. Finally, the source and recipient areas have both been there a long time, much longer than the history of the seed plants requires. I must, therefore, conclude that long-distance dispersal is more than adequate to account for the relatively insignificant, though spectacular, floristic relationships between tropical Africa and tropical America.

References

Aubréville, A. 1969. Essais sur la distribution et l'histoire des angiospermes tropicales dans le monde. Adansonia, series 2, 9:189-247.

Carlquist, S. 1966a. The biota of long-distance dispersal, I: Principles of dispersal and evolution. Quart. Rev. Biol. 41:247-270.

————. 1966b. The biota of long-distance dispersal, II: Loss of dispersibility in Pacific Compositae. Evolution 20: 30-48.

————. 1966c. The biota of long-distance dispersal, III: Loss of dispersibility in the Hawaiian flora. Brittonia 18: 310-335.

————. 1966d. The biota of long-distance dispersal, IV: Genetic systems in the floras of oceanic islands. Evolution 20:433-455.

————. 1967. The biota of long-distance dispersal, V: Plant dispersal to Pacific islands. Bull. Torrey Bot. Club 94:129-162.

Castellanos, A., and H. LeLong. 1943. Cactaceae. Pp. 49-133, *in* H. R. Descole (Editor). Genera et Species Plantarum Argentinarum. Vol. 1. Guillermo Kraft, Ltd. Buenos Aires.

Cuatrecasas, J. 1961. A taxonomic revision of the Humiriaceae. Contr. U. S. Nat. Herb. 35:25-214.

de Laubenfels, D. J. 1969. A revision of the Malesian and Pacific rain-forest conifers, I: Podocarpaceae, in part. J. Arnold Arbor. 50:274-369.

DeWolf, G. P., Jr. 1964. On the sizes of floras. Taxon 13:149-153.

Engler, A. 1964. Syllabus der Pflanzenfamilien, II. Band. Angiospermen. Edited by H. Melchior. 666 pages. Gebruder Borntraeger. Berlin.

Ernst, W. R., and J. J. Thompson. 1963. The Loasaceae in the southeastern United States. J. Arnold Arbor. 44:138-142.

Florin, R. 1963. The distribution of conifer and taxad genera in time and space. Acta Hort. Berg. 20:121-312.

Fosberg, F. R. 1948. Derivation of the flora of the Hawaiian Islands. Pp. 107-119, *in* E. C. Zimmerman. Insects of Hawaii. Vol. 1: Introduction. Univ. Hawaii Press. Honolulu.

Good, R. 1964. The geography of the flowering plants. 3rd edition. 518 pages. Wiley & Sons. New York.

Gressitt, J. L., R. D. Leech, and K. A. Wise. 1963. Entomological investigations in Antarctica. Pacific Insects 5:287-304.

Gressitt, J. L., and S. Nakata. 1958. Trapping of air-borne insects on ships on the Pacific. Proc. Hawaiian Entomol. Soc. 16:363-365.

Gressitt, J. L., and C. M. Yoshimoto. 1963. Dispersal of animals in the Pacific. Pp. 283-292, *in* J. L. Gressitt (Editor). Pacific basin biogeography. Bishop Museum Press. Honolulu.

Guppy, H. B. 1906. Observations of a naturalist in the Pacific between 1896 and 1899. Vol. II: Plant dispersal. Macmillan & Co. London.

————. 1917. Plants, seeds, and currents in the West Indies and Azores. Williams & Norgate. London.

Hepper, F. N. 1965. Preliminary account of the phytogeographical affinities of the flora of West Tropical Africa. Webbia 19:593-617.

Hutchinson, J. 1959. The families of flowering plants. Vol. I: Dicotyledons. 2nd edition. 510 pages. Oxford Univ. Press. London.

Hutchinson, J., and J. M. Dalziel. 1927-1936. Flora of West Tropical Africa. 2 vols. Crown Agents. London.

————. 1954-1968. Flora of West Tropical Africa. 2nd edition. 3 vols. Edited by R. W. J. Keay and F. N. Hepper. Crown Agents. London.

Iltis, H. H. 1967. Studies in the Capparidaceae, XI: *Cleome afrospina*, an African endemic with neotropical affinities. Amer. J. Bot. 54:953-962.

Johnson, L. A. S. 1961. Zamiaceae. Contr. New South Wales Nat. Herb. 1:21-41.

Johnson, L. A. S., and Barbara G. Briggs. 1963. Evolution in the Proteaceae. Australian J. Bot. 2:21-61.

Martius, K. F. P., von. 1840-1906. Flora Brasiliensis. 15 vols. Munich & Leipzig.

Meusel, H. 1965. Vergleichende Chorologie der zentraleuropäischen Flora. 2 vols. Gustav Fischer. Jena.

Ridley, H. N. 1930. The dispersal of plants throughout the world. 744 pages. L. Reeve & Co. Ashford, Kent.

Simon, J.-P. 1970. Comparative serology of the order Nymphaeales, I: Preliminary survey on the relationships of *Nelumbo*. Aliso 7:243-261.

Stafleu, F. A. 1954. A monograph of the Vochysiaceae, IV: *Erisma*. Acta Bot. Neerl. 3:459-480.

Steyermark, J. A. 1932. A revision of the genus *Menodora*. Ann. Missouri Bot. Gard. 19:87-176.

Thorne, R. F. 1968. Synopsis of a putatively phylogenetic classification of the flowering plants. Aliso 6(4):57-66.

van Steenis, C. G. G. J. 1962. The land-bridge theory in botany. Blumea 11:235-542.

van Steenis, C. G. G. J., and M. M. J. van Balgooy. 1966. Pacific plant areas. Vol. 2. Blumea 5, suppl. vol. Leyden.

Vester, H. 1940. Die Areale und Arealtypen der Angiospermen-Familien. Bot. Arch. 41:203-275, 565-577.

Wild, H. 1965. Additional evidence for the Africa-Madagascar-India-Ceylon land-bridge theory with special reference to the genera *Anisopappus* and *Commiphora*. Webbia 19:497-505.

Willis, J. C. 1966. A dictionary of the flowering plants and ferns. 7th edition. 1214 pp. Edited by H. K. Airy Shaw. Cambridge Univ. Press. Cambridge.

Angiosperm Evolution and the Relationship of the Floras of Africa and America

ALBERT C. SMITH

Introduction

Few scientists now doubt the reality of continental drift. Evidence from many fields points to continuing sea-floor spreading from oceanic ridges, and to the movement of continent-carrying plates of the earth's crust. The movement of South America and Africa away from one another is now so well documented that it must be accepted as a fact of the past. As research has been accelerated, and as the formation and movement of these two plates have become better understood, one of the remaining questions bears on the date of the original separation of South America and Africa. The geological and paleomagnetic evidence suggests that the South Atlantic came into being during Jurassic or very early Cretaceous time (Martin 1968). The results of recent drilling indicate that by Campanian time the South Atlantic was already some 3000 kilometers wide (Maxwell et al. 1970). A different assumption (Axelrod 1970)—that the breakup of Gondwanaland may be dated from the medial Cretaceous and that the middle Atlantic opened up only in the early Albian stage—does not seem widely held by geophysicists.

One of our objectives in this symposium is to attempt to correlate the concepts of physical and biological scientists as to the date of separation of Africa and South America. Can biological data bearing on the known distribution of different major taxa help to fix a date? Can evolutionary and distributional hypotheses, in reference to the different major taxa, be correlated with the date currently favored by geophysical opinion?

It has often been pointed out that generalizations about the past migrations of diverse major taxa of plants and animals are dangerous. The perils of drawing sweeping conclusions from hypotheses based on a single taxon are obvious; each major group of organisms had its own origin in time and space, its own dispersal mechanisms, and its own paleoecological adaptations. Biologists who have reached conflicting conclusions as to the role of continental drift as a distributional factor seem sometimes to have been speaking different languages, in that one is knowledgeable about marine invertebrates, another about insects, and a third about mammals. To be sure, there have been syntheses of a broad and scholarly nature (Simpson 1965, Darlington 1957, 1965), whose authors are understandably cautious about dating past continental movements. In the botanical field different conclusions on distributional factors have been expressed by students of bryophytes (Schuster 1969), gymnosperms (Hair 1964), and angiosperms.

My intent is to discuss angiosperm evolution as it is relevant to the modern ecosystems on the two sides of the South Atlantic. The contributions of such botanists as Plumstead (1961), Good (1964), Cranwell (1964), Hawkes and Smith (1965), Melville (1966, 1969), Axelrod (1970), and others who call upon contiguity of continents as the key to angiosperm distribution cannot here be discussed in detail, but neither can their views be ignored. The gist of their arguments has been questioned by many students of angiosperms, among whom may be mentioned Thorne (1964, 1965, and herein),

Albert C. Smith, Department of Botany, University of Massachusetts, Amherst, Massachusetts 01002.

Smith (1967, 1970), and Takhtajan (1969). Since my thesis embraces the concept that angiosperms arrived on the scene too late for their modern distribution to have been seriously affected by continental movements, it is evident that I am not going to defend proponents of alternative hypotheses.

I propose to examine the angiosperms, from the viewpoint of their time and place of origin and their primary distributional routes, for possible illumination of the similarities and differences in the present-day floras of Africa and South America.

I need not explain that angiosperms are flowering plants, the group of plants that today dominates the vegetation of most land surfaces. In terms of numbers, there may well exist 300,000 or more species of angiosperms; perhaps two-thirds or even three-quarters of thus-far described species of plants are flowering plants. (To be sure, this proportion may be altered when less obvious plants, such as soil organisms that may be so classified, become better known.) Only certain temperate areas are dominated by another plant taxon, the gymnosperms, but even in extensive coniferous forests angiosperms are by no means lacking. Conversely, in many tropical and subtropical areas angiosperms have flourished almost to the exclusion of significant gymnosperm elements. Other taxa of land plants, those composing the pteridophytes and bryophytes, are also widespread, but they do not dominate the landscape as angiosperms do, and it is apparent that they are much older taxa, phyletically dwindling in number and in degree of dominance.

In elaborating a hypothesis of the evolutionary history of a group of organisms, the biologist reaches at least tentative conclusions as to which living representatives of his group best reflect the original composition of that group in its early history. Such conclusions are not pulled out of thin air, as sometimes implied (Lam 1961, Meeuse 1965), but are based on the known sequences of the appearance in the fossil record of certain character-states, significant correlations between these and other character-states, and painstaking examination of evidence from all disciplines. Logical conclusions have been so well summarized by Cronquist (1968) and Takhtajan (1969) that I need not, at present, justify the thesis that the most primitive extant angiosperms are to be sought among the so-called "ranalean" families of dicotyledons.

Current classification outlines of angiosperms do not agree in detail as to the arrangement of the "ranalean" taxa in families and orders, even though there is general agreement that approximately 500 genera and 12,000 extant species retain some of the appreciably primitive characteristics. Estimates of the number of higher taxa concerned range from the 37 families and 4 orders suggested by Thorne (1968) to 47 families and 13 orders indicated by Takhtajan (1969). I have recently proposed to arrange the so-called ranalean genera in 60 families and 14 orders (Smith 1972). We are not now concerned with the taxonomic details, but in my subsequent discussion I propose to consider the distribution of some of the 60 families into which the putative primitive extant angiosperms may be divided.

The present-day distribution of our 60 families in three major land areas of the world may now be considered (Figure 1). In the area of eastern Asia and Australasia, 53 of the families are known to occur; in all of Africa and Madagascar and adjacent Mediterranean Europe 25 families; and in all of North and South America 40 families. How does it happen that the extant primitive angiosperms are so much more diverse in Asia-Australasia than in America and so conspicuously more diverse than in Africa? These figures alone do not necessarily indicate that the angiosperms originated in Asia-Australasia; the difficulties of distinguishing between centers of origin and centers of survival, in the absence of really convincing paleobotanical evidence, are obvious. On the other hand, a reasonably monophyletic taxon does not arise from different populations of a pre-existing taxon in widely separated areas. The angiosperms, if claimed to be monophyletic (a claim sometimes disputed but not to be defended here), must have had their origin in one region of the earth's surface and subsequently must have spread to other regions. Our question is: Did this large taxon originate within the Asian-Australasian region and then spread throughout the world, or was it already widespread when the continents were grouped and, through differential extinction, did its primitive elements survive in greater diversity in Asia-Australasia than elsewhere? As suggested, the paleobotanical record does not greatly illuminate this question. We know that many of our primitive families did occur substantially farther north and south during the Upper

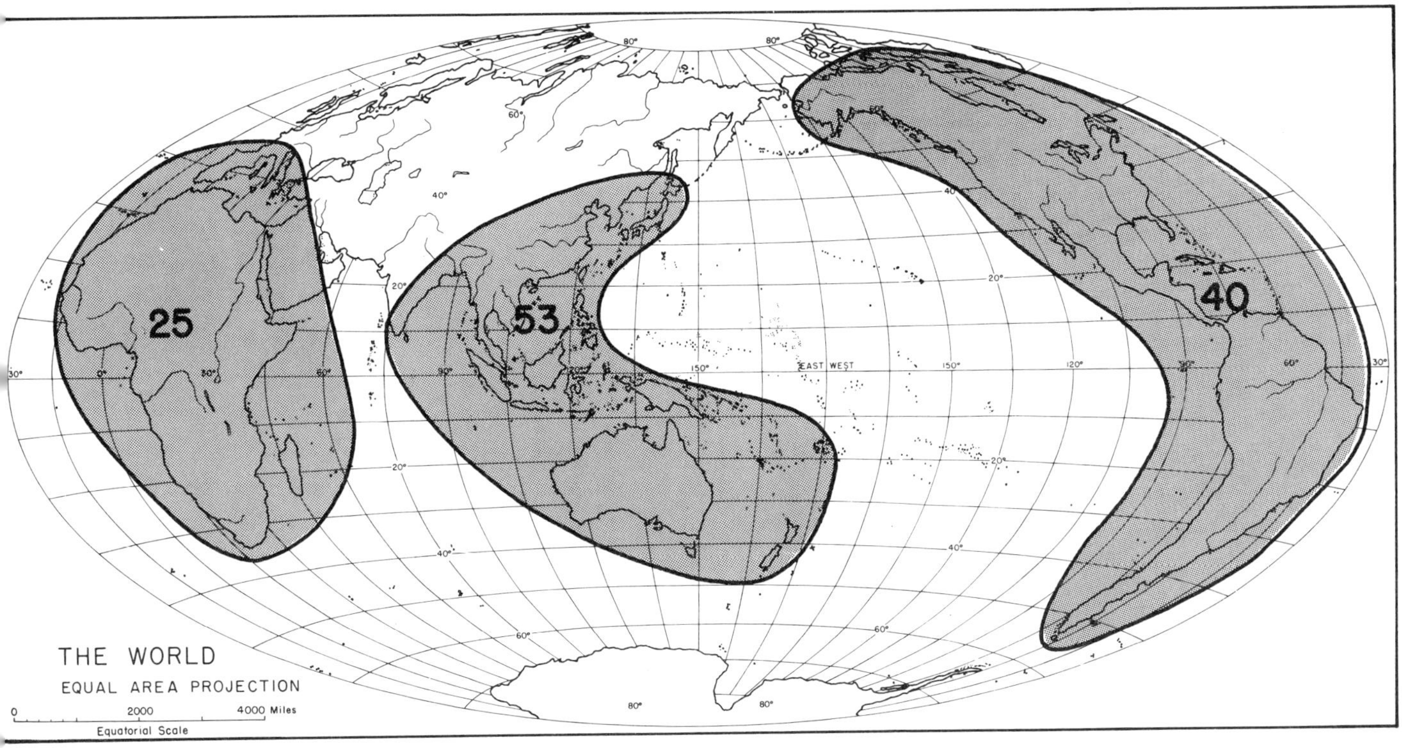

FIGURE 1. The representation of 60 primitive extant angiosperm families in three comprehensive divisions of the earth's land surfaces. (Reprinted from Smith 1970, with permission of the Harold L. Lyon Arboretum, University of Hawaii.)

Cretaceous and Tertiary periods, and it is probable that the Pleistocene glaciations played a part in narrowing their modern distributions. But we do not find in the fossil records of Africa and Middle America any convincing evidence that Asian-Australasian families lacking there today did occur there during the Tertiary. This fact suggests that the center of diversity of primitive angiosperms may have been in Asia-Australasia in the past as well as in the present.

To reach a tentative solution as to the center of origin of angiosperms we must attempt to reconstruct the evolutionary history of the group, calling upon authenticated fossil evidence wherever it exists, but in the main depending upon study of extant organisms. By the application of all disciplines and techniques that come to hand, it is indeed feasible to reconstruct evolutionary history through the phylogenetic method. Again, I do not now propose to defend this viewpoint, but for competent defenses I refer you to the recent summarizing works of Cronquist (1968) and Takhtajan (1969). By utilizing well-documented phylogenetic conclusions we may suggest a model of angiosperm origin and dispersal on a world-wide scale.

A Distributional Hypothesis

Evidence that the flowering plants originated in the Asian-Australasian area has been accumulating for many decades, and this probability has frequently been tentatively suggested (e.g., Bailey 1949). Perhaps the best documented argument for the cradle of the angiosperms having been "between Assam and Fiji" is that of Takhtajan (1969). The present writer (Smith 1967, 1970) has briefly summarized the case for an Asian-Malesian origin and has suggested the probable major migrational routes. An outline of this model of angiosperm distribution will serve to orient present readers.

On the basis of currently available evidence, we may suggest that the angiosperms originated in the Jurassic or conceivably even in the Triassic Period. Hypotheses of an early Cretaceous origin, simply because no thoroughly authenticated pre-Creta-

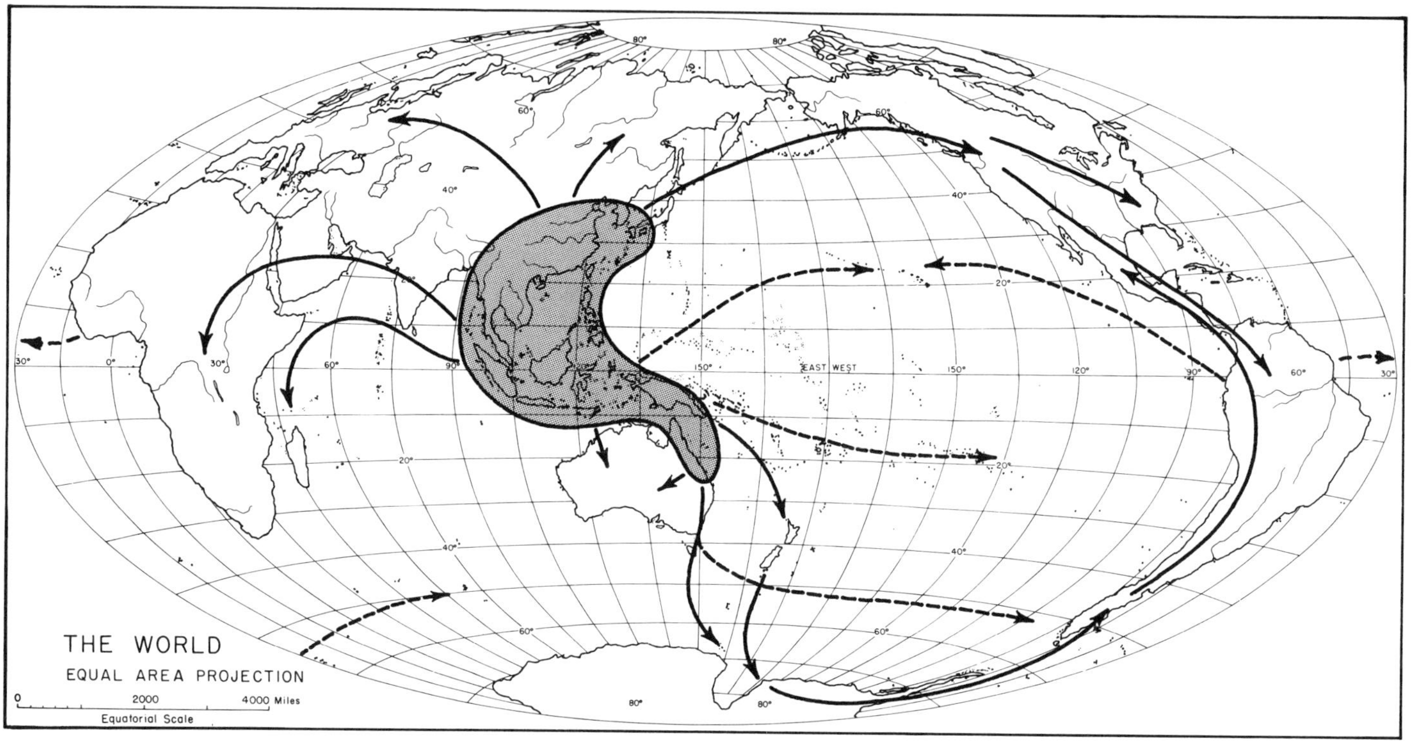

FIGURE 2. A generalized model of angiosperm distribution, indicating the probable routes of migration followed by early flowering plants over land or along insular chains (solid lines), and probable movements by long-distance dispersal (dotted lines). (Reprinted with slight modification from Smith 1970, with permission of the Harold L. Lyon Arboretum, University of Hawaii.)

ceous angiosperm fossils are at hand (Scott et al. 1960), seem to ignore the need of an extended period of pre-Cretaceous time for the evolution of the great diversity and extended distribution that we find in early Cretaceous time. We may reasonably conclude that flowering plants came into existence at about the same time that the original supercontinents, Laurasia and Gondwanaland, were beginning to be fragmented. That the angiosperms were monophyletic, that their ancestry is to be sought among the pteridosperms (Meeuse 1971), and that their center of origin lay in southeastern Asia are also suggested by increasingly well-documented evidence. We may assume, from what we know about evolutionary rates in other major taxa of plants, that for many millions of years the angiosperms must have been confined to their area of origin, only slowly becoming locally dominant over entrenched pteridophytes and gymnosperms, by means of the elaboration of quite spectacular morphological and ecological innovations. During the first few tens of millions of years of their existence, without meeting serious barriers, they could have extended their range northward into warm temperate Asia and southeastward into eastern Malesia and eastern Australia (Figure 2). You will note my assumption that by the late Jurassic or early Cretaceous Australia had drifted into essentially its present position, and that the precursors of the eastern Malesian islands had permitted intermittent passage of many angiosperm elements. This hypothesis as to the early drift of Australia is not necessarily approved by geophysicists, but I have recently (Smith 1970) suggested a possible reconciliation of the conflicting evidence as to the position of Australia.

During the early Cretaceous the crowding angiosperms, with their rapidly diversifying insect visitors, began to spread from their source area in three major streams: (1) northward into northern Asia and Europe, some of these elements crossing Beringia and then moving southward; (2) westward along Indian Ocean coasts to Madagascar and Africa; and (3) southward into Australia and along the New Caledonia-New Zealand insular chain into the West Antarctic Archipelago and thence into South America. If the suggested timing and the suggested migrational routes seem overly speculative to readers, I can only refer them to a few prior discussions (Glaessner 1962, Thorne 1964, Darlington 1965, Audley-Charles 1966, Smith 1970, and bibliographies cited therein). Superimposed on these major migrational routes we must assume movement north and south, and vice versa, along transtropical highland bridges, and we may be sure that there occurred in the past, as in the present, a random movement of floristic elements through saltational long-distance dispersal.

This distributional model satisfies the known facts about the modern patterns of angiosperms, primitive as well as advanced, and it does not conflict with the authenticated fossil record. Because the angiosperms occur in great diversity and profusion in the middle Cretaceous records of many parts of the world, we are not justified in concluding that they arose autochthonously in all those areas. They must have moved into some of them from a center of origin, diversifying as they moved into new and different terrain. If this movement took place after the components of Pangea separated, our model will explain the present distributions. If the worldwide dispersal of angiosperms occurred before the breakup of Pangea, or at least before the breakup of Laurasia and Gondwanaland, as sometimes hypothesized, certain floristic patterns would be evident in modern floras.

In the first place, all three major areas of the world's land surface would have somewhat comparable shares of primitive angiosperm elements; but this is certainly not the case. Secondly, such elements as survive in each of the three areas would be somewhat interrelated within that area; but in fact such elements are definitely disharmonic in Africa and America, and only in the Asian-Australasian sector are groups of interrelated genera and families concentrated in a harmonic manner. Thirdly, no distributional continuum of primitive angiosperms would be apparent from southeastern Asia to Malesia to New Guinea to Australia, New Caledonia, and New Zealand, since these areas would have received their original angiosperm elements from quite different directions (if angiosperms had existed on Gondwanaland before its breakup); in fact, however, Australasian areas are for the major part floristically related to Indo-Malesia, a conclusion particularly striking when one examines the distributions and relationships of ranalean families and genera. And finally, the paleobotanically better known parts of the world —North and South America, Europe, and Africa— would by this time have disclosed very early Cretaceous angiosperm fossils, if angiosperms had indeed occupied them at that time; but to the present all such putative fossils from the first Cretaceous horizons, as those from the Jurassic, have met with paleobotanical skepticism.

In reference to the third point mentioned above, the distributional continuum from Malesia to Australia, New Caledonia, and New Zealand, some mention of Wallace's Line is perhaps required. The significance of Wallace's Line has been stressed by many zoologists, and it appears to have been a barrier to the spread of certain taxa of animals (Darlington 1957, 1965). This Line indeed may reflect the collision between the Australian and Asian crustal plates, but the implication that it divides two basically different angiosperm floras (Schuster 1969) is not demonstrable. It is unrealistic to assign the surviving primitive angiosperm groups to "Asian" and "Australasian" centers of origin. In the first place, how could such basically similar groups have originated at the two extreme ends of the ancient landmass extending from Laurasia through Africa into Antarctica and Australasia? (Contiguity of Africa and Australia is no longer an acceptable hypothesis in Gondwanic reconstructions, cf. McElhinny and Luck 1970.) Secondly, both primitive and advanced angiosperm families and orders have crossed Wallace's Line in Quaternary, Tertiary, or perhaps earlier times without impediment.

The above distributional hypothesis conflicts with the concepts of those biologists who maintain that angiosperms were widespread in Gondwanaland before its breakup. Some adherents of a concept of early diffusion throughout Gondwanaland (e.g., Axelrod 1970) are convinced that records of pre-Cretaceous angiosperms are valid, although a repetition of this viewpoint without new and incontrovertible evidence seems unproductive. Others (e.g., Maguire 1970) are impressed by similarities of the angiosperm components of the Guayana Highland and the planalto of Brazil, and by their apparent floristic relationships with tropical and southern Africa. The two South American shield areas are certainly remarkable refugia and secondary centers of speciation. They inevitably contain older elements than the intervening Amazonian lowlands, which have been subject to interglacial sea transgressions during the Pleistocene (Vuilleumier 1971). Nevertheless, the unique angiosperm elements of the Guayana Highland belong to comparatively advanced families; to push back their origins to the Juro-Cretaceous or to ignore their fundamental relationship at the ordinal level with Indo-Malesian groups is to neglect pertinent evidence. Still other students of angiosperm evolution (Plumstead 1961, Melville 1966) are sure that ancestral forms are to be sought among Gondwanic glossopterids, a viewpoint not supported by our growing acquaintance with the Permian Antarctic fossil record (Schopf 1970).

PRIMITIVE ANGIOSPERMS IN AFRICA AND SOUTH AMERICA

An examination of floristic lists indicates that there are a striking number of angiosperm families in common between Africa and South America, in addition to an appreciable number of genera. In the comprehensive and very useful Appendix to his *Geography of the Flowering Plants*, Good (1964) lists 92 angiosperm genera or generic pairs occurring disjunctively between Africa (and/or Madagascar) and South America. This statistic would be somewhat more impressive if one were not also aware of the much stronger links between tropical South America and the tropical western Pacific borderlands (see Thorne, pp. 37–42 herein). It seems to me that there would be a much larger element common to Africa and America if angiosperm families had occupied those regions before their separation. Nevertheless, we must explain discontinuities of this striking nature, infrequent as they are. I would suggest that each such case be critically examined with the following possibilities in mind: (1) the groups require more study; perhaps the taxonomy is faulty and the same or closely related genera are not actually involved; (2) long-distance dispersal is occasionally effective across the Atlantic; or (3) these taxa are relicts of pantropical genera now lacking in Asia-Malesia.

To illustrate the possible effects of these alternatives, I now propose to discuss a few of the putative primitive angiosperm groups, and subsequently to mention some more advanced groups common to Africa and South America. We may first eliminate altogether the 19 families (of my 60) that today are endemic within the Asian-Australasian area, occurring in neither Africa nor America. To be sure, some of these, such as the Trochodendraceae and Cercidiphyllaceae, were present in America or Eurasia in the Tertiary; that such families are today relict in eastern Asia tells us nothing positive about their place of origin. Next, let me briefly illustrate the workings of our model to explain the distributions of a few of our primitive families that are fairly widespread but do not occur in both Africa and South America.

The functioning of the Beringian route from southeastern Asia into America is illustrated by such families as the Magnoliaceae, Illiciaceae, Schisandraceae, Saururaceae, and Calycanthaceae. The first three of these have large centers of diversification and distinctly primitive elements in the Asian-Malesian region; their American elements are comparatively limited in diversity and are phylogenetically advanced. The two latter families are more evenly distributed between the hemispheres, as to diversity and abundance, but one may also hypothesize for them an Asian origin and a simple migrational route across Beringia in Cretaceous and/or Tertiary time.

To illustrate the functioning of the West Antarctic Archipelago route from Australasia into America our best example is perhaps the Winteraceae, a family of great diversity in Australasia, with a single comparatively advanced genus in America now persisting in a broken and relict pattern from Tierra del Fuego to southern Mexico. This same family demonstrates another distributional feature of some consequence, in that a single species of an otherwise Papuasian genus, *Bubbia*, occurs in Madagascar. The significance of a few floristic elements in common between Madagascar and New Caledonia has sometimes been overemphasized (Good 1950). It may be pointed out (Thorne 1965) that such elements are often basically characteristic of New Guinea and other parts of Malesia, and that they apparently spread in both directions from Malesia, southward into New Caledonia and westward into Madagascar. Another comparatively primitive family showing the Madagascar connection is the Chloranthaceae, which has five genera distributed in the same general pattern as the seven genera of the Winteraceae. In this case the Madagascar element is an endemic monotypic genus, but its derivation from a basically Indo-Malesian population is probable.

Many of our primitive angiosperm families are tricentric, with representatives in all three major land areas. As an example we may contemplate the present distribution of the Papaveraceae, a comparatively advanced ranalean family widely distributed in the Northern Hemisphere in a broad altitudinal range from sea level to more than 5500 meters. The genus *Papaver* has extensions south of the equator in Africa and Australia, while in America the genera *Bocconia* and *Argemone* extend southward in the Andes. The family is well represented in Tertiary European records. One may suggest that it originated toward the northern portion of our putative angiosperm homeland, becoming adapted to temperate and subarctic conditions, and subsequently spread throughout the Northern Hemisphere, sending separate extensions southward across the equator. The African and South American elements are not closely interrelated. The related Fumariaceae have a somewhat similar, but more limited, range; in South Africa there are four small endemic genera, and the typically northern genus *Corydalis* is known from high elevations in East Africa. Like that of the Papaveraceae, this distribution is readily explained in terms of our model. Both families show familiar disjunctions between eastern Asia and eastern North America that point to populational exchange during the Tertiary.

But the tricentric families that especially concern us are those with a predominantly tropical distribution, an example of which is the Menispermaceae, a family with about 70 genera and more than 400 species. Although most genera are not only strictly tropical, but in addition are endemic to one or another major region, the family extends well to the north and south of the equator, and there are Cretaceous and Tertiary reports far north of the present range. Although this distribution may be explained in terms of our model, it may be premature to assume that the movement into America was exclusively northern; in this case perhaps both the northern and the southern routes were functional. An exhaustive phylogenetic study of the family at some future date may clarify this point. At any rate, the African and South American elements are no more closely interrelated than each of them is related to the large Indo-Malesian element.

To illuminate the common occurrence of a taxon in Africa and South America, we may contemplate the so-called family Monimiaceae. The word "so-called" is advisedly used, as the Monimiaceae have been a traditional catch-all for genera with a vague similarity. The basic complex with which we are here concerned, within the ranalean complex, is characterized by having comparatively primitive xylem, ethereal oil cells, monocolpate pollen or pollen derived from that type, and unilacunar nodes. (Incidentally, these are not glaringly apparent characters, but they are some of the basic characters that taxonomists concerned with angiosperm evolution must use if they seek to understand relationships.) The interrelationships of families with these particular characters, comprising the order Laurales, became apparent only with the work of Money et al. (1950). Even when such related but well demarcated families as the Austrobaileyaceae, Hernandiaceae, Lauraceae, and a few others are set aside, there exists a residue that passed as Monimiaceae until the last decade or two, and that even now is placed in Monimiaceae in most of our major herbaria. In this limited sense, the family (of 37 genera and about 536 species) still exhibits such a wide range of morphological variability that its further division is required, and it has recently been replaced by six well-marked but interrelated families. To understand the distributional features of the Monimiaceae (sensu lato) we are definitely required to examine the distributions of its six component parts (Figure 3). Without going into the documentation outlined by Money et al. (1950), Sampson (1969), Schodde (1970), and others, I proceed from the most primitive family of the over-all complex to the most advanced.

In this discussion we see another example of the pitfalls encountered by an unwary phytogeographer who would accept such a family as Monimiaceae in an outmoded sense. He could, for instance, quote it as an example of the occurrence of related genera in Africa and South America. In fact, however, when such an unnatural family is broken down on a rational basis into six natural component parts, we see that five of them are Australasian in their basic aspects. Three of these parts are endemic to the Asian-Australasian re-

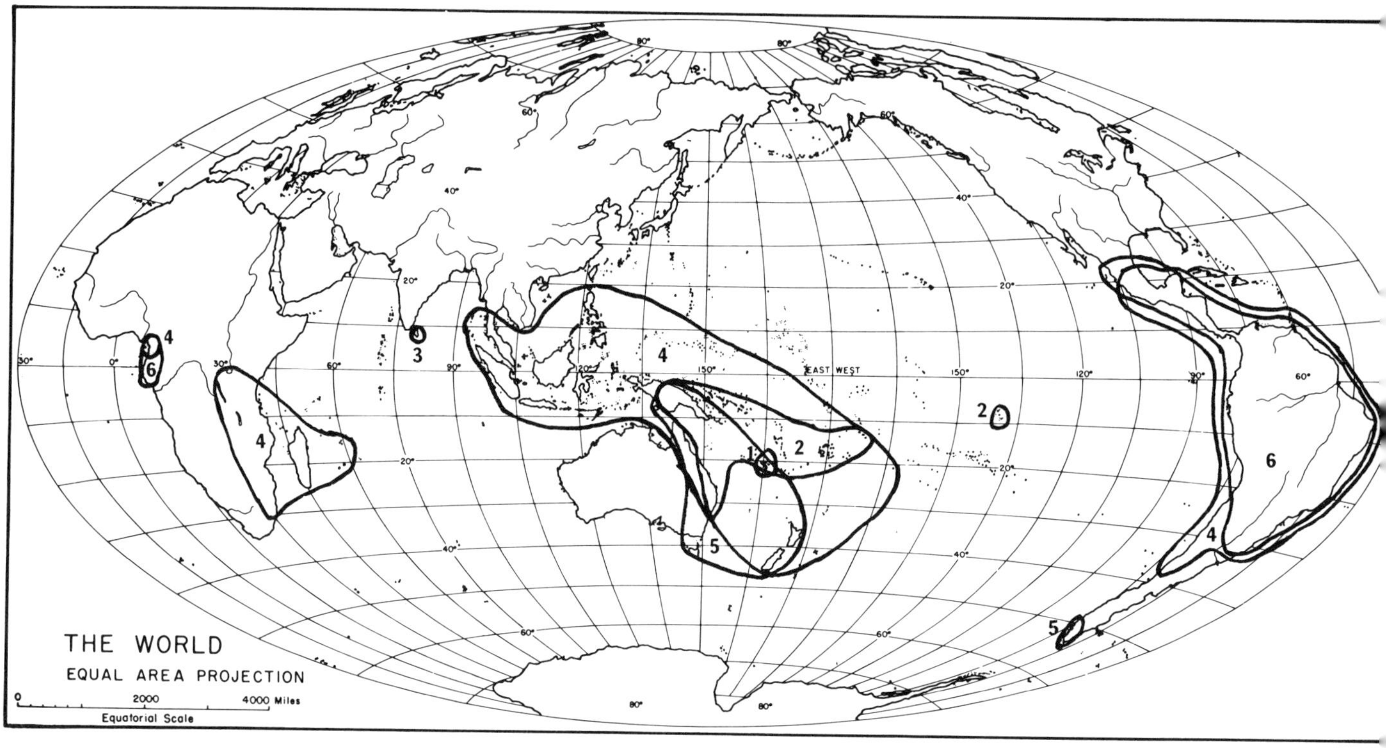

FIGURE 3. Generalized modern distribution of six families of the monimiaceous alliance. (1) Amborellaceae; (2) Trimeniaceae; (3) Hortoniaceae; (4) Monimiaceae; (5) Atherospermataceaea; (6) Siparunaceae.

gion: the monotypic Amborellaceae to New Caledonia; the Trimeniaceae, with two genera and eight species, extending from New Guinea and Australia to the Marquesas; and the unigeneric Hortoniaceae, with two or three species endemic to Ceylon. A fourth part, the Monimiaceae proper, occurs in all three tropical areas, but its center of diversity is the Asian-Australasian region (12 genera and 177 species); 6 genera with 51 species are known from the Mascarene-Madagascar area, but we note only one genus with 4 species from Africa proper; America has 5 genera and 98 species. A fifth part, the Atherospermataceae, has 6 genera and 17 species in Australasia, but one of these genera, *Laurelia,* with a single species in New Zealand, reappears in southern South America. Only the sixth part of our original complex, the Siparunaceae, is lacking from Asia-Australasia. This family currently consists of 3 genera and 174 species, all except five of them in the genus *Siparuna.* This, the most advanced family in the complex, probably originated in America, or enroute to America, from some Australasian ancestor that has now been lost to our knowledge. The Siparunaceae also have a West African genus, *Glossocalyx,* with four species, and this distribution we might be inclined to dismiss as an instance of long-distance dispersal from South America to West Africa at a time when the continents were not as far separated as they are today. But let's look again: Is this indeed the explanation of the African genus *Glossocalyx*? Money et al. (1950) seem unconvinced of the close relationship of *Glossocalyx* to *Siparuna* and *Bracteanthus,* the American elements of the Siparunaceae. In short, it is entirely possible that this African genus has acquired its superficially siparunoid characters from a quite different ancestor, and it is also possible that it reached Africa from Indo-Malesia directly rather than via America.

The Monimiaceae (sensu lato) illustrate several points basic in biogeographic philosophizing. First, the family is an unnatural group and its distribution can be understood only by analyzing the evolutionary history of its component parts; secondly, it still requires basic work (and is in fact now under intensive study) before phytogeographic conclusions can become really convincing; thirdly, upon such analysis as is currently feasible, we see that the complex is basically Asian-Australasian and that its spread to Madagascar, Africa, and America follows the pattern of our model, a fact by no means obvious prior to systematic investigations that are imperative for the soundness of any biogeographic discussion.

A ranalean complex with a distribution that may require a different explanation is the order Annonales (sensu stricto), composed of the families Annonaceae, Myristicaceae, and Canellaceae. The Annonaceae, with about 125 genera and more than 2000 species, occur in all three tropical areas, with an extension northward in eastern North America. Although generic relationships are complex, the family is sharply demarcated from its relatives. Walker (1971, 1972) concludes that the primary center of origin for the family was probably South America, possibly Africa, but certainly not Asia or Australasia. This apparent anomaly, among ranalean families, suggests a very early migration of a pre-annonalean element from southeastern Asia into Africa, a secondary evolutionary radiation there into basic tribes, and a subsequent dispersal of elements across the then narrow (or opening) Atlantic to America. An alternative explanation would call for the early Cretaceous arrival in America (via either the West Antarctic islands or Beringia), an evolutionary expansion there, and dispersal into Africa across the Atlantic. In any case, it is evident from the high degree of generic endemism that the Annonaceae are an ancient family. Similar continental generic disjunctions are shown by the two smaller families of the annonalean alliance, the Myristicaceae (Smith 1938, Sinclair 1958, Hutchinson 1964) and Canellaceae (Wilson 1960-1966). The former has four genera endemic to the Asian-Pacific region, five to Africa proper, two to Madagascar, and five to tropical America; the latter has one genus endemic to Madagascar, one to East Africa, and four to tropical America. The annonalean complex may be critical for our understanding of mid-Cretaceous angiosperms in America. Monosulcate pollen widely reported from the middle and late Lower Cretaceous as *Clavatipollenites* (Doyle 1969) may in part be referable to annonalean plants with sufficient vagility to have reached Africa and America before the major advances of other angiosperms.

The fact that some monocotyledonous families (Ayensu, p. 117 herein) show a strong Madagascar-African-South American link may suggest that they, like the annonalean families, moved into Africa earlier than the major angiosperm invasion and thus reached tropical America over a still narrow (or opening) Atlantic. This hypothesis, if borne out, would imply a very early monocot-dicot separation and would cast further doubt on the origin of monocots from any currently recognizable dicot complex like the Nymphaeales.

MORE ADVANCED ANGIOSPERMS IN AFRICA AND SOUTH AMERICA

If our basic model of angiosperm distribution should be deemed applicable to the ranalean dicotyledons, one may still ask: Can the distribution of every angiosperm family be so explained? Such a claim would indeed be difficult to substantiate. There are many families that seem definitely to have been autochthonous in either Africa or tropical America, some of them being endemic to one of these areas. We need not assume that every family originated in Asia-Australasia and followed the distributional model I have proposed. Many of our pantropical families date back in their origins at least to early Cretaceous, and perhaps the distributions of most of these follow the pattern I have described. As an example of such a chronologically old, but morphologically fairly advanced, family, I shall briefly discuss the Proteaceae. Other families, and we may mention the well-known examples of Bromeliaceae and Cactaceae, are essentially American endemics. Occasionally such families have very limited elements in West Africa. The Humiriaceae will serve as an example of this kind of family in the dicotyledons. Finally, there are certain genera of angiosperms occurring in all three tropical regions that, by features of their morphology and ecology, arouse our suspicions as to the naturalness of their distributions. In other words, do their present distributions show the influence of man? As an example of this type of distribution I propose to look briefly at the genus *Ceiba* in the family Bombacaceae.

The Proteaceae provide an illustration of how purely statistical considerations can lead to questionable conclusions as to (1) centers of origin and (2) migrational routes. The family consists of about 1400 species that are placed in 62 (Johnson and Briggs 1963) or 63 (Venkata Rao 1960) genera. The great preponderance of genera and species occurs in Australia, with about two-thirds of the species, and South Africa, with about one-quarter of the species. These areas of greatest species concentration are assumed by some authorities (Venkata Rao 1957, Beard 1959, Burbidge 1960, Good 1964) to exhibit strong evidence of a southern or at least an Australian origin of the family. The Proteaceae are one of several families (among them the monocotyledonous Restionaceae) that are sometimes discussed as an Afro-Australian phytogeographic element. Its presence has led some students to suggest that the ancestors of the modern Afro-Australian flora must have been established while Africa and Australia were interconnected by land. (In fact, however, even in the Paleozoic Africa and Australia seem to have been separated by the whole breadth of the East Antarctic continent (McElhinny and Luck 1970), from which, significantly, no angiosperm fossils are known.)

A careful analysis of evolutionary trends in the Proteaceae refutes both the probability of a southern origin and the influence of continental drift. In the words of Johnson and Briggs (1963:56): "This view [as to a southern origin] results from giving approximately equal weight to all genera irrespective of their relationships, degree of advancement, and specialized or unspecialized ecological characteristics. In statistical biogeography, argument from a summation of unweighted data has led to many misleading conclusions concerning past distributions and centres of origin." In the Proteaceae we find independent modifications of original features, and as complex a pattern of mosaic evolution as noted in ranalean families. By combining the morphological and cytological evidence, Johnson and Briggs point to a remarkable concentration of comparatively primitive, small, relict genera in the rain forests of northeastern Queensland. The suggestion of a southern connection between Australian and African elements is considered to be unsupported, and it is concluded that the Proteaceae passed from one continent to the other, or into both from outside, by a tropical route. The so-called Afro-Australian element disappears in the Proteaceae, as in the Restionaceae and similar groups, when comparative and evolutionary studies in plant geography are emphasized. Nothing in the review by Johnson and Briggs negates an explanation of proteaceous distribution in terms of our model. From an original center in Australasia, or perhaps in Malesia in the broad sense, the West Australian, African, and American elements were independently derived. The African and American genera are only remotely related, but there is a striking disjunction of four generic distributions between Australasia and South America (Sleumer 1955). In short, it would be extremely difficult to explain the tricentric distribution of the modern Proteaceae in any other terms than those of our model.

For another example of the African-South American disjunction I turn to the family Humiriaceae, which has recently been taxonomically clarified by Cuatrecasas (1961). The family consists of 8 genera and 49 species. Known fossils occur in the Tertiary of tropical South America, except for two from Europe. Both of these have now been identified as belonging to other families, one of the innumerable instances that should make botantists hesitate to accept fossil evidence without rigid confirmation. All of the extant species are South American except for *Sacoglottis gabonensis,* which is a frequent timber tree in the rain forests extending from Sierra Leone to northern Angola. The African species, one of eight in the genus *Sacoglottis,* is closely related to *S. amazonica,* which occurs in lowland forest from Costa Rica to Amazonian Brazil. Cuatrecasas (1961:36) hypothesizes that the single African *Sacoglottis* is an offspring of an ancestral form of *S. amazonica*: ". . . its ancestor at some time in the Tertiary may have found its way along the Brazilian current to establish itself on the West African coast. Furthermore, evidence has established the Amazon and Orinoco origin of the *Sacoglottis* drift endocarps found in the West Indies and British Isles, but no examples of *Sacoglottis gabonensis* have been found on American shores."

In this instance, then, we have an illustration of an African-South American discontinuity that was clearly due to the dispersal of ancestral fruits by sea. Oceanic drift in the Tertiary would have been facilitated by the then lesser distance between the continents, and there has apparently been adequate time for the two isolated populations to have evolved into quite distinct species. We can ask for no better indication of the efficacy of long-distance dispersal between continents.

Although any reasonable interpretation suggests that the Humiriaceae are an American family, with a purely fortuitous and disharmonic occurrence in West Africa, are we then to assume that the family originated in America? This is certainly quite possible, but it is necessary to examine the broader relationships of the family, which is a natural, homogeneous, and compact one. Its relationship (Hutchinson 1959, Thorne 1968) appears to be with the family Linaceae, with which it forms a cluster of taxa composed of four families. Two of these, the Linaceae and Ixonanthaceae, have tricentric and basically tropical distributions; a third, the Ctenolophonaceae, has an interesting disjunction of Malesia and tropical West Africa; and the fourth is the Humiriaceae. This group of taxa seems to have acquired its modern distribution through dispersal from Indo-Malesia along the routes we have discussed, but the last three families have acquired different types of disjunction suggesting that they are independently derived relicts evolved from a basic linaceous stock in Indo-Malesia. A superficial and statistical appraisal of the distribution of the modern Humiriaceae could be quite misleading without such an analysis. Each family and generic disjunction between West Africa and South America requires such a reexamination.

For an example of a modern species with a tricentric tropical range I turn to the paper by Baker (1965) on *Ceiba pentandra*, the well-known kapok tree of the family Bombacaceae. Its range coincides well with that of the tropical rain forest in America, West Africa, and Asia-Malesia. This perhaps unique specific distribution has caused botanists to question the naturalness of the range. Baker's conclusion is that *Ceiba pentandra* is native in the American tropics, since the other nine species of the genus are endemic there; the American population of the species is designated as var. *caribaea*. As to the West African forest representation of the species, there are minor morphological differences between it and the American forms, but it is also referable to var. *caribaea*. Nevertheless, in the savanna woodlands of West Africa, quite a distinct form has evolved, now referred to var. *guineensis*. Additionally, a strictly cultivated form is also known in West Africa, and it is demonstrated that this form arose through hybridization between the two varieties thought to be native to Africa. The cultivated form, indeed, is var. *pentandra*. This form is identical with cultivated southeastern Asian material, which was the basis of the Linnaean species and which therefore for nomenclatural purposes is the type-variety.

Thus it is concluded that the African and Asian cultivated variety must have had its origin in West Africa, the only region where both parental forms occur. There is evidence that the kapok tree was present in Asia as early as the tenth century A.D., at which time there was a well-developed trade between the Sudan and India. Baker's interesting detective work, supported by cytological and genetic experiments, indicates that *C. pentandra* owes its occurrence in Asia-Malesia to man's activities, but its occurrence in West Africa is natural, probably the result of pre-human long-distance dispersal by sea-drift.

The three examples of disjunction among comparatively advanced angiosperms that I have adduced, then, demonstrate different types of distribution in respect to their occurrence on both sides of the South Atlantic. The Proteaceae probably derive their present range from Cretaceous or early Tertiary migrations from the broad Asian-Australasian area; the African and American elements are clearly not interrelated. The Humiriaceae may be an autochthonus tropical American derivative of a basically Indo-Malesian linaceous stock, with an African species resulting from long-distance Tertiary dispersal. (To suggest that families with an unbalanced distribution, such as Bromeliaceae, Vochysiaceae, Rapateaceae, and Mayacaceae, indicate "the splitting apart of a more homogeneous flora by fragmentation of Gondwanaland" (Axelrod 1970:293) is to ignore the proven efficacy of long-distance dispersal.) The genus *Ceiba* is clearly a tropical American group; the tricentric occurrence of one of its species is due in part to carriage by man and in part to natural long-distance dispersal in the past. However, the Bombacaceae, the family to which *Ceiba* belongs, is pantropical. It is my thesis that every angiosperm disjunction, at the level of family, genus, or species, may logically be explained in terms of ultimate derivation from an Asian, Malesian, or Australasian ancestry. This ancestry is far from obvious in many instances, but it becomes clearer with every careful systematic study; and without such study, I would contend, biogeographic explanations are too often based on statistical records and on outmoded taxonomy, and can be very misleading.

In reference to the age of angiosperm genera, an

often-cited paper by Hawkes and Smith (1965) suggests that many modern genera, including such advanced ones as *Gossypium, Bromus,* and *Solanum,* were in existence by the early Cretaceous, 100–120 million years ago. It is also indicated that angiosperms, then dispersed throughout Gondwanaland, did not invade Laurasia until mid-Cretaceous. This viewpoint, needless to say, is not in accord with evidence suggesting a Laurasian (southeastern Asian) origin of angiosperms and a subsequent spread into Australasia, Africa, and America. Possibility of the persistence of modern genera for more than 100 million years, although verified by neither the macrofossil nor the microfossil record (Doyle 1969), has led various students to very detailed distributional hypotheses based on the assumed occurrence of such genera as *Gossypium* in pre-drift Gondwanaland (e.g., Johnson and Thein, 1970). When the taxonomy of *Gossypium* is analyzed in a somewhat different manner, however (Fryxell 1971), a late Cretaceous or Tertiary dispersal from an Australasian center becomes quite logical.

In summary, the present-day distribution of modern angiosperm genera provides no real evidence that they existed throughout Gondwanaland and rode on its fragments as these were dispersed by drift. On the contrary, the available evidence suggests that widespread extant angiosperm genera are Tertiary or late Cretaceous in origin, and that they acquired their present ranges along established migrational routes or by sporadic long-distance dispersal.

Conclusions

Evidence from evolutionary history indicates that Africa and South America received their initial stocks of angiosperms from different directions, from the general source area of southeastern Asia and adjacent Malesia. Inferences based on this large plant taxon suggest that the date of separation of Africa and South America most often mentioned by physical scientists—Jurassic or at the latest very early Cretaceous time—is reasonable. When these areas separated from actual contact or from contiguity they were presumably occupied by many groups of plants, but the angiosperms were not among them. A conceivable exception may be noted in the annonalean complex, the three families of which do indeed suggest a possible African or South American radiation, presumably secondary, with dispersal across a narrow or just opening Atlantic. In every other ranalean family that has been sufficiently well studied to permit analysis, the African and the South American elements seem separately related to an Asian-Australasian element which usually persists to a degree in that central area. In the vast majority of ranalean complexes we find no relationship between African and South American elements that would indicate a direct interchange of genetic components.

More modern tropical families than those of the ranalean complex usually illustrate the same type of origin and spread. However, there are also instances of disjunction in modern families and genera that imply a more direct and more recent interchange of genetic material between African and American populations. In such cases the explanation lies in the saltational long-distance dispersal of colonizing elements from one of the continents to the other. This latter pattern, when viewed against the whole picture of angiosperm distribution, is distinctly disharmonic, such groups being scattered throughout the evolutionary sequence. As the distance between Africa and South America has increased with the passage of time, the opportunities for direct interchange of floristic elements, except for such interchange as promoted by man, have been decreasing.

The similarity between the angiosperm floras of the tropical forests of Africa and America is thus superficial and represents the independent modifications of ultimately allochthonous elements. In this connection, a sentence from a recent discussion by Ehrendorfer (1970:185) is pertinent: "Groups with similar variation potential therefore will tend to produce parallel syndromes of structure and differentiation under similar environmental conditions." In summary, the modern angiosperm floras of Africa and South America, when broadly examined, refute any concept that these floras had a common origin in a shared continental area.

References

Audley-Charles, M. G. 1966. Mesozoic palaeogeography of Australasia. Palaeogeography, Palaeoclimat., Palaeoecol. 2:1-25.

Axelrod, D. I. 1970. Mesozoic paleogeography and early angiosperm history. Bot. Rev. 36:277-319.

Bailey, I. W. 1949. Origin of the angiosperms: need for a broadened outlook. J. Arnold Arb. 30:64-70.

Baker, H. G. 1965. The evolution of the cultivated kapok tree: a probable West African product. *In* Brokesha, D. (Editor). Ecology and Economic Development in Tropical Africa. Inst. Internat. Studies, U. Calif. Berk. Res. Ser. 9:185-216.

Beard, J. S. 1959. The origin of African Proteaceae. J. South Afr. Bot. 25:231-235.

Burbidge, N. T. 1960. The phytogeography of the Australian region. Austr. J. Bot. 8:75-211.

Cranwell, L. M. 1964. Antarctica: cradle or grave for its *Nothofagus*? *In* Cranwell, L. M. (Editor). Ancient Pacific Floras, 87-93. Univ. Hawaii Press. Honolulu.

Cronquist, A. 1968. The evolution and classification of flowering plants. 396 pp. Houghton Mifflin Co. Chicago.

Cuatrecasas, J. 1961. A taxonomic revision of the Humiriaceae. Contr. U. S. Nat. Herb. 35:25-214.

Darlington, P. J., Jr. 1957. Zoogeography: the geographical distribution of animals. 675 pp. Harvard Univ. Press. Cambridge, Mass.

————. 1965. Biogeography of the southern end of the world. 236 pp. Harvard Univ. Press. Cambridge, Mass.

Doyle, J. A. 1969. Cretaceous angiosperm pollen of the Atlantic Coastal Plain and its evolutionary significance. J. Arnold Arb. 50:1-35.

Ehrendorfer, F. 1970. Evolutionary patterns and strategies in seed plants. Taxon 19:185-195.

Fryxell, P. A. 1971. Phenetic analysis and the phylogeny of the diploid species of *Gossypium* L. (Malvaceae). Evolution 25:554-562.

Glaessner, M. F. 1962. Isolation and communication in the geological history of the Australian fauna. Pp. 242-249, *in* Leeper, G. W. (Editor). The Evolution of Living Organisms. Melbourne Univ. Press.

Good, R. 1950. Madagascar and New Caledonia: a problem in plant geography. Blumea 6:470-479.

————. 1964. The geography of the flowering plants, 3rd edition. 518 pp. John Wiley & Sons. New York.

Hair, J. B. 1964. Cytogeographical relationships of the southern podocarps. Pp. 401-414, *in* Gressitt, J. L. (Editor). Pacific Basin Biogeography, Bishop Mus. Press. Hawaii.

Hawkes, J. G., and P. Smith. 1965. Continental drift and the age of angiosperm genera. Nature 207:48-50.

Hutchinson, J. 1959. The families of flowering plants. 2nd edition, vol. 1: Dicotyledons. 510 pp. Clarendon Press, Oxford.

————. 1964. Myristicaceae. Gen. Fl. Pl. 1:146-153.

Johnson, L. A. S., and B. G. Briggs. 1963. Evolution in the Proteaceae. Austr. J. Bot. 11:21-61.

Johnson, B. L., and M. M. Thein. 1970. Assessment of evolutionary affinities in *Gossypium* by protein electrophoresis. Amer. J. Bot. 57:1081-1092.

Lam, H. J. 1961. Reflections on angiosperm phylogeny (I, II). Proc. Koninkl. Nederl. Akad. Wetens. Ser. C. 64:251-276.

Maguire, Bassett, 1970. On the flora of the Guayana Highland. Biotropica 2:85-100.

Martin, H. 1968. A critical review of the evidence for a former direct connection of South America with Africa. Vol. 1, pp. 25-53, *in* Fittkau, E. J., et al. (Editors). Biogeography and ecology in South America. Dr. W. Junk Publ. The Hague.

Maxwell, A. E., et al. 1970. Deep sea drilling in the South Atlantic. Science 168:1047-1059.

McElhinny, M. W., and G. R. Luck. 1970. Paleomagnetism and Gondwanaland. Science 168:830-832.

Meeuse, A. D. J. 1965. Angiosperms—past and present; Phylogenetic botany and interpretative floral morphology of the flowering plants. Adv. Frontiers Pl. Sci. 11:1-228.

————. 1971. Interpretative gynoecial morphology of the Lactoridaceae and the Winteraceae—a re-assessment. Acta Bot. Neerl. 20:221-238.

Melville, R. 1966. Continental drift, Mesozoic continents and the migrations of the angiosperms. Nature 211:116-120.

————. 1969. Leaf venation patterns and the origin of the angiosperms. Nature 224:121-125.

Money, L. N., I. W. Bailey, and B. G. L. Swamy. 1950. The morphology and relationships of the Monimiaceae. J. Arnold Arb. 31:372-404.

Plumstead, E. P. 1961. Ancient plants and drifting continents. South Afr. J. Sci. 57:173-181.

Sampson, F. B. 1969. Studies on the Monimiaceae, II: Floral morphology of *Laurelia novae-zelandiae* A. Cunn. (subfamily Atherospermoideae). New Zealand J. Bot. 7:214-240.

Schodde, R. 1970. Two new suprageneric taxa in the Monimiaceae alliance (Laurales). Taxon 19:324-328.

Schopf, J. M. 1970. Petrified peat from a Permian coal bed in Antarctica. Science 169:274-277.

Schuster, R. M. 1969. Problems of antipodal distribution in lower land plants. Taxon 18:46-91.

Scott, R. A., E. S. Barghoorn, and Estella B. Leopold. 1960. How old are the angiosperms? Amer. J. Sci. 258A:284-299.

Simpson, G. G. 1965. The geography of evolution. 249 pp. Chilton Books. Philadelphia.

Sinclair, J. 1958. A revision of the Malayan Myristicaceae. Gard. Bull. Singapore 16:205-466.

Sleumer, H. 1955. Proteaceae. Fl. Males. I, 5:147-206.

Smith, A. C. 1938. The American species of Myristicaceae. Brittonia 2:393-510.

————. 1967. The presence of primitive angiosperms in the Amazon Basin and its significance in indicating migrational routes. Atas Simpos. Biota Amaz. 4:37-59.

————. 1970. The Pacific as a key to flowering plant history. H. L. Lyon Arb. Lecture 1:1-27.

————. 1972. An appraisal of the orders and families of primitive angiosperms. J. Indian Bot. Soc. [in press].

Takhtajan, A. 1969. Flowering plants: origin and dispersal. [translated from Russian by C. Jeffrey] 310 pp. Oliver & Boyd. Edinburgh.

Thorne, R. F. 1964. Biotic distribution patterns in the tropical Pacific. Pp. 311-350, *in* Gressitt, J. L. (Editor). Pacific Basin Biogeography. Bishop Museum Press. Honolulu.

————. 1965. Floristic relationships of New Caledonia. Univ. Iowa Stud. Nat. Hist. 20(7):1-14.

————. 1968. Synopsis of a putatively phylogenetic classification of the flowering plants. Aliso 6:57-66.

Venkata Rao, C. 1957. Cytotaxonomy of the Proteaceae. Proc. Linn. Soc. N. S. W. 82:257-271.

————. 1960. Studies in the Proteaceae, I: Tribe Persoonieae. Proc. Nat. Inst. Sci. India B26:300-337.

Vuilleumier, B. S. 1971. Pleistocene changes in the fauna and flora of South America. Science 173:771-780.

Walker, J. W. 1971. Pollen morphology, phytogeography, and phylogeny of the Annonaceae. Contr. Gray Herb. 202:1-131.

————. 1972. Chromosome numbers, phylogeny, phytogeography of the Annonaceae and their bearing on the (original) basic chromosome number of angiosperms. Taxon 21:57-65.

Wilson, T. K. 1960-1966. The comparative morphology of the Canellaceae. I. Trop. Woods 112(1960):1-27. II. Amer. J. Bot. 52(1965):369-378. III. Bot. Gaz. 125(1964):192-197. IV. Amer. J. Bot. 53(1966):336-343.

Palms in the Tropical Forest Ecosystems of Africa and South America

HAROLD E. MOORE, JR.

The Role of Palms in the Ecosystem

Palms epitomize the tropics in the minds of many. Thus it seems especially appropriate to consider them in a symposium on the comparative evolution of tropical forest ecosystems in Africa and South America. I propose to treat the palms of these continents in two contexts: first to consider briefly their role in the ecosystem for whatever light this may throw on evolution; secondly to consider in some detail their diversity, adaptations, and present and past distributions.

Caveats must be registered before attempting any study of palms. Excepting the work of Read (1967) on *Thrinax* in Jamaica, studies of the biology of palms are in their infancy, though replete with possibilities. Palms are notable components of forests on poorly drained soils (see Richards 1952, for example); yet soil and water relationships of palms, to say nothing of light and temperature relationships, have been little studied. Though Walter (1964: 165) suggests that tree monocotyledons (*Pandanus*, Palmae) compete favorably with dicotyledonous trees on poorly aerated soils (as do monocotyledonous herbs with dicotyledonous herbs), no physiological basis is given. Despite the impressive diversity in morphology of inflorescence, flowers, and fruit, and the presence of internal and external protective mechanisms such as crystals, fiber sheaths (Uhl and Moore 1973), and spines (Tomlinson 1962, Janzen 1969) we may, for the most part, only speculate on the interrelationships between palms and pollinators, palms and predators, palms and agents of dispersal, palms and other plants.

Harold E. Moore, Jr., L. H. Bailey Hortorium, Cornell University, Ithaca, New York 14850. Acknowledgments: Much of the background work for this paper has stemmed from or been associated with National Science Foundation grants G-18770, GB-1354, GB-3528, GB-7758, and GB-20348X. I am much indebted to a host of colleagues and students with whom I have discussed the contents and in particular to J. Dransfield, C. H. and N. W. Uhl, and C. E. Wood, Jr., who have read and commented on the manuscript. All conclusions, however, are my own. Maps have been plotted on Goode Base Map Series No. 201HC, World, Homolosine, Henry M. Leppard, Editor, copyright by the University of Chicago.

PALMS AND ANIMALS IN THE ECOSYSTEM

Palms have often been considered to be largely wind-pollinated, though they are abundant in the rain forest which "structurally and floristically . . . is inhospitable for anemophily" (Whitehead 1969: 32). Schmid (1970) has shown in *Asterogyne martiana* the interrelationship between floral morphology, phenology, and insects which pollinate (syrphid flies) or otherwise visit flowers for food (pollen, nectar, floral tissue), a site for mating, or a site for oviposition. Essig (1971) has shown the intricate interaction among bees, beetles, and flowers in two species of *Bactris* and the probable relationship of beetle pollination to floral anatomy. Both Schmid and Essig note that stingless bees gather pollen and take nectar from staminate flowers but seldom visit pistillate flowers. Though wind is presumably or clearly the agent in pollination of some palms [e.g., *Phoenix dactylifera* (many authors), *Arenga pinnata* (Miller 1964), *Cocos nucifera* (Menon and Pandalai 1958), *Elaeis guineensis* (Hartley 1967), *Thrinax* (Read 1967)], the abundance of pollen and multiplicity of stamens or excess of staminate flowers produced by many palms of the rain forest may well be an adaptation to insect depredation rather than to wind pollination.

Palms figure prominently in the life cycle of a variety of insects, though most attention has been paid to those that affect commercially important palms such as the coconut (see Lever 1969). Lepesme (1947) devotes several pages to the various habitats offered by palms, ranging from the roots and stem to the crown, inflorescence, flowers and fruit, each tending to have its particular fauna of insects. The trunk and leaf bases may also support a flora of epiphytes from lichens to phanerogams and a varied fauna from mites to vertebrates. Some insects are host-specific, others are obligate or facultative associates of palms. Some genera of palms appear to be host to only one or a few species of insects, but *Cocos* is host to 751 (Lepesme 1947:127) of which 165 or 22 percent are host-specific; another 278, or 37 percent, are essentially restricted to palms. H. Scott (1933) devoted a section of three pages to "Insects living between leaf-bases of palms and *Pandanus*" in his account of the insect fauna of the Seychelles and adjacent islands. Leaf-axils of some palms and *Pandanus* are nearly comparable to the "aquaria" of Bromeliaceae. Scott wrote (1933:339):

> Picado showed that permanent marshes and other standing terrestrial waters do not usually exist in great tropical forests. . . . In the mountains of the Seychelles the extreme steepness of the ground, leaving scarcely any level spaces in which water can collect in standing pools, constitutes an additional factor against their [marshes] existence. But their place is taken by the accumulations of wet vegetable detritus and clear water in the axillary spaces between the overlapping bases of the leaves of monocotyledonous plants, which Picado likened to a great marsh split up into numerous little divisions, lifted to varying heights above the soil. The separate territories of marsh formed by each of these plants are further subdivided into many compartments, as the spaces between the sheathing leaf-bases do not intercommunicate and the water and detritus stand at different levels in the several compartments of a single plant. It is obvious that this separation into compartments of very limited size will, by limiting the food supply, have marked effects on the number and nature of the animals inhabiting the axillary spaces, particularly in the case of predaceous forms. Various causes combine to bring about that some members of the fauna of "reservoir-plants" display astonishing modifications adapting them to their very specialized biological *milieu*. Throughout my stay in the Seychelles I devoted special attention to this *milieu* and its inhabitants, often felling trees and cutting off the sheathing leaf-bases one by one, from the outermost inwards, carefully collecting every visible form of animal life found between them.

Of these palms in the Seychelles, *Phoenicophorium, Verschaffeltia,* and *Lodoicea* leaf-axils yielded insects of 6 orders and 62 species, 38 in the axils of *Phoenicophorium* alone, 20 in the axils of *Lodoicea.* Eighteen species were found exclusively in leaf-axils. While this may represent an unusual situation, experience in the American tropics suggests that it may not be unduly exaggerated with respect to some American palms.

Palm fruits are also an important source of food. Leck (1969) touched on the progression of animals feeding on palms as fruit ripens and noted the dietary shift of migrant birds from insects to the fruit of *Roystonea. Prestoea montana ("Euterpe globosa")* is an important parrot food in Puerto Rico (Kepler 1970). Rodents, chiroptera, and ungulates appear to be the chief vertebrate feeders apart from birds, but Janzen (1971) finds bruchid beetles associated with the seeds of *Scheelea rostrata* in Costa Rica in a relationship probably quite as complex as that in fruits of legumes which they more frequently attack (Janzen 1969). Zacher (1952) lists species referable to 18 genera of palms that are fed on by larvae of the bruchid genera *Caryobruchus* and *Pachymerus.*

Man, however, is the animal in the South American forest that utilizes palms in the greatest number of ways. Braun (1968:46) has recently given an account of the usefulness of palms in Venezuela. He lists palm products in order of importance as leaves, stems, fruits, fibers, roots, and writes: "An Indian village in the tropics is always built where a rich palm flora prevails, for here man depends on palms." The versatility of palms in the hands of man is astonishing. Houses, baskets, mats, hammocks, cradles, quivers, packbaskets, impromptu shelters, blowpipes, bows, starch, wine, protein from insect larvae, fruit, beverages, flour, oil, ornaments, loincloths, cassava graters, medicines, magic, perfume—all are derived from palms. The importance of man as a biotic factor in the tropical ecosystem has been argued (Richards 1952, 1963; C. O. Sauer 1958). However, to whatever extent man has been involved in the tropical ecosystem, palms have certainly been a major factor in making possible this involvement and even today, despite the advent of the corrugated tin roof and the rifle, they are of primary importance to many primitive American cultures.

PALMS AND THE VEGETATION

The foregoing paragraphs have touched very briefly on relationships between palms and animals as we so imperfectly understand them. It remains to consider relationships between palms and other plant components of the tropical ecosystem. I propose to do so in two ways: first, to touch on the role of palms in ecological succession; secondly, to note the place of palms in climax formations.

Palms for the most part are ill-adapted to truly xeric situations. Even those that grow in deserts depend on underground sources of water (e.g., *Washingtonia filifera,* Vogl and McHargue 1966, or *Wissmannia carinensis,* Moore 1971b). It is not surprising, therefore, to find that they are often principal components of the hydrosere. Richards (1952) has reviewed hydroseres in South America, Africa, and the eastern tropics. Except that there is no strict counterpart for the mangrove palm, *Nypa,* outside Asia (though Richards considers the American *Manicaria* analogous), a parallel entry of palms into the succession may be seen in all three parts of the world. In South America, following the establishment of herbaceous swamp, palms appear to enter as initial stages in the formation of seasonal swamp forest or of true swamp forest (following the terminology of Beard 1944, 1955). Along the Atlantic lowland and parts of the Pacific lowland of Central America, *Raphia taedigera* is able to withstand a degree of salinity (Allen 1965) and may form pure stands hundreds or even thousands of square acres in extent. Anderson and Mori (1967) suggest that in Costa Rica, the *Raphia* palm swamp is a pioneer community on imperfectly drained sites giving way to forest with more dicotyledons as drainage improves. In the Amazon estuary, *Raphia taedigera* grows at the back of the mangroves within the limits of tidal influence (Bouillenne 1930) while *Euterpe oleracea, Mauritia flexuosa, Manicaria saccifera,* and *Maximiliana martiana* (*M. regia*) succeed mangroves on coastal and river margins and may lead directly to the establishment of fresh-water swamp forest climax. *Astrocaryum jauari* replaces *Mauritia* on river margins from the mouth of the Xingú River in Brazil to Peru, with *Mauritia* then farther from the river banks at the margin of forest (Bouillenne 1930). Pure stands of *Mauritia* ("aguajales") occur in old oxbows of the rivers in eastern Peru where they are an unfailing indicator of poorly drained acid soils. Scattered individuals of the species indicate either local swampy land or rainfall of two meters per annum or more (Moore, Salazar, and Smith 1960).

In Nigeria, palm swamp, dominated by one or more species of *Raphia,* is a stage in the succession to tall fresh-water swamp forest of dicotyledonous trees (Richards 1939). Ainslie (1926) describes briefly a similar fresh-water swamp forest in Nigeria, where it is a landward extension of the mangrove, with *Raphia, Eremospatha,* and *Calamus* extending into high forest. Chipp (1927) describes a seral *Calamus-Ancistrophyllum-Raphia* association in Ghana adjoining lagoon marsh with *Raphia hookeri* sometimes in a pure consocies. *Phoenix reclinata,* in a narrow belt of swamp about 20 yards wide, separates the herbaceous swamp communities from the forest margin in the Namanve swamp on Lake Victoria (Eggeling 1935), while swamps of *Raphia monbuttorum* lie between the lake and forest on the Sese Islands of the same lake (Thomas 1941). In the eastern tropics, *Metroxylon sagu* forms swamp forests in New Guinea (Barrau 1959).

Palms appear to figure in other successions though less has been written about them. Blydenstein (1967) noted that on the Colombian llanos, gallery forest along small streams developed rapidly from formations dominated by *Mauritia minor* to denser broadleaf semi-evergreen forest; similarly, groves in dry savanna of the piedmont region develop around a nucleus of *Acrocomia* sp. with species of *Annona, Casearia,* and *Davilla* as pioneer members. *Prestoea montana ("Euterpe globosa")* is a colonizer of landslip scars in the West Indies, forming "palm brakes," a seral stage leading to montane thicket (Beard 1949). It also forms a high (75) percentage of the hurricane forest on St. Vincent (Beard 1945). Bannister (1970), however, suggests that the palm is a normal component of climax rain forest vegetation in Puerto Rico and that palm forests appear on slopes because of their ability to anchor in saturated soil and because light is available at a critical stage. *Acrocomia ierensis* in Trinidad "characterizes bush on areas of shifting cultivation and is not known in natural forest except for the fringes of the Erin savanna" (Beard

1946). Janzen (1971) notes that he has not seen *Acrocomia* except in disturbed areas in Costa Rica and the same may be said for Mexico, where I have always seen it in second growth. Beard writes that in Trinidad a deflected succession toward fire grasslands begins with increasing abundance of *Maximiliana elegans,* or more rarely *Acrocomia* or *Scheelea,* which eventually may form groves, then thin out and be replaced by grass; *Orbignya* may behave in somewhat similar fashion in Brazil (Sternberg 1968).

PALMS IN THE CLIMAX FORMATIONS OF SOUTH AMERICA

The optimum formation of mixed rain forest varies as to palm content (see Beard 1955 for categories to follow). Davis and Richards (1934) list palms, most frequently *Jessenia,* as occasional in the lower canopy in Guyana while Fanshawe (1952) states that palms are few and only occur in the shrub or undergrowth layers. Espinal and Montenegro (1963), however, remark on the abundance of palms in the rain forest of the Chocó, Colombia, which has an extraordinary assemblage of species for so limited an area. In seasonal formations, palms may play a more distinctive role. The evergreen seasonal forest characteristically has representatives of some cocosoid palms such as *Astrocaryum, Maximiliana, Scheelea,* the arecoid genera *Dictyocaryum, Euterpe, Iriartea, Jessenia, Socratea,* and the nearly ubiquitous *Mauritia,* all of which reach at least the lower story of the canopy where I know this forest in Peru, Colombia, and Venezuela. *Bactris, Geonoma, Iriartella, Hyospathe,* and *Lepidocaryum* are representative of the understorey while *Leopoldinia* is usually a riparian genus.

In the semi-evergreen seasonal forest of Trinidad, *Sabal, Coccothrinax,* and *Scheelea* occur in both stories of some associations, *Bactris* in the understorey (Beard 1946). *Sabal* and *Scheelea* occur in forests of this nature in Venezuela also. *Maximiliana elegans* is the sole palm in the deciduous seasonal forest in Trinidad. Many species of cocosoid genera of the *Syagrus* alliance are apparently associated with this as well as with thorn woodland and cactus scrub in continental South America, especially in the "caatinga" and "cerrado" of southern Brazil (Braun 1968, Bouillenne 1930, Goodland 1970, Rizzini 1963).

Among montane formations, palms figure prominently in lower montane and montane rain forest or cloud forest. They are found only rarely (*Geonoma, Ceroxylon*) in montane thicket, where *Ceroxylon* stands above the low forest at high elevations in Peru. In elfin woodland of Trinidad, Beard lists *Euterpe* and *Prestoea.* Many genera of the seasonal evergreen forest extend into the lower montane forest on the South American continent (*e.g., Dictyocaryum, Iriartea, Astrocaryum, Bactris, Geonoma, Hyospathe, Euterpe, Socratea*) though Beard considers them unimportant in this formation in Trinidad. In the cloud forest, *Chamaedorea, Geonoma, Euterpe, Prestoea, Catoblastus, Dictyocaryum, Socratea,* and *Ceroxylon* spp. predominate and may be very abundant as on the upper levels of Ptari-tepui or in the coastal range of Venezuela.

Of dry evergreen formations, Beard cited the littoral woodland, wherein *Roystonea oleracea* is an impressive element in Trinidad; presumably *Allagoptera arenaria* on the beaches of southern Brazil would fit into the category of littoral hedge. *Syagrus orinocensis* might be categorized as rock pavement vegetation, except that it occurs in a seasonal deciduous formation, judging from Braun (1968:43) who wrote:

> In the northern part of Amazon Territory there are savannas which are interrupted by the enormous granite massifs of the Maypures region. Here are the natural habitats of *Syagrus orinocensis,* an extremely hardy and tough palm which grows abundantly on the steepest granite slopes, withstanding there many months of excessive dryness and high daytime temperatures. . . . Dwarf shrubs, where they still exist, are stripped of foliage during this dry season, yet the palm remains green.

The edaphic climaxes of seasonal-swamp and swamp formations are sometimes dominated by palms. Beard (1944) recognized four variations dominated by palms: palm marsh, palm swamp, palm brake, and marsh forest. *Mauritia* spp. (especially *M. flexuosa*), *Euterpe* spp. (especially *E. oleracea*), *Manicaria, Jessenia,* and *Roystonea* are components of both formations; *Raphia, Manicaria* and *Mauritia* are perhaps most frequent in swamp forest. Myers (1933:345) says of wet savanna in the Orinoco delta that there is "an uninterrupted sea of sedges fringed by "aeta" palms (*Mauritia flexuosa*) then passing into the dominant swamp palm forest which covers most of the 12,000 square miles of the delta." *Leopoldinia, Astrocaryum* spp. and

Bactris spp. are characteristic of swamp woodland. *Copernicia tectorum* and *Mauritia* spp. are characteristic of the extensive savanna formations of Colombia and Venezuela, though *Acrocomia* and *Sabal mauritiiformis* may also figure. Equally or more extreme seasonal forests are those of *Copernicia prunifera* in northeastern Brazil and the "palmares" in the Gran Chaco of Paraguay and adjacent Argentina, Bolivia, and Brazil, where *Copernicia alba* forms solid stands hundreds of square miles in extent. The largest contain perhaps half a billion individuals and they occur on areas subject to periodic flooding followed by drainage and drying (Markley 1955).

PALMS IN THE CLIMAX FORMATIONS OF AFRICA

South American palms are represented in some 18 or 19 of the 28 formations listed by Beard (1955). The diversity of habitats in Africa, by contrast, is markedly less extensive. Excluding *Chamaerops* of the North African macchia, *Medemia* of oases in the Nubian desert, and *Jubaeopsis* in the coastal forest-savanna mosaic of South Africa, five chief habitats seem to harbor palms in the tropical belt. The most striking difference from South America is the complete lack of palms in montane regions of Africa: none is known to exceed the 1000 meter elevation line if, in fact, any attains that elevation. Since no listing of climax formations for Africa is quite comparable to that of Beard's, I use terms in part derived from Keay (1959) and in part from Chipp (1927).

Most important and extensive (as a natural formation) is swamp forest near the coast or pocketed in the rain forest, whether seral or as an edaphic climax. Lepidocaryoid palms (*Ancistrophyllum, Calamus, Eremospatha, Raphia*) are especially prominent in swamp forest in West Africa, but *Phoenix reclinata* is also a significant element. *Raphia, Phoenix,* and sometimes *Calamus* are important in Central Africa (Eggeling 1947) and in East Africa, where they may form edaphic climax formations (Bogdan 1958). *Phoenix reclinata* in West Africa may also be an element of the strand association where soil is sandy and mangroves are absent (Ainslie 1926, Chipp 1927). *Hyphaene* is largely a genus of East African tropical subdesert steppe or moist to dry coastal forest-savanna (after Keay). Engler (1908: pl. 12) for example, figures *Hyphaene* on the coast at Dar-es-Salaam.

The savanna or parkland is the habitat of *Borassus aethiopum* (sometimes referred to as *B. flabellifer* in the literature on Africa) which is often a prominent feature of the landscape in West Africa (Chipp 1927) and Central Africa. *Phoenix reclinata* is also present as a component of riparian fringing forest or in a peculiar termite mound complex in valley grassland (Thomas 1945). *Hyphaene* may also occur in a palm-grassland formation (Walter 1964). *Elaeis guineensis,* though widely distributed as a cultivated palm or persisting after cultivation, occurs also in derived savanna (Clayton 1958). According to Zeven (1967), it is probably native in a riparian habitat or in the forest fringe of the savanna where light is sufficient.

It may be judged from the foregoing, and correctly I think, that palms play an important role in the active evolution of the tropical forest ecosystem today, especially in South America. Undoubtedly they also have in the past, perhaps to an even greater extent. They pioneer in the formation of or become dominant in swamp forest; they pioneer on landslips or are the nucleus about which gallery forests develop on the llanos. They provide habitats, sometimes very special ones, and/or food for various invertebrate and vertebrate animals; they are used in many ways by man, who most actively changes the tropical forest and who, at least in Africa, has substantially modified the patterns of vegetation in the past few millennia.

Some Comparisons Between the Palm Floras of Africa and South America

We may next ask what palms occur on the two continents, where they may have come from, what their relationships may be, and why the floras differ. Some notes of introduction, however, are again required before embarking on any comparative study. Important and imposing as they are in the tropical landscape, palms have often been ignored entirely or poorly collected by field botanists because of the problems of logistics and time involved. The kinds of bases for comparison, both in terms of morphology and of distribution, that are available for more tractable subjects, for example the grasses (excepting bamboos), are mostly lacking for the palms. All too often for a given species we know only the flowers or only the fruit, but not both in association. The species, in turn, may be known from only a single collection. Although the nature of the inflorescence is of major importance in understanding relationships and the biology of palms, only infrequently is the entire inflorescence with its attendant bracts collected or described.

Similarly, the leaf, which may exceed about 20 meters in length for *Raphia* (see Anderson and Mori 1967), is often represented in the herbarium by a section with a few pinnae, but without notes on the whole blade or the petiole and sheath, the nature of which may be of importance at both generic and specific levels, as in *Thrinax* (Read 1967). Thus, when figures for numbers of species are used in text or in tables, it should be understood that, especially for larger genera, these are the best estimates that can be obtained by combining data from monographic or synoptic treatments and standard indices with those in unpublished notes and manuscripts. Though distinctive species remain to be described, the number of names now in use probably will be substantially reduced when monography is undertaken (e.g., Wessels Boer 1968 for *Geonoma,* Moore 1971a for *Synechanthus*). The distribution of palms is also difficult to work out, except in rather general terms, because of inadequate collections. Moreover, we do not know the extent of man's activity in the dispersal of many useful taxa, for example species of *Actinorhytis, Areca, Borassus, Caryota,* and *Metroxylon.*

I elect, therefore, to treat palms very largely at the generic level in the present paper, noting that even here an evaluation of taxonomic limits is as yet incomplete and that the number of genera presently accepted will doubtless decline in the future as it has in the recent past. With these further caveats registered, we may attempt an assessment of palms in Africa and in South America in the context of the palm flora of the world.

For convenience, since palms are essentially tropical, I have not attempted to eliminate those few that are represented in subtropical or warm temperate regions. I have broken the world palm flora into four geographic components for purposes of analysis: (1) the Western Hemisphere, which is further divided into South America (including Trinidad, Tobago, and the Juan Fernandez Islands), and North America (including Bermuda, Bahama Islands, Greater and Lesser Antilles); (2) Africa (including the Arabian peninsula, Mediterranean Europe, the Canary Islands); (3) the Indian Ocean region with Madagascar, the Comore Islands, Pemba, the Mascarene Islands, and the Seychelles Islands; (4) the eastern tropics including the remainder of the Old World.

Although the coconut (*Cocos nucifera*) is cultivated throughout the tropics, for purposes of analysis, following Lepesme (1947) and Child (1964), I have considered it probably of Melanesian origin, though it has been mapped for a broader Indo-Pacific region to accommodate alternatives of Beccari (1917) and J. D. Sauer (1967). *Phoenix dactylifera,* a second cultigen, is included with both the African and Asiatic floras; *Phoenix reclinata* and *Raphia farinifera* are considered both with Africa and with the islands of the Indian Ocean because of their possibly natural occurrence on Madagascar and the Comore Islands. *Elaeis guineensis* is accepted as introduced into Madagascar, following Jumelle (1945) and Zeven (1967), and is thus omitted from the listing for the Indian Ocean region. Apart from these species, none is common to more than one major region, though seven genera of continental Africa are shared with other continents—*Raphia* and *Elaeis* with South America, *Chamaerops* with Europe, *Phoenix, Borassus, Hyphaene,* and *Calamus* with Asia.

THE KINDS OF PALMS AND THEIR DISTRIBUTION

The palm order Principes [= Arecales] is unquestionably natural and ancient. Among tropical monocotyledons, it probably ranks after only the orchids and perhaps the grasses in number of genera and species. Among the world's monocotyledons it is exceeded by the Orchidales, Poales, and Liliales in number of genera, and by the foregoing and the Cyperales in number of species (based on figures from Melchior 1964). I have recently reviewed the geographic distribution and the substantial diversity in morphology of inflorescence, flower, and fruit of palms in an effort to present a preliminary and informal arrangement that would reflect relationships and levels of evolutionary advancement somewhat more clearly than has been true in the past (Moore 1973). This arrangement is by no means final; yet it is built from as much and as many kinds of information as it has been possible to assemble at present. Fifteen groups of palms have been recognized tentatively (Table 1) repre-

TABLE 1. Distribution of Genera, Endemic Genera, and Species of Palms by Groups and Major Geographic Regions.

	Western Hemisphere																			
	North America			*South America*			*American Total*			*Africa Arabia Europe*			*Madagascar Mascarenes Seychelles*			*Eastern Tropics*			*Grand Total*	
Palm Lines and Groups	*Genera*	*Endemic genera*	*Species*	*Genera*	*Endemic genera*	*Species*	*Genera*	*Endemic genera*	*Species*	*Genera*	*Endemic genera*	*Species*	*Genera*	*Endemic genera*	*Species*	*Genera*	*Endemic genera*	*Species*	*Genera*	*Species*
I Coryphoid	13	9	91	7	3	15	16	16	105	3	2	3	–	–	–	14	13	214	32	322
II Phoenicoid	–	–	–	–	–	–	–	–	–	1	0	5	1	0	1	1	0	12	1	17
III Borassoid	–	–	–	–	–	–	–	–	–	3	1	42	5	3	8	3	1	6	6	56
IV Lepidocaryoid	1	0	1	3	2	31	3	2	31	5	3	60	1	0	1	16	15	573	22	664
V Nypoid	–	–	–	–	–	–	–	–	–	–	–	–	–	–	–	1	1	1	1	1
VI Caryotoid	–	–	–	–	–	–	–	–	–	–	–	–	–	–	–	3	3	35	3	35
VII Pseudophoenicoid	1	1	4	–	–	–	1	1	4	–	–	–	–	–	–	–	–	–	1	4
VIII Ceroxyloid	–	–	–	2	2	18	2	2	18	–	–	–	2	2	12	–	–	–	4	30
IX Chamaedoreoid	4	2	103	3	1	39	5	5	141	–	–	–	1	1	5	–	–	–	6	146
X Iriarteoid	3	0	3	8	5	50	8	8	52	–	–	–	–	–	–	–	–	–	8	52
XI Podococcoid	–	–	–	–	–	–	–	–	–	1	1	2	–	–	–	–	–	–	1	2
XII Geonomoid	6	2	29	4	0	77	6	6	92	–	–	–	–	–	–	–	–	–	6	92
XIII Cocosoid	11	2	78	24	15	507	26	25	580	2	1	2	–	–	–	1	1	1	28	583
XIV Arecoid	8	1	24	9	2	91	10	10	109	1	1	3	19	19	105	58	58	543	88	760
XV Phytelephantoid	1	0	6	4	3	9	4	4	15	–	–	–	–	–	–	–	–	–	4	15
Totals	48	17	339	64	33	837	81	79	1147	16	9	117	29	25	132	97	92	1385	212	2779

senting five major lines along which palms appear to have evolved.

These major lines may be characterized in very general terms. Groups I-III may be termed the coryphoid line. This line includes all palms with leaves induplicately folded in bud, except the caryotoid palms; all palms with costapalmate or palmate leaves, except two South American lepidocaryoid genera; and all apocarpous palms except *Nypa*. The inflorescence in general is unspecialized, many genera have perfect flowers, and many genera are also adapted to more seasonal and/or drier and/or cooler conditions than most other palms. All the remainder, except the caryotoid palms, have reduplicately folded leaves. Group IV limits only the lepidocaryoid line, which are characterized by the presence of scales on the ovary and fruit. Groups V and VI include nypoid and the caryotoid lines respectively while groups VII-XV compose what may be termed an arecoid line.

The component groups of the arecoid line are diverse and not easy to characterize as a whole, but they form three subunits: (1) The pseudophoenicoid, ceroxyloid, and chamaedoreoid groups are interrelated (Moore 1969). (2) The iriarteoid, podococcoid, geonomoid, cocosoid, and arecoid groups are all characterized by a sympodial unit of two staminate flowers and a pistillate flower called a triad (Uhl 1966). (3) The dioecious phytelephantoid palms form a very highly specialized third unit. I propose now to assess these groups from the point of view of their characteristics briefly stated, their relationships, adaptations, and past and present distributions as they relate to the geographic regions of the world outlined in previous paragraphs, and particularly as they relate to the two continents with which we are here concerned.

Before embarking on this survey, it is important to note that my coverage of the fossil record is incomplete, as is the record itself. Many papers are

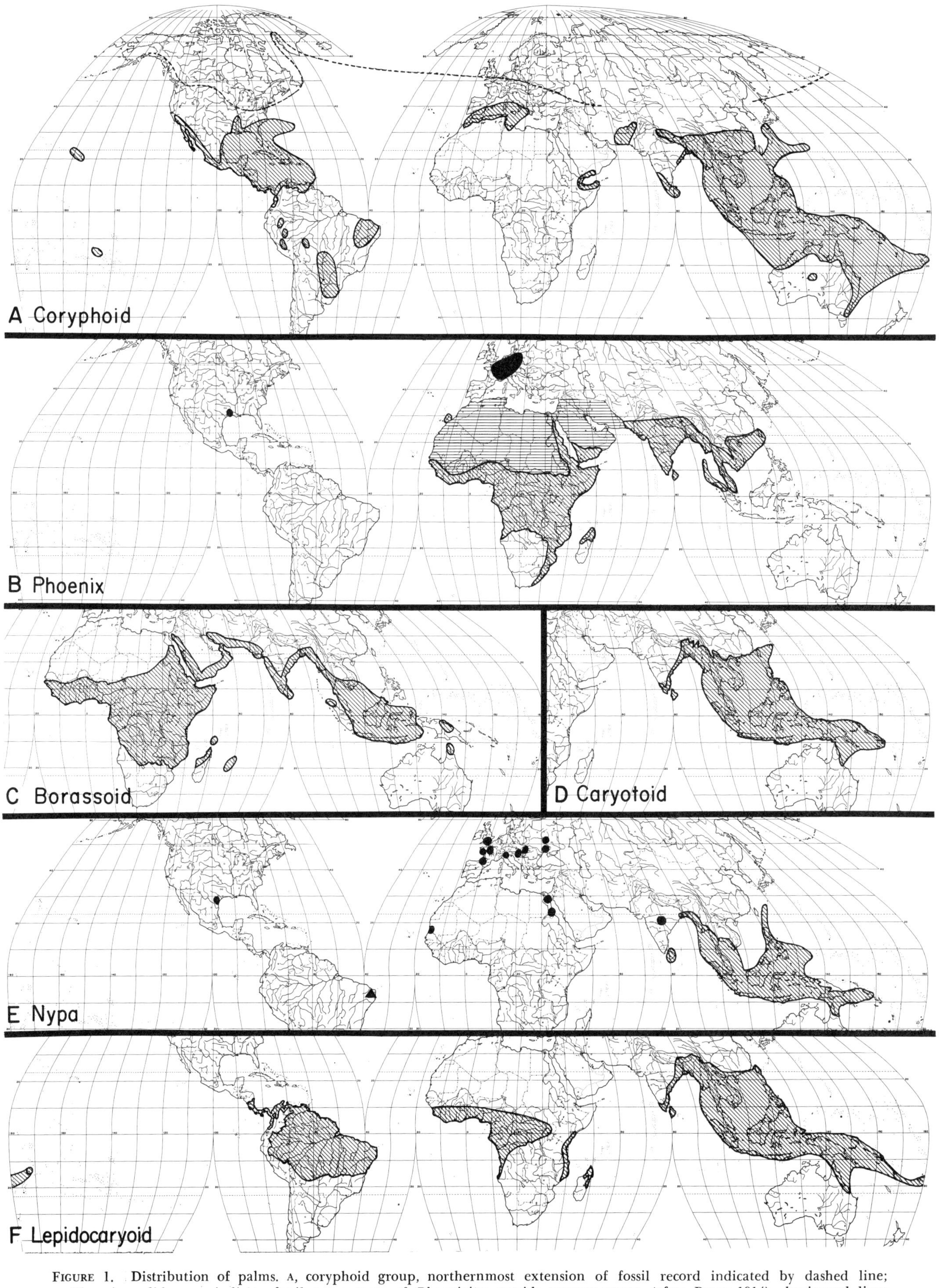

FIGURE 1. Distribution of palms. A, coryphoid group, northernmost extension of fossil record indicated by dashed line; B, Phoenix, solid areas indicate fossil occurrences of *Phoenicites* outside present range (after Berry 1914), horizontal lines indicate generalized distribution of the date palm, *P. dactylifera*; C, borassoid group; D, caryotoid group; E, *Nypa*, fossil localities in solid triangle (Paleocene) and solid circles (Eocene, Miocene) after Tralau (1964); F, lepidocaryoid group.

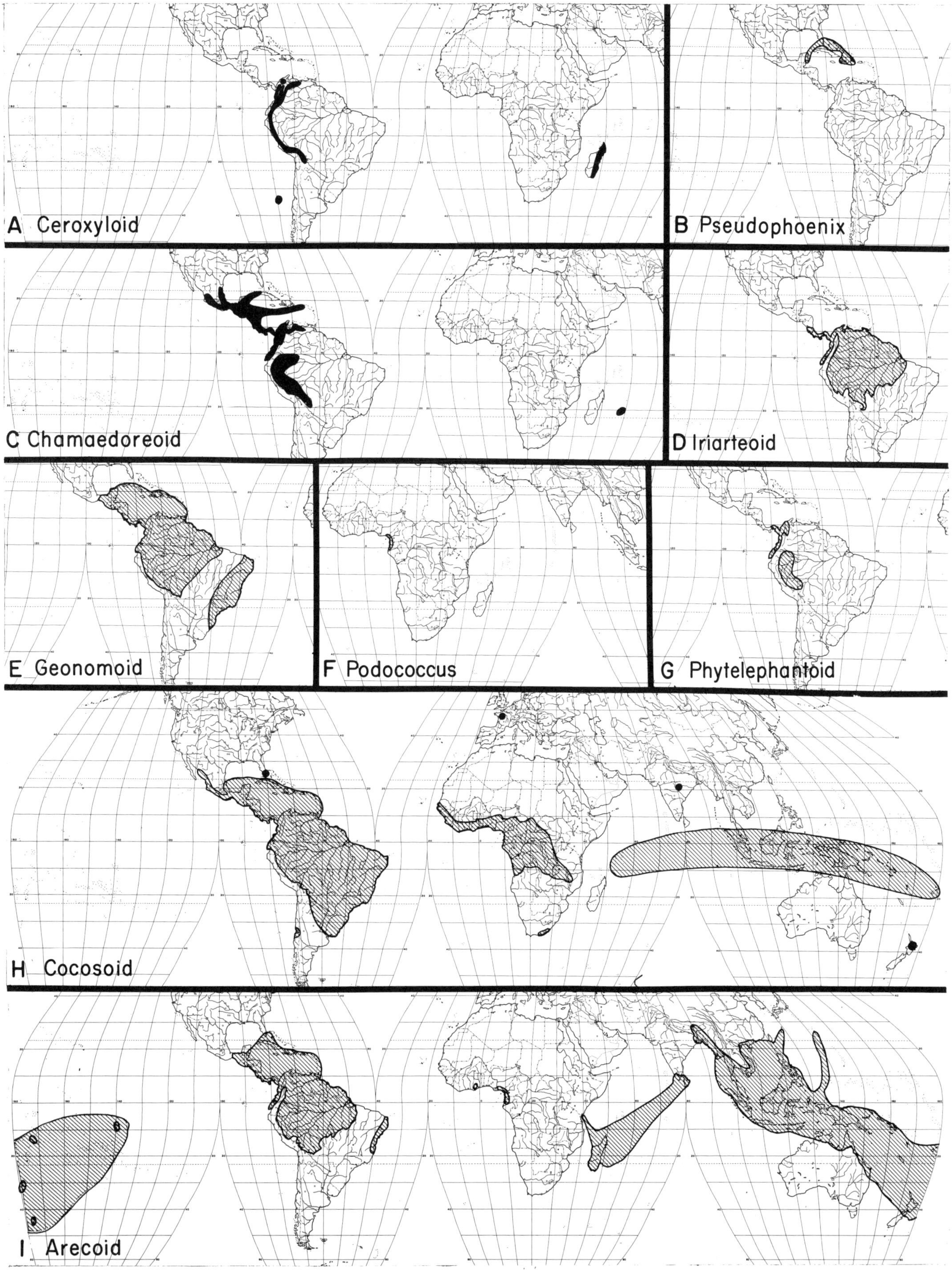

FIGURE 2. Distribution of palms. A, ceroxyloid group; B, *Pseudophoenix*; C, chamaedoreoid group; D, iriarteoid group; E, geonomoid group; F, *Podococcus*; G, phytelephantoid group; H, cocosoid group, records of fossil fruits outside present range in solid circles, *Cocos* generalized in Indo-Pacific; I, arecoid group.

cited individually but other data have been taken on the authority of summary papers on the Tertiary floras of the Old and New Worlds by Chandler (1961-1964, vol. 4) and on South America by Menéndez (1969). The diversity of fossil palm stems in the western United States (see Tuta 1967) appears to have been little studied. Even were my survey complete it would have to be accepted with care. As Ball (1931:35) has put it: "Students of fossils are well acquainted with the extraordinary difficulties which beset attempts at identification of remains of palms. Feather-palms are . . . extremely rare as fossils, and when found, are generally mere fragments of fronds." Fossil pollen is perhaps the most reliable indicator for palms, followed by seeds or fruits when these are as well preserved as are the remains identified as species of *Sabal* and *Nypa* by Reid and Chandler (1933). Fossil stems need to be reevaluated (Mahabalé 1959), and even then may not yield identification of sufficient precision to be useful. Scott et al. (1972), for example, have recently suggested that the affinities of some contested palm woods appeared to lie among genera representing three distinct groups (*Bactris, Cocos*—cocosoid palms; *Erythea, Rhapidophyllum*—coryphoid palms; *Wallichia*—caryotoid palms). Any identification of fossil palm leaves without careful and detailed cuticular or other studies must be accepted with reservation. This is particularly true of pinnate leaves which have similar patterns of venation and dissection in several groups. For these reasons, the fossil record carries less weight than it would were we studying vertebrates or some other groups of plants.

Where chromosome numbers are given, they are taken from Read (1965, 1966). Certain characteristics of phloem correlate well with groups and these, where noted, are taken from Parthasarathy (1968).

Coryphoid palms. The 32 genera and over 300 species of coryphoid palms represent the most primitive in the order (Moore 1973). They are widely distributed and occur on all continents (Figures 1A, 3). They have been divided into four alliances that are based chiefly on the nature of the gynoecium, whether apocarpous (*Trithrinax* alliance), partially syncarpous, i.e., with carpels united by the styles (*Livistona* alliance), or variously syncarpous (*Corypha* alliance, *Sabal* alliance). These alliances also correlate rather well with the nature of phloem strands in the central vascular bundles of the petiole. Many genera have a single strand, but *Chamaerops, Rhapidophyllum, Rhapis, Trachycarpus,* and the *Livistona* alliance are marked by two strands. It is only the coryphoid group that exhibits both types, all other groups being apparently constant for one or the other condition.

Coryphoid palms are noteworthy for the diversity of inflorescence, for the diversity of sexual expression from hermaphrodite to polygamous or dioecious, and for the uniform chromosome complement of $n = 18$. The fruit of many genera is colored and apparently adapted to dispersal by birds.

Distribution of alliances and of genera within the most primitive of these is plotted in Figure 3. I have recognized five units in the primitive *Trithrinax* alliance. Genera 1-4, 5-7, and 9-12 appear to represent different but more or less comparable lines within the apocarpous genera while genera 8 and 13-15 are more advanced. The *Livistona* alliance has two centers of diversity—six genera in the eastern tropics, six genera in the Western Hemisphere and one African genus, *Wissmannia,* which is very closely related to *Livistona,* differing chiefly in less connation in the corolla. Except for *Washingtonia* in western North America and *Johannesteijsmannia* in Malaya, Borneo, and Sumatra, the genera appear to be of about the same level of specialization and clearly interrelated. For syncarpous genera this is not so: syncarpy has developed in two different alliances represented by three genera in Asia (*Corypha* alliance) and the genus *Sabal* in the Western Hemisphere.

Among the most primitive genera, the *Trithrinax* unit is more clearly adapted to cooler and drier, more seasonal climates than most palms. *Chamaerops* reaches 44°N latitude in Europe; *Rhapidophyllum* and *Trachycarpus* both reach about 32°N latitude and *Trithrinax* about 32°S latitude; *Trachycarpus takil* occurs in the western Himalaya at 2400 meters elevation where the land is under snow from November to March (Beccari 1933). The *Chelyocarpus* unit is restricted to South America except the advanced *Cryosophila,* which extends north to Mexico. The South American genera are rain forest palms, but some species of *Cryosophila* withstand marked seasonality in Mexico. *Schippia,* a monotypic genus with unicarpellate gynoecium and a grade from perfect to staminate flowers on the same axis, is endemic to

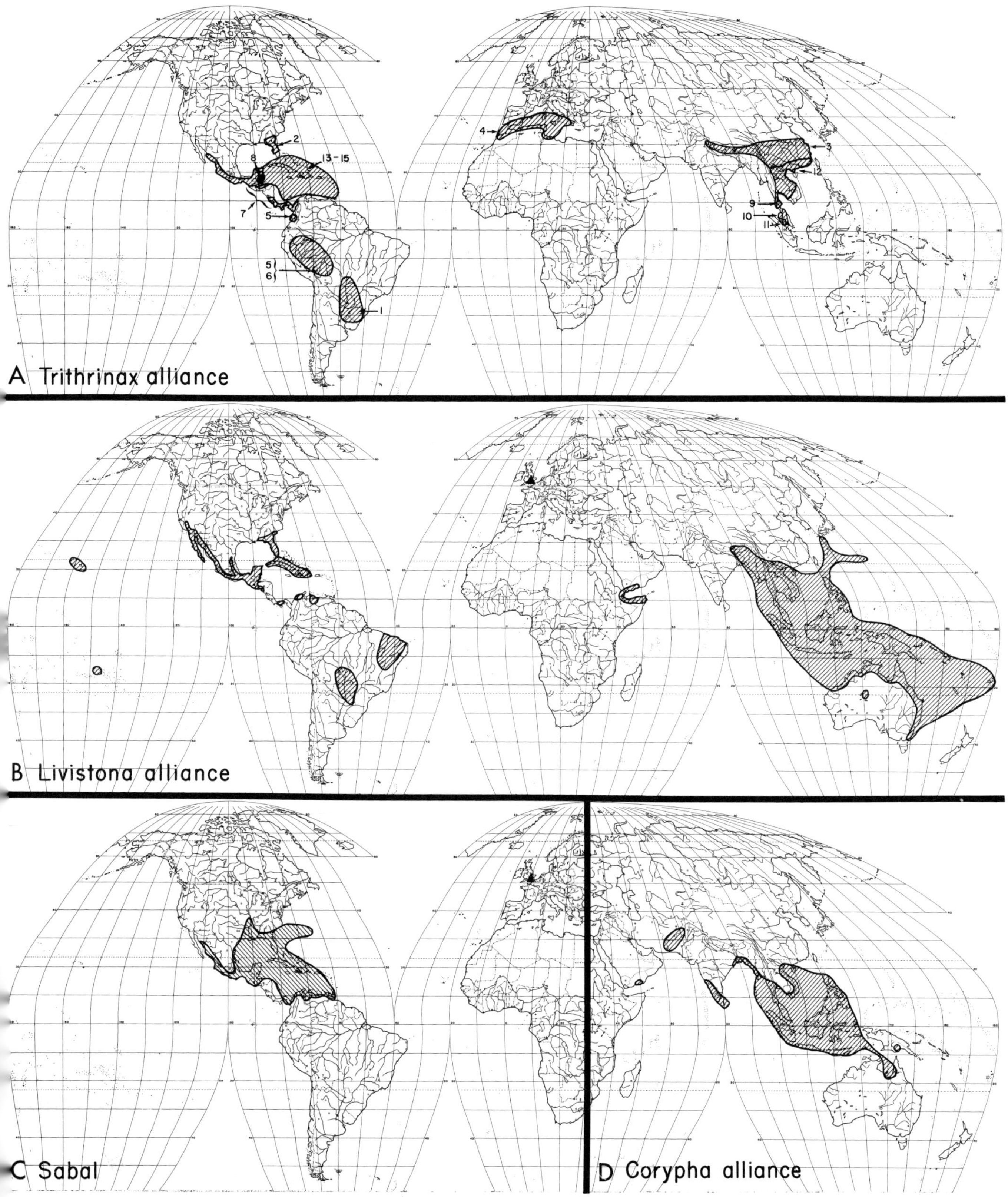

FIGURE 3. Distribution of coryphoid alliances. A, *Trithrinax* alliance: (1) *Trithrinax* unit—*Trithrinax*, (2) *Rhapidophyllum*, (3) *Trachycarpus*, (4) *Chamaerops*; (5) *Chelyocarpus* unit—*Chelyocarpus*, (6) undescribed genus, (7) *Cryosophila*; (8) *Schippia* unit—*Schippia*; (9) undescribed genus, (10) *Rhapis* unit—*Liberbaileya*, (11) *Maxburretia*, (12) *Rhapis*; (13) *Thrinax* unit—*Thrinax*, (14) *Coccothrinax*, (15) *Zombia*. B, *Livistona* alliance, solid triangle indicates fossil seeds reported for *Livistona* and *Serenoa*. C, *Sabal*, solid triangle indicates fossil seeds reported for the genus. D, *Corypha* alliance.

British Honduras and seems to show some affinity to the *Chelyocarpus* unit. The *Thrinax* unit, essentially Caribbean in distribution, is advanced with a uniseriate reduced perianth, sometimes more than six stamens and a carpel showing anatomical specialization that accords with wind pollination (Uhl and Moore 1971). Species of the three genera are strand plants or associated with sandy or calcareous soils. A third unit expressing primitive features, the *Rhapis* unit, is isolated in southeastern Asia and Malaya, where at least *Liberbaileya* and *Maxburretia* are restricted to limestone hills (Whitmore 1970, 1971).

In the *Livistona* alliance, *Serenoa* of the southeastern United States occurs well north of the 30th parallel and withstands occasional frost, and *Washingtonia* and *Wissmannia* are desert or subdesert palms, though associated with underground water (see Vogl and McHargue 1966, Moore 1971b). *Brahea* in Mexico undergoes a marked dry period, and the species of *Copernicia* in South America are drowned for half the year, and baked and seared by sun and wind for the other half (Markley 1955). Most species of *Licuala*, however, a large and a highly evolved genus in Indomalaysia and Melanesia, are undergrowth palms in rain forest. Among the syncarpous genera, *Sabal* and *Corypha* are more often associated with moist and more equable, or even hot, climates, though some species of *Sabal* are adapted to cold (*S. minor, S. palmetto*), others to more xeric conditions (*S. rosei, S. uresana*).

The fossil record for coryphoid palms is extensive, assuming that most or all of the palmate leaves, seeds and pollen assigned to *Sabal, Sabalites, Serenoa, Thrinax,* and *"Flabellaria"* belong here. The group appears to have had a substantially more northern range in Upper Cretaceous and early to middle Tertiary time (Figure 1A), having reached Kyushu and Hokkaido, Japan, and the Aleutian Islands (Kryshtofovich 1929, 1955 cited by Chandler 1961-1964, vol. 4), Cook Inlet, Alaska (Hollick 1936), western Greenland (Heer, 1883[1]), southern England (Reid and Chandler 1933, Chandler 1961-1964), Austria (Unger 1852), Dagestan and Kazakhstan, western Russia (Sterlin 1950, Budanchev 1953). I am not aware of any record for Africa, nor does Menéndez (1969) list any coryphoid genera for the Cretaceous of South America. A report of *Sabalites* in the Tertiary of Venezuela (Berry 1922b) is suspect: the material illustrated greatly resembles a characteristic divided pinna of *Iriartea*.

PHOENICOID PALMS. *Phoenix,* the sole genus of phoenicoid palms, is today restricted to Africa and Asia (Figure 1B). The species are dioecious with specialized inflorescences that bear somewhat dimorphic, probably wind-pollinated flowers. The pistillate flowers have an apocarpous gynoecium. Fruits are fleshy and sweet, apparently admirably adapted to dissemination by birds. The chromosome complement is $n=18$. Leaves are pinnate (the lower pinnae modified into spines), and the central vascular bundles of the petiole have a single phloem strand. The characteristics of the group in sum place it near but on a somewhat more specialized level than the coryphoid palms with which it must certainly have had a common origin.

One species, *P. paludosa,* is associated with littoral forests, tidal swamps, and mangroves. It tolerates a higher percentage of salt in the water than *Nypa fruticans,* and it may grow in the same general area as that species though not in the same locality in the Andaman and Nicobar Islands (Mahabalé and Parthasarathy 1963). In northern Sumatra, however, the two grow side by side (J. Dransfield, personal communication). Other species occur in dry districts where their presence indicates the presence of watercourses or underground water.

Berry (1914) noted fossil records of *Phoenicites* for Europe and described and figured a seed from the late Eocene or early Oligocene of Texas as *Phoenicites occidentalis.* This does indeed resemble the seed of a modern *Phoenix* but parts of pinnate leaves described and figured by Ball (1931) from the Eocene of Texas as *Phoenicites integrifolia* and *P. brazosensis* do not convince me, since they appear to lack any indication of an induplicate nature or of modified basal pinnae.

BORASSOID PALMS. Borassoid palms are more specialized than the foregoing but seem clearly related to them and derived from a common progenitor or perhaps even from some coryphoid stock. They are dioecious with stout inflorescences in which markedly dimorphic flowers are sometimes (pistillate) or always (staminate) borne in pits

[1] Brown (1962) noted that some writers do not accept Heer's species of *"Flabellaria"* as plants at all. One might construe *F. johnstrupii* as ripple marks, but leaf-fragments of *F. grönlandica* (probably of Lower Paleocene, see Koch 1964:544) as illustrated by Heer show cross-veinlets characteristic of some palms. Brown suggested that the taxon may pertain to *Sabal grayana* Lesquereux.

formed by connation and adnation of bracts and inflorescence-axis (Uhl and Moore, 1973). The chromosome complement, where known, is $n=18$, 17, or 14. Leaves are costapalmate. Phloem strands in the central vascular bundles of the petiole are single, and fruits are fibrous or corky. *Borassus* and *Hyphaene* are reportedly spread by elephants (Corner 1966) and Dransfield (personal communication) thinks it highly probable that *Borassodendron* in Borneo is spread by orang-utan, but other modes of dissemination seem not to be recorded.

The group is restricted to the Old World (Figure 1c) and consists of two alliances: *Borassus* and related genera; *Hyphaene* and its relatives. *Borassus,* which appears to be least specialized, occurs from Africa to Madagascar, India, and New Guinea, with a reported but somewhat questionable occurrence in northern Queensland, Australia. Three related genera are found respectively in the Seychelles Islands (*Lodoicea*), the Mascarene Islands (*Latania*), and Malaya and Borneo (*Borassodendron*). *Hyphaene* is represented by one species in India but is otherwise restricted to Africa in the company of *Medemia,* and to Madagascar in the company of *Bismarckia.*

Borassus occurs both in open savanna (in Africa) and in wetter regions (Asia) but *Borassodendron* is a palm of rain forest, *Lodoicea* of rocky hillsides, and *Latania* occurs in rocky ravines or on slopes not far from the sea. All species of the *Hyphaene* alliance occur in dry regions, both *Hyphaene* and *Bismarckia* in some abundance on the dry west coast of Madagascar.

LEPIDOCARYOID PALMS. Lepidocaryoid palms are set off from all others by the imbricate scales that cover the ovary and the mature fruit. Within the complex, there is a wide range of inflorescence type, but a uniform pattern of floral arrangement in or derived from a monopodial system. Two phloem strands occur in central vascular bundles of the petiole and usually sieve elements of stem and petiole have simple sieve plates. Plants are hermaphrodite, polygamous, monoecious, or dioecious, and are nearly always armed, sometimes very fiercely so, with spines on the trunk, on leaf-sheaths and leaves, on inflorescence, or on all parts. Habit ranges from plants with subterranean stems to large upright palms or to lianes which reach great lengths in the Old World tropics. Flowers are somewhat modified when unisexual, but not so markedly as in some of the groups to follow. Fruits are often provided with a sweetish fleshy mesocarp or seeds with a sarcotesta which may attract pigs, birds, and fruit bats. The chromosome number, where known, is $n = 14$.

The majority of the 22 genera in this group are dioecious and inhabit the Indomalaysian tropics, where *Calamus* and *Daemonorops* are large genera. Africa has five genera, two of them shared: *Raphia* (Figure 4) is restricted to Africa and Madagascar except for one species in America; about nine species of *Calamus* occur in Africa, the remainder in Asia (Figure 5). Three genera of hermaphrodite lianes are known only from the wet forests of tropical West Africa. Two dioecious genera restricted to South America have the only palmate leaves that are reduplicately folded.

The group as a whole is clearly well adapted to the wet tropics between latitudes 25° S and N. Where representatives are found in drier regions, as in the savannas of South America, they usually indicate the local presence of water.

Pollen is the chief source of information on their past history. Muller (1970) lists pollen records of *Mauritia* and *Calamus,* both advanced genera, from the Paleocene of South America and Borneo, respectively, and of two contemporary species of *Eugeissona* from Borneo dating to lower Miocene. A lepidocaryoid fruit is known from the Tertiary of Colombia (Berry 1929b). Spines (*Calamus daemonorops*) are known from the lower Tertiary of Britain and Europe (Chandler 1961-1964, vol. 2, 3).

NYPOID PALMS. *Nypa,* like *Phoenix,* is the sole genus in its group. At present it is monotypic, *N. fruticans* occurring as a monoecious mangrove palm in estuarine habitats from Ceylon and the Ganges delta to the Solomon Islands and the Ryukyu Islands (Figure 1E). The inflorescence is complex with much adnation and the staminate flowers are borne on spicate axes, the pistillate in a terminal head (see Uhl 1972). The gynoecium is apocarpous with an unusual type of carpel and the sepals and petals are very similar. The fibrous fruits commence germination on the head and are well adapted to dispersal by sea. Though the staminate flower has only three stamens, these are united. The chromosome complement is $n = 17$.

Nypa has a substantial fossil history. Pollen of *Nypa* is one of the first angiospermous pollens to be identified to genus (Muller 1964a, 1970). The

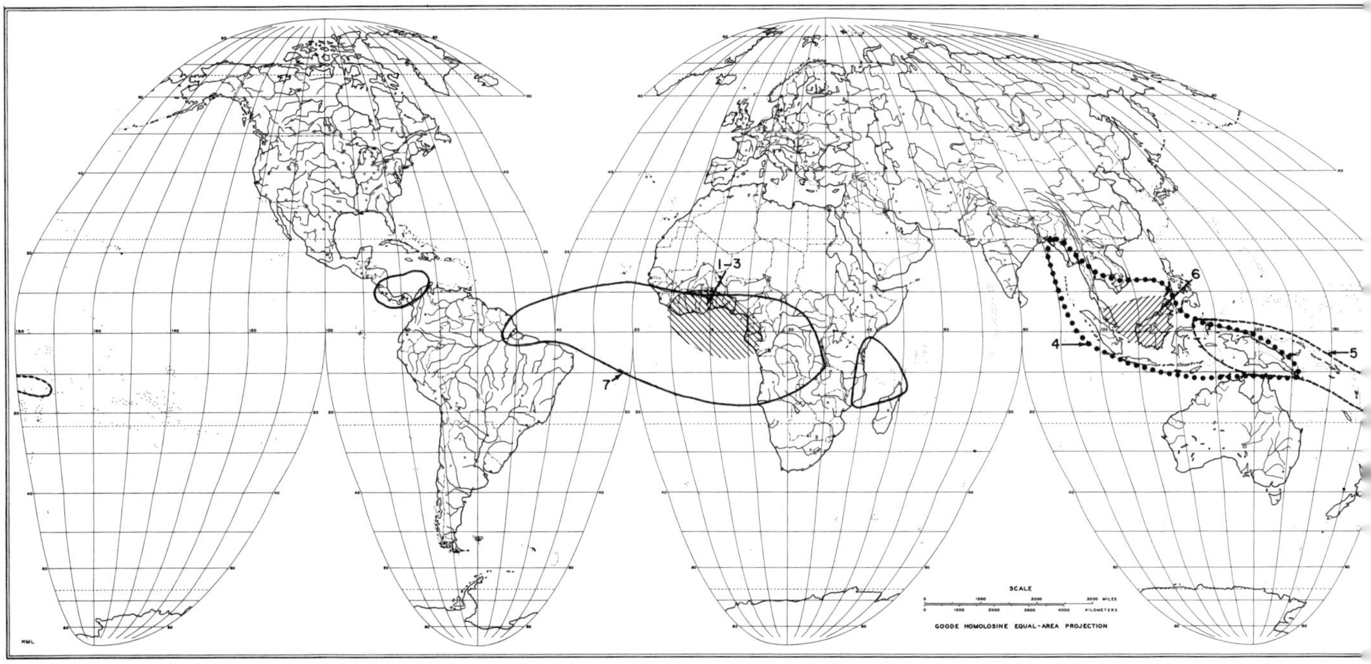

FIGURE 4. Distribution of less specialized lepidocaryoid palms. Hermaphrodite genera: (1) *Ancistrophyllum*, (2) *Eremospatha*, (3) *Oncocalamus*, diagonal lines in Africa; (4) *Korthalsia*, dotted line. Polygamous genera: (5) *Metroxylon*, dashed line; (6) *Eugeissona*, diagonal lines in Malaysia. Monoecious genus: (7) *Raphia*, solid line.

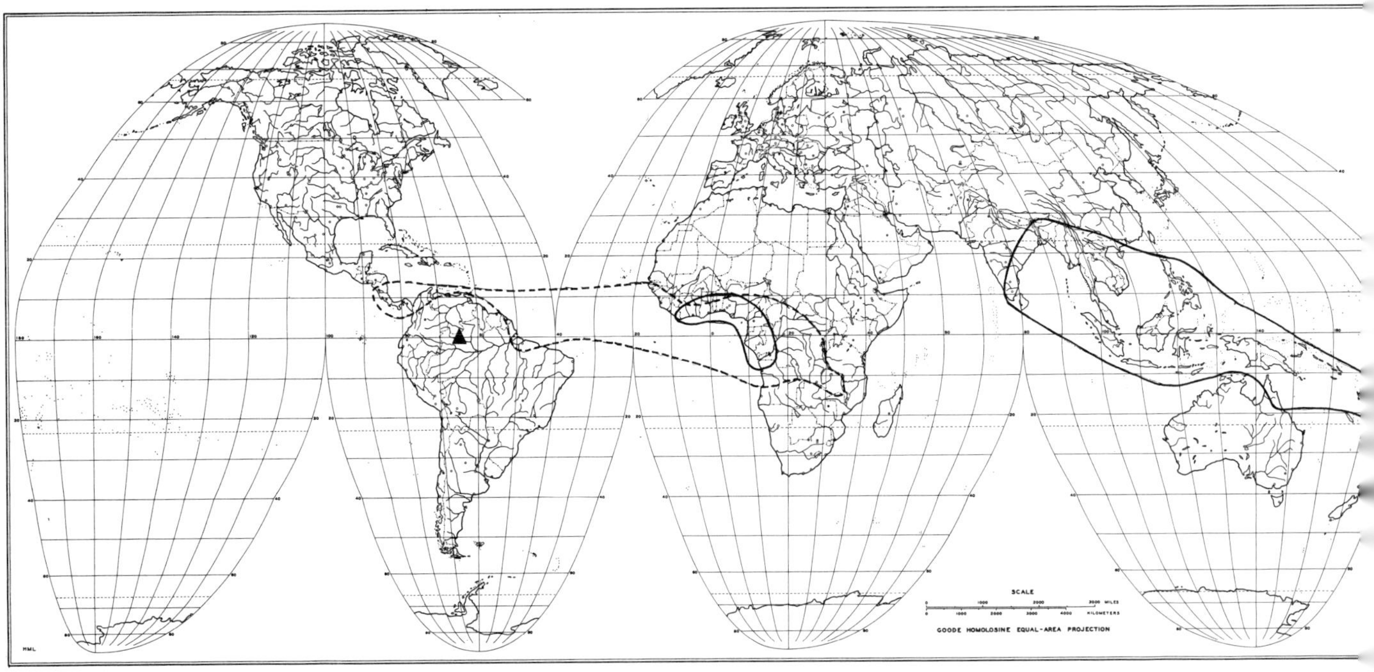

FIGURE 5. Distribution of *Calamus* (solid line), *Barcella* (solid triangle), and *Elaeis* (dashed line).

genus has occurred in Borneo since Senonian (Upper Cretaceous) times, presumably in the same habitat, and it must have been common along the margins of the Tethys Sea in early Tertiary times (see Tralau 1964, Chandler 1961-1964, vol. 4) until the Miocene (Figure 1E). A record from the Paleocene of Brazil (Dolianiti 1955) is in marine deposits and the seeds are badly worn suggesting the possibility that they may have drifted from the Tethys while that sea was still open to the Atlantic. The condition of a relatively unworn fruit in the Eocene of Texas suggested to Arnold (1952) that it might have come from the Mississippi embayment. If so, both American records may represent autochthonous palms. Several species have been described from fossil fruits either as *Nipadites* or as *Nypa* (*"Nipa"*) but their differences from *N. fruticans* are chiefly in size.

In sum, the characteristics of *Nypa* are so distinctive that the genus seems best left in an evolutionary line of its own though perhaps closer to the arecoid line than to others.

CARYOTOID PALMS. The caryotoid palms include three monoecious genera and about 35 species in the wet tropics of the Old World, from India to the Ryukyu Islands in the north and to the Solomons and perhaps the New Hebrides in the south (Figure 1D). Their distinctive characteristics set them off from other lines, though they appear closest to members of the greater arecoid line. The leaves are morphologically induplicate in vernation but are anatomically reduplicate (Tomlinson 1960). The pinnae are often narrowly to broadly cuneate, toothed along the upper margins and apex, and with nerves that diverge from the base or midnerve rather than running parallel with it. The central vascular bundles of the petiole have single phloem strands. Inflorescences have numerous bracts and dimorphic flowers are arranged in triads or in derivates of triads like those of several groups in the arecoid line. Chromosome numbers are $n=16$ and 17. Fruits are fleshy but filled with crystals which do not seem to deter fruit bats, wild pigs, and civet cats from eating and disseminating them (Miller 1964, Corner 1966).

Fossil pollen of *Arenga* is recorded from Miocene deposits of Borneo (Muller 1970) and perhaps the lower Eocene of India (Lakhanpal 1970) while fossil pollen of *Caryota* has been identified from Oligocene deposits on the Isle of Wight (Pallot 1961 cited by Chandler 1961-1964, vol. 4) and seeds of this relationship were described from the London Clay flora as *Caryotispermum* by Reid and Chandler (1933).

PSEUDOPHOENICOID PALMS. *Pseudophoenix,* a small genus of only four species, is restricted to the islands of Hispaniola, Cuba, Bahamas, Florida Keys, and to the Yucatan Peninsula (Figure 2B). In its much-branched inflorescence, solitary hermaphroditic flowers, and 1-3-seeded fleshy fruits, it is the most primitive of those palms which I place in the greater arecoid line. According to Read (1968: 170), species of *Pseudophoenix* occur naturally "only in areas of very little rainfall and on exceedingly well-drained sand or porous limestone, always where the substratum is alkaline or saline." No evident fossil record is known.

CEROXYLOID PALMS. Four genera comprise this group which is more advanced than *Pseudophoenix,* but less so than the chamaedoreoid group, and like each of these, it has single phloem strands in the central vascular bundles of the petiole. Salient characteristics are the much-ramified inflorescences bearing pedicellate or sessile, unisexual (but not markedly dimorphic) flowers singly along the axes. Fruits are usually red and fleshy. Floral anatomy indicates the homogeneity of the group (Uhl 1969).

The distribution of genera in this group is unlike that of any other except the chamaedoreoid group (Figure 2A). *Ceroxylon* is a genus of high elevations in the Andes of South America where it is usually associated with cloud forest, while *Juania* occurs only in wet forest on one island in the Juan Fernandez chain. The remaining genera are in Madagascar and the Comore Islands, where *Ravenea* occurs in swampy or sandy places in littoral forest, in wet forest to elevations of 900 meters or more, or occasionally in drier locations. Fossils assignable to this group are not recorded.

CHAMAEDOREOID PALMS. The monoecious or dioecious chamaedoreoid palms appear to represent an endpoint in an evolutionary series commencing with *Pseudophoenix* and continuing with the ceroxyloid group. Inflorescences are simplified; flowers are more markedly dimorphic though small and not easy to differentiate superficially. In the monoecious genera they are borne in lines of several or rarely only two staminate and usually a basal pistillate. The fruit is usually red or shining black on a red or orange axis and is apparently adapted to dispersal by birds. The mesocarp is fleshy but filled with irritant crystals like those found in the caryotoid group. Chromosome complements are $n=16$, 14 or 13.

Six genera are restricted to the New World, except for the genus *Hyophorbe* in the Mascarene Islands, a distribution pattern comparable to that of the ceroxyloid palms (Figure 2c). Two genera are restricted to dry limestone hills in Cuba and Puerto Rico or to the limestone region of the Petén in Guatemala and British Honduras and extending into Mexico. The remaining American genera usually occur in wet forest or cloud forest along the eastern side of the Andes in the upper Amazon Basin and along the western and northern slopes from Ecuador to Venezuela and thence north to Mexico, where *Chamaedorea radicalis* is associated with drier limestone outcrops from Hidalgo to Tamaulipas. Another northern species, *C. pochutlensis,* is associated with pines and oaks in western Mexico. *Hyophorbe* usually grows on some of the steepest and wettest slopes in the Mascarene Islands but is fully exposed on Round Island close to the sea (see Bailey 1942). It is curious that the color and scent of staminate flowers of *Hyophorbe verschaffeltii* are the same as those found in some species of *Chamaedorea.*

Fossil leaves have been described as *Chamaedorea danai* in the Eocene of southeastern United States (Berry 1916).

IRIARTEOID PALMS. The most striking feature of the palms in this group is the usual production of stout adventitious stilt roots but in other ways they also show a distinct homogeneity. Pinnae are erose-truncate with several principal ribs and there are several bracts on the peduncle of the inflorescence. Phloem strands in the central vascular bundles of the petiole are single. Most of the genera are South American, though *Iriartea* and *Socratea* reach Costa Rica, and *Wettinia* has recently been collected in the Darien region of Panama (Figure 2D). All are associated with abundant moisture, either as palms of the wet lowland forest or of montane cloud forest. Fossil materials have been ascribed to *Iriartites* (Berry 1920, 1922a, 1929c) but need to be restudied.

PODOCOCCOID PALMS. *Podococcus* (Figure 2F) is an unusual genus of one or perhaps two species restricted to the West African rain forest. Leaves appear imparipinnate and the pinnae have several divergent ribs as in the iriarteoid group. Flowers are borne in triads sunken in pits along a spicate axis, bracts of the inflorescence are four, and the fleshy orange-red fruit is stalked and horizontally oriented. Two distinct phloem strands are present in the central vascular bundles of the petiole. The genus was included with geonomoid palms by Hooker (1883). The pits resemble those of geonomoid palms, but the structure of the flowers and of the fruit is alien to that group. It seems, instead, to lie between the iriarteoid palms and the geonomoid palms, sharing certain characteristics of each group but distinctive in its own right. Nothing is known of the past history of *Podococcus.*

GEONOMOID PALMS. Geonomoid palms also are restricted to the New World. Four of the six genera occur both in South and Central America while two are either Antillean or Central American (Figure 2E). The corollas are gamopetalous throughout the group, staminodia are prominent, and styles are elongate. Flowers are borne in triads sunken in pits. Two distinct phloem strands are present in central vascular bundles of the petiole. Both chromosome complements known are $n=14$. Fruit is often black or green, rarely red, and small, apparently adapted to dispersal by birds.

Geonoma is a common component of most tropical American forests from the lowlands to the cloud forest and there at elevations exceeded only by *Ceroxylon* among the palms. Fossil leaves which may or may not belong here have been described as *Geonomites,* both from North America (see Ball 1931) and from South America (Menéndez 1969).

COCOSOID PALMS. Cocosoid palms have long been given a rank equal to a broadly defined arecoid complex. They differ in an obvious characteristic of the fruit which has a thick, hard endocarp with three or more pores each corresponding to an actual or potential embryo, and a peculiar and probably associated pattern of growth in the gynoecium. Otherwise, the group shares most features with groups X to XIV and especially with the arecoid palms in a limited sense—a prophyll and one major bract on the inflorescence, dimorphic flowers borne in triads, two distinct phloem strands in the central vascular bundles of the petiole. Chromosome complements are $n=16$ and 15.

South America, where 24 of 28 genera occur, is the great center for cocosoid diversity (Figure 2H). Only *Gastrococos* and *Rhyticocos* in the West Indies, *Jubaeopsis* in South Africa, and *Cocos* in the Indo-Pacific region do not occur there naturally. The genus *Elaeis* is shared with Africa (Figure 5), but it is interesting to note that the allied but less specialized *Barcella* is known only from a limited area on the Rio Padauiri in Brazil. Furthermore, of the two species in *Elaeis,* the American seems less specialized than the African although the latter appears to have a record back to the Miocene in the Niger Delta (Zeven 1964).

Many of the less specialized cocosoid palms appear to be adapted to cooler, drier, and more seasonal climates. *Butia,* for example, is native from southern Brazil to Argentina and is cultivated as far north as South Carolina in the United States or in Europe as far north as Firenze, Italy, where individuals withstood snow and a period of prolonged below-freezing temperatures with no visible effect (Moore 1957). The more advanced *Elaeis* and *Bactris* alliances are, in general, associated with greater heat and moisture.

The fossil record of cocosoid palms outside the present range is not extensive but it is important. Since hard endocarps of cocosoid palms are borne long distances by sea today, their presence in marine deposits may not necessarily be accepted as proof of autochthony without substantiation from other evidence. Thus the earliest macrofossils of cocosoid palms, *Astrocaryopsis* and *Cocoopsis,* from the Cenomanian (lower Upper Cretaceous) of France (Fliche 1896), may not be autochthonous and the fruit of *Attalea* in the Eocene of Florida (Berry 1929a) is also suspect for the same reason. Fossil cocosoid fruits in the Eocene of India (Kaul 1951) and fossil stems perhaps pertaining to this group (Rao and Menon 1966) are, in combination, suggestive of autochthony. So also is the presence of fossil cocosoid fruits, *Cocos zeylandica* (Berry 1926a), associated with pollen inferred to be of the same genus in the Miocene of New Zealand (Couper 1952). Within the present range of cocosoid palms, fossil fruits are known from the Eocene of Peru and Ecuador (Berry 1926b, 1929c, 1932).

Arecoid palms. Arecoid palms, even when considering the term in a restricted sense, are widely distributed (Figure 21). They are highly advanced with only a prophyll and one major bract (very rarely two) on the inflorescence, with strikingly dimorphic flowers borne in triads or in a modification of the triad, frequently with a pseudomonomerous gynoecium (Uhl and Moore 1971). They are adapted primarily to the tropical rain forest, the montane cloud forest, and only rarely extend beyond the Tropics of Capricorn and Cancer to southern Florida in the Western Hemisphere (*Roystonea*), to the Himalaya region (*Pinanga*), the Ryukyu and Bonin Islands (*Satakentia, Clinostigma*), and New Zealand (*Rhopalostylis*) in the Old World. Gametic chromosome numbers of 16 and 18 have been reported.

I have recognized 18 alliances within this complex of which five are American. Three of these five contain triovulate genera, and only one, the *Euterpe* alliance, is not unigeneric. Africa has only one genus and alliance, the highly advanced *Sclerosperma,* but Madagascar and nearby islands have 19 genera and 105 species, one of which (*Sindroa*) is surely triovulate, another (*Beccariophoenix*) possibly triovulate. Eight alliances in the eastern tropics include one triovulate genus (*Orania* alliance) and seven uniovulate alliances in one of which, the *Clinostigma* alliance, there has been exceptional evolution at the generic level (see Moore 1973). All but two of these alliances, *Oncosperma* and related genera in the Mascarene and Seychelles Islands, Ceylon to Borneo and Java, and the *Archontophoenix* alliance in New Caledonia, New Zealand, and Australia, are represented to a greater or lesser extent in New Guinea.

The macrofossil record is scant and difficult to interpret, and I place little credence in present identifications which purport to be arecoid (e.g., *Manicaria* in the Eocene of the United States; see Ball 1931). Microfossil records of *Oncosperma* in the Oligocene and *Nenga* in the lower Miocene of Borneo are reported by Muller (1964b, 1970) and of *Rhopalostylis* from the Miocene of New Zealand by Couper (1952).

Phytelephantoid palms. One or three to perhaps four genera included in this group are the most highly evolved among the palms by reason of the dioecious state, two bracts on the staminate inflorescence, reduction of inflorescence to a large staminate spike in *Phytelephas* or a pistillate head in all genera, reduction of the perianth in staminate flower, large number of stamens, extraordinary size of the pistillate perianth, and the numerous locules of the gynoecium. The chromosome complement is $n=16$ and phloem strands of the central vascular bundles of the petiole are two. Phytelephantoid palms are restricted to rain forest of the upper Amazon Basin and the northwest corner of South America from Ecuador to Venezuela and into the Darien region of Panama (Figure 2G). The fossil record is limited. A fruit from the upper Tertiary of Ecuador (Brown 1956) seems to be correctly identified; the identification of fossil stem material from the West Indies by Kaul (1943) is less certain.

A Question of Origin

We now have before us information on the kinds of palms in Africa and South America, as well as something of their relationships, general levels of evolution, adaptive nature, and past and present distribution. Figures are obtained from Table 1 that show a striking contrast between the two continents. There are 16 genera and about 117 species of palms in Africa, Europe, and Arabia representing seven groups. This representation is smaller than that in the geographically closer but much smaller islands of the Indian Ocean, where 29 genera and about 132 species from 6 groups occur. It also is substantially smaller than that for South America, where 64 genera and 837 species from 9 groups occur, and for the Eastern Tropics where we find 97 genera and about 1385 species in 8 groups. The obvious question raised by these figures is how the distribution came about. In attempting an answer I put it in the frame of continental drift, the reality of which now can scarcely be denied, and of the reconstruction of relationships and positions of landmasses since the end of the Permian attempted by Dietz and Holden (1970a, 1970b).

THE PRESENT SCENE

The more immediate history of palms in the two ecosystems we consider here is intimately involved with events of the late Tertiary and the Pleistocene and for this we have reasonable evidence from the geological record. The past of Africa has been treated at some length in a series of papers by R. E. Moreau (1952, 1963, 1966) as background for an understanding of the evolution of birds on that continent. After a long period of peneplanation, during which the land surface is thought to have been reduced to a level mostly below 600 meters, uplift of eastern and southern Africa in the Miocene, followed by volcano-building in the Pliocene and Pleistocene, changed the face of the continent greatly. Climatic changes have also occurred, most severely in the Pleistocene when several major fluctuations and more minor ones are inferred using data from several sources. The Congo Basin, site today of one of the great tropical forests, is rooted in Kalahari sand which is thought to have been redistributed between 75,000 and 52,000 years ago. The bounds of the Sahara have fluctuated even within the past 5000 years as have those of the once more extensive rain forest which today is divided into two unequal regions separated by a dry gap in Ghana, Togo, and Dahomey (Keay 1959). Moreau (1966:59) summarizes as follows: "The evergreen forests of West Africa were reduced and fragmented during some period prior to about 22,000 years ago, those of the Congo basin to an enormous extent for a period prior to about 50,000 years ago and again to a limited extent around 12,000 years ago." Indeed, Moreau (1963:416) speaks of the changes in West Africa in the last 5000 years as "almost comparable with the climatic catastrophe of Australia. . . ."

Such sweeping changes may well have been responsible for the limited palm flora we find today in Africa. The montane regions of Africa were once very extensive and probably offered suitable habitats for some palms (Moreau 1963). Assuming that such palms ever existed in Africa, their absence today may be due to the extinction of palms adapted to such habitats during the period of maximum peneplanation. Yet despite oscillations in climate, Moreau suggests that the evergreen forest, savanna, and arid regions have co-existed in series similar to those of today, though doubtless much modified in extent, over long periods of time.

Nor has South America been undisturbed by Pleistocene climatic change. Haffer (1969) and more recently Vuilleumier (1971) have suggested that marked alterations have occurred in the extent of forest and nonforest vegetation. They have reconstructed a series of retreats to and advances from forest refugia based on evidence from distribution of birds and selected plants. Our knowledge of palms at the specific level does not as yet permit an analysis comparable to those of the authors cited, except in *Pholidostachys* and *Geonoma* section *Taenianthera* (see Wessels Boer 1968) where parallels appear to be present.

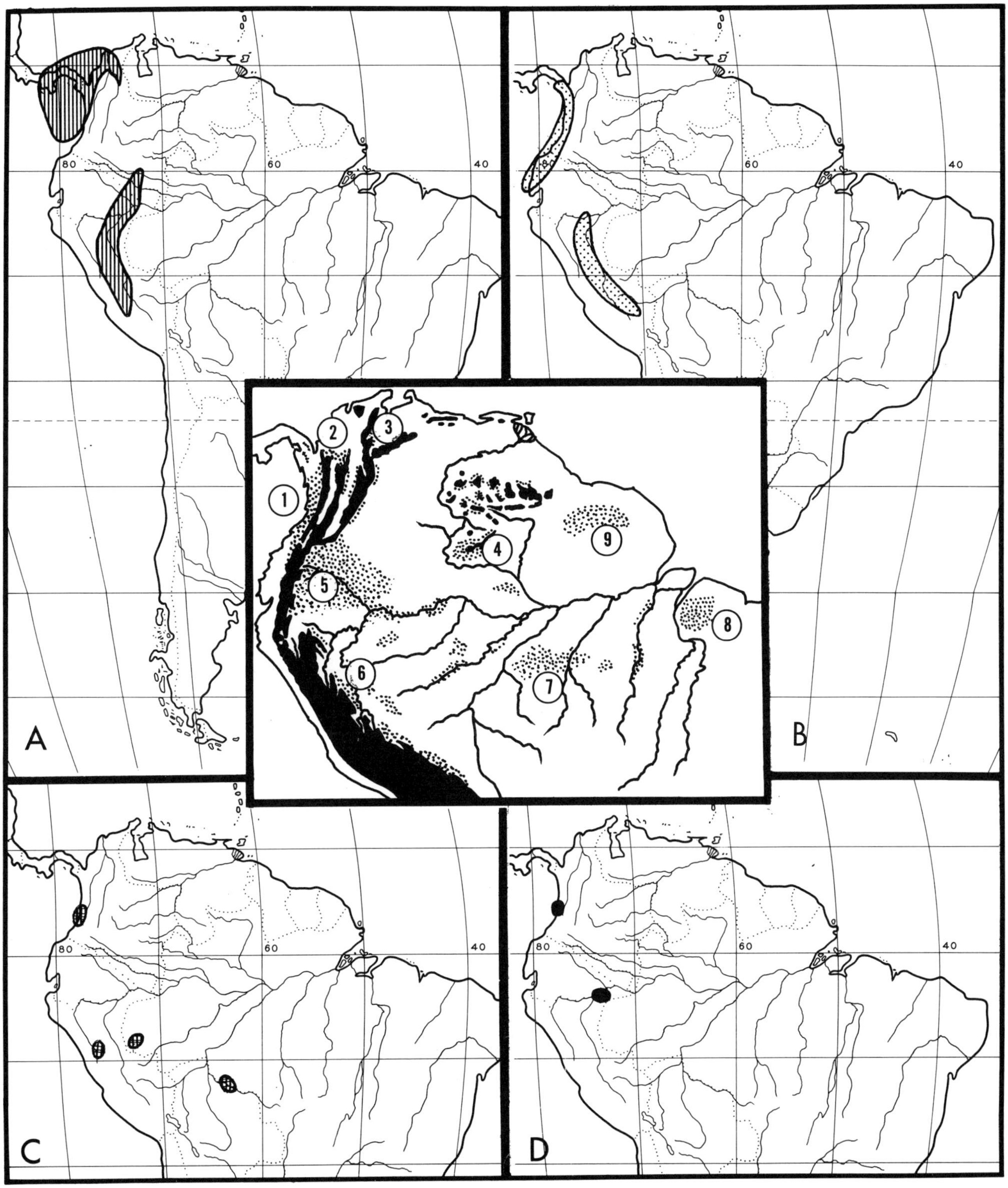

Figure 6. Disjunct distribution of some palms in South America. A, *Phytelephas* and allied genera; B, *Wettinia augusta, W. maynensis, W. quinaria*; C, *Chelyocarpus* spp.; D, *Orbignya* (section *Spirostachys*) spp. Center inset, presumed forest refuges in central and northern South America during warm-dry climatic periods of the Pleistocene, adapted from Haffer (1969, fig. 5).

There is, however, a clear and unusual relationship between two of Haffer's refugia east of the Andes—the East Peruvian and Napo refuges—and the Chocó refuge on the northwest coast of Colombia and into Ecuador. Identical or vicarious species of *Mauritia, Oenocarpus, Wettinia, Orbignya* section *Spirostachys,* and related species of *Chelyocarpus* and *Phytelephas* have disjunct distributions with no intervening populations in the lower Amazon Basin (Figure 6). At the same time, a few species have been eminently successful in persisting through change or more likely in repopulating the rain forest of South America. Prominent among these are the nearly ubiquitous *Mauritia flexuosa, Euterpe precatoria, Socratea exorrhiza, Maximiliana martiana,* and *Geonoma deversa* which, by their numbers and aspect, tend to give character to wet forest at low to moderate elevations. Conversely, it is likely that the range of some of the cocosoid palms adapted to drier situations, such as species of *Syagrus* or *Orbignya,* has been of even greater extent in the past, large as today's range may seem. But abundant as palms are, especially in South America, and intriguing as problems of present distributions may be, the greater question is where did they come from? Three alternatives are considered in pages to follow.

PALMS IN THE EASTERN TROPICS

Smith (1970) has suggested on the basis of distribution of the primitive "ranalean" families that the angiosperms originated as a monophyletic taxon during the Jurassic or Triassic periods in the general area of southeastern Asia to Malesia, and that they spread within and migrated from an Asian-Australasian center to other parts of the world. The abundance of palms in this region suggests that their distribution be examined against this thesis. When one attempts a model for palms along similar lines, but using primitive groups and genera instead of families, there is no predominance of primitiveness in the center proposed by Smith. Rather, the center of diversity and of primitiveness appears to be in South America (or in the Western Hemisphere when *Pseudophoenix* is added). Here nine groups occur, including the most primitive coryphoid genera (*Chelyocarpus, Trithrinax*) and the most primitive groups of the arecoid line, as well as most cocosoid palms and triovulate genera of the advanced arecoid group. The eastern tropics, in contrast, have two units of apocarpous coryphoid genera, but *Trachycarpus* is clearly advanced in the *Trithrinax* unit; and the *Rhapis* unit may be less primitive than either the *Trithrinax* or the *Chelyocarpus* units. Elsewhere, the palms of the region appear to be advanced. There are many dioecious genera of lepidocaryoid palms, two polygamous genera (*Eugeissona, Metroxylon*), and only one hermaphroditic genus, *Korthalsia,* which may be secondarily so through loss of the staminate flower in the pairs of perfect and staminate flowers that characterize the polygamous genera. *Nypa* may or may not have originated in the region, but it certainly occurred on other continents in the early Tertiary. The caryotoid palms are relatively specialized, wherever they may have originated. The important arecoid line is represented by the single genus *Cocos* among the cocosoid palms and by numerous genera in the advanced arecoid group which appear to have evolved in response to island isolation. *Phoenix* and the borassoid palms are shared with Africa and show no particular concentration of primitive species or genera.

This examination carries with it a further inference. Implicit in Smith's concept is the assumption of a single origin for the angiosperms, i.e., that the monocotyledons have been derived from the dicotyledons. It seems likely, however, that, like the origin of birds and mammals from different reptilian stocks, the monocotyledons and dicotyledons may have evolved separately from an already diverse but not yet recognized protangiospermous stock (Moore and Uhl, unpublished). The evidence is that, among the monocotyledons, palms had a different place of origin from that postulated for "ranalean" families by Smith.

A NORTHERN ORIGIN?

There is a relatively rich fossil palm flora in the Northern Hemisphere from lower Upper Cretaceous through early Tertiary times. [Recent reports of palms from Jurassic beds of the western United States (Tidwell et al. 1970; Tidwell, Rushforth, and Simper 1970) appear to be in error and the fossils belong to Tertiary times (Scott et al. 1972).] Can Laurasia have been the place of origin of palms? The possibility cannot be ruled out but the total evidence is not convincing. Admittedly primitive coryphoid genera—*Rhapidophyllum, Chamaerops,* and *Trachycarpus* of the *Trithrinax* alliance

—fit an "Arcto-Tertiary" pattern and fossils of presumed coryphoid palms in Alaska, the Aleutian Islands, and Japan provide evidence for the possibility of migration through a Beringian corridor. Wood (1969) noted the disjunct distribution of species between southeastern North America and southeastern South America, a pattern that might admit though not explain the presence of *Trithrinax,* the least specialized genus of its alliance, in southern Brazil, Uruguay, and Argentina. Costapalmate and palmate leaves in North America and Eurasia, seeds closely resembling those of *Sabal, Serenoa,* and perhaps *Livistona* in the London Clay flora, and pollen identified as that of *Livistona, Sabal, Serenoa,* and *Thrinax* (Reid and Chandler 1933, Chandler 1961-1964, vol. 4) are evidence for a much wider dispersal of other coryphoid palms in the past.

Pinnate leaves, fruits of *Nypa,* and possibly autochthonous endocarps of cocosoid palms in North America, and the same elements plus seeds referred to *Oncosperma* and *Caryotispermum,* spines thought to be lepidocaryoid, and pollen of *Phoenix* and *Calamus* in Europe (see Chandler 1961-1964, vol. 4) provide evidence for a more diverse flora in early Tertiary times than now exists in either region. It is noteworthy that fossils of pinnate palms lie far south of the limit of those of palmate palms, suggesting that climatic conditions that were compatible with coryphoid migration across Beringia may not have permitted the less cold-tolerant pinnate palms to follow this route.

According to Dietz and Holden (1970a, 1970b) there was no terrestrial connection between North and South America from the initiation of rifting in the early Triassic until the establishment of a connection in the late Pliocene. Migration between Africa and Eurasia presumably might have occurred intermittently over a long period of time, affording interchange of plants. Hence the African flora is less significant in the consideration of a possible northern origin. If palms originated in the north, however, then migration to South America was most likely by long-distance dispersal. We do know that fossils presumed or known to be of palms occur in South America in formations from the Cretaceous to the Pliocene (Menéndez 1969). The kinds are less clear because the evidence is restricted to pinnate leaves, seeds of undetermined affiliation, some cocosoid endocarps, and problematical fruits of *Nypa.*

It is the evidence from living palms that must be considered. Table 1 shows that there is relatively little difference between North America and South America so far as groups of palms are concerned. Each has one group peculiar to it—*Pseudophoenix* in the north, the ceroxyloid palms in the south—but all other groups are shared. Within the shared groups, however, so far as our knowledge now permits an analysis, it seems apparent that in general the less specialized genera and the predominance of species are South American. I have already mentioned *Pseudophoenix* and the ceroxyloid groups. *Chamaedorea* is certainly the most highly evolved of its group and has a center of diversity and specialization in Mexico and Central America, with apparently less specialized representation in South America.

The iriarteoid palms are scarcely found north of the Darien region of Panama—one species each of *Socratea* and *Iriartea* reaches Costa Rica. Of geonomoid palms, two specialized genera—*Calyptrogyne* and *Calyptronoma*—are restricted to Central America and Mexico, and to the Greater Antilles, respectively. The remainder are present on both continents but with what I take to be the less specialized species in South America. At most, two genera of cocosoid palms are peculiar to North America (in the West Indies); the diversity of genera and species is again South American. One monotypic genus of the arecoid group, *Neonicholsonia,* occurs in Central America. The other nine genera of arecoid palms are all represented in South America, though there are more species of *Roystonea* in the West Indies than on that continent. Finally the phytelephantoid palms are essentially South American.

The presence of fossils over a long period and the overwhelming balance of diversity and usually of primitiveness among the coryphoid group and the groups of the arecoid line in South America suggest that most of the palms in North America are relatively recent migrants. *Pseudophoenix, Sabal, Copernicia,* and perhaps the chamaedoreoid palms may be exceptions to be considered later.

WEST GONDWANA AS A CENTER OF DISTRIBUTION

A third possibility is that palms may have had an austral origin as suggested for the angiosperms as a whole by Camp (1947). We have seen that there is a substantial diversity of palms in South America and that Africa and Australia are relatively poor compared with other regions. But if the diversity of South America and Africa together (i.e., West Gondwana) is considered it becomes apparent that, when the fossil record of *Nypa* is added, 13 of 15 groups are represented there. Only *Pseudophoenix* in the Caribbean and the caryotoid palms of Asia are missing. It is also apparent that there is a concentration of primitive types. This focus of diverse and primitive palms in West Gondwana raises a question: Could palms have originated there? And if so, when?

West Gondwana had begun to rift by late Jurassic, according to Dietz and Holden's (1970a, 1970b) reconstruction, but Africa and South America were apparently still yoked or in very close proximity into the Lower Cretaceous in those portions which today correspond with the bulges of the two continents (Allard and Hurst 1969, Smith and Hallam 1970). Generally warm climates prevailed there throughout the Mesozoic and early Tertiary (Gill 1961, King 1961, Schwarzbach 1963). Large continental shields in Africa and South America were peneplaned in the Jurassic but there were periods of uplift (King 1967) and a reasonable diversity of habitats may be assumed. Thus the physical setting may easily have been conducive to the development of such palms as the more primitive types of the coryphoid and arecoid lines in equable tropical uplands with later migration to and further evolution in the lowlands, much as suggested for evolution of angiosperms by Axelrod (1952, 1970).

There is no undisputed fossil evidence of palms in the late Jurassic or early Cretaceous but all other factors appear to have been right for their evolution in West Gondwana during that time, with the establishment of stocks which would lead to evolution along major lines noted earlier. Some palms might even have evolved to nearly modern states: *Nypa* has essentially "sat still" for 70 million years and the highly evolved cocosoid palms have been extant for at least 100 million years, since the Cenomanian (lower Upper Cretaceous). The occurrence in both Africa and South America of *Elaeis* and *Raphia* (Figures 4, 5) might even be considered in this context.

Within such a center, the more adaptable groups, for example the coryphoid palms or primitive arecoid types such as *Pseudophoenix* and allied groups, may have achieved wide distribution while others developed and remained in more local patterns, e.g., iriarteoid and geonomoid palms in South America and *Podococcus* in Africa. Having achieved wide distribution, basic types may then have migrated by way of intermittent contact or island "stepping-stones" from Africa to Laurasia (see Kurtén 1969). Dispersal then followed a westward course along a Eurasian-North American corridor (see Termier and Termier 1960) as suggested by Kurtén (1966) for mammals (in the Paleocene), and eastward along the then warmer and more humid shores of the Tethys Sea to Asia and Australasia. This is a reversal of the route predicated by Reid and Chandler (1933) and Lakhanpal (1970).

In this fashion, the northern genera of the *Trithrinax* alliance or their prototypes and early members of the *Rhapis* alliance might have reached their present positions (Figure 3A), *Chamaerops* directly, *Rhapidophyllum* by westward migration, and *Trachycarpus* (with supposed fossils in England and Russia, see Chandler 1961-1964, vol. 2; Takhtajan 1958) by an eastward path, having left *Trithrinax* behind in South America. The distribution of the *Livistona* alliance is very suggestive in this regard. Many of its members are adapted to seasonal, dry, or even warm-temperate climates. The genera, except the somewhat different *Washingtonia* in western United States and *Johannesteijsmannia* in Malaya, Sumatra, and Borneo, are rather similar and evenly distributed. Five genera (*Acoelorrhaphe, Brahea, Colpothrinax, Copernicia, Serenoa*) are in the Western Hemisphere, all north of South America save three species of *Copernicia*, which may have invaded South America secondarily. Another five genera (*Licuala, Livistona, Pholidocarpus, Pritchardia, Pritchardiopsis*) are in the eastern tropics where they generally occupy wetter or more equable habitats. Finally, *Wissmannia*, very close to *Livistona* but with a slightly less specialized corolla, occurs centrally in three apparently relict populations in desert or near-desert regions of eastern Africa and southern Arabia (Moore 1971b). *Sabal* (Figure 3C), though now native to the Western Hemisphere, may have extended from England to Russia in the Tertiary (Takhtajan 1958), and its present representation in northeastern South America (one species) may be of recent origin there.

Similar patterns may be traced, at least in part, for other groups: for *Phoenix* or its antecedents, now extinct in North America and Europe though once present; for the borassoid palms, possibly of African origin and since spread to Asia; quite conceivably for *Nypa,* now in Asia but extinct in America, Africa, Europe; for the wet-adapted lepidocaryoid palms in South America and Africa, with mostly advanced genera in Asia. Perhaps *Oncosperma* or its prototype was African; if so, the present distribution of fossil and extant *Oncosperma* in Europe and Asia and of related genera in the Seychelles fits the pattern. *Pseudophoenix* (Figure 2B), adapted to marginal situations, may be conceived of as a relict persisting from a wider distribution of an arecoid line. So may the chamaedoreoid palms, though each, and more likely the chamaedoreoid group, may equally have persisted in the south and migrated northward since the establishment of late Tertiary connections. The present pattern of the arecoid group, however, suggests a different pathway noted below.

Some migration might have been effected by an austral route also. The finding of palmoid pollen in deposits at McMurdo Sound, Antarctica, dated between the Cretaceous and Oligocene (Cranwell, Harrington, and Speden 1960) suggests this alternative, as do some of the possible times of break-up and dispersal of Gondwanaland listed by Smith and Hallam (1970). The distribution of fossil cocosoid palms in New Zealand and of *Cocos* in the Indo-Pacific might be explained in this way. It is also conceivable that some original stocks of the arecoid group may thus have reached New Guinea, New Zealand, and New Caledonia, where all but one of the alliances in the eastern tropics are concentrated (see Moore 1973).

Two patterns in particular give credence to this third alternative so lengthily considered. These are the disjunct distributions of ceroxyloid and chamaedoreoid palms already mentioned (see Figure 2A, C) in America and in Madagascar and the Mascarenes, respectively. Other plants demonstrate essentially this pattern, witness *Ravenala-Strelitzia-Phenakospermum.* I find the coincidence of relationships too strong to accept chance long-distance dispersal as the agent and it is clear that parallel evolution is not involved here (see Moore 1969, Uhl 1969). A once much broader distribution of these and many other palms now interrupted in Africa by massive extinction of groups that demand conditions not present today and perhaps not present for a long time in the past seems a much more satisfactory explanation. As noted earlier, such extinction is plausible at least in the light of the Pleistocene history of Africa if not before.

Three alternatives have been provided to explain palm distribution in its broadest implications. None can be proved, but the third alternative, that of origin in and spread from West Gondwana, appears to fit present distribution best. It is of more than passing interest to note that Hallam (1967), considering implications of paleozoogeographic data on continental drift, provides examples from vertebrate distribution that may fit in part with the above pattern for palms. The distribution and adaptive nature of the more primitive palms, especially the most primitive coryphoid group, also fits admirably with Axelrod's hypothesis of angiosperm origin, while the position of landmasses in late Jurassic and early Cretaceous times satisfies the need for a unified center of origin and initial dispersal for palms. There is no undisputed evidence for an origin so early as Jurassic but there is an increasing acceptance of the possibility (Axelrod 1970, Hawkes and Smith 1965, Maguire 1970, Smith 1970). Perhaps as the fossil record becomes clearer, or as further refinements of our understanding of palms and other plants become available, other alternatives will suggest themselves. So long as the fossil record remains inadequate, only the accumulation of detailed analyses of distribution and relationship in family after family is likely to provide the kind of circumstantial evidence that may lead to a clearer understanding of that still great mystery, the origin of the angiosperms.

References

Ainslie, J. R. 1926. The physiography of southern Nigeria and its effect on the forest flora of the country. Oxford Forest. Mem. 5:1-36.

Allard, G. O., and V. J. Hurst. 1969. Brazil-Gabon geologic link supports continental drift. Science 163:528-532.

Allen, P. 1965. *Raphia* in the western world. Principes 9:66-70.

Anderson, R., and S. Mori. 1967. A preliminary investigation of *Raphia* palm swamps, Puerto Viejo, Costa Rica. Turrialba 17:221-224.

Arnold, C. A. 1952. Tertiary plants from North America, 1: A *Nipa* fruit from the Eocene of Texas. Palaeobotanist 1:73-74.

Axelrod, D. I. 1952. A theory of angiosperm evolution. Evolution 6:29-60.

———. 1970. Mesozoic paleogeography and early angiosperm history. Bot. Rev. 36:277-319.

Bailey, L. H. 1942. Palms of the Mascarenes. Gentes Herb. 6:49-104.

Ball, O. M. 1931. A contribution to the paleobotany of the Eocene of Texas. Bull. Agr. & Mech. Coll. Texas, ser. 4, 2(5): 1-173.

Bannister, B. A. 1970. Ecological life cycle of *Euterpe globosa* Gaertn. *In* H. T. Odum (Editor), A tropical rain forest. B 299-B 314. Div. Tech. Info. U.S. Atomic Energy Comm. Oak Ridge, Tenn.

Barrau, J. 1959. The sago palms and other food plants of marsh dwellers in the South Pacific Islands. Econ. Bot. 13:151-162.

Beard, J. S. 1944. Climax vegetation in tropical America. Ecology 25:127-158.

———. 1945. The progress of plant succession on the Soufriere of St. Vincent. J. Ecol. 33:1-9.

———. 1946. The natural vegetation of Trinidad. Oxford Forest. Mem. 20:1-152.

———. 1949. The natural vegetation of the Windward & Leeward Islands. Oxford Forest. Mem. 21:1-192.

———. 1955. The classification of tropical American vegetation-types. Ecology 36:89-100.

Beccari, O. 1917. The origin and dispersal of *Cocos nucifera*. Philippine J. Sci., C. Bot. 12:27-43.

———. 1933. Asiatic palms—Corypheae. Ann. Roy. Bot. Gard. Calcutta 13:1-356.

Berry, E. W. 1914. Fruits of a date palm in the Tertiary deposits of eastern Texas. Amer. J. Sci. 187:403-406.

———. 1916. Lower Eocene floras of southeastern North America. U. S. Geol. Surv. Profess. Paper 91:1-481.

———. 1920. Miocene fossil plants from northern Peru. Proc. U. S. Natl. Mus. 55:279-294.

———. 1922a. A palm nut from the Miocene of the Canal Zone. Proc. U. S. Natl. Mus. 59:21-22.

———. 1922b. Tertiary fossil plants from Venezuela. Proc. U. S. Natl. Mus. 59:553-579.

———. 1926a. *Cocos* and *Phymatocarpon* in the Pleiocene of New Zealand. Amer. J. Sci. 212:181-184.

———. 1926b. A fossil palm fruit from the middle Eocene of northwestern Peru. Proc. U. S. Natl. Mus. 70(3):1-4.

———. 1929a. A palm nut of *Attalea* from the Upper Eocene of Florida. J. Washington Acad. Sci. 19:252-255.

———. 1929b. Tertiary fossils from Colombia, South America. Proc. U. S. Natl. Mus. 75(24):1-12, pls. 1-5.

———. 1929c. Eocene plants from Restin formation of Peru. Pan-Amer. Geol. 51:241-244.

———. 1932. A new palm from the Upper Eocene of Ecuador. J. Washington Acad. Sci. 22:327-329.

Blydenstein, J. 1967. Tropical savanna vegetation of the llanos of Colombia. Ecology 48:1-15.

Bogdan, A. V. 1958. Some edaphic vegetational types at Kiboko, Kenya. J. Ecol. 46:115-126.

Bouillenne, R. 1930. Un voyage botanique dans le bas-Amazone. Arch. Inst. Bot. Univ. Liège 8:1-185, pls. 1-34.

Braun, A. 1968. Cultivated palms of Venezuela. Principes 12:39-103, 111-136.

Brown, R. W. 1956. Ivory-nut palm from late Tertiary of Ecuador. Science 123:1131-1132.

———. 1962. Paleocene flora of the Rocky Mountains and Great Plains. U. S. Geol. Surv. Profess. Paper 375:1-119.

Budanchev, L. J. 1953. New discovery of the palm *Sabal* in Kazakhstan. Dokl. Akad. Nauk SSSR ser. 2, 93:347-348, pl. 1.

Camp, W. H. 1947. Distribution patterns in modern plants and the problems of ancient dispersals. Ecol. Monogr. 17:159-183.

Chandler, M. E. J. 1961-1964. The Lower Tertiary floras of southern England. 4 vols. British Museum (Natural History). London.

Child, R. 1964. Coconuts. 216 pp. Longmans, Green & Co., Ltd. London.

Chipp, T. F. 1927. The Gold Coast forest: A study in synecology. Oxford Forest. Mem. 7:1-94.

Clayton, W. D. 1958. Secondary vegetation and the transition to savanna near Ibadan, Nigeria. J. Ecol. 46:217-238.

Corner, E. J. H. 1966. The natural history of palms. 393 pp., 133 figs., 24 pls. Weidenfeld and Nicolson. London.

Couper, R. A. 1952. The spore and pollen flora of the *Cocos*-bearing beds, Mangouni, North Auckland. Trans. Roy. Soc. New Zealand 79:340-348.

Cranwell, L. M., H. J. Harrington, and I. G. Speden. 1960. Lower Tertiary microfossils from McMurdo Sound, Antarctica. Nature 186:700-702.

Davis, T. A. W., and P. W. Richards. 1934. The vegetation of Moraballi Creek, British Guiana: An ecological study of a limited area of tropical rain forest, Part II. J. Ecol. 22: 106-155.

Dietz, R. S., and J. C. Holden. 1970a. Reconstruction of Pangaea: Breakup and dispersion of continents, Permian to Present. J. Geophys. Res. 75:4939-4956.

———. 1970b. The breakup of Pangaea. Sci. Amer. 223(4): 30-41.

Dolianiti, E. 1955. Frutos de *Nipa* no Paleoceno de Pernambuco, Brasil. Dep. Nac. Prod. Min., Div. Geol. e Min. Bol. 158:1-36.

Eggeling, W. J. 1935. The vegetation of Namanve Swamp, Uganda. J. Ecol. 23:422-435.

———. 1947. Observations on the ecology of the Budongo rain forest, Uganda. J. Ecol. 34:20-87.

Engler, A. 1908. Die Pflanzenwelt Afrikas 2. *In* A. Engler and O. Drude, Die Vegetation der Erde 9(2):222-235. W. Engelmann. Leipzig.

Espinal T., L. S. and E. Montenegro M. 1963. Formaciones vegetales de Colombia. 201 pp. Geográfico "Augustin Codazzi." Bogotá.

Essig, F. B. 1971. Observations on pollination in *Bactris*. Principes 15:20-24.

Fanshawe, D. B. 1952. The vegetation of British Guiana: A preliminary review. Imp. Forest. Inst. Paper 29:1-96.

Fliche, P. 1896. Études sur la flore fossile de l'Argonne Albien-Cénomanien. Bull. Soc. Sci. Nancy, ser. 2, 14:114-306.

Gill, E. D. 1961. The climates of Gondwanaland in Kainozoic times. Pp. 332-353, *in* Nairn, A. E. M. (Editor). Descriptive palaeoclimatology. Interscience Publishers. New York.

Goodland, R. J. A. 1970. Plants of the cerrado vegetation of Brasil. Phytologia 20:57-78.

Haffer, J. 1969. Speciation in Amazonian forest birds. Science 165:131-137.

Hallam, A. 1967. The bearing of certain palaeozoogeographic data on continental drift. Palaeogeography, Palaeoclimatology, Palaeoecology 3:201-241.

Hartley, C. W. S. 1967. The oil palm (*Elaeis guineensis* Jacq.). 706 pp., illus. Longmans Green & Co. London.

Hawkes, J. G., and P. Smith. 1965. Continental drift and the age of angiosperm genera. Nature 207:48-50.

Heer, Oswald. 1883. Die fossile Flora Grönlands, II.2: Die tertiäre Flora von Grönland. *In* Flora Fossilis Arctica 7:47-142. J. Wurster & Co. Zurich.

Hollick, A. 1936. The Tertiary floras of Alaska. U. S. Geol. Surv. Profess. Paper 182:1-185.
Hooker, J. D. 1883. Palmae. *In* G. Bentham and J. D. Hooker, Gen. Pl. 3:870-948.
Janzen, D. H. 1969. Seed-eaters versus seed size, toxicity and dispersal. Evolution 23:1-27.
————. 1971. The fate of *Scheelea rostrata* fruits beneath the parent tree: Predispersal attack by bruchids. Principes 15:89-101.
Jumelle, H. 1945. 30ᵉ Fam.—Palmiers. *In* H. Humbert, Flore de Madagascar et des Comores. 186 pp. Tananarive, Madagascar.
Kaul, K. N. 1943. A palm stem from the Miocene of Antigua, W. I.—*Phytelephas sewardii,* sp. nov. Proc. Linn. Soc. London 155:3-4, figs. 1-2.
————. 1951. A palm fruit from Kapurdi (Jodhpur, Rajasthan Desert) *Cocos sahnii* sp. nov. Curr. Sci. 20:138.
Keay, R. W. J. 1959. Vegetation map of Africa south of the Tropic of Cancer. Oxford University Press. London.
Kepler, C. B. 1970. Appendix A: The Puerto Rican parrot. *In* H. T. Odum (Editor), A tropical rain forest. E 186-E 188. Div. Tech. Info., U. S. Atomic Energy Comm. Oak Ridge, Tenn.
King, L. C. 1961. The palaeoclimatology of Gondwanaland during the Palaeozoic and Mesozoic eras. Pp. 307-331, *in* A. E. M. Nairn (Editor), Descriptive palaeoclimatology. Interscience Publishers. New York.
————. 1967. The morphology of the earth. 2nd edition. 726 pp. Oliver & Boyd. Edinburgh.
Koch, B. E. 1964. Review of fossil floras and nonmarine deposits of West Greenland. Geol. Soc. Amer. Bull. 75:535-548.
Kryshtofovich, A. N. 1929. Evolution of the Tertiary flora in Asia. New Phytol. 28:303-312.
————. 1955. Development of the phytogeographical regions of the Northern Hemisphere from the beginning of the Tertiary period. Problems of the geology of Asia 2:824-844 [in Russian, cited by Chandler 1961-1964, vol. 4].
Kurtén, B. 1966. Holarctic land connexions in the early Tertiary. Comment. Biol. Soc. Sci. Fennica 29(5):1-5.
————. 1969. Continental drift and evolution. Sci. Amer. 220(3):54-64.
Lakhanpal, R. N. 1970. Tertiary floras of India and their bearing on the historical geology of the region. Taxon 19:675-694.
Leck, C. F. 1969. Palmae: hic et ubique. Principes 13:80.
Lepesme, P. 1947. Les insectes des palmiers. 903 pp. Paul Lechevalier Éditeur. Paris.
Lever, R. J. A. W. 1969. Pests of the coconut palm. FAO Agricultural Studies 77:1-190.
Maguire, B. 1970. On the flora of the Guayana Highland. Biotropica 2:85-100.
Mahabalé, T. S. 1959. Resolution of the artificial palm genus *Palmoxylon*: A new approach. Palaeobotanist 7:76-84.
Mahabalé, T. S., and M. V. Parthasarathy. 1963. The genus *Phoenix* Linn. in India. J. Bombay Nat. Hist. Soc. 60: 371-387.
Markley, K. 1955. Caranday—a source of palm wax. Econ. Bot. 9:39-52.
Melchior, H. (Editor). 1964. A Engler's Syllabus der Pflanzenfamilien, ed. 12, 2:498-625.
Menéndez, C. A. 1969. Die fossilen Floren Südamerikas. Monogr. Biol. 19:519-561.
Menon, K. V. P., and K. M. Pandalai 1958. The coconut palm: A monograph. 384 pp. Indian Central Coconut Committee. Ernakulam, India.
Miller, R. H. 1964. The versatile sugar palm. Principes 8: 115-147.
Moore, H. E. 1957. Cold tolerance of the palms in Italy. Principes 1:77-78.
————. 1969. The genus *Juania* (Palmae-Arecoideae). Gentes Herb. 10:385-393.
————. 1971a. The genus *Synechanthus* (Palmae). Principes 15:10-19.
————. 1971b. Wednesdays in Africa. Principes 15:111-119.
————. 1973. The major groups of palms and their distribution. Principes 17: [in press].
Moore, H. E., A. Salazar C., and E. E. Smith. 1960. A reconnaissance survey of palms in eastern Peru. Agriculture Division, U.S. Operation Mission to Peru, I. C. A., Lima. [Typescript report.]
Hoore, H. E., and N. W. Uhl. (In manuscript). Palms and the origin and evolution of the monocotyledons. Presented at AIBS Symposium, 30 August 1972.
Moreau, R. E. 1952. Africa since the Mesozoic: With particular reference to certain biological problems. Proc. Zool. Soc. London 121:869-913.
————. 1963. Vicissitudes of the African biomes in the late Pleistocene. Proc. Zool. Soc. London 141:395-421.
————. 1966. The bird faunas of Africa and its islands. 424 pp. Academic Press. New York and London.
Muller, J. 1964a. A palynological contribution to the history of the mangrove vegetation in Borneo. Pp. 33-42, *in* Cranwell, L. M. (Editor), Ancient Pacific Floras. Univ. Hawaii Press. Honolulu.
————. 1964b. Palynological contributions to the history of Tertiary vegetation in NW. Borneo. Tenth Internat. Bot. Congr., Abstracts of papers, 271.
————. 1970. Palynological evidence on early differentiation of angiosperms. Biol. Rev. Cambridge Philos. Soc. 45:417-450.
Myers, J. G. 1933. Notes on the vegetation of the Venezuelan llanos. J. Ecol. 21:335-349.
Pallot, J. M. 1961. Plant microfossils from the Oligocene of the Isle of Wight. PhD dissertation. University of London [cited by Chandler 1961-1964, vol. 4].
Parthasarathy, M. V. 1968. Observations on metaphloem in the vegetative parts of palms. Amer. J. Bot. 55:1140-1168.
Rao, A. R., and V. K. Menon. 1966. A new species of petrified palm stem from the Deccan Intertrappean series. Palaeobotanist 14:256-263, pls. 1-2.
Read, R. W. 1965. Chromosome numbers in the Coryphoideae. Cytologia 30:385-391.
————. 1966. New chromosome counts in the Palmae. Principes 10:55-61.
————. 1967. A study of *Thrinax* in Jamaica. PhD dissertation. 228 pp. Univ. West Indies. Mona, Jamaica.
————. 1968. A study of *Pseudophoenix* (Palmae). Gentes Herb. 10:169-213.
Reid, E. M., and M. E. J. Chandler. 1933. The London Clay flora. 561 pp., 17 figs., 33 pls. British Museum (Natural History). London.
Richards, P. W. 1939. Ecological studies on the rain forest of southern Nigeria, I: The structure and floristic composition of the primary forest. J. Ecol. 27:1-61.
————. 1952. The tropical rainforest. 450 pp. Cambridge Univ. Press. Cambridge, England.
————. 1963. What the tropics can contribute to ecology. J. Ecol. 51:231-241.
Rizzini, C. T. 1963. A flora do cerrado. Pp. 125-177, *in* Ferri, M. G. (Editor). Simpósio sôbre o cerrado. Univ. São Paulo. Brazil.

Sauer, C. O. 1958. Man in the ecology of tropical America. Proc. Ninth Pacific Sci. Congress 20:104-110.

Sauer, J. D. 1967. Plants and man on the Seychelles coast. 132 pp. University of Wisconsin Press. Madison.

Schmid, R. 1970. Notes on the reproductive biology of *Asterogyne martiana* (Palmae), I: Inflorescence and floral morphology; phenology. II: Pollination by syrphid flies. Principes 14:3-9, 39-49.

Schwarzbach, M. 1963. Climates of the past. 328 pp. D. Van Nostrand, Ltd. London.

Scott, H. 1933. General conclusions regarding the insect fauna of the Seychelles and adjacent islands. Trans. Linn. Soc. London Zool. 19:307-391.

Scott, R. A., P. L. Williams, E. S. Barghoorn, L. C. Craig, L. J. Hickey, and H. D. Macginitie. 1972. Pre-Cretaceous angiosperms from Utah: Evidence for Tertiary age of the palm woods and roots. Amer. J. Bot. 59:886-896.

Smith, A. C. 1970. The Pacific as a key to flowering plant history. Harold L. Lyon Arboretum Lecture 1:1-26.

Smith, A. G., and A. Hallam. 1970. The fit of southern continents. Nature 225:139-144.

Sterlin, B. P. 1950. On the discovery of a leaf imprint of *Sabal major* in the Karaganskian deposits of northern Dagesten. Dokl. Akad. Nauk SSSR ser. 2, 72:113-114, fig. 1.

Sternberg, H. O. 1968. Man and environmental change in South America. Monogr. Biol. 18:413-445.

Takhtajan, A. L. 1958. A taxonomic study of the Tertiary fan palms of the U.S.S.R. Bot. Zhurn. 43:1661-1674 [English summary].

Termier, H., and G. Termier, 1960. Atlas de paléogéographie. 36 maps and legends. Masson et Cie. Paris.

Thomas, A. S. 1941. The vegetation of the Sese Islands, Uganda. J. Ecol. 29:330-353.

————. 1945. The vegetation of some hillsides in Uganda. J. Ecol. 33:10-43.

Tidwell, W. D., S. R. Rushforth, J. L. Reveal, and H. Behunin. 1970. *Palmoxylon simperi* and *Palmoxylon pristina*: two pre-Cretaceous angiosperms from Utah. Science 168: 835-840.

Tidwell, W. D., S. R. Rushforth, and A. D. Simper. 1970. Pre-Cretaceous flowering plants: Further evidence from Utah. Science 170:547-548.

Tomlinson, P. B. 1960. Anatomy of the Caryotoideae. Principes 4:118-119.

————. 1962. Essays on the morphology of palms, VII: A digression about spines. Principes 6:44-52.

Tralau, H. 1964. The genus *Nypa* van Wurmb. Kungl. Svensk. Vetenskapsakad. Handl. ser. 4, 10 (1):1-29.

Tuta, J. A. 1967. Fossil palms. Principes 11:54-71.

Uhl, N. W. 1966. Morphology and anatomy of the inflorescence axis and flowers of a new palm, *Aristeyera spicata*. J. Arnold Arboretum 47:9-22.

————. 1969. Floral anatomy of *Juania, Ravenea,* and *Ceroxylon* (Palmae-Arecoideae). Gentes Herb. 10:394-411.

————. 1972. Inflorescence and flower structure in *Nypa fruticans* (Palmae). Amer. J. Bot. 59:729-743.

Uhl, N. W., and H. E. Moore. 1971. The palm gynoecium. Amer. J. Bot. 58:945-992.

————. 1973. The protection of flowers in palms. Principes 17 [in press].

Unger, Franz. 1852. Iconografia plantarum fossilium. Denkschr. Kaiserl. Akad. Wissensch. Math-Naturw. Classe, Wien 4 (1):73-118, pls. 24-45.

Vogl, R. J., and L. T. McHargue. 1966. Vegetation of California fan palm oases on the San Andreas Fault. Ecology 47:532-540.

Vuilleumier, B. S. 1971. Pleistocene changes in the fauna and flora of South America. Science 173:771-780.

Walter, H. 1964. Die Vegetation der Erde in öko-physiologischer Betrachtung. I. Die tropischen und subtropischen Zonen. 2nd edition. 592 pp. Gustav Fischer Verlag. Stuttgart.

Wessels Boer, J. G. 1968. The geonomoid palms. Verh. Kon. Ned. Akad. Wetensch. Afd. Natuurk, Tweede Reeks, 58(1) : 1-202.

Whitehead, D. R. 1969. Wind pollination in the angiosperms: Evolutionary and environmental considerations. Evolution 23:28-35.

Whitmore, T. C. 1970. *Liberbaileya gracilis*. Principes 14:97-107.

————. 1971. *Maxburretia rupicola*. Principes 15:3-9.

Wood, C. E., Jr. 1969. Some floristic relationships of eastern North America. XI Internat. Bot. Congr., Abstracts 242.

Zacher, F. 1952. Die Nährpflanzen der Samenkäfer. Zeitschr. angew. Entomologie 33:210-217, 460-480.

Zeven, A. C. 1964. On the origin of the oil palm (*Elaeis guineensis* Jacq.). Grana Palynol. 5:121-123.

————. 1967. The semi-wild oil palm and its industry in Africa. 178 pp. Agricultural Research Report 689, Centre for Agricultural Publications and Documentation. Wageningen, Netherlands.

Leguminous Resin-producing Trees in Africa and South America

JEAN H. LANGENHEIM

Discussion of the distribution and possible evolutionary patterns of members of the Leguminosae is particularly pertinent to the purposes of this symposium for at least two reasons. First, they are prominent components of various tropical ecosystems in both Africa and South America, and secondly, certain genera display striking discontinuity in their distribution between the two continents.

Our work has been focused upon a group of resin-producing genera, several of which occur in both Africa and South America, in the Cynometreae-Amherstieae tribal complex within the subfamily Caesalpinioideae. Interest in leguminous resin-producing trees was stimulated by discovery that amber (fossilized resin) from Mexico, Colombia, and Brazil was derived from *Hymenaea* (Langenheim 1966, 1967, 1969) and that amber from Angola was derived from *Copaifera* (Langenheim 1968, 1970).

Resins consist of complex mixtures of mono-, sesqui-, di- and triterpenoids, some of which are volatile whereas others are nonvolatile. The nonvolatile fractions are generally composed of some sesquiterpenoids but predominantly di- or triterpenoids. These nonvolatile terpenoids may fossilize and certain components may be sufficiently stable that they may be studied through time (Langenheim 1964; Langenheim and Beck 1965, 1968; Langenheim 1969; Thomas 1970).

There is a long record of trees producing resins, beginning with coniferous trees in the Carboniferous Period. Although resins are still synthesized in all coniferous families today, more than two-thirds of resin-producing plants occur in tropical angiosperm families. Many members of the Burseraceae, Anacardiaceae, Guttiferae, Rubiaceae, Styracaceae, and Euphorbiaceae synthesize resins, but members of the Leguminosae and Dipterocarpaceae are particularly noted for copious resin production which often has been utilized commercially. Since leguminous resins do have stable components which fossilize, they provide an interesting opportunity to study chemical evolution.

In this presentation I shall be primarily concerned with a selected few genera, recent studies of which bring into perspective some problems relevant to the purpose of the symposium. I shall be considering: (1) similarities in morphological characters as well as in chemical ones related to resin production; (2) possible sites of origin and routes of migration; and (3) similarities in species radiation into different ecosystems on both continents.

Systematic Review

All of the known resin-secreting leguminous plants belong within the subfamily Caesalpinioideae and the Cynometreae-Amherstieae tribal complex. Delimitation of these tribes has been subject to controversy and it is a formidable task to understand

JEAN H. LANGENHEIM, Division of Natural Sciences, University of California, Santa Cruz, California 95060. ACKNOWLEDGMENTS: Grateful acknowledgment is made to Dr. J. P. M. Brenan, Keeper of the Herbarium, Royal Botanic Gardens, Kew, for stimulating discussions of the material in the manuscript and also to Professor J. Léonard, Université de Bruxelles and to Dr. Velva Rudd, Smithsonian Institution, for commenting on the text.

the relationships within the groups. Genera belonging to these tribes occur throughout the tropics but have been most intensively studied in Africa and the New World; they are least well known in Asia. In Africa the Cynometreae and Amherstieae are the most important tribes in the Caesalpinioideae, not only in numbers of species but as significant forest components (Léonard 1951, Aubreville 1961).

In defining the Cynometreae and Amherstieae one of the greatest difficulties has been the choice of the most important taxonomic characters. Bentham (1840) and Bentham and Hooker (1865) used the fusion or nonfusion of the stipe of the ovary to the receptacle and the number of ovules. Because the distinctions provided by these characters could not be universally applied, Harms (1915) questioned whether or not the two tribes shouid be united. Baker (1926) then actually united the two tribes into the Amherstieae, and Dwyer (1954a) further suggested the union of both the Cynometreae and the solely New World Sclerobieae into the Amherstieae. Léonard (1957), however, thought that two tribes should be recognized, based on the presence of either enveloping or nonenveloping bracteoles. Because this basic separation encompasses significant secondary characters, it results in a more natural, coherent classification.

I prefer to accept Léonard's analysis of the tribes as probably the best for evolutionary considerations. Using his interpretation, all African and South American leguminous resin-producing genera (Table 1) belong to the Cynometreae. With previous tribal concepts, most of these genera come within the Amherstieae. Although all of the genera listed in Table 1 secrete resins, the six starred ones have species that produce a sufficient quantity to have been of commercial value (Tschirch and Stock 1933-1936, Howes 1949, Léonard 1950a). This discussion will be centered upon four of these genera (*Hymenaea, Trachylobium, Copaifera,* and *Guibourtia*) which are exemplary of the most interesting biogeographic and evolutionary problems and which have been most recently investigated. In the past at least one species of *Cynometra* was thought to produce resin, but recent evidence indicates that this is not the case (Léonard, pers. comm.). Nonetheless, *Cynometra* will be discussed for its possible significance in understanding the origin and dispersal of the four above-mentioned genera.

TABLE 1. Resin-producing Genera in the Cynometreae (sensu Léonard 1957) occurring in Africa and/or South America

Genus	*No. of Species*	*Distribution*
Cynometra?	70	Pantropics
**Copaifera*	30 4	Neotropics Tropical Africa
**Daniellia*	7	Tropical Africa
**Tessmannia*	10	Tropical Africa
**Guibourtia*	13 1?	Tropical Africa Neotropics
Colophospermum	1	Tropical Africa
**Trachylobium*	1	Tropical Africa
**Hymenaea*	17	Neotropics
Peltogyne	28	Neotropics
Gossweilerodendron	1	Tropical Africa
Oxystigma	5	Tropical Africa
Prioria	1	Neotropics
Eperua	6	Neotropics

* indicates genera with species which produce copal of commercial value.

Resin Secretory Systems and Resin Chemistry

Resin is synthesized in parenchyma cells that usually line rounded pockets or cysts and elongated canals. These containers, into which the resin is secreted, may arise by either schizogeny, lysigeny, or both. Schizogeny involves the separation of cells, which round off and increase their intercellular spaces to produce pockets or canals, with the secretory cells as an epithelial layer. Lysigenous cavities result from the breakdown or disintegration of secretory cells. These cavities may occur in the parenchyma tissue in any part of the plant, although their location varies with different genera.

Resin production is particularly interesting among members of the Cynometreae because of the variety of anatomical arrangements for secretion within different organs, or in different stages of maturity in the same organ in the same plant (Guignard 1892, Moens 1955, Langenheim 1967, 1969). Different genera or groups of genera are also characterized by different secretory systems, a point of interest to our consideration.

Primarily because of their commercial importance some of the Cynometreae resins, commonly called "copals," have been analyzed chemically. Various physical and chemical properties, such as melting point, solubility, and saponification values, were commonly used to characterize resins until relatively recently when more modern techniques, such as gas chromatography, mass spectrometry, and infrared spectrophotometry, became available to facilitate structural analysis. Unfortunately most of the earlier types of analysis did not provide useful data for botanical purposes. Considerable work on coniferous resins, emphasizing quantitative compositional differences, has shown that these resins can be useful chemotaxonomically (e.g., Zavarin and Snajberk 1965, Mirov 1967, Zavarin, Reichert, and Tsien 1970). In these studies it was also demonstrated that resin composition may vary greatly in tissues of different organs such as leaf, wood, bark, or fruit, and must be considered in any chemotaxonomic interpretation.

The first work on leguminous resins that could be used in any chemotaxonomic context was based upon infrared spectra of Congo copals. Hellinckx (1955) demonstrated infrared spectra of carbon tetrachloride extracts of copals to be useful in determining the botanical source of various species of *Guibourtia, Copaifera, Daniellia,* and *Tessmannia.*

Infrared spectra made from pellets including the untreated resin have also been used in identifying the sources of some fossil resins (Langenheim and Beck 1965, 1968). In the polymerization of resins in fossilization, all simple functional groups are preserved except carbon-carbon double bonds, whereas skeletal frequencies are damped but not extinguished. Only major constituents of the complex of terpenoids comprising resins would be expected to give strong absorption bands, and although absolute identity of spectra can never be expected, the presence of similarities indicates structural similarity of the major constituents. The similarity in spectra between recent and fossil resins indicates that major constituents of some taxa appear to be relatively stable over millions of years of time. In the course of determining that amber from Mexico, Colombia, and Brazil was derived from *Hymenaea* and amber from Angola came from *Copaifera,* spectra were made of numerous species of *Hymenaea, Copaifera, Trachylobium, Guibourtia,* and *Daniellia.* Characteristic spectra appear to be representative of either genera or species complexes in the various groups studied (Langenheim 1968, 1969, 1970).

Although mono- and sesquiterpenoids are present in legumes, they have not been studied intensively as they have in coniferous resins. On the other hand, considerable work has been done on elucidating the structure of the diterpenoids. Most of these analyses thus far have been approached from the viewpoint of the structural chemist interested in novel compounds, although interest in biogenetic interpretation relevant to systematics has been indicated (Ponsinet, Ourisson, and Oehlschlager 1968). Harborne (1971) pointed out that the about 40 diterpenoids that have been isolated and characterized show considerable promise as taxonomic markers and that intensive systematic surveys of these compounds should be "especially rewarding." Diterpene resin acids, which Sandermann (1962) thought to have chemotaxonomic potential, have received the most study, although the data are generally too scattered in terms of specific compounds taken from different organs and from isolated species to be of systematic and phylogenetic value as yet. Excluding seed pod resin from *Trachylobium* (Hugel et al. 1965a, 1965b, 1965c, Hugel and Ourisson 1965), which consists mainly of tetra- and pentacyclic diterpenoids, these resin acids are mainly bicyclic compounds having the labdane skeletal type with the stereochemistry frequently reversed. Although diterpene resin acids have been studied in genera such as *Copaifera, Daniellia, Oxystigma, Prioria,* and *Eperua,* the most intensive studies have been made on *Trachylobium* and *Hymenaea* (Nakano and Djerassi 1961, Ponsinet, Ourisson, and Oehlschlager 1968, Harborne 1971, Martin and Langenheim 1970, Martin, Langenheim, and Cunningham 1971).

HYMENAEA–TRACHYLOBIUM

Let us now consider the relationships of several pairs of African and South American genera. I shall discuss the relationship of *Hymenaea* and *Trachylobium* first because it represents a fascinating case of two genera with a high degree of similarity in morphology and resin chemistry, yet having a striking discontinuity in distribution between the two continents.

Considerable morphological similarity between *Hymenaea* and *Trachylobium* has long been recognized. *Hymenaea courbaril* was the type-species of the genus described by Linnaeus (1737) from a Brazilian specimen without fruit. Gaertner then described the fruit of *H. courbaril* and also in 1791 described *H. verrucosa* from Madagascar. Hayne (1830) synonymized *H. verrucosa* to *Trachylobium hornemannianum*, which he had described from Mauritius. He also described several other species of *Trachylobium*, one from the Amazon Basin which was later redefined as a *Cynometra*. In 1871 Oliver put the *Trachylobium* populations of eastern Africa from coastal southern Kenya to Mozambique and the islands of Zanzibar, Madagascar, Mauritius, and the Seychelles (Figure 1) into *T. verrucosum*. Although considerable morphological variation has been recognized within populations occurring in the lowland evergreen forest, woodland, and coastal evergreen bushland, they have not been studied in sufficient detail to warrant establishment of additional species. *Trachylobium* appears more closely related to the New World *Hymenaea* than any of the extant African genera. Species of *Cynometra* appear to be the closest African relatives.

Hymenaea is not only closely related to the African *Trachylobium* but has obvious affinities in the New World with *Peltogyne*, which has 28 species distributed quite similarly to *Hymenaea* (Ducke 1935, 1938; Martinez 1960). *Peltogyne* apparently does not produce large amounts of resin but does synthesize it, at least in leaf resin pockets. *Peltogyne* also appears to be closely related to the New World *Cynometra* (Dwyer 1958).

In our present investigation of the genus *Hymenaea*, 17 species have been recognized (Langenheim, Lee, and Martin 1970, 1971), which are distributed from 23°N in central Mexico to 27°S in northern Argentina (Figures 1, 2). On the basis of seedling characteristics and differences in flowers and fruit, the genus is divisible into two groups. They tend to be segregated ecologically, one primarily comprised of species from the Amazonian and coastal evergreen forests and the other centered around *H. courbaril* and species occurring in the extra-rain forest habitats (Table 2; Lee and Langenheim 1972).

Nine *Hymenaea* species comprising Group I occur only in evergreen rain forest habitats (Table 2), although there is considerable variation in amount and seasonality of the precipitation within the "evergreen rain forest type." Also very important differences occur in the igapó, or permanently flooded swamp forest, and the várzea, or periodically inundated forest types, in the Amazon Basin which has been considered the center of distribution of the genus (Record and Hess 1943). One species, *H. torrei*, occurs in evergreen coastal scrub.

Six species comprise Group II, morphologically centering around *Hymenaea courbaril*, which has a distribution similar to that of the genus. Occurring from central Mexico, along the west coast of Central America to Panama, it cuts across to the east coast of Central America to Belize and Honduras, thence throughout the islands of the West Indies, and is found in all of the South American countries except Uruguay, Chile, and Argentina.

TABLE 2. Distribution of *Hymenaea* Species in Various Ecosystem Types

Species of *Hymenaea*	ECOSYSTEM TYPES: Evergreen Forest								
	Terra Firme	*Várzea*	*Igapó*	*Coastal*	*Subdeciduous Forest*	*Deciduous Forest*	*Savanna*	*Thorn Forest*	*Dry Littoral*
GROUP I									
H. oblongifolia	X	X							
H. palustris			X						
H. parvifolia	X								X
H. reticulata	X								
H. intermedia	X								
H. adenotricha	X								
H. davisii	X								
H. multiflora	X								
H. rubriflora				X					
H. torrei									X
GROUP II									
H. courbaril	X			X	X	X	X	X	X
H. altissima				X					
H. stilbocarpa					X		X		
H. martiana							X	X	
H. velutina								X	
H. eriogyne								X	
H. stigonocarpa							X	X	

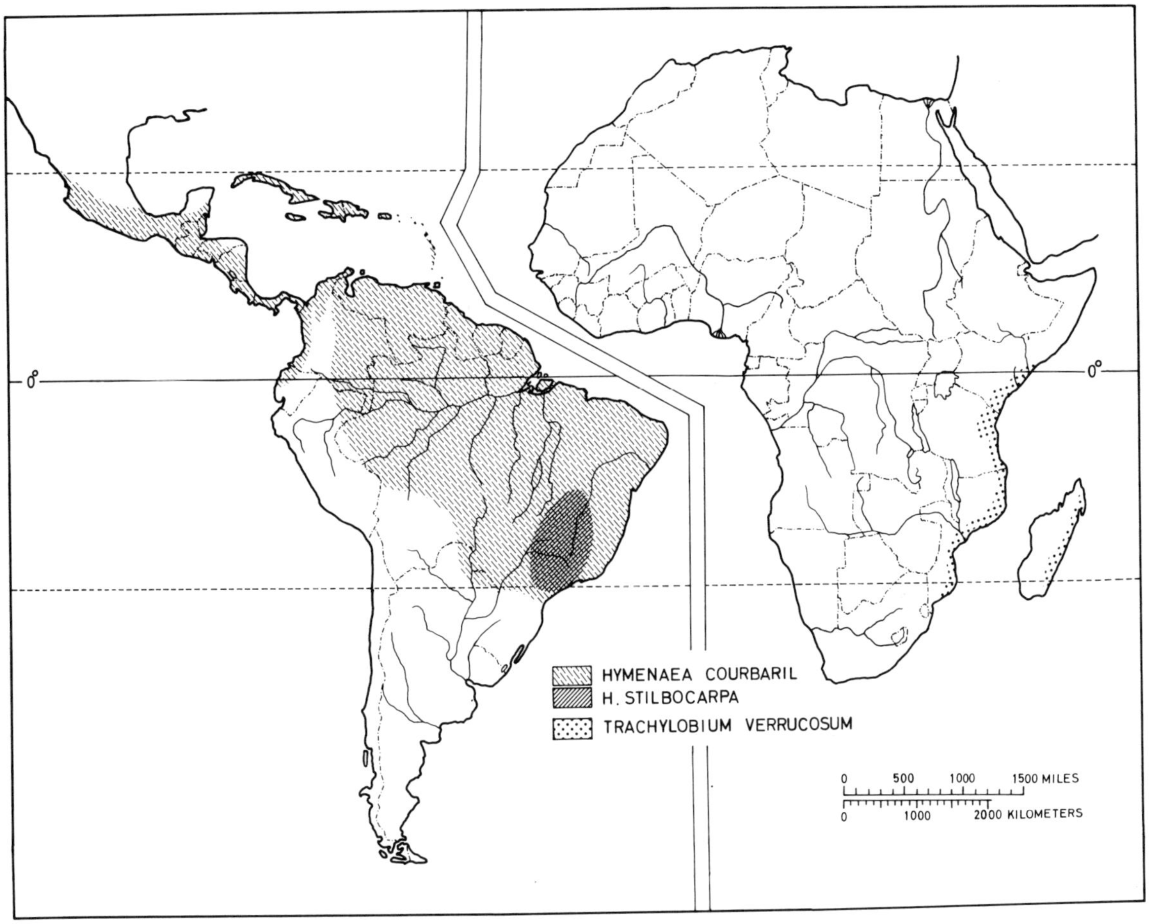

FIGURE 1. Distribution of *Hymenaea courbaris, H. stilbocarpa,* and *Trachylobium verrucosum.*

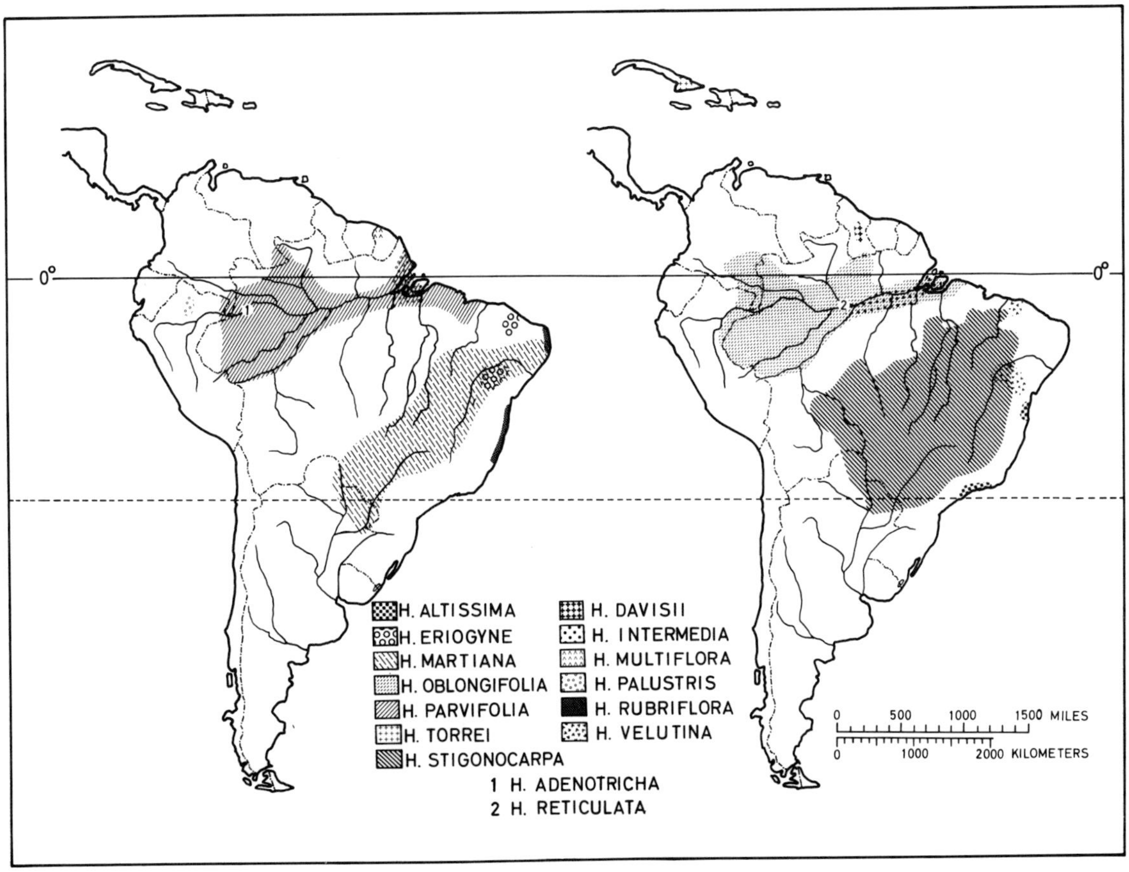

FIGURE 2. Distribution of all *Hymenaea* species, excepting *H. courbaril* and *H. stilbocarpa.*

This single species has radiated into a wide range of climatic and edaphic conditions throughout its extensive distribution, and is suspected of being replete with ecotypes. For example, it is a highly dispersed emergent tree in the terra firme forest of the central Amazon Basin, primarily where there is a 1 to 2 month dry season, but one variety (*subsessilis*) is restricted to the sandy river beaches. The species also occurs in evergreen forests of Central America (3 to 5 months dry season) and is a common component of semideciduous and deciduous forests, growing in almost pure stands in northern Mexico. It is a prominent member of diverse types of savannas, occurring, for example, along streams in the Brazilian cerrado and as scattered trees in the Venezuelan llanos. It grows in the thorn forest in Mexico and has a variety (*obtusifolia*) in the caatinga of Brazil where the dry season extends from 6 to 8 months. Other species belonging to Group II tend to be more restricted in their distribution than *H. courbaril* and occur predominantly in drier habitats such as savannas and thorn forests (Table 2).

Considering the highly disjunct distribution of *Hymenaea* and *Trachylobium,* the degree of similarity is striking and the old question as to whether they may belong to the same genus is still with us. Certainly there is no question of a closely reticulate pattern of similarity varying within different species and dependent upon the particular character. For example, the petals in *Trachylobium* are five as in *Hymenaea,* but three of the upper ones may be clawed and suborbicular with the lower ones minute and scale-like. In some populations, however, they may all be clawed and subequal. Generally in *Hymenaea* the petals are not clawed and are subequal. Nevertheless, in *H. reticulata* the petals are distinctly clawed and in *H. oblongifolia* may be subclawed. The fruit in *Trachylobium* is one or two seeded, coarsely verrucose-rugose as a result of prominently developed resin pockets. Most *Hymenaea* fruits are several seeded and the surface is relatively smooth. On the other hand, several species are one or two seeded (e.g., *H. oblongifolia, H. parvifolia,* and *H. torrei*) and the latter from Cuba has verrucose fruit extremely similar to *Trachylobium.*

Fashbender (1959), in a study of the pollen of the Amherstieae, Cynometreae, and Sclerobieae, has shown pollen from three species of *Hymenaea* (*H. oblongifolia, H. parvifolia,* and *H. stilbocarpa*) to be quite similar to her sample of *Trachylobium verrucosum.* In fact, *H. oblongifolia* and *Trachylobium* are virtually identical.

In preliminary analysis of epidermal patterns of the leaves, a similarly close relationship has been demonstrated between *H. oblongifolia* and *Trachylobium* (J. A. Wolfe, U.S.G.S., pers. comm.).

The secretory structures in *Hymenaea* and *Trachylobium* appear essentially the same, and have much in common with those in *Guibourtia* (Moens 1955). In *Hymenaea courbaril* schizogenous pockets form in cortical and medullary tissue of the epicotyl and stem (but not the root) portion of the hypocotyl soon after germination (Langenheim 1967). These pockets persist until bark formation finally obliterates them. In the older plant, lysigenous cavities appear in the cambial zone of the trunk or root. Extensive cavities are probably formed by coalescence of smaller ones along the cambial zone.

The resins of *Hymenaea* and *Trachylobium* have been more intensively studied than any of the other genera. The initial work on *Trachylobium* was done by organic chemists principally interested in elucidating structures of diterpene acids (Hugel, Oehlschlager, and Ourisson, 1966). Our work on both *Hymenaea* and *Trachylobium,* on the other hand, has been biologically oriented with investigation not only of the diterpene acids from trunk resins but also sesquiterpene hydrocarbons in both the exuded trunk resin and leaf resin pockets (Martin and Langenheim 1970, Martin, Langenheim, and Cunningham 1971). Our attention was first directed toward a comparison of *Hymenaea* and *Trachylobium* because of the virtual identity of infrared spectra of their trunk resins and their known close botanical relationship. As might be suspected from the similarity of the spectra, further chemical analysis reveals the presence in each of several of the same diterpene acids (e.g., copalic and eperuic acids). We have analyzed trunk resin of 10 *Hymenaea* species, finding significant differences in the proportions of these compounds despite qualitative similarity. We have noted similar quantitative differences in two Kenyan populations of *Trachylobium verrucosum* and find these also quantitatively different from the data published for *T. verrucosum* growing as an ornamental in Borneo (Hugel, Oehlschlager, and Ourisson 1966).

Sesquiterpene hydrocarbons in both trunk resins

and leaf resins of 11 *Hymenaea* species and 2 Kenyan populations of *Trachylobium* show further similarities, with many compounds common to all, but with interesting quantitative variations. In every case, whatever the class of resin compound being examined, we find no greater differences between *Trachylobium* and any of several *Hymenaea* species than between the latter and other members of the genus *Hymenaea*. This chemical evidence further supports the morphological data showing a very close relationship between these two genera.

The most striking distinction thus far discovered between the two genera is the chromosome number. The 11 species of *Hymenaea* examined have a diploid number of 24 (Lee 1972), which is the number reported for the majority of the Cynometreae. *Trachylobium,* on the other hand, has a diploid number of 16 (Atchison 1951). This smaller chromosome number might point to *Trachylobium* as being more primitive and thus ancestral to the New World *Hymenaea*. Unfortunately the paleobotanical record does not provide further information about the early relationship between *Hymenaea* and *Trachylobium*.

The first fossil evidence reported for either genus is leaf impressions of four species considered to be *Hymenaea* from Upper Cretaceous (Cenomanian to Turonian) beds in North America (Knowlton 1919). Recent analysis, however, shows that these superficially leguminous-appearing leaves assigned to *Hymenaea* belong to another source (J. A. Wolfe, pers. comm.). All leaves sufficiently well preserved to be identified are simple rather than compound and there also is no evidence of cross-hatching on the petiolule, a characteristic of all compound-leaved Caesalpinioideae. In addition, no caesalpiniaceous pollen has been discovered from over 100 samples of pollen floras, ranging in age from Cenomanian to basal Paleocene, from the same geographic and stratigraphic succession in which these supposed leaf types were found.

Thus the first well-authenticated record of *Hymenaea* appears to occur as amber and its enclosed leaf, sepal, and pollen remains from the Oligo-Miocene Simojovel formation of Chiapas, Mexico (Langenheim 1966). The leaf and sepals in the amber were identified as probably being *H. courbaril,* although some characters were close to *H. intermedia* (Faustino Miranda of Universidad Nacional Autónoma de México, pers. comm.). The identity of the amber was determined by close similarity of the infrared spectrum with that of resin from *H. courbaril*. Amber with similar spectra also came from the early Miocene Pirabas formation from Pará, Brazil, and from two Colombian formations near Girón and Medellín. The age of occurrence of the Colombian amber is unknown but thought to be younger than Miocene. A leguminous leaflet similar to *Hymenaea* (*Leguminosites hymenacoides*) was reported from Pliocene beds in Bahia, Brazil (Hollick and Berry 1924).

Fossil resin from Pleistocene beds on Zanzibar is the only known paleobotanical record of *Trachylobium*. Zanzibar was isolated from the African Coast during the Quaternary (Furon 1963) and probably the *Trachylobium* populations were continuous before that time.

COPAIFERA–GUIBOURTIA

The confusion in systematic interpretation of *Hymenaea* and *Trachylobium* is considerably less than that concerned with *Copaifera, Guibourtia,* and allied genera. A history of some of the nomenclatural changes will be detailed to exemplify the problems, particularly as they are relevant to explaining American-African distribution patterns.

The genus *Copaifera* was described by Linnaeus in 1762, the American *C. officinale* being the type. In 1857 Bennett established the genus *Guibourtia* based on the African *G. copallifera*; in 1862 Bolle described the African genus *Gorskia*. Bentham and Hooker (1865), finding the distinctions unclear, then enlarged the concept of *Copaifera* to include both *Guibourtia* and *Gorskia*. In subsequent studies, however, Léonard (1950a, 1950b, 1957) was able to relatively clearly separate *Copaifera* and *Guibourtia* on the basis of a variety of characters including those of the inflorescence, bud, leaflets, seedling, and anatomy of the secretory system.

Approximately 30 species of *Copaifera* are presently recognized to occur in the New World (Dwyer 1951) and 4 in Africa (Léonard 1957). In the neotropics most species are distributed in the eastern part with the majority occurring in Brazil. Since many species are known from a single locality with poor habitat data recorded, it is difficult to make generalizations regarding distribution among ecosystem types. In Africa three species occur in the western African rain forests but one (*C. baumiana*) is present in the Zambezi savannas of Angola and Northern Rhodesia.

In *Copaifera* resin is secreted in pockets which occur in the pith of young stems, in the parenchyma of the leaf and flower bud, and in bark. Concentric circles of canals, which anastomose tangentially, occur in the secondary wood. Resin pockets also are present in the fruit. The resin system in *Copaifera* appears to be quite similar to that of *Daniellia* (Guignard 1892).

Considerable difference in resin composition occurs between the African and South American species of *Copaifera.* The African species produce a viscous complex of terpenoids that hardens into lumps of copal whereas the New World species produce a free-flowing, liquid oleoresin. Despite commercial importance of both the oleoresin and copal, little is known about the chemical structure of the constituents (Cocker, Moore, and Pratt 1965).

When Léonard (1950a, 1950b) made clear the distinctions between *Guibourtia* and *Copaifera,* he recognized 14 species of *Guibourtia* as occurring in Africa and 4 in the New World. In subsequent work African species have been little changed, only one species having been dropped from the genus. In contrast, the nomenclatural problems which reflect the difficulties in assessing relationships in the New World species have continued.

The four New World species which Léonard (1950a, 1950b) had transferred from *Copaifera* to *Guibourtia* were: *G. hymenaeifolia, G. chodatiana, G. confertifolia,* and *G. fissicuspis.* Dwyer (1954b) agreed with Léonard that the first two species should be considered *Guibourtia.* Léonard in 1957, however, decided that these two species should be referred back to the genus *Psuedocopaiva* based upon analysis of additional characters. He also indicated that *G. confertifolia* probably was more closely related to "*Peltogyne* or some neighboring genus" than *Guibourtia.* The status of *G. fissicuspis* is now equivocal but it is most frequently considered *Cynometra fissicuspis* (Dwyer 1958). The distributions of these taxa according to the systematic views just expressed are presented in Figure 3.

This survey of the systematic confusion surrounding *Guibourtia* emphasizes the occurrence of a complex of closely related but variable populations on both sides of the Atlantic Ocean. An immediate clarification of the interrelationships is complicated by relatively few available collections of complete material. Although question now exists as to the occurrence of *Guibourtia* in the New World, perhaps the most significant problem posed is to explain the degree of similarity within such genera as *Guibourtia, Psuedocopaiva, Peltogyne,* and *Cynometra.*

Despite the considerable commercial value of copal from species such as *G. demeusei* and *G. copallifera,* little useful chemical information is available on resins of the genus. Infrared spectral analysis has indicated compositional differences between species of *Guibourtia* and *Copaifera,* as well as complexes of species within *Guibourtia* (Hellinckx 1955, Langenheim 1968, 1970).

Chromosome numbers are interesting in these genera. The three species of *Copaifera* examined (two New World and one African) show a diploid number of 24, the most frequent number for the tribe (Atchison 1951, Manginot and Manginot 1957, Turner and Irwin 1961). On the other hand, polyploidy, rarely recorded among these leguminous genera, occurs in *Guibourtia. Guibourtia ehie*

TABLE 3. Distribution of African *Guibourtia* Species in Various Ecosystem Types

African Species of *Guibourtia*	ECOSYSTEM TYPE				
	Evergreen Forest	*Subdeciduous*	*Woodland*	*Savanna*	*Dry Littoral*
SUBGROUP GUIBOURTIA					
G. copallifera	X				
G. demeusei	X				
G. carrissoana					X
G. sousae					?
SUBGROUP GORSKIA					
G. arnoldiana	X				
G. ehie	X				
G. dinklagei	X				
G. conjugata			X	X	
G. schliebenii		X	X		
SUBGROUP PSEUDOCOPAIVA					
G. coleosperma			X	X	
G. tessmannii		X			
G. leonensis	X				
G. pellegriniana		?			

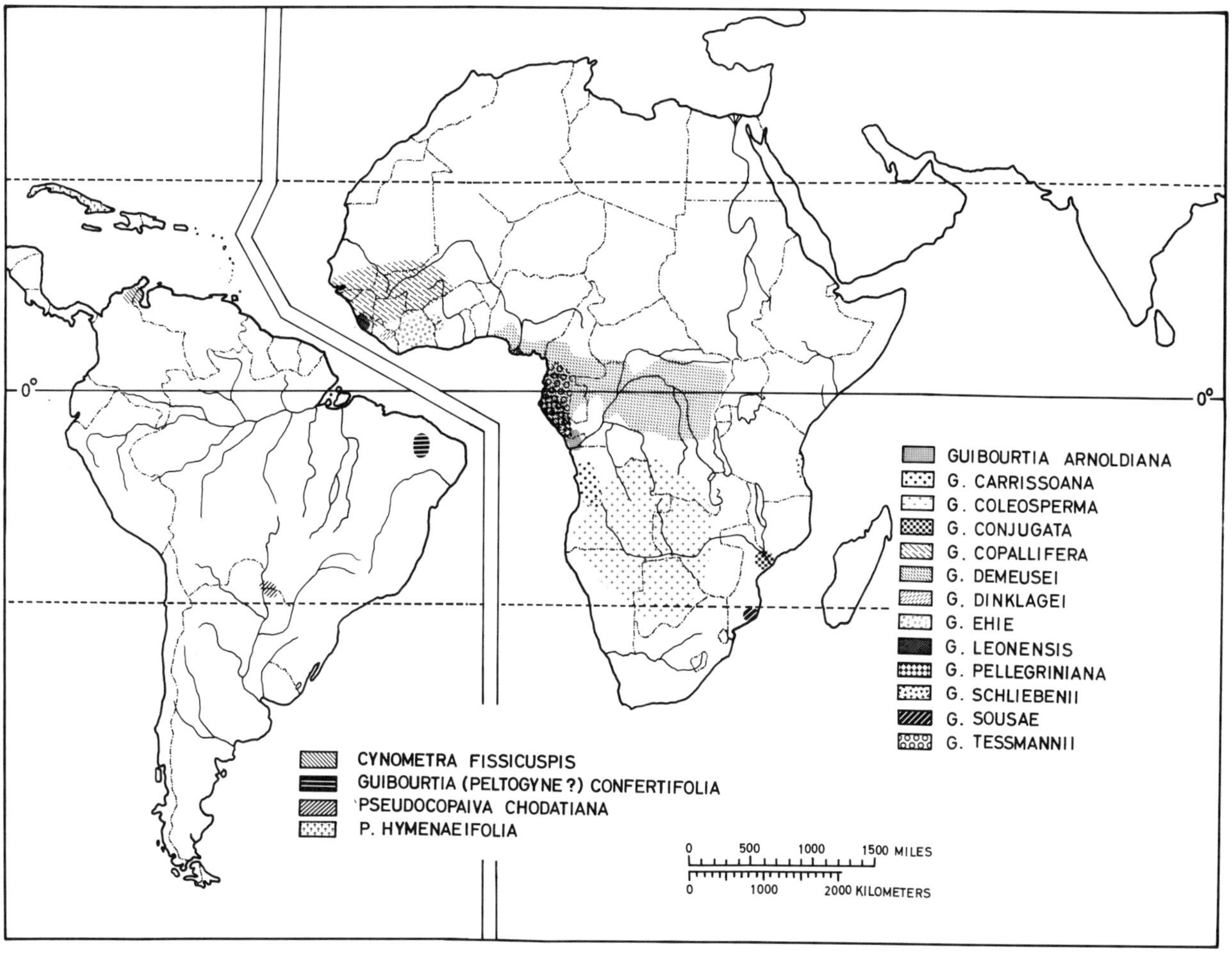

FIGURE 3. Distribution of *Guibourtia* species, including *G.* (*Peltogyne?*) *confertifolia, Cynometra fissicuspis, Pseudocopaiva chodatiana,* and *P. hymenaeifolia,* which were recently included in *Guibourtia.* Ranges of species according to Léonard (1950a).

has the usual diploid number of 24 (Manginot and Manginot 1957) whereas *G. coleosperma* has 48 (Turner and Fearing 1959).

Few habitat data have been recorded for species of the New World *Copaifera.* Some information is available for the better studied *Guibourtia* partially as a result of its capacity to produce copal. Although species of *Guibourtia* occur primarily in the evergreen forest, some have radiated into the drier forests and savannas (Table 3). This dispersion, however, is not as great as that occurring in the coastal, drier evergreen forests of eastern Africa (Figure 3). Some of these habitats appear to overlap those of *Trachylobium verrucosum.*

Few reliable fossil records are reported for either *Copaifera* or *Guibourtia.* Doubtful leaf impressions of *Copaiva* (*Copaifera*) occur in middle Oligocene beds from Puerto Rico (Hollick 1928). Relatively recent amber (2830–1650 B.P.) from northeastern Angola has been attributed to *Copaifera mildbraedii* (Langenheim 1968, 1970).

CYNOMETRA

Cynometra is a pantropical genus which was described by Linnaeus in 1753 based on *C. cauliflora* and *C. ramiflora* from India. Today, 70 species are generally recognized, with about 25 each from Africa (Léonard 1951) and the New World (Dwyer 1958), and 20 from Asia (Knaap-Van Meeuwen 1970).

Both the Asiatic and New World cynometras represented homogeneous assemblages. All of the

Indo-Pacific species occur in evergreen forests predominantly along streams or in swamps such as mangroves (Knaap-Van Meeuwen 1970). *Cynometra* in Indo-Malesia is closely related to *Maniltoa.*

In the New World *Cynometra* has classically been considered allied to *Copaifera* and the pantropical genus *Crudia,* but Dwyer suggests that it is more closely related to *Peltogyne.* The ranges of *Cynometra* and *Peltogyne* also are similar in extending from Mexico to southeastern Brazil; however, *Peltogyne* is absent from the West Indies. Despite the relatively wide distribution of *Cynometra* in the New World, half of the species occurs in the Amazon Basin, which may be the center of their distribution (Dwyer 1958). The remarkable floral stability of New World cynometras suggests that they diverged from the main line evolving in Africa and were able to maintain themselves in a stable fashion in the New World (Dwyer 1958).

The greatest diversification within the genus apparently has occurred in Africa. Taxonomic confusion resulting from interpretation of this diversification has additionally been increased by *Cynometra* having been made a depository for species difficult to classify within the entire Cynometreae (Léonard 1951). Léonard in 1951 distributed 60 species previously assigned to *Cynometra* among 11 genera. Even with this clarifying revision, *Cynometra* is difficult to distinguish from other genera such as *Scorodophloeus, Plagiosiphon,* and *Hymenostigia.*

In Africa *Cynometra* has a distribution similar to *Guibourtia,* occurring primarily in the western evergreen forests (particularly in the Congo and Gabon), but also having several species in coastal forests of Tanzania and Kenya and a doubtful one in Mozambique.

Previously it has been thought that *Cynometra sessiliflora* produced copal, in fact, that it rivaled *Guibourtia demeusei* in quantity produced (Léonard 1950a). However, Moens (1955) was unable to find secretory tissue in samples of *C. sessiliflora* collected by Léonard. Hellinckx (1955) also found a sample of resin thought to be collected from *C. sessiliflora* to give an infrared spectrum identical with that of *G. demeusei.* A solution to this dilemma has recently been suggested by Léonard (pers. comm.). *C. sessiliflora* grows under *G. demeusei* in the forest, and resin from *Guibourtia* could thus have fallen onto *Cynometra* where it was collected. Thus, the listing of *Cynometra* as a resin-producing genus has probably been a mistake.

Despite *Cynometra* now not being thought to produce resin, this pantropical genus still appears on morphological bases and geographical distribution to be the most likely parental stock for such genera as *Trachylobium, Hymenaea, Guibourtia,* and *Copaifera,* as well as other African and Asiatic resin producers not discussed here.

The fossil record of *Cynometra* unfortunately is scanty. A leaf record, along with *Copaiva,* is reported from middle Oligocene beds from Puerto Rico (Hollick 1928). An early fossil record of the pantropical *Cynometra* might be extremely useful in interpreting the evolution of these resin-producing genera.

Interpretation of Distribution Patterns

In the previous paragraphs I have tried to trace some of the systematic history of these genera to illustrate the confusion created by names and to display important similarities and differences useful in interpreting the present distribution patterns and dispersal routes.

In any interpretation of past migrational routes paleobotanical evidence is desirable. Few authenticated paleobotanical records are available either for these genera or for the entire subfamily Caesalpinioideae (Archangelsky 1968, Muller 1970). There appears to be no authenticated evidence for the Cynometreae in the Cretaceous when the group might be expected to have evolved. The Paleocene pollen occurrence of *Crudia* is the earliest accredited record of the tribe (Muller 1970). The only other well-authenticated fossil record for the Cynometreae is that of *Hymenaea* in the Oligo-Miocene and Miocene.

This paucity of paleobotanical information for such a large and important tropical group is thought-provoking. Partial explanation may be provided by the very fact of the tropical distribution. Good tropical megafloras are few because of extreme weathering and inaccessibility of outcrops. Therefore, the most extensive knowledge of tropical paleofloras has been obtained from pollen (Germeraad, Hopping, and Muller 1968, Muller 1970). Many tropical trees are either self- or animal-pollinated. In *Hymenaea,* for example, the role of bats and bees as pollinators (Vogel 1968, H. G. Baker and G. Frankie, both of the University of California, Berkeley, pers. comm.) may preclude the shedding of any significant amounts of pollen that would become incorporated into sediments. Perhaps similar pollinating mechanisms exist in other genera in the Cynometreae, with the result that the rare pollen record might be attributable to methods of pollen dispersal.

Another difficulty in obtaining paleobotanical data for members of the Cynometreae is that 67 percent of the Caesalpinioideae are endemic to Africa (Brenan 1961). Relatively few Cretaceous and Tertiary beds occur in Africa, with at least one half of the surface being covered with pre-Cambrian outcrops (Furon 1963). Thus, the lack of availability of sediments which would contain paleobotanical material is added to the problem presented by the tropical environment.

The lack of fossil record for Caesalpinioideae in Cretaceous formations naturally poses the question as to whether this subfamily had evolved by this time. Very little has been written about the phylogeny of the Leguminosae and their expected time of appearance in the geologic record. Generally considered primitive floral characteristics have led some to speculate that this subfamily probably evolved relatively early (Brenan of the Royal Botanic Gardens, Kew, pers. comm.). Others, such as Andrews (1915:358), have believed the Leguminosae to be composed of highly developed types, suggesting that present members "had not been outlined until the earlier forms of the dicotyledons had been well established and differentiated." This latter view accords better with the present geological record than the former. Absence of fossil evidence in a situation such as this, however, simply poses further questions as to why it may be lacking, as one can never be sure when the hitherto undiscovered evidence will be found.

The geologic record, thus, is of relatively little direct help in understanding the present distribution of the genera we are discussing, although it produces sobering thoughts. My present hypothesis is that these genera may have been derived from a pantropical stock which probably originated in the Indo-Pacific region. The pantropical *Cynometra,* because of its extraordinary morphological diversity, could be envisioned as the ancestral group. Certainly there is evidence for "close reticulate affinities" that Knaap-Van Meeuwen (1970) mentions for many genera in the Cynometreae. *Cynometra* could well have migrated westward from Malesia along the Indian Ocean to Africa where the greatest diversification seems to have occurred.

The knottiest problem, and the one central to this symposium, is the explanation for the amphi-Atlantic distribution of the genera. Three ideas which must be considered are (1) continental drift, (2) southern island hopping and (3) oceanic dispersal of disseminules.

Generally speaking, the idea of the unity of the African and South American continents sometime in the past now is a widely accepted view. The question centers about the timing of the separation. As Martin (1969:48) states in a recent summary, "none of the lines of evidence, which have been adduced to prove the hypothesis of continental drift (or global expansion or ocean spreading), is able to establish the former existence of a Gondwana continent comprising South America and Africa convincingly, but the whole array of facts still gives a considerable degree of probability to this theory". Although Cretaceous faunal and facies relationships between the northeastern part of Brazil and the African coast bordering the Gulf of Guinea strongly suggest a connection in this area lasting well into the Cretaceous, paleomagnetic evidence contradicts this idea, suggesting that the South Atlantic came into being during Jurassic or early Cretaceous, or even during the Paleozoic. Therefore, most currently acceptable dates for possible union of the two continents appear too early to explain the distribution of our leguminous genera, even if they evolved previous to their presently known appearance in the geologic record.

Smith (1967) has suggested the possibility of migration between the two continents via land connections, perhaps insular, which existed between Antarctica, South Africa, or Madagascar. Margolis and Kennett (1971) have just presented evidence for glaciation in Antarctica during the Eocene. Thus this migrational route would be precluded for the Cynometreae unless ancestral types existed at an earlier time.

We are then left with the third migrational possibility, that of oceanic dispersal. Despite *Hymenaea* and *Trachylobium* being recognized as two genera, the morphological and chemical similarities are greater between them than between the African and South American species within the genera of *Copaifera* and *Cynometra*. The tantalizing problem posed by the *Hymenaea-Trachylobium* distribution is that *Trachylobium* occurs only along the coast of eastern Africa. It seems quite possible that fruits of *Trachylobium* and *Hymenaea* have been successfully dispersed across oceanic barriers. The indehiscent pods float easily in water; in fact, *Hymenaea* is commonly observed to be dispersed by streams. The pods remain intact after wetting for long periods of time and the seeds remain viable for many years. *Hymenaea* fruits also have been picked up along the southeastern Florida coast where presumably they have been carried in the Gulf Stream from the West Indies (Gunn 1968). However, the oceanic transport hypothesis is complicated by the direction of the currents as they are known today. With the distribution of *Trachylobium* presently restricted to eastern Africa, fruits would tend to enter either the Aguilhas or the South Equatorial currents which get deflected eastward. It would thus be a difficult feat for pods to reach the eastern coast of Brazil. It also would be difficult for *Hymenaea* fruits to reach eastern Africa via the Brazil Current unless they took an extremely long trip into the West Wind Drift and finally entered the South Equatorial Current. Looking at the present distribution patterns the only explanation would appear to be Simpson's (1952) probability theory that even the most improbable, isolated event can occur if sufficient time elapses!

However, another possible means of allowing oceanic dispersal of *Trachylobium* to South America does exist. Both *Guibourtia* and *Cynometra* in Africa are distributed primarily in western evergreen forests but have outlying populations in the east African coastal forests. A more continuous distribution of these genera probably occurred across the African continent. Genera from other families, such as *Mansonia* and *Cola* in the Sterculiaceae, are generally evergreen forest components that are scattered from east to west across central Africa in a pattern that suggests a former more continuous distribution of this forest type (Chatterjee and Brenan 1950). The evergreen forests of West Africa apparently were reduced and fragmented about 22,000 years ago, and, in fact, there is increasing evidence for the rapid succession and recent date (primarily Pleistocene) of the ecological changes that determine the present vegetational patterns throughout Africa (Moreau 1966). During the Eocene Period tropical rain forest vegetation reached its maximum world-wide development, extending to at least 50°N and probably attaining almost double its present distribution (J. A. Wolfe 1971). At such a time when rain forest vegetation had its widest geographic distribution because of maximum extent of favorable climate, we can infer that dispersal probabilities were highest. Individuals of tropical species would generally be more numerous, and with increased population size, a greater number of seeds would be supplied to the environment for dispersal. In addition, environments suitable for colonization were much larger in area and long-distance migration would be favored. It may well be that *Trachylobium* is a relict of a taxon that was more widely distributed in the past. If it once had also been distributed in western Africa (or had relatives that were), as is the case today with *Cynometra* and *Guibourtia,* current transport of fruit to the Amazon region via the Equatorial Current or to more southern coastal areas via the Brazil Current would appear to represent a fairly simple solution to the distributional problem. Moreover, species of *Hymenaea* apparently most closely related to *Trachylobium* occur either in the Amazonian or coastal evergreen forests. In fact, the center of distribution of *Hymenaea* is considered to be the Amazon Basin, as also is true of *Cynometra* and possibly *Peltogyne.* Additionally, these taxa could have been widely distributed in Central and South America during the Eocene, when tropical vegetation and associated faunas were common as far south as central Patagonia (Menéndez 1969, Simpson 1969). Of considerable interest in this respect is the presence in the relictual rain forest area in central coastal Bahia of undescribed *Hymenaea* species with affinities to those occurring today in the Amazon Basin (Langenheim, Lee, and Martin 1971). These species possibly are remnants of the time when an Amazonian-type evergreen forest was more extensive than today. It is also probable that during this time *Hymenaea,* as well as *Cynometra* and *Peltogyne,* could have migrated northward into Central America. A continuous water barrier existed between Central and South America from Cretaceous to Pliocene (Weeks 1947, Harrington 1962) and faunal exchange was disrupted between the two areas. For plants with fruits capable of oceanic dispersal, this relatively narrow band of ocean would not be insurmountable. Geological evidence from the Antilles also reveals the presence of larger and more continuous landmasses during the early Tertiary when climatic conditions were extremely favorable to long range dispersal of rain forest species (Woodring 1954, Graham and Jarzen 1969). Graham and Jarzen have developed a model of two possible Tertiary migration routes for extension of tropical species into northern Central America—the Isthmian-coastal Mexico route and the Antillean arc. The present distribution of *Hymenaea* (Figures 1, 2) suggests both routes may have been utilized, depending in part upon the availability of populations producing fruits along the coasts of Colombia and Venezuela. *Hymenaea* was known to occur during later Tertiary time from the Girón and Medellín areas of central Colombia and had been present in southern Mexico earlier in the Oligo-Miocene.

The distribution of *Copaifera* on both continents appears more simple to explain than that of *Hymenaea* and *Trachylobium* because the African species are concentrated in the western tropical zone. With the presently available systematic data, it seems improbable that *Guibourtia* has representatives in the New World, although the relationship to *Pseudocopaiva* appears relatively close.

Convergent evolution provides an alternative explanation to that of oceanic dispersal in explaining the similarity displayed between *Hymenaea* and *Trachylobium*. Groups with similar variation potential may tend to produce parallel syndromes of structures and differentiation under similar environmental conditions (Ehrendorfer 1970). If no direct genetic interchanges were possible between *Hymenaea* and *Trachylobium*, the parallel differentiation has been striking. Even more startling than the morphological similarities are those of the chemistry of the resins. It would seem incredible that isolated populations such as these could develop at the same time similarities in floral and fruit characters, anatomical mechanisms for the secretion of the resin, and the same pathways in production of such complicated secondary products as the complex of terpenoids comprising resins.

Even though convergent evolution does not seem to answer the distributional questions posed by these leguminous genera, it appears that a somewhat parallel radiation of species from Eocene rain forest to drier forest ecosystems probably occurred on both continents with what is thought to be world-wide climatic deterioration in the Oligocene (Wolfe 1971). Brenan (1965) has pointed out that 76 percent of the genera of the Caesalpinioideae in Africa occur exclusively in evergreen forests and that only 19 percent (10 genera) also occur in savanna ecosystems. At least six of the resin-producing genera in the Cynometreae occur in a diversity of ecosystem types. This radiation is tabulated for *Hymenaea* (Table 2) and *Guibourtia* (Table 3) as these genera have the most complete habitat data available. The same general pattern, however, is also indicated for *Cynometra, Copaifera, Daniellia,* and *Tessmannia* (Léonard 1950a, 1957). In *Hymenaea* it appears that species within Group II may well have evolved with progressive drying trends probably initiated during the Oligocene. This morphologic-ecologic division within the genus does not seem as clear-cut for *Guibourtia,* but a similar analysis of all of these genera would be interesting, if data were available. Despite the apparent parallel radiation into ecosystem types, it should be remembered that the same ecosystem types display not only floristic but even physiognomic differences on the two continents (Aubreville 1961).

Conclusions

Analysis of the presently accepted systematic relationships, distribution patterns, resin chemistry, and secretory mechanisms of leguminous resin-producing genera indicates close affinities between those occurring both in South America and Africa. Genetic interchange appears likely, but probably occurred after the two continents had separated. In fact, paleobotanical evidence suggests evolution within the Cynometreae did not gain momentum at least until the Eocene epoch although the group may have evolved at an earlier time. It appears that *Hymenaea, Peltogyne, Guibourtia, Trachylobium* and *Copaifera* (as well as other resin-producing genera) might have evolved from a pantropical stock such as *Cynometra.* The genera, however, have been too little studied as yet from an evolutionary viewpoint to establish phlogenetic lines among them.

Long distance oceanic dispersal appears to be the most probable explanation for the present distribution patterns and, therefore, tends to emphasize their chance occurrence. A major migration may well have occurred during the Eocene when tropical moist climates and, correspondingly, rain forest ecosystems had their maximum development, although it is never necessarily to be assumed that all genera had to migrate simultaneously. From basic rain forest stock, species in these genera may have radiated into drier habitats in a relatively parallel manner on both continents. I do not interpret these taxa, particularly *Hymenaea* and *Trachylobium*, to have resulted from convergent evolution under similar environmental conditions on the two continents. Despite highly disjunct distribution patterns, the similarities in both morphological and chemical characters seem too great to accept this conclusion. Even such genera as *Cynometra* and *Copaifera*, which display more diversity than *Hymenaea* and *Trachylobium* between the two continents, seem most reasonably to have had genetic interchange. The distributional patterns of *Copaifera* and *Cynometra* seem relatively simply solved by oceanic dispersal. In considering the *Trachylobium-Hymenaea* relationship, the existence of an extinct West African relative of *Trachylobium verrucosum* needs to be assumed to make oceanic dispersal plausible. Distributional evidence from *Guibourtia* and *Cynometra*, as well as from genera in the Sterculiaceae, Tiliaceae, etc., indicate the possibility of *Trachylobium* once having a somewhat continuous distribution across Central Africa. The presently highly disrupted patterns of distribution probably have resulted from relatively recent Pleistocene ecological changes.

More detailed studies are obviously needed to understand evolutionary relationships among these related South American and African taxa. They represent a fascinating example of a classic problem; interest in these particular genera is enhanced by the additional potentiality of unraveling the physiological and chemical story associated with the capacity to produce resin.

References

Andrews, E. C. 1915. The development and distribution of the natural order Leguminosae. Jour. and Proc. Roy. Soc., New South Wales (for 1914):333-407.

Archangelsky, S. 1968. Paleobotany and palynology in South America: A historical review. Rev. Palaeobot. Palynol. 7:249-266.

Atchison, E. 1951. Studies in Leguminosae, VI: Chromosome numbers among tropical woody species. Amer. J. Bot. 38:538-546.

Aubreville, A. 1961. Étude écologique des principales formations végétales du Brésil et contribution à la connaissance des forêts de l'Amazonian brésilienne. Centre Technique Forestier Tropical Nogent-sur-Marne (Seine). France.

Baker, E. G. 1926. The Leguminosae of tropical Africa. Erasmus Press. Ghent.

Bennett, J. J. 1857. Jour. Linn. Soc. Bot. 1:149.

Bentham, G. 1840. Contributions toward a flora of South America. Hooker's J. of Bot. 2:72-103.

————. 1870. *In* Martius (Editor). Flora Bras. 15:247.

Bentham, G., and J. D. Hooker. 1865. Gen. Pl. 1(2):562-600.

Bolle, C. 1862. In Peters, Reise Mossamb., Bot. 1:15.

Brenan, J. P. M. 1965. The geographical relationships of the genera of the Leguminosae in tropical Africa. Webbia 19 (2):545.

————. 1967. Leguminosae—subfamily Caesalpinioideae. *In* E. Milne Redheed and R. M. Polhill (Editors). Flora of tropical East Africa. Crown Agents for Oversea Govts. and Administration.

Chatterjee, D., and J. P. M. Brenan. 1950. *Mansonia dipikae* Parkayastha. Hooker's Icones Pl. 35. pl. 3484.

Cocker, W., A. L. Moore, and A. C. Pratt. 1965. Dextrorotary hardwickiic acid, an extractive of *Copaifera officinalis*. Tetra. Lett. 24:1983-1985.

Ducke, A. 1935. As especies brasileiras de jatahy, jutahy ou jatobá (genero *Hymenaea* L., Leguminosas Caesalpiniaceas). Ann. Acad. Bras. Sc. 7:203-211.

————. 1938. Notes on the purpleheart woods of Brazilian Amazonia. Tropical Woods 54:1.

Dwyer, D. 1944. The genus *Pseudocopaiva* Britton & Wilson. Trop. Woods 80:7-10.

————. 1951. The Central American, West Indian, and South American species of *Copaifera* (Caesalpiniaceae). Brittonia 7 (3):143-172.

————. 1954a. Rapports entre stipe et coupe réceptaculaire dans la classification des Amherstieae. Proc. VIII Intern. Bot. Congress, Paris.

————. 1954b. Further studies on New World species of *Copaifera*. Bull. Torr. Bot. Club 81 (3):179-187.

————. 1958. The New World species of *Cynometra*. Annals Missouri Bot. Gar. 45:303-345.

Ehrendorfer, F. 1970. Evolutionary patterns and strategies in seed plants. Taxon 19 (2):185-195.

Fashbender, M. V. 1959. Pollen grain morphology and its taxonomic significance in the Amherstieae, Cynometreae, and Sclerobieae (Caesalpiniaceae) with special reference to the American genera. Lloydia 22 (2):107-162.

Furon, R. 1963. Geology of Africa. Hafner. New York.

Gaertner, J. 1791. De fructibus et seminibus plantarum. 2:306.

Germeraad, J. H., C. A. Hopping, and J. Muller. 1968. Palynology of Tertiary sediments from tropical areas. Rev. Paleobot. and Palynology 6:189-348.

Graham, A., and D. M. Jarzen. 1969. Studies in Neotropical paleobotany, 1: The Oligocene communities of Puerto Rico. Ann. Missouri Bot. Gard. 56:308-357.

Gunn, R. 1968. Stranded seeds and fruits from southeastern shore of Florida. Garden Journal (March/April) 43-54.

Guignard, L. 1892. L'appareil sécréteur de *Copaifera*. Bull. Soc. Bot. France 39:253.

Harborne, J. B. 1971. Terpenoid and other low molecular

weight substances of systematic interest in the Leguminosae. Pp. 257-283, *in* Harborne, J. B., D. Boutler, and B. L. Turner (Editors). Chemotaxonomy of the Leguminosae. Academic Press. New York.

Harms, H. 1915. Caesalpinioideae. Der Pflanzenwelt Afrikas. *In* Engler and Drude (Editors). Vegetation der Erde 93 (1):423-520.

Harrington, H. J. 1962. Paleogeographic development of South America. Amer. Assoc. Pet. Geol. Bull. 46:1773-1814.

Hayne, F. G. 1830. Arzneikunde gebrauchlichen Gewachse XI: 6-16. Berlin.

Hellinckx, L. 1955. Les propriétés des copals du Congo Belge en relation avec leur origine botanique. Institut National pour l'Étude Agronomique de Congo Belge. Publications Series Technique no. 44, Bruxelles: 8-25.

Hollick, A. 1928. Paleobotany of Porto Rico. Sci. Surv. Porto Rico and Virgin Islands 7:177-393.

Hollick, A., and E. W. Berry. 1924. A later Tertiary flora from Bahia, Brazil. Johns Hopkins Univ. Studies Geol. 5.

Howes, F. N. 1949. Vegetable gums and resins. Chronica Botanica. Waltham, Massachusetts.

Hugel, G., L. Lods, F. M. Mellor, D. W. Theobald, and G. Ourisson. 1965a. Diterpènes de *Trachylobium,* I: Introduction générale, Isolement du kauranol et de huit diterpènes nouveaux. Bull. Soc. Chim. France 1965:2882-2887.

————. 1965b. Diterpènes de *Trachylobium,* II: Structure des diterpènes tétra- et pentacycliques de *Trachylobium.* Bull. Soc. Chim. France 1965: 2888-2894.

————. 1965c. Diterpènes de *Trachylobium,* III: Réactions des dérivés trachylobaniques. Bull. Soc. Chim. France 1965:2894-2902.

Hugel, G., A. C. Oehlschlager, and G. Ourisson. 1966. The structure and stereochemistry of diterpenes from *Trachylobium verrucosum* Oliv. Tetrahedron, Suppl. 8 (Pt. I) : 203-216.

Hugel, G., and G. Ourisson. 1965. Diterpènes de *Trachylobium,* IV: Structure et stéréochemie de l'acide zanzibarique. Bull. Soc. Chim. France 1965:2903-2908.

Knaap-Van Meeuwen, M.S. 1970. A revision of four genera of the tribe Leguminosae-Caesalpiniodeae-Cynometreae in Indomalesia and the Pacific. Blumea 18 (1):1-52.

Knowlton, F. H. 1919. A catalogue of the Mesozoic and Cenozoic plants of North America. U. S. Geol. Surv. Bull. 696.

Langenheim, J. H. 1964. Present status of botanical studies of ambers. Harvard Bot. Mus. Leaflets 20 (8):225-287.

————. 1966. Botanical source of amber from Chiapas, Mexico. Ciencia 24:201-211.

————. 1967. Preliminary investigations of *Hymenaea courbaril* as a resin producer. J. Arn. Arb. 48:203-230.

————. 1968. Infrared spectrophotometric study of resins from northeastern Angola: Appendix, Observations on forest destruction in the Congo Basin in prehistoric times with special reference to northeast Angola. Publicações Culturais, Companhia de Diamantes de Angola 78:151-154.

————. 1969. Amber: A botanical inquiry. Science 1633: 1157-1169.

————. 1970. Botanical origin of amber and its relation to forest history in northeastern Angola. Amer. J. Bot. 57:759.

Langenheim, J. H., and C. W. Beck. 1965. Infrared spectra as a means of determining botanical source of amber. Science 149:52-55.

————. 1968. Catalogue of infrared spectra of fossil resins (ambers), I: North and South America. Harvard Bot. Mus. Leaflets 22 (3): 65-120.

Langenheim, J. H., Y. T. Lee, and S. Martin. 1970. Amazonian species of the resin-producing genus *Hymenaea.* Amer. J. Bot. 57:754.

————. 1971. Brazilian species of the leguminous resin-producing genus *Hymenaea.* Amer. J. Bot. 58:466.

Lee, Y. T. 1972. Chromosome number in the genus *Hymenaea* (Leguminosae, Caesalpiniodeae, Cynometreae) [Unpublished ms.]

Lee, Y. T., and J. H. Langenheim. 1972. Systematic studies of the genus *Hymenaea* (Leguminosae, Caesalpiniodeae, Cynometreae). [Unpublished ms.]

Leonard, J. 1949. Caesalpiniaceae-Amherstieae Africanae Americanaeque: Notulae Systematicae IV. Bull. Jard. Bot. Bruxelles 19:373-408.

————. 1950a. Étude botanique des copaliers en Congo Belge. Public. I.N.E.A.C. Ser. Scient. 45.

————. 1950b. Nouvelles observations sur le genre *Guibourtia* (Caesalpiniaceae): Notulae Systematicae IX. Bull. Bot. Gard. Bruxelles 20:270-284.

————. 1951. Les *Cynometra* et les genres voisins en Afrique tropicale: Notulae Systematicae XI. Bull. Jard. Bot. Bruxelles 21:373-450.

————. 1957. Genera des Cynometreae et des Amherstieae Africaines. Memoire Classe des Sciences de l'Académie Royale de Belgique 30 (2):1-314.

Linnaeus, C. 1737. Genera Plantarum. 1st ed. 366 pp.

————. 1762. Species Plantarum. 2nd ed. 557 pp.

Mangenot, S., and G. Mangenot. 1957. Nombres chromosomique nouveaux chez diverses dicotylédones et monocotylédones d'Afrique occidentale. Bull. Jard. Bot. Bruxelles 27:639-654.

————. 1958. Deuxiéme liste de nombres chromosomique nouveaux chez diverses dicotylédones et monocotylédones. Bull. Jard. Bot. Bruxelles 28:315-329.

Margolis, S. V., and J. P. Kennett. 1971. Cenozoic paleoglacial history of Antarctica recorded in subantarctic deep-sea cores. Amer. J. Sci. 271:1-36.

Martin, H. 1969. A critical review of evidence for a former direct connection of South America with Africa. Vol. 1, pp. 25-53, *in* Fittkau, E. F., et al. (Editors). Biogeography and ecology in South America. Dr. W. Junk Publ. The Hague.

Martin, S. S., and J. H. Langenheim. 1970. Studies of resins from *Hymenaea* and *Trachylobium.* Amer. J. Bot. 57:756.

Martin, S. S., J. H. Langenheim, and A. Cunningham. 1971. Resin acids in *Hymenaea* (Leguminosae). Amer. J. Bot. 58:479.

Martinez, M. 1960. Una especie de *Peltogyne* en México. Anales del Instit. de Biol. 31:123-131.

Menéndez, C. A. 1969. Die Fossilen Floras Sudamerikas. Vol. 2, pp. 519-561, *in* Fittkau, E. F., et al. (Editors). Biogeography and ecology in South America. Dr. W. Junk Publ. The Hague.

Mirov, N. T. 1967. The genus *Pinus.* Ronald Press. New York.

Moens, P. 1955. Les formations sécrétrices des copaliers congolais. Cellule 57:35-59.

Moreau, R. E. 1966. The bird faunas of Africa and its islands. Academic Press. New York.

Muller, J. 1970. Palynological evidence on early differentiation of angiosperms. Biol. Rev. 45:417-450.

Nakano, T., and C. Djerassi. 1961. Terpenoids XLVI: Copalic acid. J. Org. Chem. 26:167-173.

Oliver, D. 1871. Flora of tropical Africa, II. L. Reeve Co. London.

Ponsinet, G., G. Ourisson, and A. C. Oehlschlager. 1968. Systematic aspects of the distribution of diterpenes and triterpenes. Vol. 1, pp. 271-302, *in* Mabry, T. J., R. E. Alston and V. C. Runeckles (Editors). Recent advances in phytochemistry. Appleton-Century-Crofts. New York.

Record, S. J., and R. W. Hess. 1943. Timbers of the New World. Yale University Press. New Haven, Connecticut.

Sandermann, W. 1962. Terpenoids: metabolism. Pp. 591-630, *in* M. Florkin and H. S. Mason (Editors). Comparative biochemistry. Academic Press. New York.

Simpson, G. G. 1952. Probabilities of dispersal in geologic time. Bull. Am. Mus. Nat. Hist. 99:163-176.

————. 1969. South American mammals. Vol. 2, pp. 879-909, *in* Fittkau, E. F., et al. (Editors). Biogeography and ecology in South America. Dr. W. Junk Publ. The Hague.

Smith, A. C. 1967. The presence of primitive angiosperms in the Amazon Basin and its significance in indicating migrational routes Atas do Simpósio sôbre a Biota Amazônica 4 (Botánica):37-59

Thomas, B. R. 1970. Modern and fossil plant resins. Pp. 59-78, *in* Harborne, J. B. (Editor). Phytochemical phylogeny. Academic Press. New York.

Tschirch, A., and E. Stock. 1933-1936. Die Harze. 3rd edition. 2 vols. in 4. Berlin.

Turner, B. L., and D. S. Fearing. 1959. Chromosome numbers in the Leguminosae, II: African species, including phyletic interpretations. Amer. J. Bot. 46:49-57.

Turner, B. L., and M. C. Irwin. 1961. Chromosome numbers of some Brazilian Leguminosae. Rhodora 63:16-19.

Vogel, S. 1968. Chiropterophilie in der neotropischen Flora. Flora, Abt. B., Bd. 157:572-602.

Weeks, L. G. 1947. Paleogeography of South America. Bull. Amer. Soc. Petr. Geol. 31:1194-2141.

Wolfe, J. A. 1971. Tertiary climatic fluctuations and methods of analysis of Tertiary floras. Paleogeography, Paleoclimatology, Paleoecology 9:27-57.

Woodring, W. P. 1954. Caribbean land and sea through the ages. Bull. Geol. Soc. Amer. 65:719-732.

Zavarin, E., and K. Snajberk. 1965. Chemotaxonomy of the genus *Abies,* I: Survey of the terpenes present in the *Abies* balsams. Phytochem. 4:141-148.

Zavarin, E., T. Reichert, and E. Tsien. 1970. On the geographic variability of the monoterpenes from the cortical blister oleoresin of *Abies lasiocarpa*. Phytochem. 9:377-395.

Phytogeography and Evolution of the Velloziaceae

EDWARD S. AYENSU

Introduction

In recent years my interest in the monocotyledons has centered on the distribution and the evolution of the Velloziaceae, a family which is one of the most interesting and puzzling links between the floras of tropical America, on the one hand, and Africa-Madagascar (Malagasy Republic) on the other.

Most phytogeographers are quite familiar with the distribution of, for example, the Bromeliaceae. Except for one West African endemic species, *Pitcarnia feliciana*, this family is centered in tropical America. Another family exhibiting a similar distribution is the Rapateaceae with most of its species distributed in South America except for the genus *Maschalocephalus* which is endemic in Liberia, West Africa. Similarly, the family Vochysiaceae has most of its genera distributed in South America except for the genus *Erismadelphus* which is found in Cameroon, West Africa.

The relationships in the Velloziaceae are not very well understood. The family generally is believed to have originated in Africa-Madagascar, later to have emigrated to South America; however, some of the reasoning generally assigned to this conclusion, in my opinion, is not very convincing. For example, Takhtajan (1969:187-188) recently stated, ". . . one of its [Velloziaceae] two genera—*Vellozia* (sensu Hutchinson 1959)—is found in South Arabia (Hadhramaut), tropical and South Africa, Madagascar and also in tropical South America; the other—*Barbacenia*—is found only in tropical South America. *Vellozia* is characterized by free or almost free perianth-segments, while in *Barbacenia* they are united into a tube. Hence, we may conclude that the Velloziaceae are more likely to have migrated from the Old World to the New than vice versa." While I am in agreement that the Velloziaceae (might) have originated in the Old World, Takhtajan's rationale that the formation of the perianth-segments into a tube is confined to *Barbacenia* is not a convincing argument.

Another controversy that is currently being discussed by others studying this interesting family was precipitated by Maguire (1969:36) who noted after studying the pollen grains of some African specimens that none has pollen similar to the American *Vellozia*: This situation led him to state that "it is not unreasonable, therefore, to suspect that all African elements belong neither to *Barbacenia* nor to *Vellozia*." Maguire described the African specimens as having small ellipsoid, unisulcate grains and concluded that they are thus similar to American *Barbacenia*. "But because of other profound morphologic differences the African elements do not appear to be assignable to *Barbacenia*." Again, while I generally support Maguire's analyses which have in fact been stated earlier by Erdtman (1963), three of the six named African species that Maguire described as having simple pollen grains that are ellipsoidal, 1-sulcate, and with reticulate exine, were based on faulty taxonomy. My recent publication on the African Velloziaceae (Ayensu 1969a) and those on the American material (Ayensu 1968, 1969b, 1973), together with unpublished studies on some 200 American species and the accounts on the

Edward S. Ayensu, Department of Botany, National Museum of Natural History, Smithsonian Institution, Washington, D.C. 20560.

Velloziaceae by Warming (1893), show conclusively that histological characters of *Vellozia* and *Barbacenia* are indeed represented in the African species. From my morphological and anatomical studies I can only conclude that there remains much to be done on the alpha taxonomy of the family, even at the generic level. Lyman B. Smith and I are currently coordinating our efforts to present a key that can be used to cover both the Old and New World species.

The taxonomic problems encountered in the Velloziaceae have historical roots. During my survey of the anatomy of the family it became quite clear that whereas the taxonomy of the American species, especially as outlined in the work of L. B. Smith (1962b), seemed to support the anatomical characters used in distinguishing the genera that constitute the family, the taxonomic treatments of the African species showed very little evidence of corroboration with morphological and anatomical characters. My recent publication (Ayensu 1969a) has shown conclusively that more serious basic taxonomic work is needed on the Old World material.

The problem of resolving the generic entities of the Old World Velloziaceae recently has come to the fore because of the various attempts to coordinate the taxonomy of the Old and New World species. The difficulties facing us today are twofold. The first involves nomenclatural problems, especially of the naming of the genera and species. The second centers around the apparent variation in the floral characteristics of the Old World species.

Perrier (1946) pointed out, when preparing for the revision of the Velloziaceae of Madagascar, that botanists are not all in agreement on the taxonomy of the family. He could recognize three genera however, *Xerophyta, Vellozia,* and *Barbacenia.* Earlier Baillon (1895) had united all the species of Velloziaceae into the genus *Barbacenia.* The first serious attempt to correct this situation was the publication of a revision of the Old World species of *Vellozia* by Greves (1921). In this publication she suggested that the genera of the family could be separated into natural groupings by the emergences on the external walls of the ovary. Perrier (1946) found such classification untenable, at least in the separation of the Madagascan species. In 1880 Baker separated the family by the presence or absence of the perianth-tube. He considered all species with free perianth-segments as *Vellozia* and those with united perianth as *Barbacenia.* Pax (1887) distinguished the genera by the number and position of the stamens. He placed all species with six stamens under *Barbacenia,* and species with more than six stamens under *Vellozia.*

Another taxonomic work which is worthy of note is the treatment of the family in the conspectus of the African Flora by Durand and Schinz (1895). In this work all the species of the family were placed under the genus *Xerophyta* in the tribe Xerophyteae of the family Amaryllidaceae. As Perrier (1946) aptly pointed out in his discussion on this family, the differences in the classification resulting from the assemblage of characters, which at best are strictly artificial, have led to the great confusion currently expressed in the literature. Hutchinson (1959:680) was indeed right in suggesting that "the family needs a careful monograph, and possibly more than two genera should be recognized, a common solution to a difficulty such as this." I must state emphatically that Perrier's (1946) recognition of three genera viz. *Xerophyta, Vellozia,* and *Barbacenia,* is strongly supported by the vegetative histology of all the species of this family.

I hope that the taxonomic work Lyman B. Smith and I are attempting to do will help in further elucidation of the problem. In the absence of a very careful taxonomic treatment of the family as a whole, however, this presentation will rely primarily on histological information for facts and arguments to support my inferences on possible origin, evolutionary sequences, and migratory routes of the family.

I am very grateful to all those who have taken interest in this study especially Dr. Lyman B. Smith. His taxonomic experience with the Velloziaceae has been of immense help in the formulation of my own ideas on this family.

General Morphology

The Velloziaceae consists of aborescent and frutescent members with mostly simple but sometimes of complex bifurcating stems (Figures 1-5). The plants are woody and fibrous and often they are covered with persistent, imbricate sheathing, the tops of which have fallen with age. The blades are rather narrow, coriaceous, and mostly terminate in a sharp point. The blades are almost articulated at the point from which the top parts are detached

FIGURE 1. *Xerophyta pinifolia* (habit) growing on gently sloping exfoliating granite, 3 km north of Anaviavy, 11 km west of Ankaramena (elevation 800 m), Fianarantsoa Province, Madagascar. [Photo by F. R. Fosberg.]

from the sheath. The tuft-like appearance of the branches is due to the arrangement of the young leaves at the ends of the branches.

The flowers are lily-like, solitary, in the axils of the upper leaves, occurring on either a short or long scape; few are sessile, cf. *V. markgrafii*. Some of the flowers are small, but generally they are large and brilliantly colored. They commonly come in very attractive white, blue, yellow, or red. The flowers are actinomorphic, bisexual, and epigynous. The perianth segments are of six equal (or sometimes unequal) parts. The petals are separate or coalese to form a tube. The stamens are six or more in *Vellozia,* but are always six in *Barbacenia*. The stigma is broadly 3-lobed in *Vellozia,* but slightly lobed in *Barbacenia*. The anthers shed their pollen by means of longitudinal slits. The ovary is inferior, trilocular with numerous ovules. The placentation is axile and stalked. The style may be either entire, attenuate, capitate, or trilobed at the summit. The stigma is capitate or it is divided into three short branches. The fruit is a loculicidally dehiscent capsule. The seeds are hard and numerous, with a small embryo and copious endosperm.

The following is the tentative key of the Veloziaceae which Lyman B. Smith and I have so far been able to construct to cover both the Old and New World genera using principally the floral characters.

1. Stigmas vertical, narrow; stamens always 6.
   2. Anthers basifixed, in line with filament; filament free from the tepals; leaf-blade persistent. Africa . . *Talbotia*
   2. Anther dorsifixed, more or less oblique to the filament.
      3. Flowers perfect, anthers appearing subsessile on a lobed tube.
         4. Stamen-tube almost wholly adnate to the tepals; leaf-blades deciduous. Africa, Madagascar, Arabia *Xerophyta (Schnizleinia) ("Barbacenia"* Africa) (*"Vellozia"* Africa)
         4. Stamen-tube free from the tepals; leaf-blades persistent. South America . . . . *Barbacenia*
      3. Flowers functionally or completely unisexual; filaments adnate to the tepals for nearly all their length; stamens appearing subsessile on the tepals; leaf blade persistent. South America . . *Barbaceniopsis*
1. Stigmas horizontal, suborbicular; stamens usually more than 6; anthers basifixed; leaf-blades deciduous. South America . . . . *Vellozia*

I must emphasize again that, while I recognize some staminal difference between the Old and New World species and the regularly persistent leaf blades and adnate filaments of many of the Old

FIGURE 2. *Vellozia minima* Pohl (habit), Maguire et al 49099.

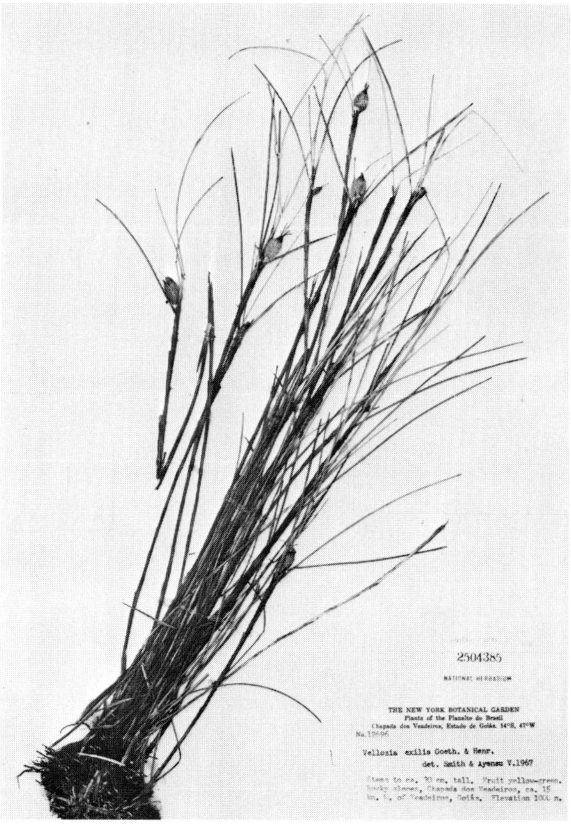

FIGURE 3. *Vellozia exilis* Goethart and Henrard (habit), Irwin et al 12696.

FIGURE 4. *Barbacenia stenophylla* Goethart and Henrard (habit), Irwin et al 9901.

FIGURE 5. *Barbacenia longiflora* Martius (habit), Archer 4089.

World species which have influenced the above tentative key, I do not consider these traits adequate to assess the taxonomy of the family at this time. It would be a serious mistake to ignore the generic delimitations that vegetative anatomy of the leaves has to offer.

In order to eliminate any confusion, I will continue to refer to the Old World species under the genera *Xerophyta, Vellozia,* and *Barbacenia* in contrast to the arrangement in the tentative key that lumps all the African and Madagascan species under *Xerophyta* and one species under *Talbotia.*

Habitat

The Velloziaceae generally occur in more or less similar environmental conditions. It is interesting to note that environmental differences with respect to rainfall, light intensity, and temperature have exerted fairly uniform influence on the morphology of the species within this family. The plants being essentially lithophytes, thrive on exposed granite rocks, as well as in dry plains of alpine regions. In Brazil, the species range in altitude from 1000–2000 meters. Species such as *Vellozia candida* and *V. plicata* thrive on barren soil while *Barbacenia purpurea* grows on the humus on rocks. Several members of the family are encountered in clefts of rocks and pebbles where the ground cover is sparse and the roots are feebly attached to the ground.

In describing the origins of the flora of southern Brazil, L. B. Smith (1962a:224) clearly described the habitat of the Velloziaceae that occur in Minas Gerais. He pointed out that "the plants grow on dry barren slopes and crests and the species in most cases have small ranges." He further indicated that in the coastal rain forest the Velloziaceae occur on exposed granite peaks rising above the forest.

In a letter on the vegetation of West Equinoctial Africa addressed to W. W. Saunders by Friedrich Welwitsch (1859) from S. Paulo de Loanda on 10 February 1858, we read the following: "But now, on the steep walls of the higher rocks, what sort of viscid shrubs with scaly stems and blue flowers do we perceive? Two species of Vellosi[ac]eae! Which, in conjunction with several fruticose Orchids that grow even on the barest rocks, cover all the mountains of Pungo Andongo." Likewise, Engler (1905) found about fifteen Velloziaceae species often in groups by themselves or together with cactus-like Euphorbias in the steppe region of tropical Africa especially in the rocky mountain steppe of the dry regions on the north slope of the West-Usambara mountains. L. B. Smith (1962a:224) has also said that "incidentally this family also illustrates the fallacy of some plant geographers. Its world range has been shown as a line enclosing the whole Amazon basin, while its true occurrence there is limited to a few bare peaks near the edge of it." Other species of the family occur in groups in South Africa in the dry and sandy campo or prairie lands. In Madagascar, the species of *Xerophyta* also grow on granite rocks.

The diurnal climatic conditions under which the Velloziaceae thrive vary somewhat. During the daytime in the dry seasons the species are subjected to intense solar radiation. Occasionally the sunny condition gives way to violent but short rainstorms to offset the intensity of solar radiation. At night, the situation is somewhat reversed. There are often cloud covers, resulting from the cooling due to the movement of the warm air upwards from the land surface into the upper atmosphere, that overshadow the mountain tops and the landscape and thus provide enough water droplets to dampen the plants and the substrate.

Morphological Adaptation

Species which grow in areas that are subjected to high solar radiation and rapid water runoff are often adapted to hold as much water as possible during rainy days. Since it is generally known that the abundance of plant species, especially those that grow on rocky surfaces, depends heavily on their capacity to reduce loss of water to the air, we can conclude that the vegetative organs of the Velloziaceae are adapted to cope with such problems.

The most obvious morphological adaptation of this family to its environment is the general reduction of leaf surface area which directly relates to low transpiration rate. For example, species such as *Vellozia pilosa, V. minima, V. ornata, Barbacenia tricolor,* and *B. rubra* have leaf blades that are 1, 1.5, 3-4, 4, and 5 millimeters wide, respectively. But we often expect that plants in such xeric environment will develop succulent leaves, which will also lead to diminution of the transpiration rate. This is, in fact, the case in a number of plants such as

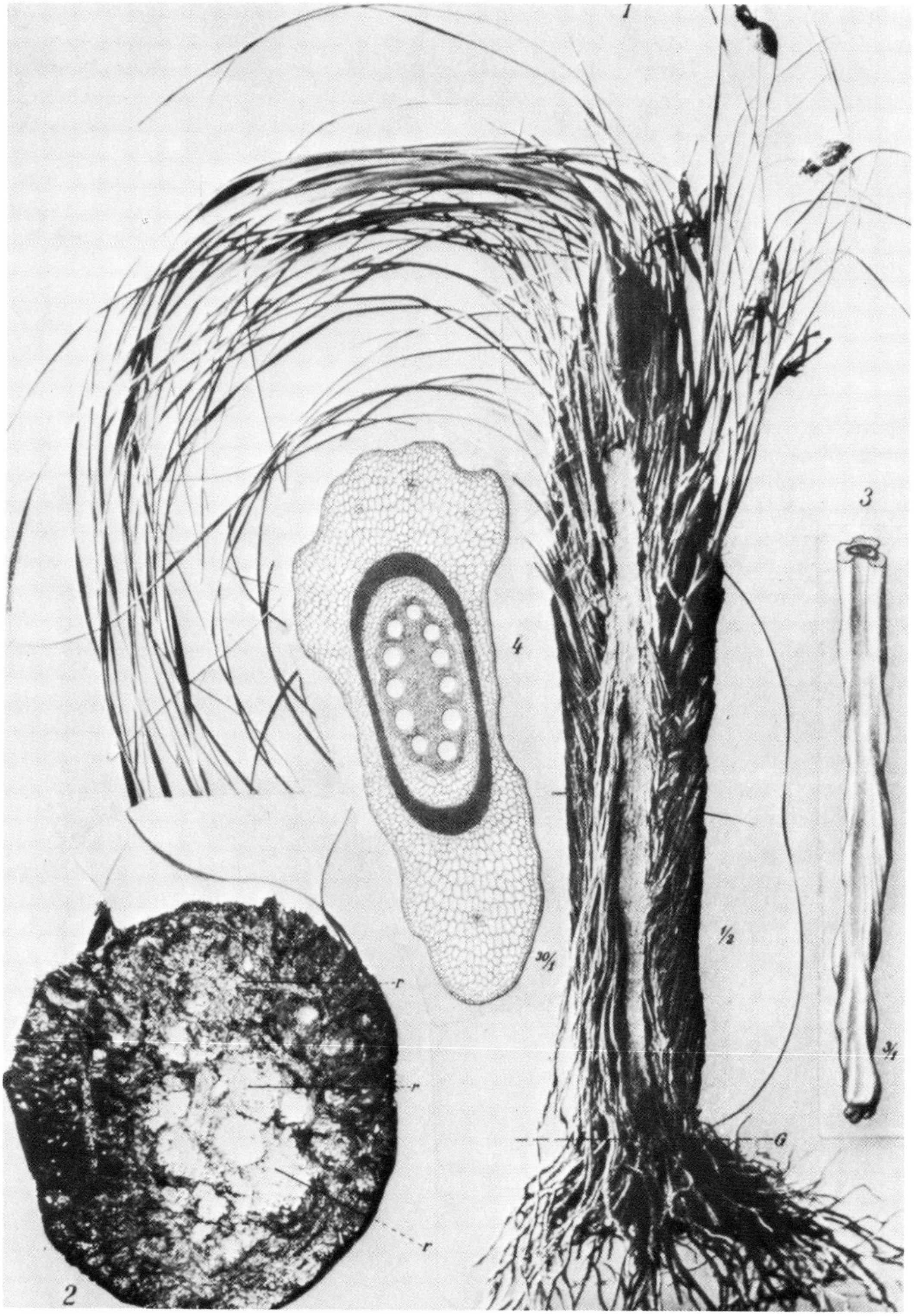

FIGURE 6. *Barbacenia retinervis* (Baker) Engler (1) Plant, the stem in longitudinal section, ½ natural size. $G =$ Surface of ground. (2) Transverse section of stem, natural size. $r =$ bundles of roots. The white circular spot in the center is the real stem. (3) Lower end of a root, 3/1. (4) Transverse section of root, 30/1. [After Marloth 1915.]

the cactus-like euphorbias that thrive in the same habitat as the Velloziaceae. The Velloziaceae, on the other hand, have developed an ingenious system for checking water loss. The slender woody stems are surrounded with foliar sheaths. A cursory look at the cylindrical stem does not give a full view of the actual structure of the real stem. A longitudinal section through a stem clearly indicates that the bulk of the mass is made up of persistent leaf sheaths (Figure 6). Apart from the slender main stem there are lateral branches of the stem as well as long aerial roots that are completely encircled by the leaf bases. The roots develop along the entire length of the main stem until they reach the ground whereupon they spread in all directions and assume the normal anchorage role of roots. It is postulated that while the roots that reach the ground cling to clefts in rocks and predominately provide a holding function, the long aerial roots, often covered with root hairs, serve as appropriate water conveyers for the plant. We should also bear in mind that perhaps the reason why the aerial roots develop so vigorously and abundantly is because they are in return supplied with the necessary humidity by the ability of the leaf bases that encircle them to hold as much water as possible.

Apart from the role that roots play in maintaining the water balance of the Velloziaceae the leaves have also developed other mechanisms for regulating transpiration. This is done by either (1) the reduction of the rate of water loss per unit area and/or (2) by the contraction of the transpiring surface area. Generally, the rate of transpiration is reduced because of the protection that is provided the stomata by the abaxial furrows in the leaves. An examination of the transverse sections of many leaves will reveal that the stomata are confined to the furrows. Furthermore, the furrows also permit the contraction, inrolling, outrolling, and lengthwise folding of the leaves against water loss. Diago (1926), in his paper on the leaves of the vellozias and their mechanism in regulating transpiration, has cited many examples of the effectiveness of this phenomenon.

Because of the unavailability of freshly pickled material for histological studies, I have often relied mostly on dried specimens. If dried leaf sheaths are placed in water or any wetting agent they immediately absorb an immense amount of the medium. One can engage in a simple experiment by releasing droplets of water on a leaf sheath. It is easily observed that the drops of water disappear instantly as if the leaf sheaths were blotting paper.

As I alluded to earlier, during the warm rainy season the foliar sheaths are able to absorb enough water to serve the needs of the plants during dry periods. Conversely, the plants are well provided for even during the long dry seasons. Clouds that lead to dew formation at night supply some water to serve the needs of the plant during the intense solar radiation in the daytime.

Important Anatomical Characters

MAJOR SCLERENCHYMA PATTERNS. There are three major sclerenchyma patterns associated with the vascular bundles, viewed in the transverse section of the leaf, that have proved significant in the study of all the species of this family. In the Madagascan species alone (designated Type A) the vascular bundles are partially surrounded with sclerenchyma on the adaxial side in the form of an inverted crescentiform or V-shaped cap. The abaxial side is either also an inverted crescentiform or V-shaped cap, or in the case of the marginal vascular bundles, crescentiform. The African species, on the other hand, exhibit two types of sclerenchyma patterns. In the first case (designated Type B), the adaxial sclerenchyma is generally inverted crescentiform or cap-shaped. The abaxial sclerenchyma is either U- or Y-shaped. In the second (designated Type C), the adaxial side of the vascular bundle exhibits an inverted Y-shaped sclerenchyma girder. The abaxial girder is either Y-shaped or three-pronged (c.f., Ayensu 1969a). The New World species, on the other hand, exhibit the sclerenchyma distributions described for Types B and C. Type A is confined to the genus *Xerophyta*.

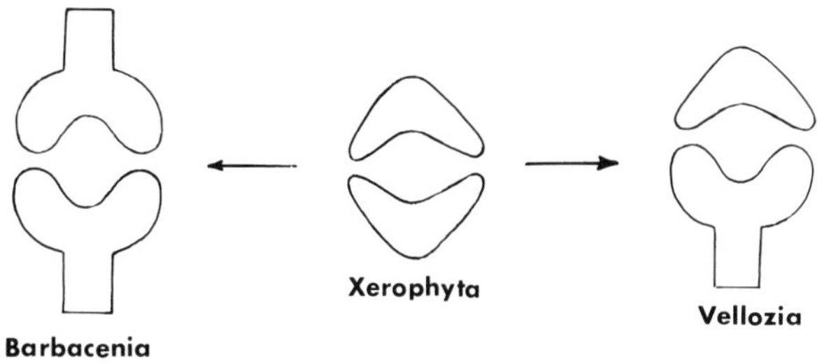

FIGURE 7. The possible results that can be obtained when *Barbacenia* and *Vellozia* leaf sclerenchyma patterns are derived from *Xerophyta*.

Type B occurs only in the genus *Vellozia* and Type C in *Barbacenia-Barbaceniopsis* (Figure 7).

The sclerenchyma changes that led from the *Xerophyta* type to the *Vellozia* type, on one hand, and to the *Barbacenia-Barbaceniopsis* type, on the other hand, seem to correlate nicely with what I envisage as an evolutionary sequence in the family (Figure 9). I consider the *Xerophyta* type as the most primitive, probably most similar to the patterns in the Hypoxidaceae. From *Xerophyta* two separate lines were developed, *Vellozia* being the first and closely followed by *Barbacenia* (Figure 7). It would appear therefore that when these two genera were brought under a certain suitable environmental condition further sclerenchyma modifications did not take place as the species spread from Africa until they reached their South American limits. Another way of assessing the sclerenchyma pattern in this family is to consider *Vellozia* as the most primitive type from which *Xerophyta* was derived by the elimination of the tail of the Y-shaped abaxial sclerenchyma girder. *Barbacenia*, on the other hand, could have been developed by the extension of the base of the inverted adaxial sclerenchyma girder subjacent to the adaxial epidermis (Figure 8). It seems very possible that both the *Xerophyta* and the *Vellozia* types may have developed from the *Barbacenia* type by the loss of sclerenchyma girders. The loss of a character is more easily achieved than the gain. Current studies will undoubtedly shed more light on this problem.

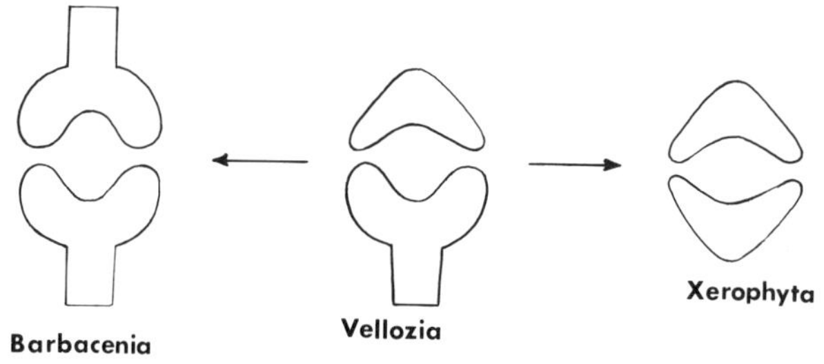

FIGURE 8. The possible results that can be obtained when *Barbacenia* and *Xerophyta* leaf sclerenchyma patterns are derived from *Vellozia*.

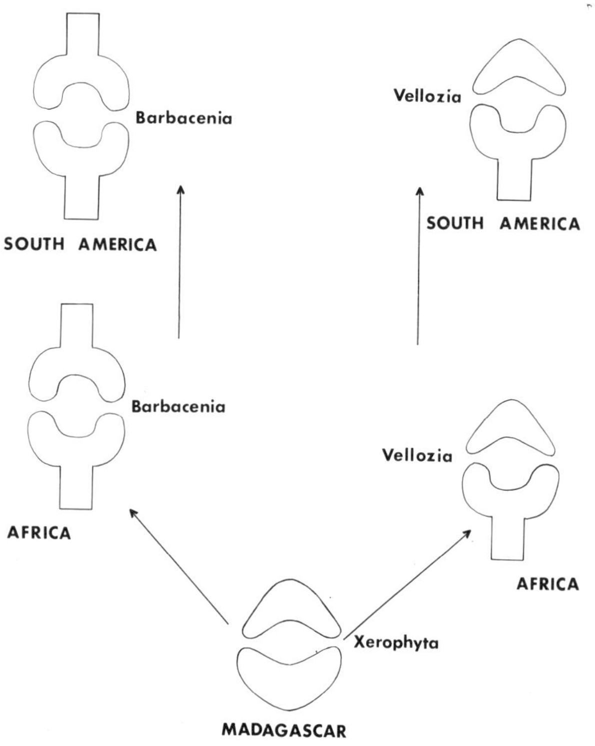

FIGURE 9. The evolutionary sequence in the leaf sclerenchyma patterns in both the Old and New Worlds.

While this alternative is entirely possible, I am inclined to favor my original suggestion that, of the three recognized genera, *Xerophyta* is the most primitive not only from the morphological standpoint (e.g., woody and aborescent forms; Figure 1) but from the fact that it is the only genus of the three represented in Madagascar (Perrier 1950), a region with a very high concentration of endemic genera (Good 1964).

MESOPHYLL PATTERNS. Two major types of mesophyll patterns occur in the Velloziaceae: (1) dorsiventral—this includes species with distinct palisade and spongy tissues (Figure 10*a-d*)—and (2) isolateral—those species with a mesophyll that has not been differentiated into palisade and spongy tissues (Figure 10*e-h*). There seems to be an interesting correlation between the mesophyll pattern and the type of sclerenchyma exhibited in the genera of this family. On the basis of the old African classifications (c.f., Perrier 1946), the dorsiventral mesophyll occurs in the genera *Xerophyta* and *Vellozia*, while the isloateral type is confined to *Barbacenia-Barbaceniopsis*. In addition to these differences, there is a slight but distinct difference between the Madagascan-African and the American vellozias in the palisade tissues. In the Madagascan-African species the palisade cells run more or less solidly across the upper side of the adaxial inverted-crescentiform sclerenchyma. The American vellozias, on the other hand, show large trans-

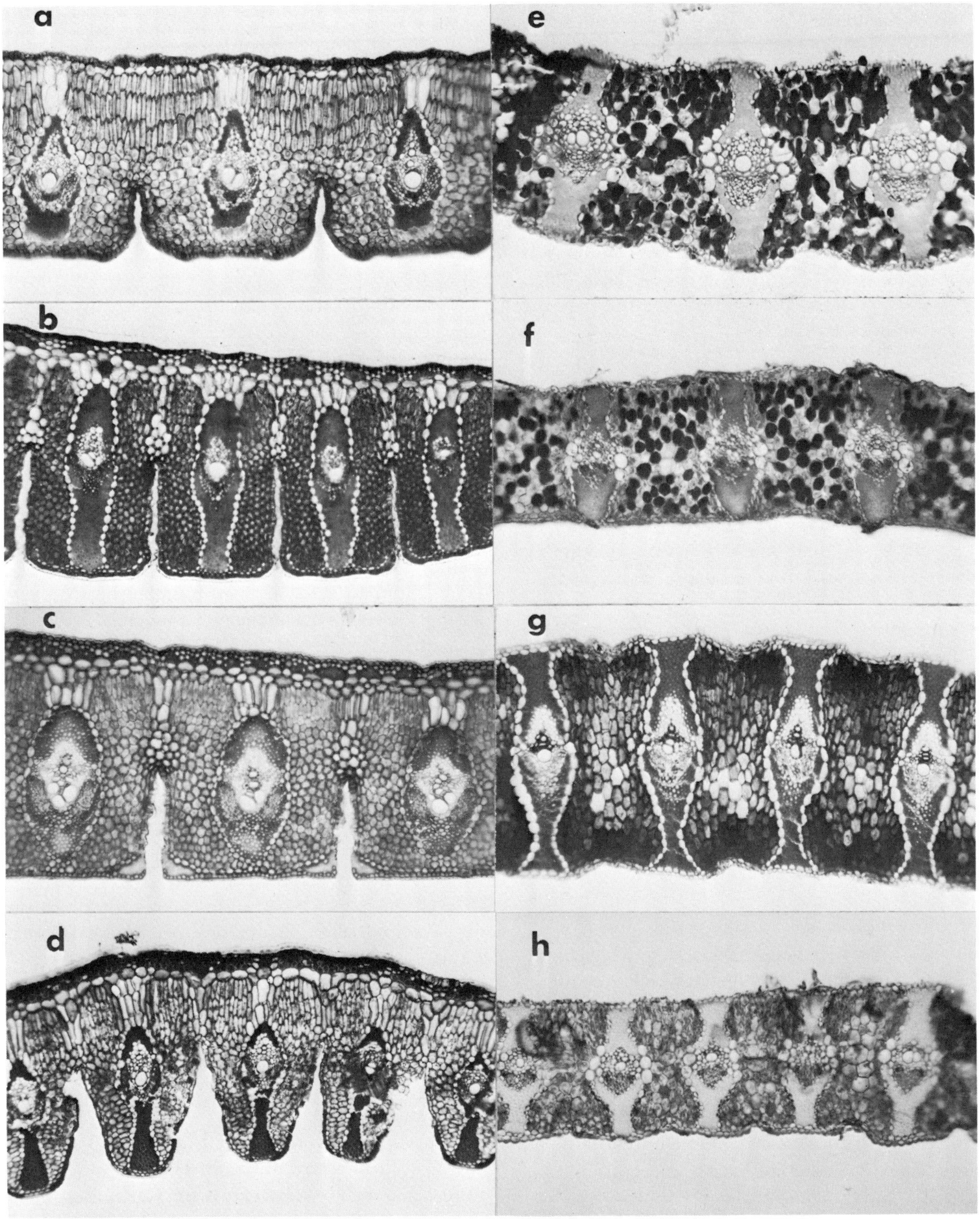

FIGURE 10. Transverse sections of leaves showing sclerenchyma patterns and mesophyll. × 115 *a, Vellozia candida* (Caranta 1868); *b, Vellozia machrisiana* (Irwin et al 12697); *c, Vellozia compacta* (Irwin et al 2407); *d, Vellozia tubiflora* (Thassen 1777); *e, Barbacenia schwackei* (Heringer 5263); *f, Barbacenia williamsii* (Williams and Assis 6386); *g, Barbacenia seubertianna* (Strang 239); *h, Barbacenia longiflora* (Archer 4089).

lucent cells immediately above the adaxial inverted crescentiform or cap-shaped girders or they alternate with the vascular bundles (c.f., Ayensu 1969a).

I view the mesophyll differentiation in the New World *Vellozia* species as a condition that occurred when Old World species spread into South America and encountered microclimatic changes that necessitated the development of large translucent cells capable of holding appreciable quantities of water. This shift from one environment to another was seemingly drastic enough to exert a significant selective pressure on the species thus resulting in the above histological changes.

Mode of Migration and Evolution

The fossil record of the Velloziaceae is nonexistent; therefore, any deductions or inferences on the origin and colonization should be made in the light of the general origins and plant exchanges that took place during the Tertiary Period between tropical floras of Africa and South America. Takhtajan (1969:187) indicated that although direct connections between the tropical floras were severed very early, isolation was probably not complete. He went on to suggest that "it is quite possible that there was still some connection between South America and Africa at the beginning of the Tertiary period, a supposition which is supported by the highly interesting disjunct distributions involving tropical America and tropical Africa to which Engler's (1905) interesting work was devoted." I am personally inclined to lean towards the hypothesis that South America and Africa were separated before the development of any but the earliest prototypes of the flowering plants.

Based on current geological and paleomagnetic evidence, A. C. Smith (1970) stated that the angiosperms must have originated in the Jurassic or Triassic periods (150-200 million years ago) in the general vicinity of Southeast Asia and spread in various directions by "saltatory long distance dispersal" during the Cretaceous or Tertiary periods. Furthermore, Thorne (see p. 45 herein) and A. C. Smith (see p. 60 herein) have emphasized that on the basis of current geological information the separation of the African and South American continents must have taken place several million years before the present distributional patterns of the angiosperms.

Be that as it may, the phenomena of long distance dispersal, as eloquently espoused by Carlquist (1966a, 1966b, 1966c, 1966d) and the papers he cites, seem to me the crucial point in the discussion of the Velloziaceae, rather than any actual connection of the continents. Furthermore, I submit that the most important factor affecting the distribution of the Velloziaceae was the availability of macro- and microenvironments that permitted the colonization and adaptation of the species, thus assuring their survival.

Two considerations have influenced my reasoning. First, all the species are adapted to xeric habitats in Madagascar, South Africa, and South America. As intimated earlier, xeric plants, as all other plants, are often modified, especially morphologically and physiologically, to cope with their environment. In the Velloziaceae modifications include reduced surface area of the leaf, revolute margins, water storage cells in the mesophyll, compactness of such tissues as the palisade cells and spongy mesophyll, strong development of the vascular bundles with distinct bundle sheaths, and strongly developed sclerenchymatous cells. Although increased thickness of leaf blade, for example, is not observed in this family, the waterholding capacity of the plant has been relegated to the leaf sheath bases that absorb water like blotting paper.

Secondly, none of the species in the family has ever been discovered on the islands between Africa and South America, such as on Tristan de Cunha, St. Helena, and Ascension. The most logical explanation is that the Velloziaceae development depends on the availability of a bare rocky substrate that is almost always granite, gneiss, or quartzite (c.f., Axelrod 1960). Plants have been observed very rarely on calcarious soils, but never on volcanic rocks such as gabbro, labradorite, or basalt. I therefore surmise that the fact that these islands are composed of relatively young volcanic rocks made their colonization by this family difficult, if not impossible.

We can quickly examine the question whether Madagascar is a continental island or an oceanic one like Tristan de Cunha, St. Helena, and Ascension. I am inclined to agree with Carlquist (1965: 382) who explains the situation thus:

"Geologically, the evidence on whether Madagascar is a continental island is curious. There are sedimentary rocks of Jurassic, Cretaceous, and Tertiary ages—but the

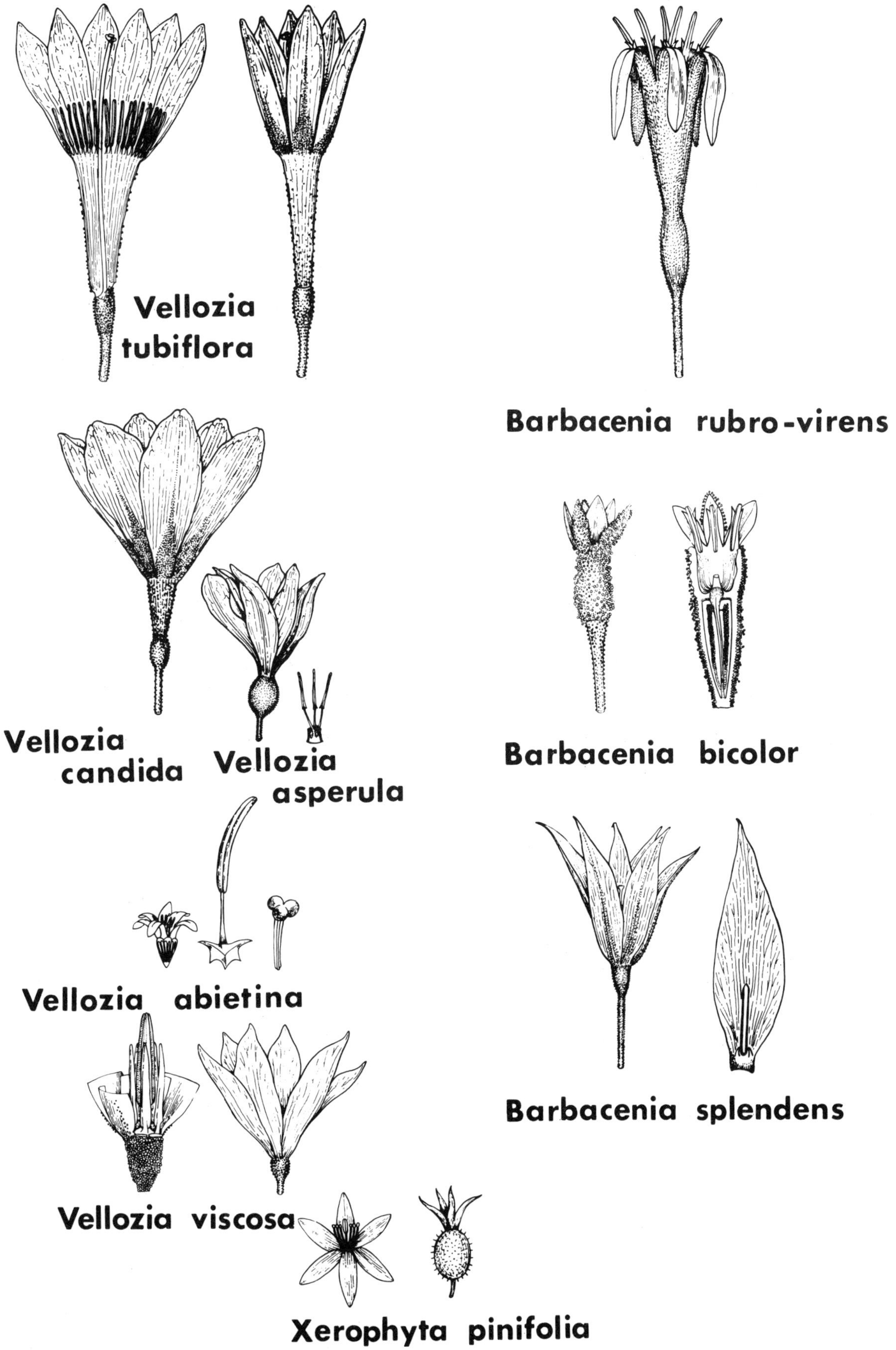

FIGURE 11. Phylogenetic progression and evolutionary specialization in the Velloziaceae.

most recent rocks face the Mozambique Channel. These and other facts hint that the Channel has existed since Triassic time, an interpretation seemingly supported by the plants and animals. The easiest explanation to accept at present seems to be that Madagascar is a continental island, but that its separation from Africa dates from about Triassic time."

Axelrod had earlier (1960:282) defined islands within oceanic regions as being of three types:

(a) *Continental islands,* composed of continental rocks, are parts of old continental areas but are now separated from them, owing to subsidence in the intervening region (Wallace, 1895). Some of these connections were severed recently (i.e. British Isles from Europe, Sunda Islands from Asia, Tasmania from Australia, Japan from Asia, etc.), but others have considerable antiquity (i.e. New Zealand—Fiji, Madagascar, Seychelles, Philippines, South Georgia, etc.); (b) *Oceanic islands* have been built up as volcanos or as volcanic ranges from the floor of the ocean basin (Wallace, 1895). Included ranges are (I) the present high volcanic islands, (II) low coral archipelagos built on the foundation of old, beveled, subsided volcanic islands, and (III) submerged archipelagos which were the sites of high islands, during the Mesozoic and Early Tertiary. . . . (c) *Composite islands,* a type not previously defined—have had the dual history, first as continental islands and then as oceanic islands. They occur today chiefly on the ridges of the Atlantic and west Indian oceans.

In reviewing the various criteria generally accepted by botanists as reflecting concepts of primitiveness and advancement in flowers (Thorne 1963, A. C. Smith 1967), I was struck by the evolutionary trend in the petals of the species in the Velloziaceae. The genera in this family clearly illustrate a phylogenetic progression and an evolutionary specialization from choripetaly to sympetaly (Figure 11). The Madagascan species of the genus *Xerophyta* are the most primitive members in the family and, as illustrated by *X. pinifolia,* the petals are clearly separate. Examination shows the two other species—*X. eglandulosa* and *X. dasylirioides*—are also choripetalous. The genus *Vellozia,* on the other hand, exhibits characteristics as exemplified by *V. viscosa.* The species *V. abietina, V. asperula,* and *V. candida* are distinctly coalescent, but the range is from slight tubing at the base to almost half the distance of the perianth. The longest perianth-tube (45-80 mm) is encountered in *V. tubiflora.*

A similar range is exhibited in *Barbacenia.* The species *B. splendens* is not completely choripetalous but slightly sympetalous. *B. bicolor,* on the other hand, is distinctly sympetalous and *B. rubro-virens* has a significantly long perianth-tube (70-75 mm).

It is significant to point out that the highly evolved members of the Velloziaceae are all South American. Whether the evolutionary development of the petals is associated with adaptation to any kind of pollination mechanisms is not known. Knuth (1904-1905) has remarked that on the basis of the floral morphology of the family it seemed likely that they are insect pollinated. It also seems likely that the large showy species of the family may be pollinated by birds. Critical field observations will undoubtedly shed some light on this subject.

I am inclined to believe that one of the major reasons for the apparent difficulty in the resolution of the genera in the Old World Velloziaceae is related to the fact that all of the most primitive members of this family occur in Madagascar and the ones in Africa are only slightly more advanced. These two areas are the regions that I and others consider as the place of origin of the family. The variations in the characters of the species—such as the shape of the stigma, the adnation of the stamens to the tepals, whether the stamens are fasicled or free—have led Velloziaceae taxonomists to conclude that there are more than three genera in the family, a conclusion I do not share. I submit that these floral characters within the African species are indications of variability and diversity that often occur in organisms undergoing differentiation and morphological adaptation to various pollinators. It seems very possible that after early phylogenetic progression of the Old World *Vellozia,* certain species migrated to South America and with rapid adaptation to the microenvironment, as well as new pollinators, were able to establish themselves. The apparent chaos in the systematics of the Old World species is not found in the New World. It is likely therefore that a "proto-*Barbacenia*" was able to establish itself in South America and possibly from a single introduction.

The taxonomic difficulties of the Old World Velloziaceae also stem from the fact that whereas the vegetative anatomical information correlates very well with the floral characteristics of the New World species, the floral characteristics of the Old World species exhibit no correlation with that of the New World species, probably because the pollinators and/or breeding systems are completely different.

DISPERSAL PHENOMENON. I have alluded to the idea that the only conceivable natural method of dispersal of fruits and seeds of the Velloziaceae from the Old World to the New World must have been through the phenomenon of long distance dispersal. Specifically I am referring to "air flotation" or "bird dispersal" as mechanisms. In his highly stimulating and well-illustrated book on *Island Life*, Sherwin Carlquist (1965) has described the various mechanisms through which small seeds can be carried to great distances. The seeds of the Velloziaceae are rather ideal for wind or bird dispersal, being about 1.5 millimeters long, flattened, cylindrical, hard, numerous and with a small embryo, but they are endowed with copious endosperm and get sticky when wet. Unfortunately, I do not know of any studies on the longevity of these seeds, though this may not be important because of the relatively short time that the wind or a bird needs to cross the Atlantic from Africa to South America. In her morphological studies in the postseminal development of the embryo of monocotyledons, Boyd (1932:153) stated that "at germination [of *Vellozia elegans*] a considerable quantity of endosperm remains. The cotyledon consists of a sheath less than 1 mm long, terminating in a minute, blunt sucker. Elongation of the cotyledon sheath and hypocotyl carries the light seed above ground, and within a few days the first leaf appears."

The arrival of seeds from one locale to the next, regardless of the great distances involved, is not the only factor that we should consider. Even more important is whether the seeds, once at their destination (in a new environment), will be capable of establishing themselves. The species of the Velloziaceae are represented in more or less similar ecological contexts wherever they occur today. We can, therefore, surmise that this family is adapted to similar environments in both the Old and New World.

It seems highly significant that the Velloziaceae migration has proceeded unidirectionally, i.e., from Africa to South America. Evidence gained from this study strongly suggests that a reverse trend has not taken place. This apparent irreversibility cannot be considered unequivocal, however, until studies on the chromosome constitution of the family have been undertaken. If these should show, for example, that the African species are diploid and the American ones are polyploid, the evidence for the unidirectional migration pattern of this family would be strengthened.

In the Introduction I made mention of some plant families that are widely distributed in tropical America with one genus and few species in Africa. Engler (1905), Hepper (1965), Iltis (1967), Axelrod (1970), and Thorne (1971) have elaborated on these kinds of distribution. There is a general tendency on the part of many phytogeographers and evolutionists to assume that the apparent wide representation of the species and genera of a family in tropical America and its meager representation in Africa means that the origin of the family is necessarily American.

The Velloziaceae is made up of some 47 species in Africa, three in Madagascar and about 200 in South America of which 124 belong to *Vellozia* and 76 to *Barbacenia*. It may be pointed out that the ratio of Old World and New World species in the Velloziaceae is not as one-sided as, for example, in the Bromeliaceae, Rapateaceae, or Vochysiaceae. The African-Madagascan species represent one-fifth of the total number of species in the Velloziaceae. While there is a temptation for some to conclude that the wide distribution of this taxon in America implies this place of origin or dispersal, it must be emphasized that all morphological and anatomical characters of the Velloziaceae quite clearly indicate that the Old World species are the most primitive members of the family. It appears, therefore, that Madagascar-Africa represents the region of original evolution and the primary center rather than South America, which constitutes a secondary center of adaptation and radiation of more recently evolved species of this family. Hence, we can conclude that the high preponderance of species of any taxon in any geographical region does not necessarily represent place of origin.

Relationship With Other Taxa

Taxonomically the Velloziaceae is considered closest to the Hypoxidaceae (Figure 12). In his treatment of the Liliales, Takhtajan (1969) considers Velloziaceae closely allied to the Hypoxidaceae. Similarly, Cronquist (1968) considers both

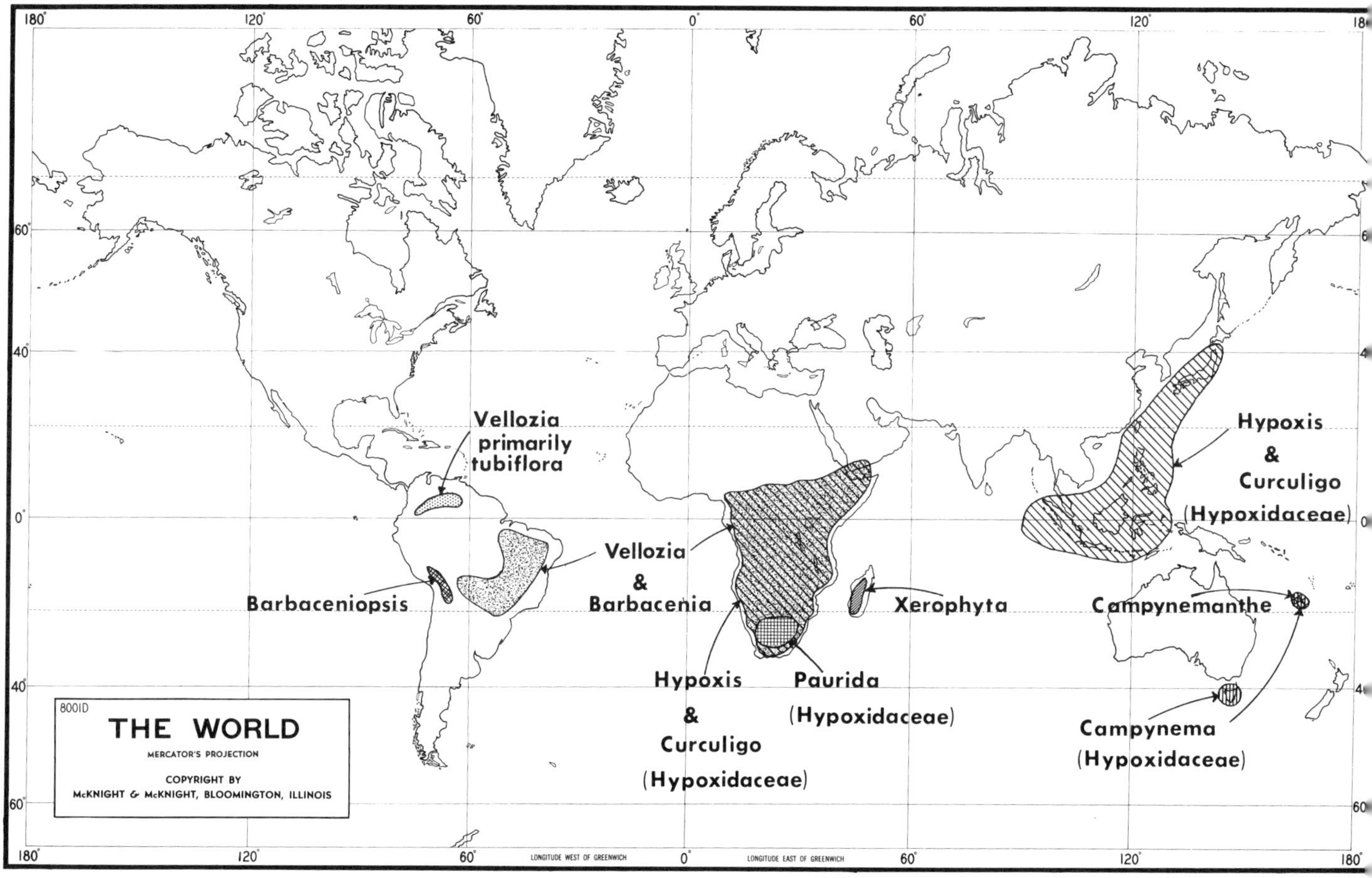

FIGURE 12. Geographic distribution of Velloziaceae and Hypoxidaceae.

families in the Liliales. However, he places the Hypoxidaceae within the Liliaceae complex excluding, inter alia, the Velloziaceae. In the Velloziaceae the ovary is 3-chambered with many ovules in each chamber while in the Hypoxidaceae, also with a 3-chambered ovary, there are 4-20 ovules in 2 series in each chamber. The habit of both families is similar and their relationship is further substantiated by the fact that, except for the placentation, their morphology is almost identical.

The Hypoxidaceae comprise about 90 species (mainly *Hypoxis*) most of which are African. About 15 are American and 5 Asiatic. There is, however, one additional genus, *Curculigo*, widespread in the Malay Archipelago, Japan, and Formosa, whose 10 species are common in forests both in the lowlands and at high altitudes. Other genera of the family (c.f., Hutchinson 1959) include *Campynema* (2 subspecies in Tasmania and New Caledonia) and *Paurida* (1 subspecies in South Africa).

The flowers and habit of *Curculigo* and *Hypoxis* show a striking resemblance especially to *Xerophyta*. A preliminary comparison of the flowers of the two families, however, has led me to conclude that the Hypoxidaceae is certainly more primitive than the Velloziaceae. If this deduction is accepted, I suggest further that there is a strong likelihood that the Velloziaceae and the Hypoxidaceae have a common ancestry in the Southern Hemisphere. Considering that in the New Caledonian-Australian-Madagascan area (c.f., A. C. Smith 1970) a large number of relictual taxa have survived, it is very likely that detailed morphological and phytogeographical studies of these small genera, especially *Campynema* and *Campynemanthe*, will shed substantial light on the whole evolution of the Haemordorales.

General Remarks

This review of the current state of our knowledge of the phytogeography and evolution of the Velloziaceae leads me to make the following closing remarks. First, it seems obvious from past works on this family that the heavy dependence on the classical usage of floral, and, to some extent, vegetative morphological characters alone in delimiting genera within this family have not worked out well in view of the results obtained from the analysis of such characters in the treatments of Baillon (1895), Greves (1921), Pax (1887), Durand and Schinz (1895), and Perrier (1946). Secondly, the introduction of a broad anatomical survey into the assessment of the systematics has presented consistent character-states covering species in the Old World as well as the New World. Thirdly, the lack of a comprehensive pollen analysis of the family, as well as similar coverage of chromosomes, render any final conclusions impossible at this point. Fourthly, because of the high preponderance of Velloziaceae species in, especially, South America where adaptive radiation has been most noticeable, it is necessary that we take advantage of that environment to study the population dynamics of this family. Such a study will undoubtedly permit us to obtain a better understanding of speciation in South America as well as in Africa.

It is only when we have been able to accumulate the missing information that we shall be in a position to formulate firm ideas on the evolutionary systematics of this family.

References

Axelrod, D. I. 1960. The evolution of flowering plants. Pp. 227-305, *in* S. Tax (Editor). Evolution after Darwin, 1: Evolution of life. University of Chicago Press. Chicago.

————. 1970. Mesozoic paleogeography and early angiosperm history. Bot. Rev. 36(3):277-319.

Ayensu, E. S. 1968. The anatomy of *Barbaceniopsis*, a new genus recently described in the Velloziaceae. Amer. J. Bot. 55:399-405.

————. 1969a. Leaf-anatomy and systematics of Old World Velloziaceae. Kew Bull. 23(2):315-335.

————. 1969b. The identity of *Vellozia uaipanensis*: Anatomical evidence. Mem. N.Y. Bot. Gard. 18(2):291-298.

————. 1973. Leaf anatomy and systematics of New World Velloziaceae. (In manuscript.)

Baillon, H. 1895. Histoire des plantes. 13:22-24.

Baker, J. G. 1880. Vellozieae. *In* Bentham and Hooker f. Gen. Pl. 3:739-740.

Boyd, L. 1932. Monocotylous seedlings. Trans. Bot. Soc. Edinburgh 31:1-224.

Carlquist, S. 1965. Island life. The Natural History Press. New York.

————. 1966a. Principles of dispersal and evolution. Quart. Rev. Biol. 41:247-270.

————. 1966b. Loss of dispersibility in Pacific Compositae. Evolution 20:30-48.

————. 1966c. Loss of dispersibility in the Hawaiian flora. Brittonia 18:310-335.

————. 1966d. Genetic systems in the floras of oceanic islands. Evolution 20:433-455.

Cronquist, A. 1968. The evolution and classification of flowering plants. Houghton Mifflin Co. Boston.

Diago, J. C. 1926. As folhas das Vellozias e seu apparelho regulador da transpiração. Arch. Mus. Nac. Rio de Janeiro 28:19-41.

Durand, T., and H. Schinz. 1895. Conspectus Florea Africae. Xerophyta 5:270-272.

Engler, Von A. 1905. Uber floristische Verwandkschaft zwischen dem tropischen Afrika und Amerika, sowie über die Annhme eines versunkenen brasilianischäthiopischen Continents. Sitzungsber. Konigl. Preuss. Akad. Wiss. Berlin 6:180-231.

Erdtman, G. 1963. Palynology. *In* R. D. Preston (Editor). Botanical Research. 1:149-208.

Good, R. 1964. The geography of the flowering plants. 3rd edition. John Wiley & Sons. New York.

Greves, S. 1921. A revision of the Old World species of *Vellozia*. J. Bot. 59:273-284.

Hepper, F. N. 1965. Preliminary account of the phytogeographical affinities of the flora of west tropical Africa. Webbia 19:593-617.

Hutchinson, J. 1959. The families of flowering plants, II: Monocotyledons. 2nd edition. Clarendon Press. Oxford.

Iltis, H. H. 1967. Studies in the Caparidaceae XI *Cleome afrospina* n.sp., a tropical African endemic with neotropical affinities. Amer. J. Bot. 54(8)953-962.

Knuth, P. 1904-1905. Die Bisher in aussereuropaischen Gebieten gemachten blütenbiologischen Beobachtungen. *In* Handbuch der Blütenbiologie. III Engelmann. Leipzig.

Maguire, B. 1969. Velloziaceae: Botany of the Guayana Highland. Mem. Bot. 55:399-405.

Marloth, Rudolf. 1915. The flora of South Africa with synoptical tables of the genera of the higher plants, IV: Monocotyledons. Darter Bros. and Co., Capetown; William Wesley and Son. London.

Pax, F. 1887. Velloziaceae. Pp. 125-127, *in* A. Engler and K. Prantl, (Editors). Die natürlichen Pflanzenfamilien. Teil 2, Abt. 5. Leipzig.

Perrier, H. 1946. On the subject of the systematics of the Velloziaceae and of genus *Xerophyta*. Natulae Systematicae 12(3 and 4):146-148.

————. 1950. Velloziacées. *In* Humbert (Editor). Flore de Madagascar 42:1-15.

Smith, A. C. 1967. The presence of primitive angiosperms in the Amazon Basin and its significance in indicating migrational routes. Atas Simpós. Biota. Amaz. 4:37-59.

————. 1970. The Pacific as a key to flowering plant history. Pp. 1-26, *in* University of Hawaii, H. L. Lyon Arboretum Lecture Number One. U. of Hawaii. Honolulu.

Smith, L. B. 1962a. Origins of the flora of southern Brazil. Contrib. U.S. Nat. Herb. 35:215-249.

————. 1962b. A synopsis of the American Velloziaceae. Contrib. U.S. Nat. Herb. 35:251-292.

Takhtajan, A. 1969. Flowering plants: Origin and dispersal. [Translated from Russian by C. Jeffery.] Smithsonian Institution Press. Washington, D.C.

Thorne, R. F. 1963. Some problems and guiding principles of angiosperm phylogeny. Am. Naturalist. 42(896):287-305.

Wallace, A. R. 1895. Island life. 2nd revised edition. Macmillan & Co. London.

Warming, E. 1893. Note sur la biologie et l'anatomie de la feuille des Vellosiacées. Overs. K. danske Vidensk. Selsk. Forh. pp. 57-100.

Welwitsch, F. 1859. Letters on the vegetation of West Equinoctial Africa. Proc Linn. Soc. (Bot.) 111:150-157.

Variation in the Bottle Gourd

CHARLES B. HEISER, JR.

Introduction

The bottle gourd (*Lagenaria siceraria*) was one of the most widely distributed plants used by man in prehistoric times. It is known archeologically from early levels in both Mexico (7000 B.C.) and Peru (ca. 3000 B.C.) (Whitaker and Cutler 1965). A recent report also places it in Thailand at about 7000 B.C. (Gorman 1969). The earliest known archeological report from Africa, which appears to be the homeland of the gourd, comes from Egypt in the fourth millennium B.C. (Schweinfurth 1884). The gourd may not have been actually cultivated at all of these sites at these early dates, but it was eventually to become an extremely important cultivated plant throughout much of the world. Today its use is dwindling with the coming of tin cans, and glass and plastic containers, but it is still extensively cultivated in many tropical areas for such uses as vessels or musical instruments. As a food plant it is still of some importance in the Far East.

The wide distribution of the bottle gourd in early times is probably easily explained. Whitaker and Carter (1954) have shown that gourds will float for prolonged periods in sea water and that their seeds retain viability. Thus we may assume that a gourd could have floated from one continent to another. Since it is not a strand plant its establishment after reaching a new continent poses some problems, but I have seen herbarium records of apparently volunteer plants growing on coastal sites in the Americas. Whether its original homeland was Africa may be considered unproven, but all wild species of *Lagenaria* are indigenous to Africa, and the bottle gourd itself has been found there apparently in the wild state (see Heiser 1969). However, it may also occur in the wild state in the Americas, and early botanists (e.g., Watt 1908) considered it wild in India, although information that I have from a number of Indian botanists indicates that it is not found there in the wild state today.

In the present paper, I propose to discuss variation in the bottle gourd, although my small sample of material of such a widely distributed plant is hardly adequate. For the past several years I have attempted to secure seeds of the bottle gourd from throughout its range. Over 180 different collections have now been grown. In the present work I have eliminated from consideration those collections from seed companies whose original geographical source could not be ascertained. While I know the geographical source of the other collections, I have no way of knowing how long they have been grown there and some of them, of course, could represent recent introductions. Seeds were started in the greenhouse in early May and transplanted into the experimental field in Bloomington, Indiana, at the end of the month for the past six summers. Three to six plants have been grown for each accession and are the source of the measurements to be referred to later. It would, of course, have been desirable to have grown a larger number of plants but the large size of the individual plants and a limited amount of field space made this impossible.

Sources of plants included in the morphological analysis, with name of collector and my accession number or letter are listed below. Further details concerning the source of the seeds, where known, are included with the herbarium collections deposited at Indiana University.

CHARLES B. HEISER, JR., Department of Botany, Indiana University, Bloomington, Indiana 47401. ACKNOWLEDGMENTS: To the National Science Foundation for research support; to Dr. Thomas W. Whitaker, whose researches on the Cucurbitaceae stimulated my interest in the group; to Jack Humbles, for his help in this study; and to the many people who have supplied me with seeds of *Lagenaria*.

AFRICA.—*Egypt*: Ashyan Yousif Zaki 10. *Ethiopia*: Taddesse Ebba 164, 165. *Ghana*: Jorge Soria 14; H. Z. Chakravarty 102-105; A. A. Enti 139-146; G. K. Noamesi 147. *Nigeria*: J. Soria 15; T. W. Whitaker 167; G. A. Akinotola 168. *South Africa*: Paul Nel 16-18. *Tanzania*: Richard Koila K; Jack Humbles 3, 117, 135; Leonard Msaki 11; S. A. Robertson 29-34; 63-65; J. Godrod 120. *Uganda*: Leonard Msaki 12; Barry Wakeman 13. *Country unknown*: Park's Seed Co. 7.

AMERICA.—*Argentina*: Heiser 35a, 35b. *Brazil*: J. Soria 57, 58; James Coleman 51-55. *Colombia*: Heiser 25. *Costa Rica*: J. Soria 6; Heiser 59. *Ecuador*: Heiser 26, 27. *Mexico*: David Nelson 39; J. Rzedowski 73-79; Antonio Marino Ambrosio 95-98, 155, 156; T. W. Whitaker 152-153; Flo Chapman 162, 163. *Peru*: B. Pickersgill 36-38; Heiser 56; T. W. Whitaker 166. *Trinidad*: E. A. Tai 60, 92; A. A. Hubbard 114. *United States*: Nelson Cross 91; Kathrine Bartlett 93, 44; Frank Chapella 99, 100.

ASIA AND PACIFIC AREA.—*Caroline Islands*: M. M. Sproat 154. *India*: Lucknow Botanical Garden E, F.; Koshy Abraham 21; M. Krishna Rao 22-24, 41-45; United States Department of Agriculture 46-48, 67; H. P. Bezbaruak 61, 62; C. P. Sreemadhaven 100, 101; T. Saxena 159. *Indonesia*: Roekmowati Hartono 20. *Lebanon*: S. Abu-Shakra 158. *Japan*: N. Tahaki 70, 71; S. Akiyama 72; Yukio Huziwara 82-86; Sinske Hattori 112, 113. *Malayasia*: R. B. Lulofs 110, 111; Anne Johnson 161. *Pakistan*: University of Karachi 68, 69. *Philippines*: Harold Conklin 149, 157. *Thailand*: Sakorn Trinandwan 40.

Comparison of Characters

An earlier review of gourds by Kobiakova (1930) provides some basis for comparison. Her study was based on 300 accessions. She does not give the details of her sources, but from her account we learn that she had no material from Indo-China or the Pacific islands and very little material from the Americas. Both India and Africa appear to be fairly well represented, and the greatest number of her collections apparently came from the Near East, an area from which I have but a single collection.

As a result of her study she concluded that two major geographical races could be recognized (Table 1), *L. vulgaris* ssp. *asiatica* and ssp. *afrikana*. The former was divided into three varieties based on fruit shape and the latter into two varieties based on seed characters. The varieties appear to be artificial groupings, but the subspecies seem to be natural. She does not give detailed measurements and little indication of the range of variability within the subspecies.

TABLE 1. Comparison of African and Asian Bottle Gourds (Kobiakova 1930)

	African	*Asian*
Stem	10–15 m in length, short pubescent	3–10 m in length, long pubescent
Leaf	kidney-shaped, entire, dark green, nerves prominent, seldom pubescent	pentagonal, laciniate, grayish green, nerves not prominent, densely pubescent
Flowers	small	large
Fruit	medium-sized, dark green, usually mottled, fruit wall thin	large, light green, usually not mottled, fruit wall thick
Seeds	bitter, rectangular with course fringed pubescence, dark brown	sweet, somewhat triangular with soft, velvety pubescence, light gray-brown, often white
Growth Period	6–8 months	4–6 months

STEMS. Since most of my plants were pruned to keep them within bounds I made no measurements of stem length. However, from plants not pruned it is apparent that the longest stems are found in African plants, although medium lengths also occur in African collections. There is considerable variation in the pubescence of stems, as well as of other parts, but my preliminary analysis indicated no significant geographical correlations.

LEAVES. There may be considerable variation in leaves on a single plant, not only in size but in degree of lobing. However, it is clear that deeply 3-5 laciniate lobed types are characteristic of Asian plants, whereas a more rounded type of lobing or unlobed leaves are mainly found in Africa and the Americas (Figure 1). The margin of the leaf tends to be somewhat serrate in Asian types, and crenate in plants from the other two continents. In spite of difficulty in scoring leaf shape precisely I have used this character in my evaluation (Figure 5). There also is variation in leaf color, the darkest leaves coming from Africa, but I failed to come out with consistent results in repeated runs of scoring color and hence do not include this character. I failed to recognize any great differences in the venation, and, as pointed out above, there is variation in pubescence with no clear cut geographical pattern evident.

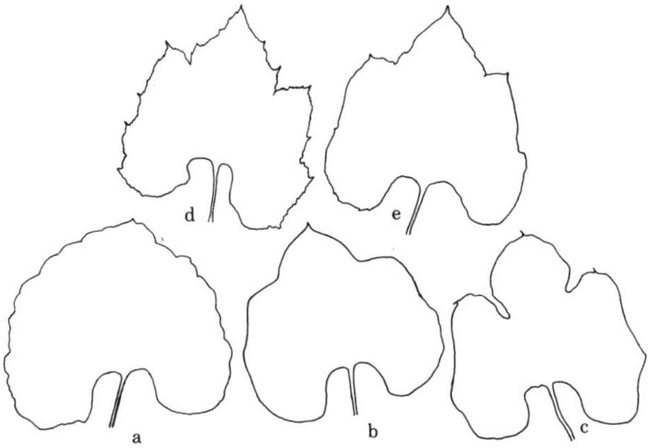

FIGURE 1. Leaf tracings of *Lagenaria siceraria. a*, Colombia (25); *b*, Mexico (95); *c*, Arizona (91); *d*, India (24); *e*, India (F). Highly reduced.

FLOWERS. My observations agree with those of Kobiakova; the largest flowers come from Asia. I have used the length of the male corolla as a measure (Figure 5). There is, as might be expected, a correlation of size characters—those plants with large flowers tending to have large leaves, large seeds, and large fruits, with some exceptions. Since most of my gourds from India are medium to large fruited, the large flowers from this country are not unexpected. However, if plants with fruits of the same size from Africa and the Americas are compared with those from India, the latter will have the larger flowers.

The calyx also offers a character showing geographical correlation. The length of the calyx lobe varies greatly—those from male flowers from Asia ranging from 6 to 20 millimeters, whereas those from Africa and the Americas range from 2 to 10 millimeters. Since the size of the calyx lobe is influenced by the corolla length I have used an index (calyx lobe length divided by corolla length) in order to make meaningful comparisons (Figure 5).

FRUITS. As one might expect, the fruit, which would be subject to strong human selection, shows great variation. My observations on fruit characters differ somewhat from those of Kobiakova. Gourds from India are mostly medium to large, whereas there is a range from small to large in Japan. In both Africa and the Americas there is a considerable range in size with the majority being medium-sized in Africa while slightly smaller ones are common in the Americas. The largest gourd that I have grown (66 inches in circumference) comes from Nigeria (Whitaker 167). I find more or less the same range of colors and patterns in the three major regions, although the lighter colors are more common in Asia than are the darker ones. One distinctive type of color pattern with somewhat wavy longitudinal lines I have seen only from Peru (Pickersgill 36). I have one fruit from Africa (Nel 16) with prominent "worts" and I have seen gourds from Mexico with this character but none from Asia. Ribbed fruits and the "Maranka" or ridged types are apparently indigenous to Africa, although the latter is now commonly offered by North American seed companies.

Kobiakova recognizes eight different classes of fruit shape. All types are found in Asia, with club types dominating in India and pyriform forms in Japan. In Africa all types except the snake are known, with bottle types being the most common. Neither the snake nor cylindrical types are represented in my samples from America, and bottle and pyriform shapes are by far the most common. I have not made a detailed study of the fruit wall or of the consistency of the flesh, both of which show considerable variation. I have observed, however, that both relatively thick- and thin-walled types occur on all three continents. For additional discussion of fruits the reader is referred to Whitaker (1948).

SEEDS. The seeds (Figures 2, 3, 4) are as variable, or even more variable, than are the fruits. This may seem somewhat surprising in view of the fact that, at least at present, the seeds are little used by man. Perhaps in earlier times they had some significance as food and there was selection for larger seeds. Such selection, however, would hardly account for the great variation in shape, coloration, and design. One wonders if there could have been a deliberate selection by man for amusement or for some ritualistic use.

Only one seed that I tested (Soria 6, Costa Rica) can be classed as bitter, and I believe that the bitter taste sometimes assigned to the seed comes fro portions of the fruit pulp adhering to the seed. In the samples grown I found 4 bitter fruits and 35 sweet from Asia, 14 bitter and 20 sweet from Africa, and 15 bitter and 25 sweet from the Americas. Seven collections gave progeny with both bitter and sweet fruits. Two studies have been made on the inheritance of bitter flesh (Pathak and Singh 1950, Joubert, cited by Watt and Breyer-Brandivijk 1962) and both found bitterness to be dominant and controlled by a single gene. However, in seven different crosses that I have made in which the parents were both sweet I secured bitter F_1's. This result would suggest a type of inheritance different from that previously reported. The fact that there are degrees of bitterness suggests that more than one gene is involved. One would expect human selection for nonbitter types, not only if the gourds were to be used for food but also if they were to be used for containers for water, beer, and the like, since the bitter fruits need lengthy treatment to remove the bitter principle from the shell.

Kobiakova's generalization about the seed shape seems to hold. I have found it more convenient to record actual measurements, and in general, Asiatic gourds have seeds more than twice as long as broad whereas the great majority of seeds from Africa and the Americas are less than twice as long as broad. This can be expressed in the form of an index (width divided by length; Figure 5). Length of seed ranges from 8.3 millimeters (Ambrosio 98, Mexico) to 30.5 millimeters (Chakravarty 102, Ghana).

The seeds from Asia as a rule have two projections or ears at the top, and these may be present or absent in seeds from the other continents (Figure 5). One African form (Whitaker 167) that lacked ears entirely gave some segregates which were prominently eared.

The surface of the seed may have prominent longitudinal lines or it may be completely smooth, as in some African and American collections (Figure 5). The lines at times, chiefly in Asian gourds, may be finely pubescent. Seeds from Africa, on the other hand, often show a prominent corky wing-like projection on the sides which I take to be the "fringed pubescence" of Kobiakova (Figure 5).

Seed color in my samples agrees with Kobiakova's observations. American accessions are similar to the African ones, and in addition to brown colors, black, olive, and rust colors have been noted from the Americas.

Seeds from the African samples appear to be most variable in shape and size, whereas in color the American ones show greater variability. The seeds, as can be seen, offer a number of good characters for analysis. Since seeds of gourds are sometimes preserved in archeological deposits it would be of interest to have comparisons of them made with modern types.

GROWTH PERIOD. Blooming dates have been recorded for all plants grown. Varieties from India and Japan are the earliest, whereas some from Africa and the Americas are the latest. However, there is considerable variability and further generalizations are unwarranted.

Geographical Variation

For presentation of my results I have chosen to use polygonal graphs (Figure 5) in which the characters are represented on radii. Each line represents the average values of plants grown from one accession. While it may not be possible to follow all individual lines on the graphs, some idea as to the central tendency and variability of the characters can be readily grasped. The complete set of data is on deposit in the herbarium of Indiana University.

Certain patterns of variability in connection with geography have already been noted. Some additional observations may be made.

The general similarity of gourds from mainland Asia (Figure 5A), Japan (Figure 5B), and Southeast Asia including the Pacific islands (Figure 5C) can be observed. It can also be seen that those from Japan show some with small flowers, less deeply lobed leaves, and seeds with ears lacking. Gourds are known to have been introduced into Japan from other regions in recent times and this perhaps may account for these differences. Very few gourds were available for study from the Pacific islands, and obviously the gourds of this area require study.

The three graphs (Figure 5D, E, F) from Africa, while indicating considerable variability, also make it apparent that the gourds of this area are for the most part quite distinct from the Asian gourds. The number of samples available is too small to make any definite statements regarding regional variation in Africa, although it may be noted that there tend to be some differences in the leaves and seeds of the samples from Tanzania and Ghana.

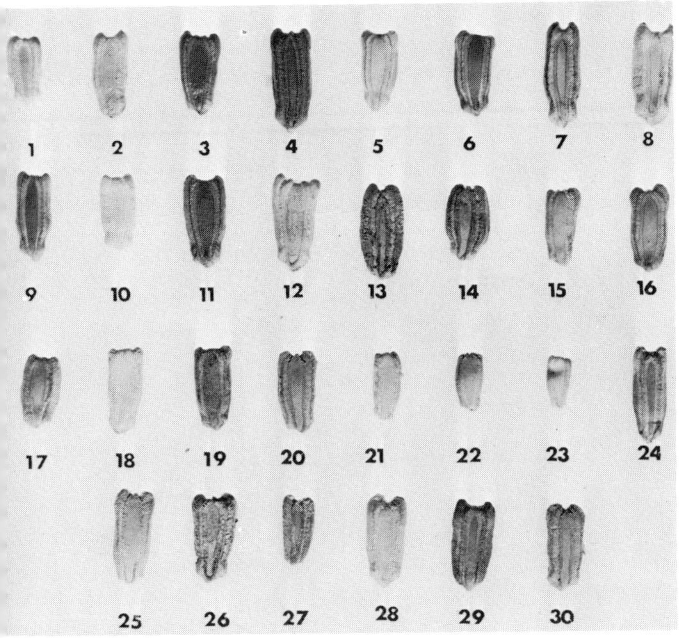

FIGURE 2.

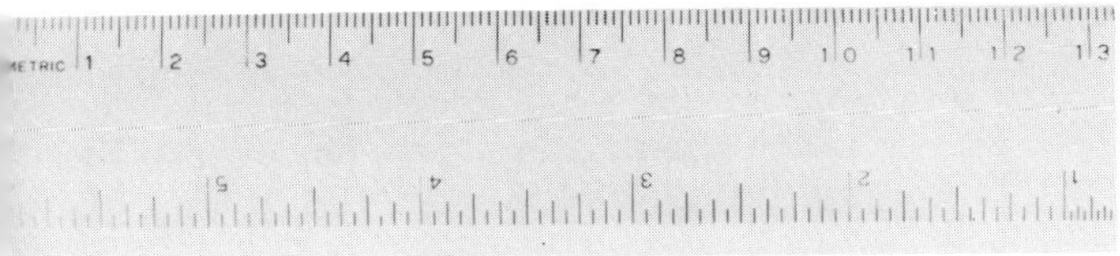

FIGURE 3.

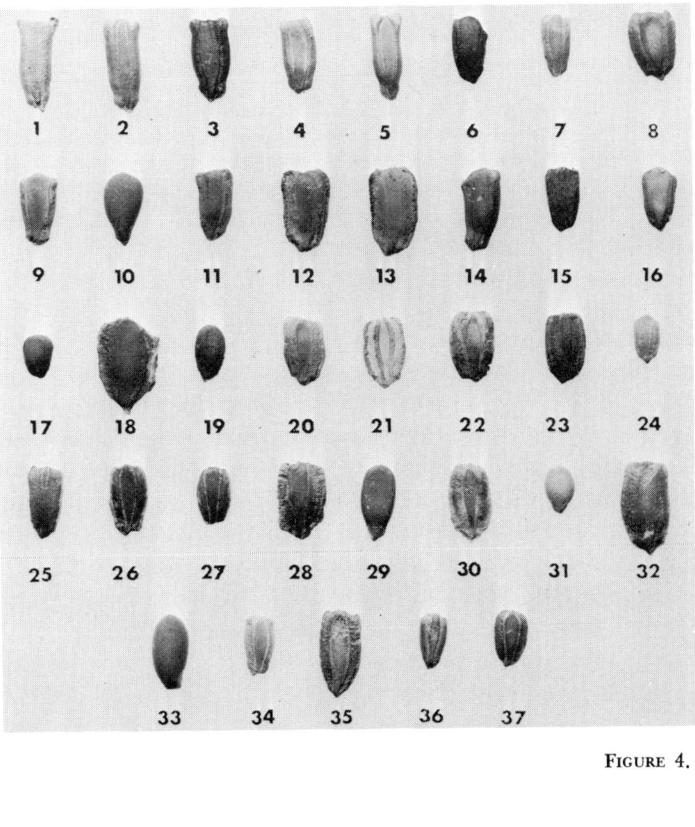

FIGURE 4.

FIGURE 2. Seeds of *Lagenaria siceraria* from Asia. Number 1-16, India; 17, Thailand; 18-19, Pakistan; 20, Malaysia; 21-25, Japan; 26-27, Philippines; 28, Caroline Islands; 29, Hawaii; 30, Malaysia.

FIGURE 3. Seeds of *Lagenaria siceraria* from Africa. Number 1-15, Tanzania; 16, Egypt; 17-29, Ghana; 30-31, Uganda; 32-36, South Africa; 37, Nigeria.

FIGURE 4. Seeds of *Lagenaria siceraria* from the Americas. Number 1-5, United States (Arizona); 6-17, Mexico; 18-19, Costa Rica; 20-22, Trinidad; 23, Colombia; 24-25, Ecuador; 26-29, Peru; 30-36, Brazil; 37, Argentina.

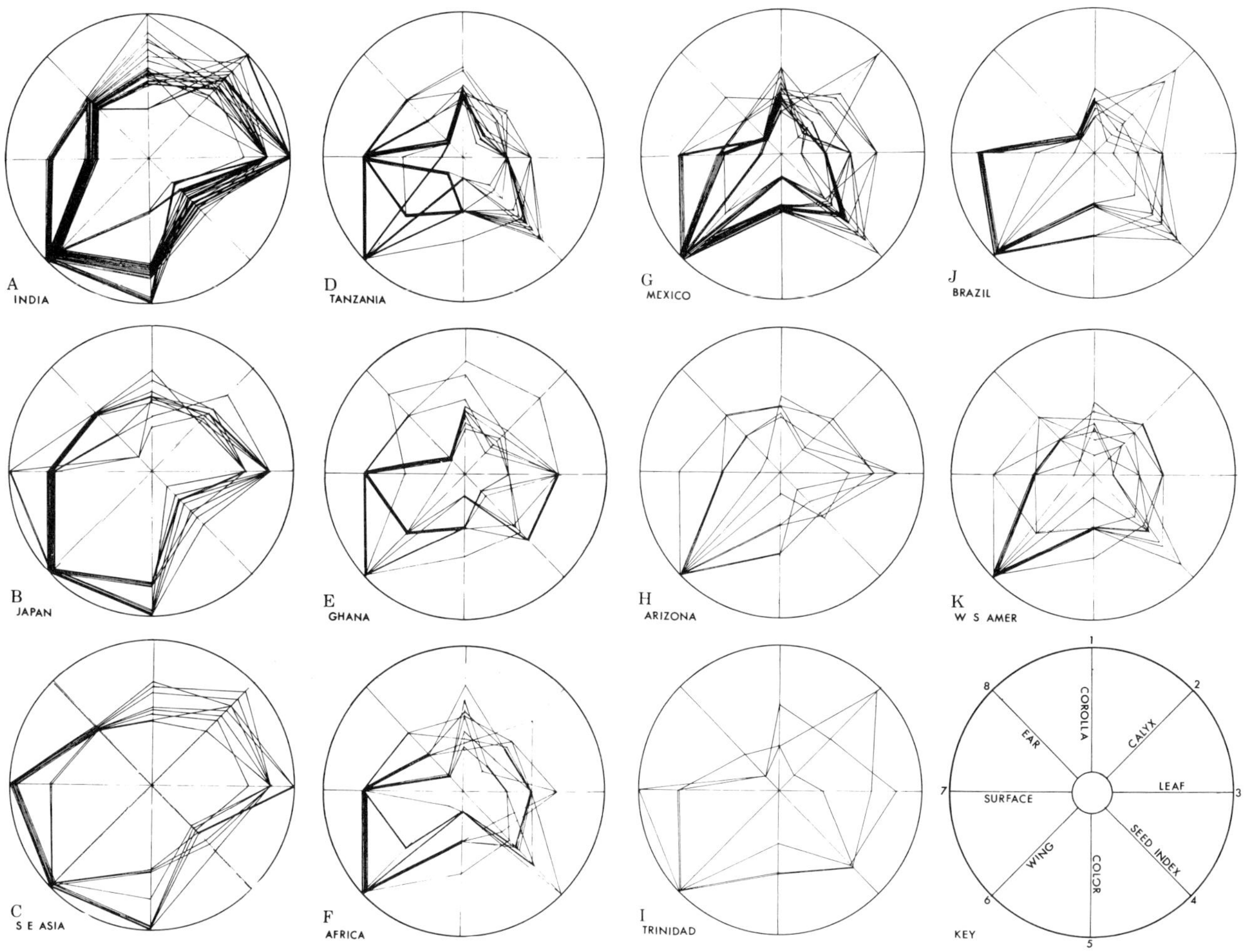

FIGURE 5. Geographical variation in *Lagenaria siceraria*. The various characters shown on the radii of the key and the range of measurements or scores, from the center to the outside, are as follows: 1. Male corolla length, from 18 to 55 mm; 2. Calyx lobe index, from .07 to .42; 3. Leaf margin, strongly crenate to deeply laciniate-lobed with four intermediate conditions; 4. Seed index, from .35 to .75; 5. Seed color, from dark to light, with 3 intermediate conditions; 6. Seed wings, from well developed, to slightly developed, to lacking; 7. Seed surface, from smooth, to lines poorly developed, to lines well developed, to lines well developed and pubescent; 8. Ears, from lacking, to inconspicuous, to well developed.

Sources and number of accessions included: A, India and mainland Asia, 22. B, Japan, 10. C, Southeast Asia and Pacific islands, 8. D, Tanzania, 14. E, Ghana, 9. F, Other African, 10. G, Mexico and Central America, 16. H, Arizona, United States, 5. I, Trinidad, 3. J, Eastern South America (mostly Brazil), 8. K, Western South America, 8.

The samples from the Americas (Figure 5G, J, K) excepting those from Arizona (Figure 5H) and Trinidad (Figure 5I), although showing considerable variability, do not vary significantly from each other. It is also apparent that the American samples are on the whole rather similar to some of the African types and quite different from most Asiatic gourds.

The great variability in the small sample from Trinidad (Figure 5I) and the similarity of some characters to the Asian gourds probably has a very simple explanation. Trinidad has a large human population derived from India, and it perhaps is not unreasonable to assume that some of the people brought gourd seeds with them. The gourds of Trinidad, therefore, could well include forms derived from hybridization between American and Asiatic gourds.

No such ready explanation is available, however, for the variation in the Arizona gourds (Figure 5H). Some of these gourds display an unusual type of leaf lobing (Figure 1, not indicated in the polygonal graphs) and all of them have relatively to extremely narrow seeds. It is possible that the gourds from this area represent a regional differentiation that occurred in early times after an original introduction from Mexico or elsewhere. It is, of course, also possible that they are contaminated by recent hybridization. Seeds purchased from seed stores are now grown by the Indians of the southwestern United States. Unfortunately, Indian varieties of gourds were not available from other parts of the United States. Thirty years ago these were not uncommon in the southeastern United States (Speck 1941).

Discussion

The early taxonomic work on *Lagenaria* was summarized by Kobiakova (1930). She used the name *L. vulgaris* Seringe for the bottle gourd. The name *L. leucantha* (Duchesne) Rusby has also received considerable use, but it is now clear that this species must bear the name *L. siceraria* (Molina) Standley. Since the appearance of Kobiakova's work two additional names have been added. Millan (1946) created a variety (*L. siceraria* var. *laevisperma*) for the smooth seeded gourd, and Chakravarty (1968) has recognized a new species (*L. bicornuta*). I have grown material secured from Chakravarty (102, Ghana) and except for the very long, prominently eared seeds it falls well within the range of other material from Africa. I have made crosses of this gourd with three other varieties and the F_1's were all fertile. Therefore, neither on morphological nor genetic grounds is there any justification of recognizing this material as a distinct species. Whitaker (1970) has independently reached the same conclusion. The prevailing opinion in this century has been that there is but a single species of *Lagenaria* in cultivation. I feel that this view is supported by the present morphological study, and also by the results from artificial hybridizations. A total of 53 crosses have been made, using morphological extremes and representatives from all three continents. The F_1's were all fertile, and the F_2's grown of seven of the combinations showed no significant breakdown in fertility or vigor.

It seems desirable to continue to recognize subspecies.

Lagenaria siceraria ssp. *siceraria*. Leaves entire, crenate or rounded-lobed; flowers generally small to medium sized, calyx lobes usually short and broad; seeds generally less than twice as long as broad, with or without corky wings; longitudinally lined or smooth, lines seldom pubescent, with or without ears, generally dark in color. Africa and the Americas. (*L. vulgaris* ssp. *afrikana* Kobiakova).

L. siceraria ssp. *asiatica* (Kobiakova) Heiser, comb. nov. Leaves rounded-lobed or laciniate-lobed; flowers usually large, calyx lobes long and narrow; seeds generally more than twice as long as broad, without corky wings, longitudinally lined, lines sometimes pubescent, usually eared, gray, gray-brown, or white in color. Asia and the Pacific region. (*L. vulgaris* ssp. *asiatica* Kobiakova).

At the present time at least, I see nothing to be accomplished by setting up a formal series of varieties for different morphological types within the subspecies.

A study of morphological variation cannot give answers as to how and where morphological differentiation occurred. The difference between the Asian gourds and those from the other continents might suggest to some that the separation of the two types was of ancient origin. However, degree of morphological divergence is not simply a measure of time, for under some circumstances morphological differentiation can proceed very rapidly following geographical isolation, whereas other examples are known in which there has been little or no morphological differentiation after prolonged isolation (Stebbins 1950). In spite of this, however, I feel that this morphological study does provide some basis for discussion, although much of it must be highly speculative. In the following discussion, I shall make two assumptions: (1) *Lagenaria siceraria* was originally indigenous to Africa, and (2) my samples reflect rather accurately the nature of the gourds of the various geographical regions at a fairly ancient date and have not been strongly modified by recent introductions that may have occurred in the last few hundred years.

Bottle gourds are not reported from archeological sites in the Near East (those gourds mentioned in the Bible are not *Lagenaria*). In view of the extensive archeological record of plants now known from the Near East, one might suppose that gourds were not present in this region in prehistoric times.

Thus one could postulate that gourds somehow reached India or Southeast Asia directly from Africa at an early date. If the identification and dating of the archeological material from Thailand are correct, we would have to assume transport by natural means of a wild gourd, since agriculture was as yet unknown in Africa. The exchange of cultivated plants between Africa and India is not known to have occurred prior to the Christian era (cf., Murdock 1959), although Burkill (1951) has postulated a movement of plants, possibly including the gourd, across the Sabaean lane before 2000 B.C. However, the clustering of the Asiatic gourds around certain morphological means suggests to me that there may have been but a single introduction and the current variability stems from mutation and selection of a single original type.

Kobiakova (1930) points to Afghanistan and Persia as being a center of variability of fruit forms. Whether or not it is also a center of variability for other characters is not known. The variability in the area could result from a fairly recent introduction of gourds both from India and Africa and would not necessarily reflect an ancient pattern.

The American gourds, as Kobiakova supposed, appear to be derived from Africa. Since the gourds were known in very early times in the Americas, we again may assume the original entry of a wild type. We also must account for a fairly extensive distribution in the Americas to explain their cultivation in both Peru and Mexico at early dates. How much of this distribution was natural is very difficult to say, for we know almost nothing about the dispersal and establishment of gourds in nature. Certainly some of its dispersal could have been by man, even previous to its cultivation, and we might think of it being carried from Mexico to Peru by this means, although it is easier to think of gourds arriving in Brazil from Africa and moving from there to other areas. There is, of course, no reason to assume that gourds did not arrive by drift from Africa on more than one occasion and perhaps in Mexico as well as Brazil. In fact, some of the parallel variation in African and American gourds might be most readily explained on this basis.

Overall, the greatest variation of gourds appears to be in Africa. This probably does not reflect a greater age in that continent but simply that there has been selection for adaptation to a greater diversity of habitats. Another possible source of variability needs to be considered, however. Although hybrids which both Dr. Whitaker and I produced between *L. siceraria* and *L. sphaerica* were sterile, hybrids showing some fertility have been reported (Cogniaux and Harms 1924). The possibility that natural hybridization occurs and is a source of variability in Africa has yet to be investigated.

References

Burkill, I. H. 1951. Habits of man and the origins of the cultivated plants of the Old World. Proc. Linn. Soc. London 164:12-42.

Chakravarty, H. L. 1968. A new species of African *Lagenaria* (Cucurbitaceae). Ann. Mo. Bot. Gard. 55:69-72.

Cogniaux, A. and H. Harms. 1924. Cucurbitaceae–Cucurbiteae–Cucumerinae. Das Pflanzenreich. 88. Heft. (IV. 275. II).

Gorman, Chester F. 1969. Hoabinhian: A pebble-tool complex with early plant associations in Southeast Asia. Science 163:671-673.

Heiser, C. B. 1969. Systematics and the origin of cultivated plants. Taxon 18:36-45.

Kobiakova, J. A. 1930. The bottle gourd. Bull. Appl. Bot., Gen., and Pl. Breed. 23:475-520.

Millan, Roberto. 1946. Nuevo mate del Uruguay (*Lagenaria siceraria* var. *laevisperma*). Darwiniana 7:194-197.

Murdock, G. P. 1959. Africa, its peoples and their culture history. McGraw-Hill. New York.

Pathak, G. N., and B. Singh. 1950. Genetical studies in *Lagenaria leucantha* (Duchs.) Rusby: (*L. vulgaris* Ser.). Indian J. Gen. and Pl. Breed. 10:28-35.

Schweinfurth, G. 1884. Neue Funde aus dem Gebiete der Flora des alten Ägyptens. Botanische Jahrbucher 5:189-201.

Speck, Frank G. 1941. Gourds of the southeastern Indians. New England Gourd Society. Boston.

Stebbins, G. L., Jr. 1950. Variation and evolution in plants. Columbia University Press. New York.

Watt, George. 1908. The commercial products of India. Murrey. London.

Watt, John M., and Maria G. Breyer-Brandwijk. 1962. The medicinal and poisonous plants of southern and eastern Africa. 2nd edition. Livingstone. Edinburgh.

Whitaker, T. W. 1948. *Lagenaria*: A pre-Columbian cultivated plant in the Americas. Southwest. J. Anthropol. 4:49-68.

———. 1970. *Lagenaria* in Ghana. Gourd Seed 31:28.

Whitaker, T. W., and George Carter. 1954. Oceanic drift of gourds: Experimental observations. Amer. J. Bot. 41:697-700.

Whitaker, T. W., and H. C. Cutler. 1965. Cucurbits and cultures in the Americas. Econ. Bot. 19:344-349.

Growth Habits of Tropical Trees: Some Guiding Principles

P. B. TOMLINSON AND A. M. GILL

Introduction

Our understanding of the growth of woody plants is currently based on studies of north temperate trees, particularly those of western Europe and eastern North America. A classical textbook, such as that of Büsgen and Münch (1929), therefore, deals largely with temperate species and gives no detailed information about tropical trees. The article by Kozlowski (1964) provides a similar but more recent example of the distribution of our knowledge. Yet it is obvious that the richest tree floras, in numbers of species, are tropical: the lowland tropical forests of Africa, Indo-Malaya, and South America are good examples of this tree-species diversity.

To what extent, therefore, is our understanding of the growth of woody plants unbalanced, due to our preoccupation with only a small fraction of the total tree flora of the world? It is our object to discuss this and, with reference to particular examples, suggest that temperate woody plants are not representative of trees as a whole and that we should be cautious in making generalized statements about tree growth on the basis of present knowledge.

In overall construction, of course, trees as represented by most dicotyledons and conifers are alike, since they have the same basic method of radial growth by development of secondary vascular tissues from a cambium, although the periodicity of this growth is very variable and still needs critical study. On the other hand, methods of extension growth are very diverse and discussion will refer chiefly to these. We have deliberately excluded many other kinds of tree (mostly tropical) such as the several types of monocotyledonous tree, tree-ferns, cycads, succulent Euphorbiaceae, Cactus-trees, etc., although these should all be considered in detailed descriptions of the tree habit. There are many well-known and obvious morphological peculiarities of tropical trees such as cauliflory, development of aerial roots, extensive buttressing, which indicate that tree growth in the tropics is at least more diverse than that in the temperate regions. In addition there are many tropical vegetation types, dominated by trees, for which there is no temperate equivalent. One of the most striking of these tropical types is the mangrove community. Growth responses and adaptations on the part of trees to this environment have no obvious counterpart in temperate regions so it is not surprising to find unique features, of which the familiar marked tendency towards vivipary in mangrove species and their pronounced peculiarities of root growth are both good examples. Since we ourselves have studied certain mangrove species in detail, we will use them as examples in subsequent description. From this we can conclude that there are aspects of tree physiology which are simply not available for study by botanists based in temperate latitudes. The reverse, of course, may be equally true; cold response of temperate trees may have no counterpart in tropical trees. It is, therefore, reasonable to expect fun-

P. B. TOMLINSON and A. M. GILL, with Fairchild Tropical Garden, Miami, Florida. Present addresses: P. B. Tomlinson, Harvard Forest, Petersham, Massachusetts 01366. A. M. Gill, Division of Plant Industry, C.S.I.R.O., Canberra, Australia 2601.

damental differences between tropical and temperate trees in growth expression.

We suggest, therefore, that quite elementary and generalized concepts of tree growth may be currently misinterpreted. Some factual basis for our statement comes from a study of methods of extension growth of woody plants of the tropics based on measurement, but most derives from the examination of shoot morphology.

We emphasize again that existing generalized concepts of tree growth refer almost entirely to *north* temperate species. Apart from observations made in South Florida, our casual observations in Australia (A.M.G.) and New Zealand (P.B.T.) suggest that the temperate woody flora of the southern hemisphere is as much "tropical" as "temperate" in growth response. This perhaps reflects the absence from the southern hemisphere of large land masses at high latitudes, together with the different past geological history and present climate of the southern hemisphere, and these consequent effects on the evolution and distribution of plants. We do not ascribe entirely different growth processes to tropical woody plants, or discuss tropical independently of temperate trees, because this would merely invert the procedure we have lamented. Rather, it is our contention that the subject of tree growth and the influence of climate (whether seasonal or nonseasonal) on tree growth should be considered as a single cosmopolitan topic. A beginning could be made by regarding any feature of growth of a woody plant in its natural environment as normal, rather than, as is the current tendency, to look at aspects of shoot morphology in tropical trees as "unusual" or even "abnormal" because they do not occur in conventionally "familiar" temperate trees. An example is provided by the tendency to distinguish second flushes of growth during the growing season at temperate latitudes by special terms (e.g., "lammas shoots") because it is regarded as normal for trees to exhibit one flush of growth per season. Lammas shoots are, in fact, the normal expression of a widespread tendency on the part of trees to flush at intervals, often independently of climatic influence. It is only in north temperate trees that this feature has become restricted to an annual event by the rigors of a climate with a short growing season. Another unfortunate generalization is to assume that morphological features which have an apparent adaptive function in an unfavorable environment must be lacking in an environment which is seemingly continuously favorable to plant growth. It has been suggested, for example, that bud-scales are rare in tropical trees (Borchert 1969) because their presumed "protective" function is not necessary. In fact, there has never been any careful survey of the incidence of bud-scales in the trees of any tropical flora. Casual observation suggests that they are common. In addition, this suggestion overlooks the fact that some temperate trees have "naked" buds. The mechanism whereby leaf primordia mature as scales rather than foliage leaves is a morphogenetic problem as much worthy of study in tropical as in temperate floras.

Modern work has demonstrated the marked growth response (bud break, leaf abscission, shoot growth) of north temperate trees to photoperiod. Can we assume that within the tropics, where seasonal changes in day length are smaller, that trees are insensitive to this effect? It may be that the reverse is true because smaller effects would be needed to induce the same response. A change in day-length of as little as 15 minutes has been shown to have a morphological effect in the herbaceous *Corchorus* (Njoku 1958). This suggests that there is unlimited scope for studies of photoperiodic responses in tropical trees. There is similar scope for studies of cold response (e.g., in relation to flowering) and the effect of drought on flowering, leaf fall, and radial growth.

When one thinks of the vast number of man-hours which have been spent on research on temperate trees, the mammoth task involved in providing a comparable body of information for the much larger woody floras of the tropics is self-evident. We are being realistic and not modest when we say that our approach at this stage can only be very superficial and all we hope to do is to provide some guiding principles using a few illustrative examples.

HISTORICAL ASPECTS

The literature on the growth of tropical trees is considerable, although it is small in relation to the scope of the subject. Some general principles have been established. Taxonomic literature by itself is not helpful, since it depends to a large extent on herbarium specimens alone, wherein aspects of growth cannot be interpreted easily. Some of the morphological principles we outline may be helpful, however, in understanding the previous growth pattern of an existing shoot as seen on a herbarium sheet.

Studies specific to tropical tree growth in older literature largely refer to comprehensive investigations of phenology (see Büsgen and Münch 1929, Klebs 1915, Volkens 1912). This is a necessary preliminary. There has been little attempt, however, to study the detailed relation between the growth of a shoot and its morphology. One considerable exception to this generalization is the relatively intensive study of tropical tree crops, notably citrus, coffee, cocoa, rubber, and tea, which has been undertaken. In tea, for example a knowledge of flushing behavior is of obvious economic importance since the crop is a portion of each flush. In subsequent discussion special reference will be made to some of this work.

In addition there are several detailed reports on trees of specific areas. Of particular value are those carried out in relatively nonseasonal tropical climates by Coster (1927-1928), Holttum (1953), and Koriba (1958). These have shown that tree growth in the tropics can be (apparently) continuous or obviously intermittent. In those trees with intermittent growth, the periodicity may or may not be related to slight seasonal variation in climate. Where tree growth is periodic but apparently independent of climatic influence, it has been concluded that trees exhibit endogenous rhythms in their growth (Koriba 1958). We might ask, on the basis of this conclusion, to what extent are periodic features of tree growth in temperate regions due to endogenous rhythms?

The historical phase of generalized observation is now passing over to one of detailed analysis of individual species. over lengthy periods of time. It is only by accumulation of well-documented information of this kind that meaningful concepts can be established. Examples of this approach are found in the work of Borchert (1969, on *Oreopanax*); Gill and Tomlinson (1971, on red mangrove); Greathouse and Laetsch (1969) and Greathouse et al. (1971, both on cocoa); Hallé (1966, on Gardenieae-Rubiaceae), Hallé and Martin (1968, on rubber), Holdsworth (1963, on mango), Taylor (1970, on mango). Hallé and Martin's work on rubber, because of its total union of anatomy, morphology and physiology, is a model for the type of work which is needed. In all these studies a careful analysis of one or more aspects of growth is related to shoot morphology together with attempts to correlate this with climatic events. An earlier and very important example of this kind of analysis is that of Bond (1942, 1945) on tea. These studies also differ from much of the early, more generalized, classical work in that they have been done on populations of established trees, often in natural environments. This is tedious work, as can be vouched for by anyone who has attempted it, but it is crucial and a very necessary corrective to more generalized studies. For instance, Koriba presents a wealth of information (for over 500 species) about trees in Singapore, but some of his observations may not be very meaningful because it is evident that his conclusions largely rest only on a morphological examination of shoots from which the method of growth is inferred, but not measured. Many of his "evergrowing" trees may in fact exhibit periodic growth which induces no morphological change he could have recognized. A more critical view of the term "evergrowing" is needed.

Another possible source of confusion is the failure to distinguish between kinds of shoots on one individual and the behavior of trees at different stages in their life cycle. This point is emphasized by Hallé and Oldeman (1970) since the growth of the sapling profoundly influences overall tree shape. Phase changes with age are a familiar feature of trees (e.g., Robinson and Wareing 1969) and this must be recognized in comparative study. Erect and horizontal shoots often differ in their morphology and this usually reflects marked differences in methods of growth (see later and e.g., Greathouse and Laetsch 1969, Hallé 1966).

ARCHITECTURE

The previous general introduction had been prepared prior to the appearance of the monograph on the architecture and dynamics of growth of tropical trees by Hallé and Oldeman (1970). This book deserves special mention. It puts forward interesting new interpretations of the growth of tropical trees supported by a wealth of new information. Undoubtedly this publication will influence considerably the whole subject of growth of tropical woody plants and we ourselves will have occasion to refer to it repeatedly in later sections of this paper. Our present survey does not conflict in any way with these European authors' very extensive one, as we do emphasize certain features of growth not considered by them.

Hallé and Oldeman use the term "architecture" to refer to the construction of tropical trees and this allows them to distinguish certain "types" referred to as "modèles," each of which is named after a botanist who has contributed to the knowledge of tropical trees. The term architecture would appear to be simply a synonym for "growth-habit" and suggests rather, at least to us, a more theoretical concept aiming at the principle of construction of a tree (determined by the genotype) as distinct from growth habit (the phenotypical expression of this principle). Perhaps this distinction is neither intended nor necessary, and need not be a point for dispute.

Methods of Extension Growth

In the following description we will restrict ourselves to vegetative growth and make no detailed reference to branching. The reason for this approach will be made clear later.

ARTICULATE GROWTH

The great majority of north temperate trees are of this kind. Shoot extension proceeds by "annual" increments (the actual period of extension growth may be very brief) with a distinct morphological discontinuity or articulation between each increment which is usually due to the scars of bud-scales. In vegetative growth all buds on current increments are inhibited by strong apical dominance but they may develop in the second year as branches which more or less repeat the pattern of growth of the parent shoot (Figure 1). Examples of this kind are the ashes, oaks, beeches, maples, of northern forests. Textbooks largely refer to this type with descriptions of interesting variations such as the difference between monopodial trees (with terminal buds) and sympodial trees (either without terminal buds or with terminal inflorescences), or the degree of specialization of some of the lateral branches as short-shoots. Of interest in this kind of tree is the extent to which the elements of a flush of growth are performed so that one can distinguish "determinate" shoots, in which all organs are preformed, from "indeterminate" shoots, in which new organs are initiated and expanded during the period of extension growth (e.g., Critchfield 1960, 1971). This topic is presented later in a comparative way.

The frequency of this articulate type of growth and construction in vegetative shoots of temperate woody plants is suggested by the observations of

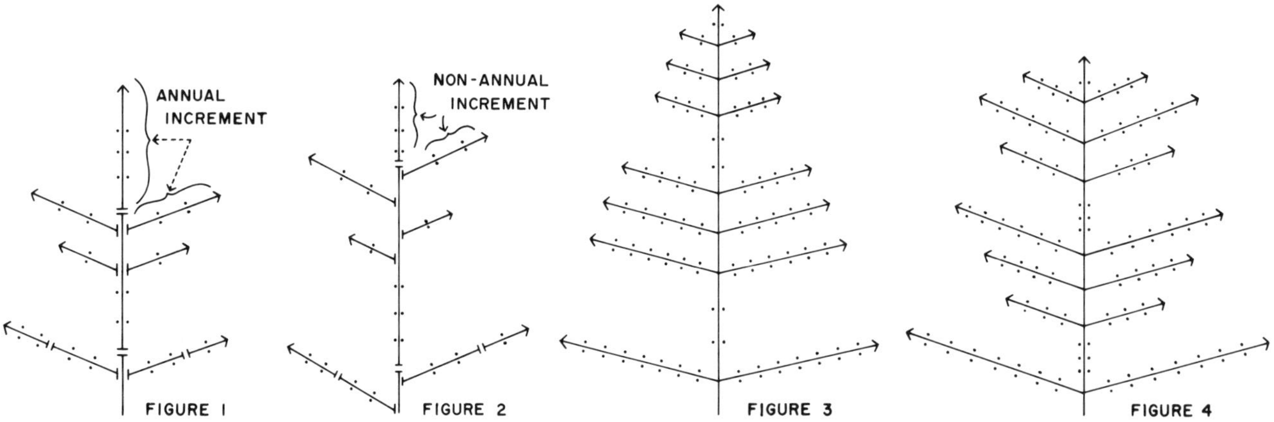

FIGURE 1. Articulate temperate type of woody plant shoot, represented by *Fraxinus americana*, without branching in the current increment. Increments are one per year. (Paired dots = nodes without branches; paired bars = positions of a previous dormant bud, i.e., regions of bud-scale scars in actual specimens.) FIGURE 2. Articulate tropical type of woody plant shoot, represented by *Swietenia* ssp., differing from Figure 1 in that increments may be more than one per year. (Paired dots = nodes without branches; paired bars = positions of a previous dormant bud, i.e., regions of bud-scale scars in actual specimens.) FIGURE 3. Nonarticulate tropical type of woody plant shoot, represented by *Rhizophora mangle*, with periodic "precocious" branch production. (Paired dots = nodes without branches.) FIGURE 4. Nonarticulate tropical type of woody plant shoot, represented by *Avicennia germinans*, like Figure 3, but branch production is more clearly correlated with fluctuations in rate of growth of leader. (Paired dots = nodes without branches.)

one of us (A.M.G.) in an area in central Massachusetts where, with two possible exceptions, all of the 120 species present belong to this general category. The type is not uncommon in tropical trees (e.g., Figure 2: *Swietenia*, and the many examples cited by Hallé and Oldeman under their "modèle de Rauh"). In a tropical species there are often several growth increments of the same kind per year so that the interval between growth flushes is much shorter than in temperate trees where there is usually only one flush per year (cf., Figures 1, 2). Repeated growth flushes during the summer months are common in trees of subtropical and warm temperate regions. Some kinds of lammas shoot are simply the infrequent expression of this type of regrowth in trees which normally have one flush per year. In this type of growth one may distinguish different kinds of flush, as when one is associated with flowering and others are wholly vegetative.

In considering the part of the cycle of shoot growth in trees during which time there is no visible extension growth, one has to distinguish carefully between different kinds of dormancy (Romberger 1963:75). In temperate trees there may be "rest" from which a shoot can only be released by some external trigger (photoperiod or chilling). In contrast, the inactivity may only be apparent, as when a bud produces new initials without undergoing extension growth. In this sense some tropical trees may never be inactive even though they have articulate growth, since flushing growth follows leaf initiation without rest. In tea, for example, leaf initiation may proceed continuously (no studies of mitotic rate have been carried out), but flushing, i.e., extension growth, is periodic (Bond 1945). Consequently the number of developing primordia in the bud fluctuates between a maximum at the end of a period of dormancy and a minimum at the conclusion of a period of extension. In the sapling rubber tree, on the other hand, there is a period between each flush of extension growth without mitotic activity in the shoot apex (Hallé and Martin 1968). The important conclusion of these and other authors is that periodicity of growth is determined endogenously, i.e., independently of obvious climatic events.

NONARTICULATE GROWTH

Red mangrove (*Rhizophora mangle*) serves as a contrasted example (Figure 3). Here there are no distinct articulations separated by bud-scales and in this sense the tree may be an example of the "evergrowing" category of Koriba (1958). The tree, however, does show periodic branch production, with the development of "precocious" branches (Figure 11). These appear at up to three successive nodes, followed by one or more successive branch-free nodes (Gill and Tomlinson 1969). This periodicity is apparently not annual and can occur in a nonseasonal climate, since it is found in *Rhizophora* species even in Singapore. In this tree, therefore, the unit of growth should be regarded as the internode itself rather than an "annual" shoot increment with many internodes as in temperate trees. Measurements show that the rate of expansion of these units is not constant, and that there is a correlation between rate of expansion and branch production; in simple terms more vigorous shoots have larger numbers of branches. Red mangrove in fact shows a complex interrelationship of growth processes with no known equivalent in north temperate trees (Gill and Tomlinson 1971). *Avicennia* and *Laguncularia* are also of this simple type, but the shoot may approach the articulate type (in South Florida) in that one short internode may mark the position of the shoot apex during winter, which is a period of slower growth. During this season, branch production is at a minimum and this contrasts with a maximum of branch production in summer when shoots are most active (Figure 4). If shoots are able to maintain high vigor throughout the year then branches may form at every node. This has been observed on vigorous stump sprouts. Data in Table 1 illustrate the relationship between plastochron and branch expression in *Avicennia* saplings in Miami, Florida.

The most extreme examples of this "precocious" branching contrast markedly with "temperate" types. In many tropical tree saplings (e.g., *Coffea*, *Drypetes*, *Trema*, many Dipterocarpaceae) apical dominance is apparently completely lacking so that each node on the erect shoot produces a precocious branch (Figures 5A, B). This corresponds to Hallé and Oldeman's type distinguished by them as the "modèle de Roux." Morphologically this class would suggest itself as truly evergrowing in Koriba's sense, but we have no growth measurements to substantiate such a belief. On the other hand, we do know that growth of a seedling can proceed periodically without effect on branch expression, because in sapling rubber trees (Figure 6) the shoot apex goes through cycles of activity and inactivity, leading to alternate zones of foliage and scale leaves, which include internodes of varying length, all carried through as an axis that retains complete apical dominance at all times (Hallé and Martin 1968). Distally, of course, the tree eventually branches essentially in the manner of Figure 2 (Hallé and Oldeman 1970).

This apparent failure of the "normal" mechanism for apical dominance will be discussed further below. It can, however, occur in shoots with articulate growth and this leads to a type of shoot construction intermediate between the two we have so far described.

INTERMEDIATE GROWTH

This may be defined as articulate growth without complete apical dominance. Units of shoot growth are clearly separated by regions with bud-scale scars, with branches on current shoots (Figure 7). This type of growth is found in many trees of the tropics and subtropics and there is a high incidence of it in warm temperate floras, as in the southeastern United States, e.g., *Liquidambar, Liriodendron* (Brown et al. 1967). It is rare in north temperate trees, but occurs, for example in *Sassafras* (Lauraceae) in eastern North America. Otherwise the branches on the current shoot may be represented by thorns (e.g., *Crataegus*). Vigorous sucker shoots ("water shoots") of temperate trees may also show this branching where it is not otherwise found in the majority of shoots. Even on different parts of tropical trees, the degree of apical dominance varies considerably. In avocado, for example, vigorous branches, especially on young trees, have incomplete apical dominance resulting in "precocious" branching, whereas adult distal shoots tend to lack them. This suggests that we are dealing with a generalized feature of tree growth which is expressed differently in temperate and tropical trees. We will suggest a hypothetical mechanism later.

RELATION BETWEEN LEAF INITIATION AND EXPANSION

Leaf growth in trees may be considered as involving two distinct phases; first, initiation when the leaf forms part of the bud; second, expansion when the leaf undergoes its final enlargement and matures. It is suggested that there are an unlimited

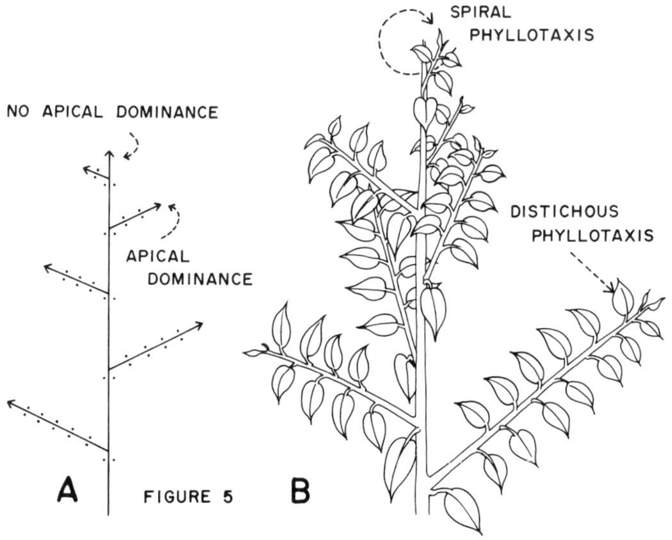

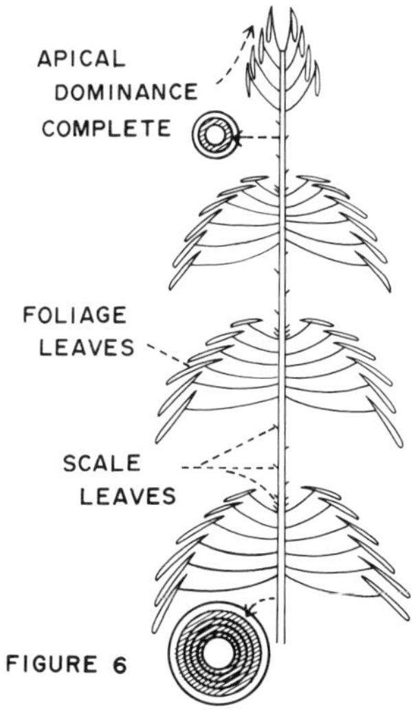

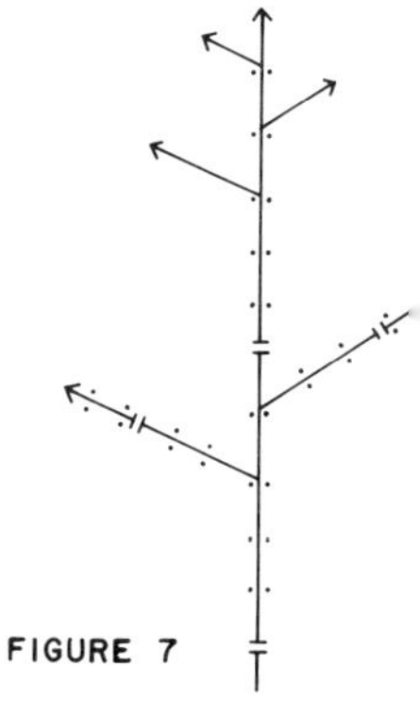

FIGURE 5. Dimorphic shoot system of woody plant shoot, represented by *Trema floridana*: A, no apical dominance of erect (orthotropic) shoots, complete apical dominance of horizontal (plagiotropic) shoots. B, leaves arranged spirally on erect shoot with a plagiotropic branch at each node, the leaves on the branches arranged distichously. FIGURE 6. *Hevea brasiliensis*, diagram of sapling axis (after Hallé and Martin 1968). Growth proceeds in flushes (at intervals of 42 days) producing alternate series of scale and foliage leaves; apical dominance complete throughout each cycle of growth. Each flush is associated with an increment of wood leading to distinct growth rings in the secondary xylem (insert diagrams). FIGURE 7. Articulate warm-temperate type of woody plant shoot, represented by *Liriodendron tulipifera*, with branching in the current flush. Increments may or may not be one per year.

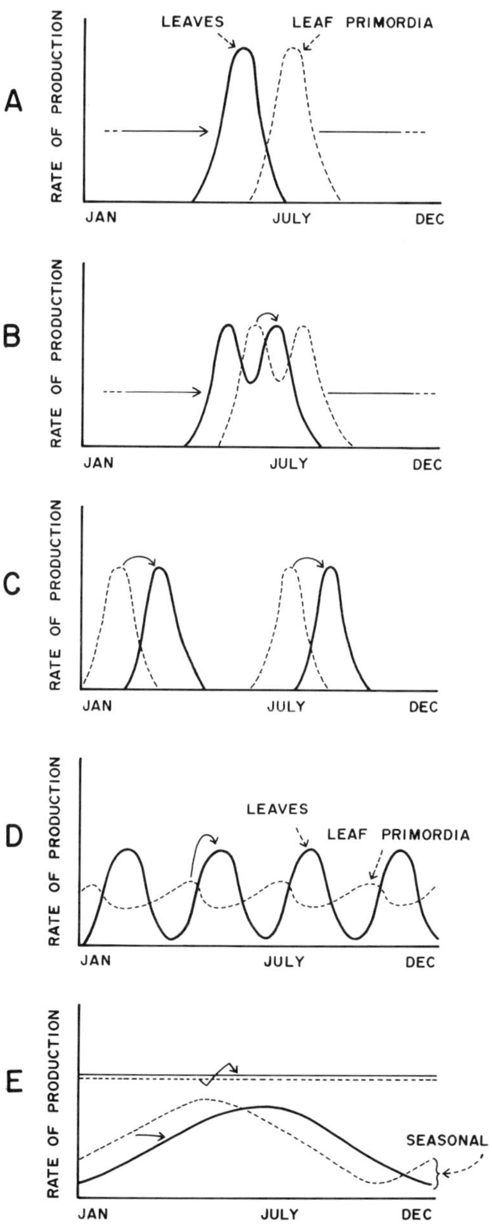

FIGURE 8. Relationship of leaf initiation and leaf expansion in various trees. The data are derived from the relevant papers cited although the diagrams shown here are not found in the original publications. To some extent, therefore, these diagrams involve interpretation of the data. In each diagram the abscissa represents some measure of rate of production of leaves, the ordinate is one calendar year. The continuous line represents leaf expansion, the dotted line leaf initiation. Arrows between curves represent the continuity of development. A, *Fraxinus americana* (Gill 1971a) U.S.A. Leaves are initiated at the end of one growing season but do not expand until the following year. The number of leaves initiated in one year equals the number expanding in the following season (determinate shoot). B, *Populus trichocarpa*, "long" shoot (Critchfield 1960) U.S.A. As in the previous example but some of the leaves initiated expand in the same year they are produced so that there is a double flush of summer growth. The number of leaves expanding in the summer is therefore not determined by the number in the resting bud (indeterminate shoot). C, *Oreopanax* sp. (Borchert 1969) Colombia. Most foliage leaves are initiated shortly before they expand, with approximately two flushes of growth per year. This means that the resting bud includes a minimum of leaf primordia, not a maximum as in A. D, *Camellia sinensis* (Bond 1942, 1945) Ceylon. Continuous leaf initiation, but not at a uniform rate, associated with periodic leaf expansion, usually of four flushes per year. E. *Rhizophora mangle* (Gill 1971b) Florida, U.S.A. Continuous leaf initiation and expansion, with a seasonal change in rates correlated with the seasonal change in climate. A suggested uniform rate for this species throughout the year in a nonseasonal climate is indicated as the simplest possible situation. This situation probably also exists in many palms.

number of ways in which these phases may be interrelated. Some interrelationships are suggested diagrammatically for a number of known types in Figure 8.

In temperate trees, where buds overwinter in a dormant state there are two main methods of growth which we have previously distinguished as "determinate" and "indeterminate. In the former, exemplified by *Fraxinus americana* (Gill 1971a) the total number of leaves expanded in the growing season equals the number initiated at the end of the previous growing season and overwintering. Normally no leaves are initiated and expanded in the same season. This relationship is expressed diagrammatially in Figure 8A. In the second category, exemplified by *Populus trichocarpa* (Critchfield 1960) there are two peaks of leaf expansion and initiation in any one season because a second series of leaves is initiated by a "long" shoot to expand during the same growing season (Figure 8B). Leaves of the two growth types can often be distinguished morphologically and are one cause of heterophylly in trees.

In tropical trees, the same situations may occur but the time scales may be different in that the cycle is gone through several times in one year. Cocoa, recently described by Greathouse et al. (1971), provides a known example. On the other hand, the example of *Oreopanax* sp. (Borchert 1969) suggests the relationship shown in Figure 8C, in which most foliage leaf primordia are initiated shortly before expansion, following a period of inactivity, in contrast to *Fraxinus* where initiation precedes expansion by several months. Another fundamental difference is that in *Oreopanax* the behavior of individual trees is nonsynchronous. The double flush per year is only a trend suggested by Borchert's results. A further "peculiarity" of *Oreopanax* is that, although bud-scales are produced in great numbers, they are conspicuous elements of a *growing* and not a *resting* shoot.

Tea provides yet another example in which leaves expand in flushes repeatedly throughout the year, but in which there may be a more or less constant initiation of primordia (Figure 8D). Thus the number of leaves in a bud fluctuates between maximum and minimum values. *Rhizophora,* in our experience, is probably the simplest possible example (Figure 8E) in which production and expansion of primorida are exactly in step, but with a change in the rate of production according to the season, at least in southern Florida (Gill and Tomlinson 1971). A theoretical situation (Figure 8E) would be a pair of superposed horizontal lines. This might be expected in a nonseasonal climate and would provide the simplest concept of an "evergrowing" tree. This, however, still remains to be demonstrated. Likely candidates for this simple relationship would be palms.

Methods of Branching

In the preceding section, a good deal has been said about the relation between vigor of growth and branching, since shoot construction cannot be discussed entirely without reference to branching. Now we speak in more detail of the morphology of branching especially as it raises topics rarely emphasized by previous workers.

BRANCH MORPHOLOGY

In the above description a distinction has been made between "precocious" branches which grow out in the current flush and those which grow out only after a period of dormancy. No previous attempt seems to have been made to name these two distinct types of branch except with reference to the so-called abnormal development of lammas shoots. This was done by Späth (1912) in his monograph on the subject. Although he distinguishes the two types clearly, his nomenclature is seen to be misleading if one considers the growth of trees on a cosmopolitan basis. With reference solely to lammas shoots which represent a second cycle of growth within one growing season, making them "abnormal" to a German forester, he distinguished: (1) *proleptic branches,* from buds which were formed on the current shoot (first flush) and had undergone a period of dormancy before expanding "abnormally" in the current year (they were "abnormal" because they should have overwintered but did not do so); and (2) *sylleptic branches,* from buds which were formed on a lammas shoot (second flush) and expanded without dormancy (the development of branches in the current year was regarded as abnormal).

With a more comprehensive understanding of tree growth we can now transpose these terms and apply them to the two kinds of shoot which occur on trees, proleptic for branches developed after some period of dormancy, sylleptic for branches developed without dormancy. This separation is pos-

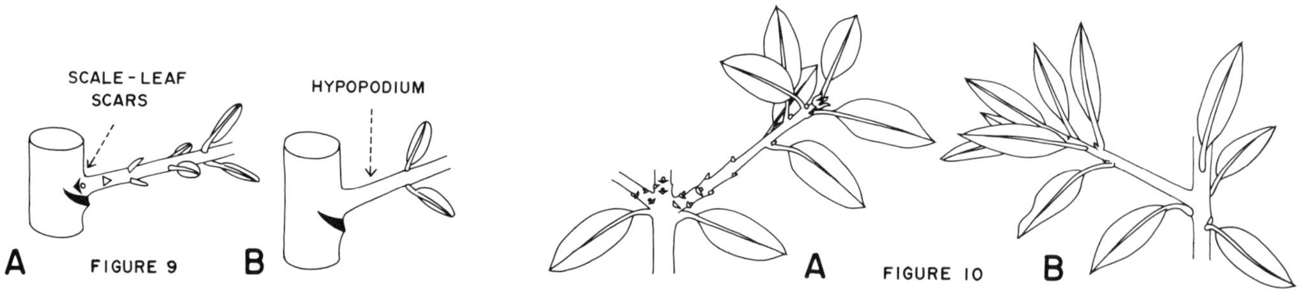

FIGURE 9. Types of branch development in woody plants: A, Proleptic branch, from a dormant bud. Subtending leaf represented by its scar (solid black). Branch has one or more series of scales and gradual transition to foliage leaves. B, Sylleptic branch, grows out without a dormant period. Subtending leaf scar (solid black) usually persists. Abrupt transition to first foliage leaves, usually associated with a long basal internode (hypopodium). FIGURE 10. Examples from avocado (*Persea americanna*), both shoots from the same tree: A, Proleptic branch. B, Sylleptic branch.

sible because the two are generally distinguished morphologically (Figures 9, 10A, B). Proleptic shoots (Figure 9A) normally produce at least one pair of basal scales (prophylls) before they enter a dormant state. Upon elongation this is reflected in a pair or more of basal scale-scars. In addition there is usually a series of transitional leaves, often separated by short internodes before normal foliage leaves are produced on the distal part of the branch. In sylleptic shoots (Figure 9B), on the other hand, there are no basal scales, the first leaf or leaf-pair has a more or less normal morphology (often it is rather smaller than normal) and it is separated from the branch insertion by a long (usually a very long) basal internode (the *hypopodium* of German authors). It should be noted that in monocotyledonous trees similar differences occur between early and late developing lateral shoots (e.g., Fisher and Tomlinson 1972). Actual examples of the two types of shoot taken from a single avocado tree are shown in Figure 10 A, B. The range of morphology expressed in sylletpic shoots in tropical trees is somewhat greater than we have thus briefly indicated, but this does not obscure our generalization.

On this basis it is often possible to distinguish the two types of growth, even in herbarium specimens. We have found exceptions, for example in *Citharexylum*, where branches may have sylleptic morphology though they have been seen to develop by the proleptic method. These examples appear to be rare, however, and it still seems safe to generalize that branches which develop without dormancy form no modified basal leaves. We should avoid making rules, however, on the basis of our present understanding. It is sufficient for our purpose to draw attention to the situation.

The etymology of Späth's useful terms produces an unfortunate anomaly. *Proleptic* comes from the Greek for "precocious" and was chosen by Späth with special reference to lammas shoots, although in fact such shoots are not "precocious" in a general sense since they have undergone dormancy. *Sylleptic* is from the Greek for "together," presumably referring to the contemporary development of main axis and branch, though on first principles this branching is more correctly "precocious," as we have described it earlier in this article, and elsewhere (Gill and Tomlinson 1969). We will now continue to use these terms, both morphologically and developmentally, because they are so useful, while realizing that they were coined with a rather restricted view of tree growth and their strict meaning is false. We can also discount the rather specialized use of the word "proleptic" with regard to *Pinus* (e.g., by Rudolph 1964) on account of this tree's distinctive shoot morphology. In our terminology both long and short shoots are developmentally proleptic in *Pinus*.

We have so far restricted the use of the terms to the development of lateral meristems. Clearly one could extend them to the development of terminal buds especially as Späth may have coined them with terminal buds in mind. Terminal buds in trees with articulate shoots could be described as having proleptic development; those in trees with nonarticulate shoots as having sylleptic development. The concept, however, is best applied to types of branching. Some special consequences of sylleptic branching will be described later.

APICAL DOMINANCE

We have dealt to some extent with branch expression in relation to various types of shoot growth. Now that we have distinguished morphologically the two main kinds of branching shown by trees it becomes possible to discuss in a meaningful way the physiological relation between extension growth and branching. In particular, our observations throw interesting light on theories of apical dominance. The first interpretations of apical dominance in hormonal terms had assumed the action of auxin as an inhibitor and that where auxin is produced in greatest quantity by vigorously growing shoots inhibition is strongest, as in current-year shoots of temperate trees (see Brown et al. 1967). It was further assumed that in the initial regrowth from buds after rest, lateral buds could grow out

TABLE 1. Relation Between Plastochron and Branch Production in a Population of *Avicennia germinans* (Miami, Florida 1970; 32 shoots observed over a 2 month period)

Plastochron (weeks)	*Frequency of Branches at a Node* None	One	Two
1 to 2	0	1	12
2 to 3	6	6	17
>3	13	2	0

because they were not immediately subjected to inhibiting quantities of auxin.

The observations of Champagnat (1961) on the "rameaux anticipés" of *Alnus* cast immediate doubt on this interpretation. He observed that branching on the "water shoots" (which we would now call sylleptic branching) was correlated in a positive way with vigor of growth, instead of a negative way as required by the auxin theory. The more vigorous shoots produced more numerous sylleptic branches. This is in fact the normal expression of tree growth in many woody plants of the tropics and subtropics as we have just pointed out (Figures 9B, 10B). The results in Table 1 for *Avicennia* further demonstrate this relation. That branching in woody plants is not determined simply by auxin levels is also evident from the studies of Brown et al. (1967), who indicate that a much more complex hormonal balance is involved. This situation seems now to be generally appreciated (e.g., Sachs and Thimann 1967) although an examination of a wide variety of shoot types in angiosperms could have prevented unnecessary oversimplification in the first place.

The failure or success of a lateral meristem on a main axis to mature precociously may be the result of some "threshold value" which has to be exceeded before the meristem can develop without undergoing dormancy. This is a purely hypothetical construct but as expressed diagrammatically in Figure 12 it serves as a unifying concept for the growth of all trees, without making an artificial distinction between tropical and temperate species. It is assumed that a certain "threshold" has to be exceeded before a branch can grow out sylleptically. Whether this threshold is exceeded is determined by the "vigor" of the individual, expressed in some quantitative way by the abscissa. In *Avicennia* (Figure 12B) the threshold is exceeded seasonally,

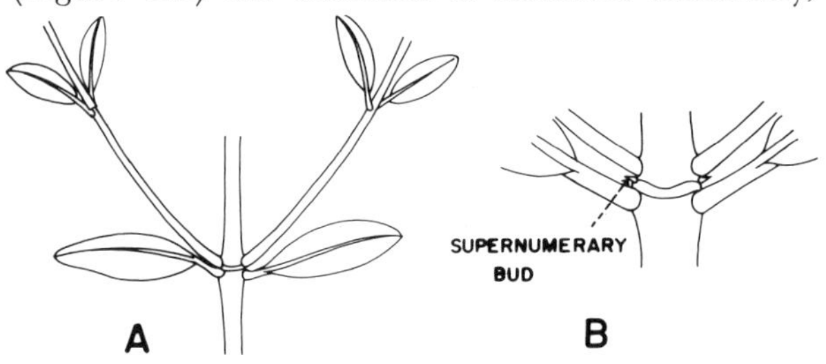

FIGURE 11. *Rhizophora mangle,* sylleptic branches: A, Pair of branches. B, Details of branch insertion showing supernumery bud below sylleptic branch. Note that two types of branch could occur at the same node.

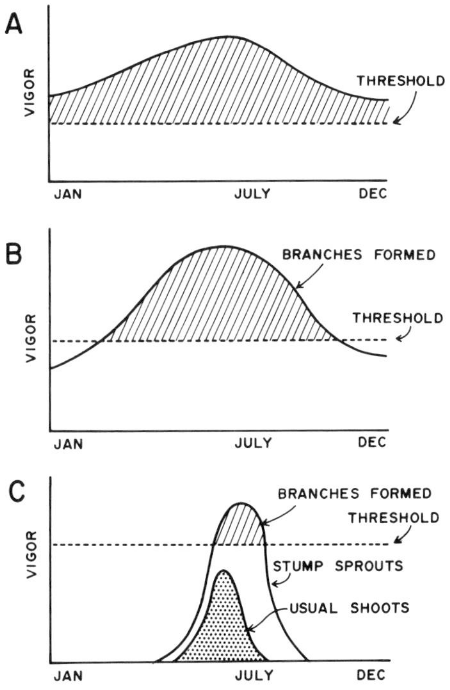

FIGURE 12. Hypothetical interpretations of branch expression in various trees: A, Tropical example, threshold exceeded continuously (see Figure 5, *Trema floridana*). B, Tropical example, threshold exceeded seasonally (see Figure 4, *Avicennia germinans*). C, Temperate example, threshold never exceeded, e.g., normal *Alnus* shoots (dotted), current increment remains unbranched (Figure 1). Exceptionally vigorous shoots such as stump sprouts may exceed the threshold and branch sylleptically (hatched).

resulting in the characteristic periodicity of its sylleptic branching (Figure 4). In tropical trees in nonseasonal climates the threshold may be exceeded periodically so that several tiers of branches can develop in one year. Here vigor may be determined endogenously. Other tropical trees, exemplified by *Trema* (Figure 12A), may grow with sufficient continued vigor so that the threshold is continuously exceeded and all lateral meristems grow out. In contrast, the current shoots of most temperate trees are continuously below the threshold, as shown in Figure 12C for *Alnus,* except for vigorous suckers or stump sprouts which may for a brief period exceed the threshold.

We readily acknowledge that this hypothesis is an oversimplification. Examination of shoot morphology in tropical trees shows that very often the leaf axil is the site of not one but several potential branches. It is very common to find at a node where a sylleptic branch has already developed that there are one or more additional dormant buds. Should these grow out they would become, of course, proleptic shoots. They represent an important source of "reserve" buds. *Rhizophora* serves as one example (Figure 11B). Another is Trema which has a pair of minute buds at the base of each sylleptic branch (these are not shown in Figure 5B). Quite clearly any theory of branching in woody plants must account for the ability of growth centers with contrasted developmental potential to originate in close proximity. Simple anatomical investigation of nodes bearing these contrasted types of buds would obviously be valuable. A survey of the frequency of multiple buds in tropical trees would be of interest, as it may be that this situation occurs more often than the "normal" single bud per leaf axil. Multiple buds are, of course, also quite common in temperate trees (e.g., *Fraxinus* spp.)

BRANCH DIMORPHISM

Branches in temperate trees with articulate growth repeat the construction of the parent axis, in a general way. In many tropical trees the situation is very different due to marked branch dimorphism with contrasted methods of growth between erect (orthotropic) and horizontal (plagiotropic) shoots. A common situation is one in which the terminal bud of the leader has a spiral phyllotaxis and lacks apical dominance. This produces an orthotropic axis with spirally arranged sylleptic branches, one at each node, separated by long internodes. These plagiotropic laterals are in marked contrast because they have alternate (distichous) phyllotaxis and more or less complete apical dominance so that they branch sparingly or not at all. This produces a very symmetrical sapling and the growth form is found in many tropical trees from different families. Numerous examples of this kind of dimorphism are given by Hallé and Oldeman (1970). Many species of the widely distributed *Trema* (Ulmaceae) show it very well (Figure 5).

A rather specialized example of this growth habit is shown by cocoa (Greathouse and Laetsch 1969, Charrier 1969). The habit also occurs in several other quite unrelated trees, as has been shown by Hallé and Oldeman (1970). The seedling axis in cocoa is orthotropic with spiral phyllotaxis and growth limited by the abortion of the apex. Lateral branches are mostly distichous, plagiotropic and have unlimited growth (i.e., the apex does not abort), except for one or more orthotropic shoots which can take over the growth of the original leader. Laterals to a large extent are predetermined as to their potentiality to become one or the other kind of shoot. It is of interest that sylleptic or proleptic development is correlated with type of branch orientation. Plagiotropic lateral shoots have sylleptic morphology whereas orthotropic shoots are proleptic, an important correlation which seems to have been overlooked, largely we believe because the difference between these types of shoot is not widely appreciated.

The subject of branch dimorphism and distribution of shoots, with or without unlimited growth, forms an important part of Hallé and Oldeman's analysis of tree growth. Many trees, for example, have leaders of limited growth, in which the apex either aborts or is converted directly into an inflorescence. The method of substitution of this leader takes place in a number of ways. That shown by cocoa (and other plants) is a peculiarly specialized one. In general, Hallé and Oldeman refer to three main categories of tree, in terms of shoot specialization, viz., those with one type of shoot in which orientation is independent of underlying morphology but with a distinction between trunk and branch; those with two types of shoot—orthotropic and plagiotropic—where orientation is closely related to morphology; and a "mixed" type in which a shoot is initially erect but later becomes horizontal so that there is no prime distinction between trunk and branch. Many leguminous trees are of this latter kind. This aspect of tree growth is dealt with very fully by Hallé and Oldeman. In the following section we deal with one very distinctive and specialized type of shoot morphology.

TERMINALIA BRANCHING

Numerous authors (e.g., Corner 1940) have drawn attention to a distinctive type of habit which is common in tropical trees, but very rare in temperate trees. Since it is particularly striking in species of *Terminalia* (e.g., *T. catappa*) it is rather colloquially referred to as "*Terminalia*-branching" which is appropriate because the generic name is derived from the habit (Figure 13). The main axis is usually monopodial with tiers of branches produced by a modified type of articulate growth. This development of tiers of branches is a general feature of tree growth, as we have seen, and is not an essential corollary to *Terminalia*-branching, although it contributes to the distinctive aspect of *Terminalia* itself. More unusual is the repeated replacement of the terminal bud of plagiotropic branches by one or more vigorous laterals. The terminal bud does not abort, but persists for a lengthy period as an erect short shoot, bearing a tuft of leaves separated by short internodes. Flowers are normally borne on these short shoots.

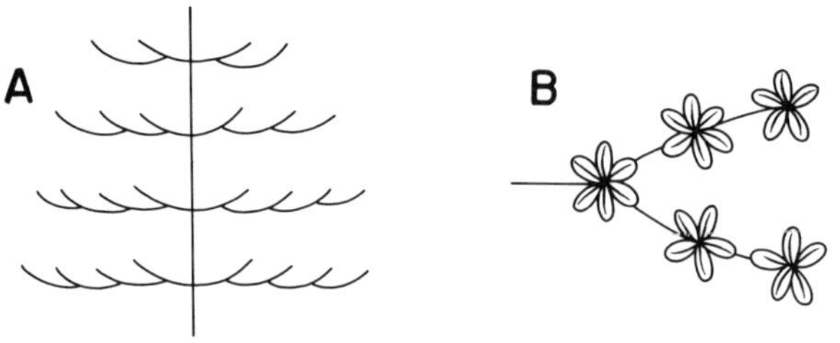

FIGURE 13. Specialized shoot in woody plant, *Terminalia* habit: A, General appearance of tree. B, Surface view of a single lateral branch.

The habit is widespread in many tropical families as has been amply demonstrated by Hallé and Oldeman, who enumerate 13 families in which it occurs. Perhaps other examples could be included since we ourselves would add *Rhizophora* (Gill and Tomlinson 1969) and *Bruguiera* of the Rhizophoraceae to the list. Apparently these genera are excluded by Hallé and Oldeman because they lack a morphological dimorphism between erect and lateral shoots, a distinction which may be artificial in physiological terms.

Explanation of *Terminalia*-branching is not readily accomplished because the terminal bud of the plagiotropic branch is evicted while it is still growing. A simple explanation would be that the lateral shoots of these tropical trees are more sensitive to gravimorphic effects than those of temperate trees. Experimentation and accurate observation of development are needed. Quite clearly, however, one cannot discuss the problem in terms of just one shoot and its immediate derivatives, as there is a distance or positional effect also which has been demonstrated, at least in *Terminalia*, in the experiments of Attims (in Hallé and Oldeman 1970: 67). These show that the main leader in some way controls growth of the horizontal branches. The propensity of tropical trees to branch sylleptically may also be a contributory factor, since the renewal shoot seems usually to have sylleptic morphology.

Cornus species in temperate forests provide a rare exception to the generalization that *Terminalia*-branching is a tropical phenomenon (Figure 14). They are interesting because growth is articulate, with several orders of branches produced by sylleptic development within each growing season. *Cornus* is an exception which may provide further evidence that *Terminalia*-branching depends on the production of sylleptic laterals.

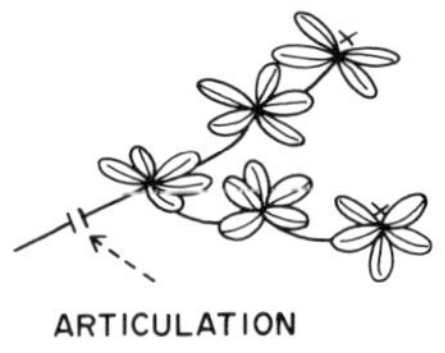

FIGURE 14. *Cornus* sp., a temperate example of specialized shoot in woody plant, but with articulate growth. Several successive orders of sylleptic shoot may be produced in one season.

SYMPODIAL BRANCHING

The *Terminalia*-habit is one specialized instance of sympodial growth in tropical trees. Sympodial growth by abortion of the leader is common in the tropics, as we have seen. The ability to produce sylleptic laterals may complicate the simple picture of substitution growth which occurs in temperate trees. In temperate trees of articulate growth the aborted terminal bud is replaced by a proleptic lateral bud after a period of dormancy (e.g., willows,

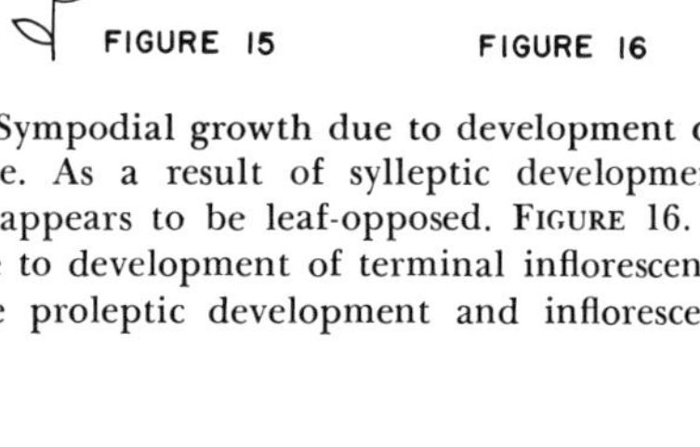

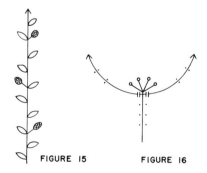

FIGURE 15. Sympodial growth due to development of terminal inflorescence. As a result of sylleptic development the inflorescence appears to be leaf-opposed. FIGURE 16. Sympodial growth due to development of terminal inflorescence. Lateral shoots have proleptic development and inflorescence is not displaced.

elms, beeches). This more or less restricts this type of branching to a particular phase in the cycle of growth of each segment. In tropical trees, replacement of a terminal shoot by a sylleptic lateral is not so restrictive. Good examples of this are seen where the terminal bud directly becomes a reproductive shoot, either an inflorescence or a single flower. Renewal growth is therefore sympodial. The continuing lateral may develop sylleptically and so produce a shoot without a major articulation at the level of branching. This accounts for seeming leaf-opposed flowers or inflorescences (Figure 15) in a number of tropical families (e.g., many Annonaceae, Apocynaceae, Asclepiadaceae). Where lateral growth is only resumed below an inflorescence after dormancy, the branch is proleptic and a major articulation is usually present, as in many species of *Psychotria* (Figure 16).

Methods of Radial Growth

This topic is largely outside the scope of our investigation, but a few comments are justifiable. The correlation between periodic extension and diameter increase leading to the production of annual rings of growth in north temperate trees is not in doubt. This has led to the assumption that there must be such a correlation in all woody plants. The situation in tropical species is very diverse, however, and the subject is much obscured by a constant faith in what is only an assumed correlation. The further assumption is made that (1) all growth rings are alike and (2) they reflect cessation of cambial activity. For most trees there is no evidence for this. One source of confusion is the tendency to observe and count growth rings in a stem sample at a polished surface, whereas in fact microscopic examination is always necessary to distinguish structural discontinuities from those produced by variation in cellular deposits such as tannin (Tomlinson and Craighead 1972). Even under the microscope it is not easy to distinguish different types of discontinuity in the wood which may have different causes. Some rings may be influenced indirectly by extension growth, whereas others may be the secondary result of climatic effects, notably drought. The term "growth ring" needs precise definition, and it may never be possible to arrive at a satisfactory one which applies to all trees.

Nevertheless, there are examples of tropical trees with growth rings correlated with shoot flushes. In rubber, for example, Hallé and Martin (1968) indicate that for each unit of shoot extension there is a corresponding growth ring, although they do not say how growth rings are recognized (Figure 6). It is equally clear, on the other hand, that other kinds of response occur. Where trees have growth or even annual rings, this may seem to be independent of extension growth. In multicyclic (multinodal) species of *Pinus*, where several (up to 5) successive spurts of growth can occur in one year, each spurt marked by a whorl of branches, there is only one ring per year in the wood. False rings, which can easily be distinguished anatomically from true rings, may have quite independent causes, notably water stress, as is probable in *Pinus elliotti* var. *densa* in southern Florida. Repeated flushing which has no effect on the production of an annual ring of wood is probably very common in tropical and subtropical trees. This may explain why lammas shoots in temperate trees normally induce no additional growth rings (Späth 1912), a phenomenon that has puzzled foresters.

The independence of cambial activity from shoot extension is further suggested by observations on the wood anatomy of the native trees of southern Florida. Here there is a marked seasonal climate. Extension growth is expressed in a great variety of ways. Despite this, most trees lack recognizable growth rings (Tomlinson and Craighead 1972). A special example is provided by *Avicennia* which has pronounced growth rings, although these are anomalous because they result from concentric bands of internal phloem produced by successive cambia. Results obtained by one of us (Gill 1971b) show that up to six rings can be formed in any one year, the number of rings depending on the vigor of a shoot. Consequently ring number is correlated with diameter rather than age. This is shown clearly in Figure 17, which represents measurements from a single plant. The fact that *Avicennia* must have nonannual rings can be demonstrated by simple examination of its shoot system, something which never appears to have been done. This example, although it represents an admittedly specialized type, is useful in demonstrating a situation for which knowledge of temperate trees provides no guidance; it may even, in fact, mislead an observer with limited facilities.

Bud Protection

A consequence of the apparent continuous growth of many tropical trees is that they lack definite terminal buds. Even those which undergo dormancy may lack bud-scales (the buds are then usually described as naked). This has led a number of observers (e.g., Potter 1891) to seek, in a teleological manner, structures which substitute for the (presumed) protective function of bud-scales. These can be found readily in the form of stipules of great variety, together with enveloping leaf bases, hairs, and specialized secretion of mucilage or resin. The value of these in protection against excessive insolation or drought still remains to be demonstrated, however, especially in the understorey of the rain forest. In tea, the manner of shoot development suggests that the apex is least enclosed when it is at its most vigorous stage of leaf initiation (Bond 1945). The foregoing discussion of leaf initiation demonstrates that bud morphology cannot be discussed independently of some knowledge of growth periodicity. The subject is dynamic not static; buds are manifestations of growth.

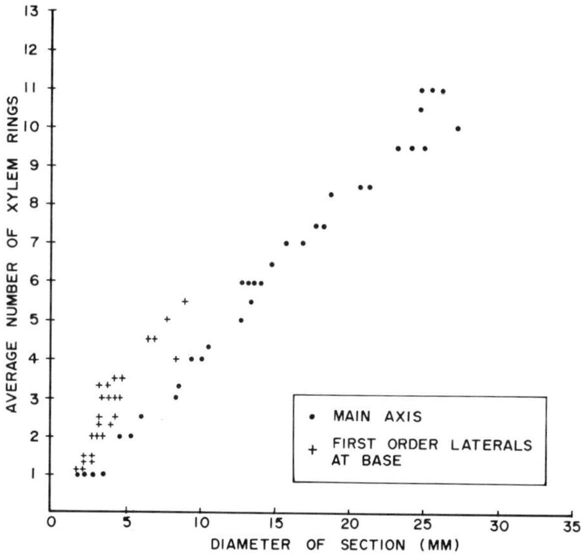

FIGURE 17. *Avicennia germinans.* Relation between anomalous ring development and shoot diameter. There is a linear correlation, indicating that ring initiation may be under endogenous control.

The failure of many tropical trees to develop bud-scales should be viewed as a morphogenetic problem, not simply as one of imaginary adaptation. In the same way the function of stipules, not merely as presumed protective structures but as organs with specialized physiological properties, still needs to be investigated. The woody vegetation of the tropics provides an endless wealth of subjects for research.

Conclusions

To anyone familiar with tropical trees the above account will be obviously superficial, but it has served to emphasize the variety of growth expression in trees as a whole and to demonstrate that trees of north-temperate latitudes comprise a minority that in many respects appears to be unrepresentative. We have recommended a reversal of contemporary attitudes about tree growth, which we think is necessary if the detailed studies of individual tropical species are to be carried out without unfortunate preconceptions. The task of detailed growth analysis is scarcely begun but offers unlimited scope for future work. It is hoped that more botanists will take up the challenge, difficult as it is, and so add to a body of information which will allow worthwhile generalizations about tree growth. Trees, after all, are undoubtedly the most important group of terrestrial organisms and we should have a clear understanding of their growth and development.

References

We do not attempt to give a full bibliography of this subject. General references in the following list which provide a more extensive guide are Alvim 1964, Büsgen and Münch 1929, Coster 1927-1928, Hallé and Martin 1968, Hallé and Oldeman 1970, Koriba 1958, Romberger 1963.

Alvim P. de T. 1964. Tree growth periodicity in tropical climates. *In* M. H. Zimmerman (Editor). Formation of wood in forest trees. Pp. 479-495. Academic Press. New York.

Bond, T. E. T. 1942. Studies in the vegetative growth and anatomy of the tea plant (*Camellia thea* Link.) with special reference to the phloem, I: The flush shoot. Ann. Bot. (Lond.) n.s., 6:607-629.

———. 1945. II: Further analysis of flushing behaviour. Ann. Bot. (Lond.) n.s., 9:183-216.

Borchert, R. 1969. Unusual shoot growth pattern in a tropical tree, *Oreopanax* (Araliaceae). Amer. J. Bot. 56:1033-1041.

Brown, C. L., R. G. McAlpine, and P. P. Kormanik. 1967. Apical dominance and form in woody plants: A reappraisal. Amer. J. Bot. 54:153-162.

Büsgen, M., and E. Münch. 1929. The structure and life of forest trees. 3rd edition. John Wiley. New York.

Champagnat, P. 1961. Dominance apicale. Tropismes, épinastie. *In* W. Ruhland (Editor). Handbuch der Pflanzenphysiologie. Vol. 14: Growth and growth substances. Pp. 872-908. Springer Verlag.

Charrier, A. 1969. Contribution à l'étude de la morphogenèse et de la multiplication végétative du Cacaoyer (*Theobroma cacao* L.). Café Cacao Thé 13:97-115.

Corner, E. J. H. 1940. Wayside trees of Malaya. 2 vols. Govt. Printer. Singapore.

Coster, C. 1927-1928. Zur Anatomie und Physiologie der Zuwachszonen und Jahresringbildung in der Tropen. Ann. Jard. Bot. Buit. 37:49-160 and 38:1-114.

Critchfield, W. B. 1960. Leaf dimorphism in *Populus trichocarpa*. Amer. J. Bot. 47:699-711.

————. 1971. Shoot growth and heterophylly in *Acer*. J. Arnold Arbor. 52:240-266.

Fisher, J. B., and P. B. Tomlinson. 1972. Morphological studies in *Cordyline* (Agavaceae), II: Vegetative morphology of *Cordyline terminalis*. J. Arnold Arbor. 53:113-127.

Gill, A. M. 1971a. The formation, growth, and fate of buds of *Fraxinus americana* L. in central Massachusetts. Harvard Forest Paper 20. 16 pp.

————.1971b. Endogenous control of ring formation in *Avicennia*. Forest Science. 17:462-465.

Gill, A. M., and P. B. Tomlinson. 1969. Studies on the growth of red mangrove (*Rhizophora mangle* L.), I: Habit and general morphology. Biotropica 1:1-9.

————. 1971. Studies on the growth of red mangrove (*Rhizophora mangle* L.), Phenology of the shoot. Biotropica. 3:109-124.

Greathouse, D. C., and W. M. Laetsch. 1969. Structure and development of the dimorphic branch system of *Theobroma cacao*. Amer. J. Bot. 56:1143-1151.

Greathouse, D. C., W. M. Laetsch, and B. O. Phinney. 1971. The shoot-growth rhythm of a tropical tree, *Theobroma cacao*. Amer. J. Bot. 58:281-286.

Hallé, F. 1966. Étude biologique et morphologique de la tribu des Gardéniées (Rubiacées). Thesis Univ. Abidjan. 141 pp. Mem. O.R.S.T.O.M. 22. Paris.

Hallé, F., and R. Martin. 1968. Étude de la croissance thythmique chez l'Hevea (*Hevea brasiliensis* Mull.-Arg. Euphorbiacées–Crotonoidées). Adansonia n.s., 8 (4) :475-503.

Hallé, F., and R. Oldeman. 1970. Essai sur l'architecture et dynamique de croissance des arbres tropicaux. Masson et Cie. Paris.

Holdsworth, M. 1963. Intermittent growth of the mango tree. J. West African Sci. Assoc. 7:163-171.

Holttum, R. E. 1953. Evolutionary trends in an equatorial climate. Symp. Soc. Exp. Biol. 7:159-173.

Klebs, G. 1915. Uber Wachstum und Ruhe tropischer Baumarten. Jb. Wiss. Bot. 56:734-792.

Koriba, K. 1958. On the periodicity of tree-growth in the tropics. Gard. Bull. (Singapore). 17:11-81.

Kozlowski, T. 1964. Shoot growth in woody plants. Bot. Rev. 30:335-392.

Njoku, E. 1958. The photoperiodic response of some Nigerian plants. J. West African Sci. Assoc. 4:99-111.

Potter, M. C. 1891. Observations on the protection of buds in the tropics. J. Linn. Soc. (Bot.) 28:343-352.

Robinson, L. W., and P. F. Wareing. 1969. Experiments on the juvenile-adult phase change in some woody species. New Phytol. 68:67-78.

Romberger, J. A. 1963. Meristems, growth and development in woody plants. Tech. Bull. U.S. Dept. Agric. no. 1293. 213 pp.

Rudolph, T. D. 1964. Lammas growth and prolepsis in Jack Pine in the Lake States. Forest Sci. Monographs. 6.

Sachs, T., and K. V. Thimann. 1967. The role of auxins and cytokinins in the release of buds from dominance. Amer. J. Bot. 54:136-144.

Späth, H. L. 1912. Der Johannistriebe. Parey. Berlin.

Taylor, F. J. 1970. Some aspects of the development of mango (*Mangifera indica* L.) leaves, 1: Leaf area, dry weight and water content. New Phytol. 69:377-394.

Tomlinson, P. B., and F. C. Craighead. 1972. Growth ring studies on native trees of sub-tropical Florida. Pp. 39-51, *in* Ghouse and Yunus (Editors). Research Trends in Plant Anatomy (Chowdhury commemoration volume). McGraw Hill. New Delhi.

Volkens, G. 1912. Laubfall und Lauberneuerung in den Tropen. Gebruder Borntraeger. Berlin.

Evolutionary Relationships Between Flowering Plants and Animals in American and African Tropical Forests

HERBERT G. BAKER

Introduction

It is a commonplace of ecology that flowering plants and animals interact at all stages of their life-cycles. If we were to attempt an examination of all the kinds of interactions which take place in the tropical forests of Africa and the Americas, a shelf full of volumes would be too small to hold the unreadable effusion that would result. Consequently, in this paper only some of the complex interactions that occur during two phases of flowering plant reproduction—pollination and seed-dispersal—will be considered, and no pretense of completeness will be made for even these discussions. Before the interactions can be treated, however, it is necessary to put the flowering plants into an appropriate historical context so that we may better appreciate the evolutionary problems that will become evident.

Historical Angiosperm Geography

Although opinions differ as to when angiosperms first became recognizable as a group distinct from their putative pteridosperm ancestors, there appears to be general agreement that the great angiosperm expansion began before the middle of the Cretaceous Period and led to a prodigious efflorescence of taxa and individuals during the Upper Cretaceous and the early part of the Tertiary Period. Many botanists believe that the center of diversity of the early angiosperms was in Southeast Asia, including some of Australasia and, consequently, have looked there for the origin of the angiosperms (Smith 1970 and pp. 50–51 herein).

Against this background, the growing volume of information about continental drift has great significance. The separation of South America from Africa (beginning in the south) appears to have begun between 120 and 200 million years ago (most estimates vary between these limits), probably during the Jurassic Period. It is not easy yet to reconcile differing statements on the timing of the final separation of Africa and South America; according to Tarling and Tarling (1971) this occurred 92 million years ago, and Heirtzler (1969) gives his opinion that "most of the opening of the South Atlantic Ocean has occurred since the Upper Cretaceous." On the other hand, Maxwell et al. (1970) suggest that by 80 million years B.P., still in the Upper Cretaceous, the South Atlantic Ocean had a width of 3000 kilometers. Smith (1970 and pp. 51–53 herein) accepts that much of the spread of the plants whose descendants make up the tropical forests of Africa and the Americas has taken place subsequent to the appearance of an Atlantic Ocean barrier. An opposite view is taken by Axelrod

HERBERT G. BAKER, Department of Botany, University of California, Berkeley, California 94720. ACKNOWLEDGMENTS: Much of the recent first-hand information in this paper has been acquired with the aid of grants (GB 7805; GB 25592) from the National Science Foundation through the Organization for Tropical Studies. In this work, I have enjoyed collaboration and discussion with several O.T.S. colleagues, particularly Drs. Gordon W. Frankie, Paul Opler, and Kamaljit Bawa.

(1970), who believes that an origin for the angiosperms in the Triassic or even late Permian times would have permitted the dispersal even of some "advanced" families before the separation of Africa from South America. Whatever the truth, it seems likely that much of the stocking of the angiosperm-dominated tropical forests was achieved *after* geographical impediments began to appear.

However, land connections or stepping-stone routes in the north from Asia through Beringia and probably across the North Atlantic Ocean from Europe must have persisted into Tertiary time, as did the southern route from Asia through Antarctica to South America. Migration across these higher latitude connections, supplemented by long-distance dispersal of seeds by wind, sea-currents and bird-flights (as well as by human carriage in recent times) must have been responsible for most of the population of the American continent by flowering plants. By contrast, Africa was receiving its share of the flowering plants by a westward migration along the coasts of the Indian Ocean, which may have involved the overcoming of fewer obstacles, at least prior to the onset of drought in South and Southwestern Asia.

The animal groups with which the flowering plants have developed relationships may not have achieved their present distributional patterns contemporaneously with the plants, and the consequences of any such discrepancies will be important matters for discussion later in the paper. However, for all considerations of the evolutionary relationships between plants and animals in the tropics, the early separation of the present-day continents is of the greatest importance. Similarities in reproductive features of the plants in neotropical and paleotropical floras, respectively, where they are to be seen, must always be considered to be possibly the products of parallel or convergent evolution rather than indications of recent (or even ancient) diffusion of the products of a single evolutionary line to more than one continent.

PARALLEL EVOLUTION AND CONVERGENT EVOLUTION

An example of parallel evolution appears to be provided by the development of pseudocopulation pollination methods by members of separate subfamilies in the Orchidaceae in at least three areas of the world (Baker and Hurd 1968). As an example of convergent evolution in the production of a character possessing reproductive significance we may take the development in separate tribes of the Rubiaceae of a semaphore signal to volant animals in the form of a vastly enlarged and brightly colored (often bright red) sepal. This is exemplified by the paleotropical genus *Mussaenda*, of the tribe Mussaendeae, and by the neotropical genus *Warscewiczia* of the tribe Rondeletieae. In both cases, butterflies seem to be the animals chiefly attracted by these brilliant signals, but sunbirds (in Africa) and hummingbirds (in the Americas) are also drawn by them.

LONG-DISTANCE DISPERSAL

Although for any particular plant family features of reproductive significance like those described above may have evolved separately in the two longitudinally demarcated hemispheres, the possibility remains that some of the reproductive features were carried from one hemisphere to the other by a plant which successfully overcame the hurdles obstructing long-distance dispersal. Provided that these pioneers are able to establish themselves, their presence and the "training effect" that they might have on the indigenous fauna could make it easier for subsequent immigrants to succeed (at least until such time as the niche is completely filled, if this ever occurs).

We do not know how the Loranthaceae (including the Viscaceae) achieved their present worldwide distribution, which includes a striking restriction of genera to either the neotropics or the paleotropics, respectively. However, this restriction suggests that long-distance dispersal across the Atlantic Ocean (and, certainly, *recent* long-distance dispersal) was not involved. Balle (1960) has suggested that there is a bipodal pattern to be observed with Indo-Malaya and Australasia at the focus and the American and African tropics at the extremes. Nevertheless, the shorter-distance dispersal-system whereby seeds are implanted on a branch of a host tree by a bird which has fed on the sticky pulp of the berries wiping its beak, regurgitating or defecating thereon (Kuijt 1969) may have provided a model utilizable by more recent transatlantic voyagers like the epiphytic cactus *Rhipsalis baccifera* (J. Miller) W. T. Stearn (= *R. cassutha* Gaertner). Although it is uncertain whether this cactus was carried from South America to the African continent by birds, there can be little doubt that its subsequent spread through Africa and into Asia as far as Ceylon has been largely by the agency of frugivorous birds operating in the Loranthaceae manner. This spread has been assisted by the fact that isolated plants of *Rhipsalis cassutha* are quite capable of setting seed (Baker 1972b).

CLIMATIC CHANGES

Climatic changes have occurred in both hemispheres with the passage of time and, presumably, have followed approximately similar patterns. At all times there must have been climatic zonation and the true rain forest in both the Americas and Africa was and is accompanied by a range of forest types adapted to the regular annual or biannual incidence of dry seasons of greater or lesser severity. Without doubt, we must attribute biological responses such as the tendency for the flowering and fruiting of woody plants to be concentrated in the dry season to parallel evolution on the American continent (Janzen 1967, Croat 1969, Frankie, Baker, and Opler 1972) and the African continent (Hopkins 1969). Both the plants and the anthophilous animals find advantage in flower visitations taking place at this time and Janzen (1967) has listed and considered these advantages in detail. In addition, the maturation of fruits and the dispersal of seeds near the end of the dry season means that the most favorable time for wide dispersal is utilized and the seeds are placed on the ground in time for germination in the oncoming wet season (Frankie, Baker, and Opler 1972). It goes without saying that among the intense biotic activity of the forest floor it is undesirable for seed to be left lying around very long before germination can take place.

Statistical studies now being made in Costa Rica (Baker, Frankie, and Opler 1972) suggest that the proportion of zoophilous pollination mechanisms to those that depend upon anemophily, the proportion of sarcochory to the various kinds of anemochory, the proportions of heavy to lighter seeds, and the proportion of cryptocotylar to phanerocotylar germination all decline as the forest becomes less dense with increasing severity of the dry season (data on germination also cullable from Duke 1969). There are some data on these matters for African forest types (if one uses such sources as Jones 1956, Keay 1957, Voorhoeve 1967), but comparable syntheses to those we are atte not come to our attention.

In the interpretation of any statistical st of great advantage to know the primitive condition so that the direction of evolutionary change may be ascertained. In these cases, there is good reason to believe that early (if not the earliest) angiosperms were insect-pollinated (particularly by beetles, Baker 1961, Baker and Hurd 1968), that sarcochory is also of immense age (van der Pijl 1969; Corner 1949, 1964) but that phanerocotyly is more primitive than cryptocotyly (Eames 1961).

Responsive or Reciprocating Evolution

The point has been made elsewhere (Baker and Hurd 1968) that the term "co-evolution," so freely used nowadays, may not be the most apt in describing the correlated development of flowering plants and flower-visiting animals over a period of time. The evolutionary pattern must rather be one of *demand et réponse* which, if prolonged, may produce a "reciprocating" evolution, wherein changes in one kind of organism stimulate the selection of alterations in the morphology or behavior of another kind of organism, to be followed by a return influence on the first kind of organism. In such responsive or reciprocating evolution (at least insofar as it involves pollination relationships), it appears to be the animals that more often make the initial selective demands and the plants that show the most profound responses (Baker and Hurd 1968).

The papaya, *Carica papaya*, appears to be a native of the neotropical forests, but this dioecious species was able to become feral very soon after it was introduced into Africa because its pollinators, which I have evidence for believing are mosquitoes, are also found there. The same is true for taxa which have been carried recently from one hemisphere to another and are successful because they can be autogamous. On the other hand, where there has been time, adaptation of flowers to locally available animals has taken place. Thus, many years ago, van der Pijl (1937) pointed to the subtle differences between the flower and inflorescence arrangements that accommodate flowers to the hovering hummingbirds (Trochilidae) in neotropical habitats and the arrangements that make paleotropical plants appropriate to the visits by acrobatic but not hovering sunbirds (Nectariniidae). Exposure to the two kinds of pollinator has resulted in evolutionary divergence within the single genus *Erythrina*. In the African *E. senegalensis*, for example, the flowers are upright like the Asiatic ones described by van der Pijl (1937), whereas in the American *E. crista-galli* the flowers are inverted and exposed on a long peduncle, making their parts more accessible to hovering birds that approach from below.

Perhaps the most precise adaptations of flowering plants and birds to each other occur between the mistletoes of the Loranthaceae and hummingbirds in the Americas, sunbirds in Africa and Asia and, particularly, the "flower peckers" or "mistletoe birds" (Dicaeidae) in the Indo-Malayan region (Kuijt 1969).

Always we are faced with the problem of deciding which characteristics of organisms are directly adaptive and which are merely the visible nonadaptive morphological results of the operation of adaptive physiological processes. Even where a character is directly adaptive this may not be appreciated if we fail to understand the function it serves. The history of the investigation of seed-dispersal mechanisms provides a case in point. I think it is fair to say that until recently we measured the effectiveness of a seed-dispersal system only in terms of how far seeds could be dispatched after each burst of flowering. But the majority of species are not now in a colonizing phase, and replacement of individuals that die is at least as important to them as range expansion. Now, an important item in replacement (at least for woody plants) is escape from predators on the seeds (Janzen 1969, Smythe 1970) and seedlings (Janzen 1971a, and Joseph Connell of the University of California, Santa Barbara, pers. comm.). These predators are most prevalent in the vicinity of existing plants, so that dispersal of seeds to an appropriate moderate distance may be important in the *retention* of a natural distribution, let alone expansion of it.

On this basis, Janzen (1972) has provided a solution of the puzzling case of the significance of irritant hairs on the *inside* of the fruit wall in *Sterculia apetala* in Central and South America. The seeds of this tree are preyed upon most destructively by insects that are always to be found in abundance near the parent trees. Survival of any seeds at all is only likely if an *undehisced* fruit is carried some distance away from the tree by an animal (which would have to be of the size of a squirrel or a monkey). These animals bite open the fruits and may manage to consume one or a few seeds (which are already ripe) before the irritant hairs cause them to drop the remainder to the ground while they clean their lips and paws. Irritant hairs on the *outside* of fruits would preserve immature fruits from robbery by fruit-plucking animals so that they might dehisce naturally (as would be advantageous in the case of wind-dispersed seeds), but in the case of *Sterculia apetala* the situation is reversed and ensures not only the desirable removal of *immature* fruits from attack by the insect predators but their dropping before all the seeds are consumed.

This mechanism of avoiding predation should be looked for elsewhere. The indehiscent fruits of *Adansonia digitata* (Bombacaeae), which are carried away by African baboons and crunched open at some distance from the parent tree, with undamaged seeds showing better germination after passage through the digestive tract of the animal (Baker, in Rick and Bowman 1961), represent another method of removing the seeds away from insect predators (*Dysdercus* bugs). Also, the entanglement of seeds in hairs so that they blow away in the wind is another means of achieving the same effect and is practiced by other members of the same family (e.g., *Ceiba* spp.) exposed to the same predators.

POLLINATION INTERACTIONS

Despite our growing realization of the inadequacy of the information available to us and the excessive simplicity of accepted explanations of plant and animal interactions, enough studies have been made to give us a glimpse of the influences that are at work at the levels of the individual species and of the community. As usual, we see the interplay of opposing forces. Thus, concerning ourselves only with pollination biology, at the level of the individual species there are clearly *cooperative* and clearly *antagonistic* interactions with animals.

The provision of nectar or of food bodies by flowers that are utilized by particular animal visitors to those flowers and the resultant cross-pollination of the flowers (which may be essential if they are self-incompatible, as Dr. Kamaljit Bawa, pers. comm., has shown to be the case for a majority of the species in a sample area of dry forest in Costa Rica) is a clear case of cooperative interaction—technically at the level of *protocooperation* or even *mutualism* (Odum 1959). Equally, the robbing of nectar from flowers by insects with powerful mouthparts, such as carpenter bees, that bite holes in the corollas and never contact stamens or stigmas, is clearly antagonistic—technically *predation* or *parasitism*. But these extremes seem to be rare compared with the mass of interactions that are neither purely cooperative nor purely antagonistic.

At one time, Jaeger (1954b) thought it possible to distinguish "pollinateurs" from "ravageurs" among the visitors to tropical flowers, but how can trigonid bees of the neotropical forests that tear flower parts to pieces in short order (but often pollinate those flowers in the process) be accommodated in such a classification? In early writings about the visits of bats to flowers there was much discussion as to whether the bats are simply destructive or not. But what does it matter if, immediately after a flower of *Bauhinia megalandra* has been visited by a bat, the corolla falls to the ground (Hart 1897), as long as the pollinated style remains?

Another complicating factor is the behavior of anthophilous animals *between* visits to flowers. Janzen (1971b) has drawn an important distinction between the foraging behavior of female euglossine bees (and, probably, other large solitary bees) in the lowland forests of the neotropics and that of smaller, often social bees found in the same habitats. The former engage in a pattern of flower-visitation that takes them from plant to plant while the latter fly from flower to flower on the same plant. In conversation, Dr. Janzen has aptly referred to the former behavior as "trap-lining" and to the latter as "opportunism." Apparently, the distances that can be covered by a "trap-lining" bee are little short of fantastic judged by our previous standards; distances of 20 to 30 kilometers each day seem to be well within their capacity.

As might be expected, there is a syndrome of characters that differentiate plant species usually visited by "trap-liners" from those visited by "opportunistic" bees. The former are characterized by the production of one or a very few flowers each day over a prolonged period, with just enough pollen or nectar content to make the bee's visit worthwhile, but far too little to satisfy its needs without visiting many more plants. By contrast, the plants visited by opportunistic bees provide abundant food material and can monopolize the attentions of the bees with a lavish display of flowers.

Trap-lining may also be practiced by some hummingbirds (data obtained by H. G. Baker) and hawkmoths in neotropical forests, and evidence will be presented later that bats also follow this pattern in the same New World forests.

In genetical terms, the significance of trap-lining and opportunism is, of course, that they promote cross-pollination and geitonogamy (the equivalent of selfing), respectively. In view of the low density of any particular species of plant in the high density tropical forests, the influence of these behavior patterns on the evolution of the taxa concerned is likely to be very great, and it is most important that they should be studied as closely in African forests as in the neotropical ones. Although euglossine bees are not present in Africa, there are other potential trap-liners among the bees, sphinx moths, birds, and bats whose behavior patterns should receive attention.

Intermediate in effect between the selfing promoted by the opportunists and the wide cross-pollination made possible by the trap-liners is the influence of feeding and mating territoriality in the pollinators. So far investigated quantitatively only for some hummingbirds in neotropical forests, this restriction of activity to a territory which (according to the relative sizes of animal and plant) may range from a part of a tree to several shrubs or large herbs, restricts wide crossing and encourages sib-crossing, if not actual selfing (Wolf 1969, Wolf and Stiles 1970, Yan B. Linhart of University of Colorado, pers. comm.). Not all hummingbird species are territorial, usually only the male is the defender of a territory, and territories are not always respected perfectly or set up with exactly the same boundaries, so there is some flexibility in the pollen dispersal opportunities. But clearly the possible existence of territoriality phenomena with an influence on pollen-dispersal must be looked for elsewhere and with other kinds of anthophilous animals.

All of these "complications" in the simple picture of cooperation between flowers and their visitors need to be looked at from the community or ecosystem point of view. Features that appear not to be at all cooperative at the level of the individual plant species may prove to be vital to the continued functioning of the ecosystem containing that species. Thus, *Ceiba acuminata* (Bombacaceae), which is a constituent of dry forests as well as of thorn scrub in the Central American tropics, is pollinated by bats (*Leptonycteris sanborni*) and by hummingbirds. However, sometimes flowering almost alone at the end of the dry season, it is visited for nectar by at least two kinds of bees, by wasps, and by skippers, none of which make contact with the stigma and, consequently, are irrelevant to its pollination. The same insects, however, that *Ceiba* provides with sustenance at a time when little else is available are probably extremely important to the pollination of other species that flower at a different season (Baker, Cruden, and Baker 1971).

Another aspect of the same phenomenon is to be seen when a number of species of flowering plant with sequential flowering times together makes possible the sustained nutrition of a large pollinator in a geographical region. First proposed by Allen (1939), and then by van der Pijl (1956) with valuable suggestions of genera to investigate, this has been backed up for some bats in West African forests (Baker 1963, 1970a, and unpublished data) and is being investigated in the neotropics by several workers. In addition to bats, birds, and large bees, social bees and wasps tend to be active almost all year round (Janzen 1967) and the synecological study of their nourishment should be revealing of ecosystem organization in both hemispheres. Van der Pijl (1969:86) has indicated that even beetles and flies may show this bond since their larvae as well as they, themselves, "must subsist on the living or decaying material of plants and animals in the whole community." The same must be true with regard to the highly specific host-preferences of the caterpillars of Lepidoptera.

SEED-DISPERSAL

A similar range of relationships is to be seen in regard to seed-dispersal. Once again, the "pure extremes" are much less frequent than the cases in which there is a balance between cooperation and antagonism in the relationship between plants and animals. Clear cases of cooperation involve the presentation of ripe fruit to birds and other frugivorous animals that carry them away, consume the fleshy parts and reject the seeds (which are often bitter), undamaged. At the opposite extreme is the predation of seeds by insects, such as bruchid beetles, that have nothing but a destructive effect. As much as 99 percent of the seed-crop of some trees of the Leguminosae may be destroyed by these beetles and the proportion may not be much less in the case of the palm *Scheelea rostrata* (Janzen 1969, 1971c). J. Connell (pers. comm.) and Janzen (1971a) have accumulated evidence that predation of seedlings may be no less important than the destruction of seeds in the synecology of tropical forests.

Just as clearly antagonistic may be the response of the plant to predation of the sort just described. In neotropical *Hymenaea courbaril* (as well as in some African relatives of this leguminous tree) protection is achieved by the welling-out of a sticky resin from the point of attack on the fruit by the female beetle. Chemicals also protect the seeds of many other leguminous plants; see Janzen (1969) for a table listing 31 different ways by which plants may cut down or avoid predation of their seeds.

In fact, three different "strategies" for predator-evasion may be recognized. First, the plant may produce seeds of such small size that they are unworthy of the attention of predators or are too small to be inhabited by them if the predators normally operate from within (Bridwell 1918, Burtt 1970). Secondly, so many seeds may be produced that predator-satiation occurs and some of the necessarily small seeds escape. Thirdly, there may be chemical protection of the seeds. Advantages in each strategy are accompanied by disadvantages, most notably the restrictions upon seedling establishment that are imposed if only small seeds, with little food reserve, are produced, as they are in the first two strategies (Janzen 1969, Baker and Hatheway 1970, Baker 1972a). However, more complicated interactions seem to be the rule (Janzen 1970) and, in many cases, the escape from predation is intimately connected with seed-dispersal, sometimes even involving the predators as dispersal agents. In any case where seeds are destroyed in the operation of a dispersal system (and this must be the case when seed-eating birds are involved or where the fruit is consumed along with the seeds it contains, even though some of them will be voided later), there is a balance struck between destruction and dispersal. The case of *Sterculia apetala* already demonstrates this.

Some years ago, Snow (1965) made a study of the fruiting of species of *Miconia* (Melastomataceae), in Trinidad. He concluded that the sequence of fruiting seasons shown by the various species had been arrived at by natural selection and serves the function of reducing competition for the attention of dispersal agents (in this case birds that eat the berries and transport the seeds endozootically). This case may also be looked upon slightly differently as increasing community efficiency by contributing to the feeding of the birds throughout the year, keeping them available for seed dispersal.

Smythe (1970) made a careful study of the fruits and seeds that fell over a period of 17 months in the humid forest of Barro Colorado Island, in the Panama Canal Zone. Although the variation in climatic factors may play the most important role in determining the fruiting seasons of the plants here, he also observed the impact of biotic factors. Small-seeded fruits, whose seeds pass through the guts of animals, ripen in sequence through the year so that the agents of dispersal are kept fully occupied and competition for their services is minimized. Where the fruits contain large seeds, however, the chances of damage to the seeds from frugivorous animals is increased and, for them, synchronous fruiting within and between species has an advantage in "saturating the market," especially when scatter-hoarding animals like agoutis (*Dasyprocta punctata*) are involved. These animals fail to find many of the fruits they have buried and these then become available for germination.

History of Animal-Flower Interactions

In the forests of the tropics a greater range of animal visitors to flowers and fruits is to be seen than in any other part of the world. Any treatment of the probable history of the interactions between the flowering plants and their animal visitors would have to be extremely lengthy and here we can do no more than consider some generalizations about pollination history and then one particular problem involving the association between flowers and bats.

The fossil record of the animals that visit flowers may be no more complete than that of the flowering plants themselves, but it shows the Coleoptera entering the picture in the Permian Period. It is widely believed (Baker and Hurd 1968) that the beetles were the first anthophilous insects and their great antiquity, matched only by the Diptera which enter in Upper Permian time, is appropriate. Lepidoptera, Hymenoptera, and Aves (birds) are represented from the Jurassic Period onwards and the Chiroptera (bats) only come on the scene at the beginning of Tertiary time. It is not suggested that the first appearance of any of these animals was as a pollinator so, if the record is in any way revealing, it simply shows that (with the exception of the beetles and flies) the animals involved in pollination were up against the same problems of interhemispheric dispersal as the angiosperms. Some, such as the "higher" Lepidoptera and Hymenoptera and the syrphid flies, only came into prominence at the beginning of the Tertiary Period (Baker and Hurd 1968.

As a result of this timing, the development of differences in emphasis and pattern between the pollination pictures in the two hemispheres was inevitable. This is evidenced in its most dramatic form by the restriction of hummingbirds (Trochilidae) to the New World where they have prospered with the consequence that, although the nonhovering sugar-birds (Coerebidae) are also found in the neotropics, the emphasis in bird pollination there is on hovering birds. In Africa, as elsewhere in the paleotropics, adaptation is to nonhovering sunbirds (Nectariniidae) and other smaller groups of birds that have developed acrobatism rather than a hovering ability.

Sapromyophilous flowers (at least among the Orchidaceae) are more common in the drier parts of Africa than they are in the neotropics, and van der Pijl and Dodson (1957:102) suggest that "the switch-over to Diptera in Africa must have some relation to dominance of flies." On the other hand, it seems likely that large, solitary bees may be more important in the pollination spectra of American forests than they are in African forests, although social bees (e.g., *Apis* in the Old World, *Melipona* and *Trigona* in the New) cannot be ignored in either hemisphere. However, the problems inherent in the study of long-range evolutionary developments in the area of interactions between pollinators and plants are well illustrated by reference to one group of animals that have flower-visiting proclivities—the bats (Chiroptera). Consequently, these are treated here in some detail.

Interaction with Bats

BAT FLOWERS. The general features of bat-pollinated flowers and the corresponding adaptations of the bats that visit them have been reviewed in recent years (e.g., van der Pijl 1936, Jaeger 1954b, Baker and Harris 1957, Baker 1961, Vogel 1958, 1968–1969). The plants themselves are often trees or vines, less often shrubs and epiphytes. Herbs (of conventional size) scarcely enter the picture as bats will not fly very close to the ground. The flowers are borne on the plant in such a manner that they may be visited freely by the bats (which may have wing-spans from less than a couple of decimeters up to a meter). Thus, some bat-pollinated trees are leafless at the time of flowering and often have a pagoda habit of branching, as in *Ceiba pentandra* (Baker 1961). Others exhibit penduliflory (e.g., *Kigelia* spp.), flagelliflory (*Mucuna*), cauliflory (*Crescentia*) or have flowers standing out prominently from the foliage (*Pachira* spp.).

There are two major syndromes of flower characters: the "shaving brush" or "stamen ball" syndrome and the "campanulate" syndrome. In the former, a brush or ball of stamens is exposed to the breasts of clasping bats (Figure 1); in the latter, the gamopetalous flower is bell-shaped (Figure 2) and the bat visitors insert either just their heads or the whole of the forward parts of their bodies (receiving pollen either on their backs or the undersides of their bodies from the stamens). In both kinds of flower, the stigmas are appropriately placed to pick up pollen brought from another flower. A few cases, notably *Lafoensia punicifolia* (Lythraceae), combine characters of both types (Vogel 1968–1969). Sometimes the "shaving brush" or "ball" syndrome of characters is made up from an inflorescence of small flowers, particularly in the Mimosaceae (Figure 1 illustrates a ball of flowers produced by *Parkia clappertoniana*). In all cases, the flowers tend to be dull in color, nocturnal in opening, and produce both a sour smell and a moderate to copious supply of rather thin nectar.

The "shaving brush" syndrome appears to be more frequent in the paleotropics. Among the bat-flowers with a campanulate inflorescence, Vogel (1968–1969) has pointed out that a subtype with a large "bell" (offering a landing place for a bat) is also more common in the paleotropics, whereas the small "bell" subtype (where only the face or head of the bat can be thrust in in search of pollen or nectar) is characteristic of the neotropics. Both circumstances can be correlated with the presence of larger, less active bats in the paleotropics.

BAT VISITORS. The bats form the Order Chiroptera, the only mammals to develop true flying abilities. Romer (1966:212–213) considers the paleontological evidence indisputable that they are derived from arboreal insectivorous ancestors and that the earliest bats were, themselves, insectivores. There are two suborders of extant bats, the Microchiroptera and the Megachiroptera. The Chiroptera as a whole make their appearance in the fossil record in the early Tertiary, and by the end of the Eocene and the beginning of the Oligocene many of the modern families of Microchiroptera have become distinguishable in Europe and North America. The earliest record of a clearly megachiropteran bat is from Oligocene deposits in Italy (Meschinelli 1903) (although the puzzling case of the fossil bat *Icaronycteris index* from an earlier deposit in North America will be considered later).

The Megachiroptera, consisting of the single family Pteropidae, (Pteropodidae in the classification by Koopman and Jones 1970) are restricted to tropical and subtropical regions of the Old World and are almost exclusively vegetarian—chewing soft fruits or flower parts or lapping nectar. In the Old World, the Microchiroptera are insectivorous; however, they are also inhabitants of the New World and there, in the neotropical regions, some of the members of this suborder have discovered the fruit- and flower-visiting niches and have exploited them.

In the New World the Microchiroptera contains nine families (eight according to Koopman and Jones 1970); one blood-lapping (the true vampires —Desmodontidae), seven carnivorous (mostly insectivorous) and one, the Phyllostomatidae, with a diversity of feeding habits. In the Phyllostomatidae there are seven subfamilies, of which the Chilonycterinae appear to be insectivorous, the Glossophaginae both insectivorous and nectarivorous, and the Phyllostomatinae and Phyllonycterinae insectivorous, nectarivorous and frugivorous. The remaining three subfamilies (Carollinae, Sturnirinae, and Stenoderminae) are primarily frugivorous (Matthews 1971). Koopman and Jones (1970) include the vampires (Desmodontidae) in the Phyllostomatidae and the Sturnirinae in the Stenoderminae, but these taxonomic modifications do not change the general picture significantly.

The usual interpretation of the present-day distribution of the two suborders of bats and their patterns of feeding in the two hemispheres has been that both suborders originated in the Old World from insectivorous ancestors but that only the Microchiroptera migrated to the New World, doing so by way of temperate regions in which insectivory was the only feasible means of feeding. When the Microchiroptera reached the neotropics, radiation into the flower- and fruit-feeding niches became possible because appropriate trees, shrubs, vines and epiphytes were available (Baker and Harris 1957, Baker 1961, Baker and Hurd 1968.)

BAT BEHAVIOR. If, as the history given above suggests, the habit of flower visitation developed independently in the two hemispheres, it would be surprising if there were not significant differences in the patterns. There is evidence that this is the case.

Thus, very obvious differences in the behavior of the bats are to be observed by anyone who attempts to make photographic records of their visitations to flowers. I have done this with both suborders of bats. The Megachiroptera, as illustrated by *Epomophorus gambianus, Nanonycteris veldkampii, Eidolon helvum* and *Micropteropus pusillus* in West Africa, lose no time in making their way to appropriate flowers at an appropriate time soon after sunset (which is not the same for each species) (Baker and Harris 1957, 1959, Harris and Baker 1958, 1959, Baker 1961). The crepuscular activity may be related to their dependence on visual and olfactory cues to the location of flowers; except for

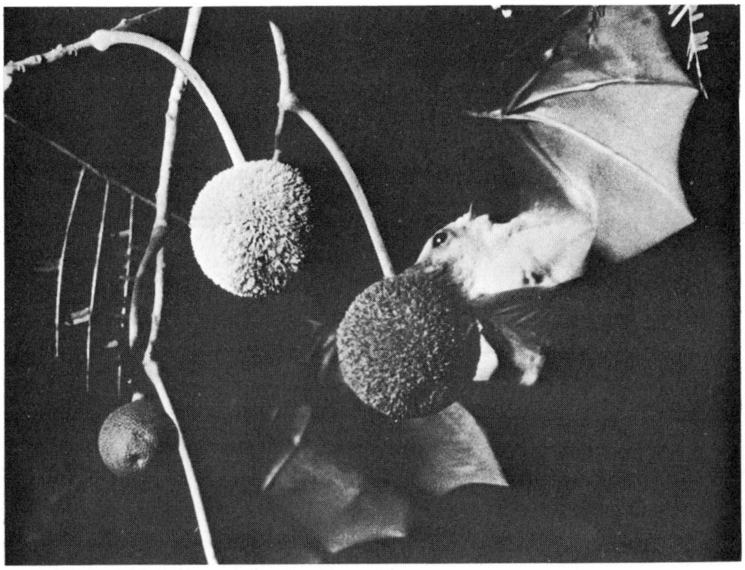

FIGURE 1. *Epomophorus gambianus* (Pteropinae, Pteropidae, Megachiroptera) visiting an inflorescence of *Parkia clappertoniana* (Mimosaceae). Achimota, Ghana, West Africa.

FIGURE 2. *Glossophaga soricina* (Glossophaginae, Phyllostomatidae, Microchiroptera) flying up to the single open flower of *Trianaea* sp. (Solanaceae). Finca La Selva, Heredia Province, Costa Rica, Central America.

Roussetus, no Megachiroptera possess a sonar system. By contrast, neotropical Microchiroptera, in my experience, delay their first visits to flowers for as much as one hour after sunset (Baker 1970b, Baker, Cruden, and Baker 1971, see also Brown, 1968). Whether this delay is due to their prior occupation with insect-catching is uncertain, for the full patterns of their nocturnal habits are, as yet, inadequately studied. *Glossophaga soricina* (Glossophaginae) takes insects as well as visiting flowers (Matthews 1971), and insects are included in the stomach contents of *Leptonycteris sanborni* (Glossophaginae) but seem to be less important in its protein nutrition than are pollen grains from the flowers (Donna Howell of University of Arizona pers. comm.). *Leptonycteris sanborni* has been photographed biting at anthers of *Ceiba acuminata* (Baker, Cruden, and Baker 1971). *Brachyphylla nana* (Phyllonycterinae), from Cuba, appears to be primarily a pollen-eater, but also takes nectar and laps the pulp from soft fruit (Silva and Pine 1969). *Artibeus jamaicensis* is interested in fruit and insects (Tuttle 1968) as well as nectar: *Micronycteris, Phyllostomus* and *Vampyrum* (which is not a vampire!) eat fruit and catch insects (Matthews 1971). *Phyllostomus* may even take lizards and other bats as well as fruit and nectar (Glass 1970).

However, even after the flower-visitations by neotropical Microchiroptera begin, the pattern is different from that of the paleotropical Megachiroptera. This is not necessarily due to any peculiarities in the flowers because trees of *Durio zibethinus*, which are attractive to megachiropteran bats in Southeast Asia, are equally attractive to Microchiroptera when planted in the neotropics (Baker 1970b). A similar conclusion may be drawn from bat-visits to *Kigelia aethiopica* planted in Panama (Vogel 1958).

It appears that Megachiroptera usually employ the "opportunistic" pattern of flower-visitation referred to earlier in this paper. Thus, on *Parkia clappertoniana*, in West Africa, individuals of *Epomophorus gambianus* and *Nanonycteris veldkampii* spend as long as 20 minutes in and around a tree even though visits to a particular inflorescence (Figure 1) always last less than a minute (Baker and Harris 1957).

In the case of *Ceiba pentandra* (also in West Africa), individuals of the same two species, as well as *Eidolon helvum*, clamber around the flower-bearing branches of the tree, visiting numerous flowers in turn (Baker and Harris 1959). Both *Parkia* and *Ceiba* are "mass flowerers" in which many inflorescences, containing multitudes of flowers, will be open at the same time on the tree. Many bats will be involved with the same tree at any one time.

By contrast, there is growing evidence of trap-lining by Microchiroptera in the neotropics. Characteristically, the visits of these bats to flowers are of much shorter duration—sometimes being mere hovering or no more than a dart at flowers while flying past (Vogel 1968–1969, Baker 1970b, Baker, Cruden, and Baker 1971). Even with an Old World tree like *Durio zibethinus*, which produces copious supplies of nectar in each of its many flowers, the phyllostomatid bats that visit it after it is planted in the New World stay only a few seconds before flying away from the tree. They may not return for as long as 20 minutes (Baker 1970b).

Similar fleeting visits by *Leptonycteris sanborni* to *Ceiba acuminata* in Mexico (Baker, Cruden, and Baker 1971), by *Artibeus jamaicensis* to the tree *Crescentia alata* (Baker, unpublished data), by *Glossophaga* sp. to the vine *Mucuna andreana* (Baker 1970b) and by *Glossophaga soricina* to the epiphyte *Trianaea* sp. (Solanaceae) (Figure 2) have been photographed. In the last case, at Finca La Selva, Heredia Province, Costa Rica, in lowland wet forest, the visits occurred at fairly regular 15 minute intervals, suggesting that the bat was following its rounds in between visits to the plant under investigation. The bat always approached from one direction and always left in a constant, different direction. Vogel (1968–1969) has also commented upon the apparent following of regular rounds by the neotropical Microchiroptera that he has observed.

Vogel (1968–1969) has also described the fleeting nature of microchiropteran flower-visits and correlates this with the "braking flight" that photographs show them to adopt in the vicinity of a flower as well as the sonar system that they possess (by contrast with their megachiropteran cousins). All of these characteristics which fit with a trap-lining habit could have developed from the agility, obstacle-avoiding and prey-detecting abilities required of an insect-catching bat.

In *Trianaea*, the amount of nectar available at any one time (indeed, the amount capable of staying by capillarity in the inverted bell of a flower) is extremely limited and is not in any way comparable to the 5 milliliters of nectar that may be produced by a single inflorescence of *Parkia* each evening. In fact, *Trianaea* exemplifies beautifully the plant contribution to the trap-lining syndrome, for no more than one flower is produced each night by each inflorescence, often by each plant. *Mucuna andreana* similarly tends to open only a single flower, with a limited supply of nectar, on each pendant inflorescence each night and the photographs (Baker 1970b) show that the bats do not always land. The explosive nature of the flower of *Mucuna* can also be regarded as an adaptation to the placing of pollen on a visitor which does not stay long at a flower before flying on.

Vogel (1968–1969) points to the appropriateness of the shallow bell-shape of a number of neotropical bat-flowers. If a bat is likely to make only a fleeting visit to a flower, it is more efficient to present the nectar so that it can be collected by a thrust of the head than to require the bat to climb into the flower to reach it.

Another difference between bat-pollination pictures in the neo- and paleotropics is provided by the extensive migratory habits of the Megachiroptera, even in the vicinity of the equator. *Eidolon helvum*, in particular, among West African flower-visiting bats, is noted for its migrations in large flocks, following the flowering and fruiting of trees (Allen 1939, Baker and Harris 1959, Harris and Baker 1959). This behavior also fits with an opportunistic syndrome.

However, more work is needed in both hemispheres on the behavior patterns of the bats and their relation to the morphology, phenology, physiology, and breeding systems of the plants that they visit. Bats are surely responsible for cross-pollination as well as self-pollination in the paleotropics (and there is indirect evidence of self-incompatibility in *Parkia*, Baker and Harris 1957). Equally, not all neotropical bat-flowers have the bell shape, and it must not be assumed without observation that all flower-visiting Microchiroptera are trap-liners.

DEMOLITION OF DOGMA

In the interpretation of the growing (but still not entirely adequate) array of observations made in the last decade or so, we must be careful not to be too closely bound to old dogmas that were largely promulgated when the information available was utterly inadequate. Some have been disposed of already. The first is that among nocturnal flowers, only those with a sweet odor are attractive to moths, while those with a sour odor will be visited only by bats. *Kigelia africana* has an unpleasant odor (and dark puple flowers as well) but in Ghana, at least, it is just as attractive to hawkmoths as to bats (Harris and Baker 1958, Baker 1961). In the neotropics, flowers of *Crescentia alata* and *C. cujete* attract moths as well as bats (Baker, unpublished data) and the same is true of planted *Durio zibethinus* (Baker 1970b).

Another dogma that has now been demolished is that there is a recognizable, taxonomically significant boundary between fruit-eating and nectar-lapping bats. All four flower-visiting genera that we have observed in West Africa (*Eidolon, Epomophorus, Nanonycteris,* and *Micropteropus*) belong to the subfamily Pteropinae (= Pteropodinae) in the Megachiroptera. This subfamily is designated the characteristically fruit-bat subfamily by Eisentraut (1945). It is true that individuals chew soft fruit, but they also indulge in nectar-lapping activity when the opportunity is presented. Similarly, in Fiji, my wife and I have photographed *Pteropus tonganus,* an acknowledged fruit bat (also a member of the subfamily Pteropinae), visiting flowers of planted *Ceiba pentandra* at a time when fruit was in short supply. In the neotropics I have seen and photographed *Artibeus jamaicensis,* a well-known member of the largely fruit-feeding subfamily Stenoderminae, visiting flowers. Conversely, Silva and Pine (1969) have shown that *Brachyphylla nana,* of the predominantly nectar-lapping Phyllonycterinae, will feed on soft fruit when this is presented to it. (Koopman and Jones, 1970, transfer *Brachyphylla* to the Stenoderminae.)

Even the most nearly sacred dogma, that in the paleotropics only the Megachiroptera are concerned with flowers and fruits as sources of nutritious and refreshing juices, is open to suspicion of being based on too little evidence. When one sees the facility with which Microchiroptera have taken to flower and fruit visitation in the neotropics, it is hard to believe that competition from the Megachiroptera has kept them completely away from this source of food in the paleotropics. In particular, if they should be operating a trap-lining system in the paleotropics, as they appear to do in the neotropics, fleeting visits of these small bats with long intervals between might well pass unnoticed in the dark.

THE PROBLEM OF PARKIA

All of these uncertainties that were once accepted dogmas must be taken into account when assessing the probable evolutionary history of bat pollination on both sides of the Atlantic Ocean, for there are very real problems that have so far defied explanation. At the center of these is the genus *Parkia* (Mimosaceae).

Species of *Parkia* occur in all three of the major tropical forest regions of the world. In Southeast Asia, Danser (1929), van Heurn (1929), and Docters van Leeuwen (1933) all saw bats visiting the pendant inflorescences of species in this genus. Later, Baker and Harris (1957) photographed other bats visiting *Parkia clappertoniana*, a West African species. We predicted that South American species would also prove to be visited by bats and this has been confirmed by the observations and photographs of Carvalho (1960) and Vogel (1968–1969). Thus, the genus *Parkia* is chiropterophilous in two areas of the paleotropics (with Megachiroptera) as well as in the neotropics (with Microchiroptera). Although there are slight differences in the manners of visitation by the two suborders of bats, the basic pattern of bat-attraction by *Parkia* is elaborate but common to both hemispheres.

There is a possibility, backed up by morphological evidence presented by Baker and Harris (1957), that the genus *Parkia* is of neotropical origin, but, in any case, in view of its pantropical distribution, it must already have been present in the American tropics very early in the Tertiary Period and probably was there considerably before that time. There are about nineteen morphologically very diverse species of *Parkia* known from South America and the West Indies, compared with eleven much less diverse species in Southeast Asia and five in the whole of Africa. Their heavy seeds are not adapted for transport over oceans and it is most likely that the genus became dispersed during the Cretaceous Period.

On the other hand, evidence available to us in 1957 (Baker and Harris, 1957:456–457), and apparently acceptable to Vogel in his review of the subject as recently as 1968–1969, suggested that the Microchiroptera have had a relatively short history of flower-visitation in the neotropics, beginning no earlier than the Tertiary (and, probably nearer the end rather than the beginning of that period). There are only three or four subfamilies of a single family in the suborder believed to contain species involved in flower-visitation and only two or three more are added when the picture is widened to include fruit consumption. In none of these subfamilies have the bats entirely lost their taste for animal protein. With the exception of a single fossil phyllostomatid (*Notonycteris*) from the Miocene of Colombia (Savage 1951), no evidence of this family has been found earlier than the Pleistocene —and there is no indication that *Notonycteris* had any interest in plantfood sources. No members of the Megachiroptera are found (living or fossil) in the American continent. Consequently, it appears that the inflorescence-visiting relationship between *Parkia* and its microchiropteran associates has developed quite recently, leaving unexplained what might have been the pollinators of *Parkia* (with its "chiropterophilous" syndrome of characters) before bats were available to it in South America.

Possible explanations for this seemingly unsatisfactory state of affairs have been offered (Baker and Harris 1957:458–459, Baker and Hurd 1968:406). The first of these, advanced only so that it might be disposed of, was that there was a purely fortuitous preadaptation of *Parkia* to a pollinator that was not yet available. We pointed out (Baker and Harris 1957:458) that

> almost inevitably this [preadaptation hypothesis] raises the doubt that, despite the deductions which seem to be most reasonably made from the systematic, geographical and paleontological data, we are putting the cart before the horse. Either the history of nectar-lapping bats goes farther back than the Eocene [probably the last opportunity for a member of a tropical forest flora to migrate by a northern or southern land-route between the continents] . . . or else the genus *Parkia* had its origin in the Old World rather than in the New (and the diversity of forms in South America is to be attributed to [recent] radiative evolution in a new environment).

In the latter case, the transportation of *Parkia* from the Old World to the New World would have to involve long-distance dispersal, an unlikely matter for this genus, as already pointed out.

Baker and Hurd (1968:406) put forward another possible explanation.

The previous pollinators of the New World Parkias may have been birds; the differences in syndromes of bat and bird pollination are not very great, and the red color of the inflorescences may, indeed, be a clue to the originally ornithophilous condition.

In connection with the interchangeability of ornithophily and chiropterophily, it may be pointed out that another problem-posing tree, *Ceiba pentandra*, is also almost certainly of neotropical origin though naturally dispersed to and established in West Africa (Baker 1965). It is pollinated by bats in West Africa (Jaeger 1954a, Baker and Harris 1959) and in the East Indies (van der Pijl 1936) although introduced by human carriage to the latter area. We have circumstantial evidence from Colombia (Baker and Harris 1959:9) that it is also visited by bats in the neotropics and this has been confirmed by Carvalho (1960) for Brazil and by Baker, Frankie, and Opler (1972) for Costa Rica. But *Ceiba pentandra* has been seen to be visited freely by hummingbirds in Mexico (Baker, unpublished data). Furthermore, the genus *Mucuna* (Papilionaceae) has bird-pollinated species as well as bat-pollinated species in the neotropics (Vogel 1968–1969, Baker 1970b) and paleotropics (van der Pijl 1941).

Vogel (1968–1969) has reviewed the historical situation more recently, adding new material that he has collected. For the *Parkia* problem he has two alternative schemes. In both, *Parkia* would have become established in the neotropics and in the paleotropics in Cretaceous time. At the beginning of the Tertiary Period, the Atlantic Ocean would have appeared as a barrier between the *Parkia* populations. (Parenthetically, we may remark that this is too late—the barrier was already there!) Then, his first scheme requires the independent development of chiropterophily by *Parkia* on both sides of the Atlantic Ocean with Megachiroptera and Microchiroptera accepting the challenge in the paleotropics and neotropics, respectively. His second scheme differs in that chiropterophily is assumed to have evolved in association with the Megachiroptera before the Atlantic Ocean barrier came into existence so that Megachiroptera and *Parkia* were cut off on both sides of the Ocean. However, the Megachiroptera then became extinct in the neotropics and their place was taken by Microchiroptera.

Arguments may be raised against both of these schemes but there may be significance in his suggestion that yet another of our dogmas—that there never have been Megachiroptera in the New World—may have to be modified.

Although they have not been considered by any botanical reviewer before now, recent reports by Jepsen (1966, 1970) on the remains of an unusually well-preserved fossil bat from Paleocene lake-bed deposits in southwestern Wyoming may be important. These deposits had been laid down in a subtropical climate and, although now standing at 2200 meters above sea-level, were at less than 300 meters elevation at the time of deposition. The remains of a young male bat were associated with fecal matter and other materials that might suggest it was a fish-eater as well as an insect-catcher. From our point of view, however, the most important feature of this bat is that despite being sufficiently close to present-day Microchiroptera to be placed in that suborder, it has the clawed index finger and several other characters now regarded as typical of the Megachiroptera.

Because there are no other satisfactorily preserved fossils of this age from the Americas, Jepsen (1966, 1970) is cautious about interpreting the phylogenetic position of this fossil bat (which he calls *Icaronycteris index*), but its intermediate morphology and its standing as the oldest extant fossil bat make it possible that it represents a stage prior to the full differentiation between Microchiroptera and Megachiroptera. If this is true, the Old World origin of the bats and the differentiation of the two suborders in that hemisphere, followed by the migration of the Microchiroptera to the New World is open to question, although not by any means ruled out because we do not know that the *Icaronycteris* line did not become extinct. Indeed, so many morphological, developmental, physiological, and ethological differences exist between present-day Megachiroptera and Microchiroptera that doubts have even been cast on their common ancestry (cf., Slaughter 1970).

We still know far too little of the history of the flower-visiting bats—and the *Parkia* mystery remains.

Perhaps the answer to these problems will not be found, even if we should be lucky enough to build up a more complete fossil history for the angiosperms and for the bats as a whole, until we understand better the ecosystems in which bats of both suborders operate. How much has the presence of small, hovering hummingbirds, some of which also operate a trap-lining system in the neotropical forests, influenced selection of a similar system for the bats and thereby prejudiced matters in favor of the neotropical Microchiroptera? How much is the ecological replacement in the paleotropics of hummingbirds by sunbirds (lacking the flight characteristics of the hummers) responsible for the success there of the "opportunistic" Megachiroptera? And are the structures of the forests as completely comparable in the two hemispheres as we, with our inadequate comparative information, think they are? Very little will be discovered in the future by studying any one group in isolation; there is now no boundary line between ecological study and evolutionary investigation. Adaptation within the full context of the ecosystem is at the heart of both. Interdisciplinary cooperation, preceded and succeeded by symposia such as this one, is essential.

References

Allen, G. M. 1939. Bats. Harvard Univ. Press. Cambridge, Mass.

Axelrod, D. I. 1970. Mesozoic paleogeography and early angiosperm history. Bot. Rev. 36:277-319.

Baker, H. G. 1961. The adaptation of flowering plants to nocturnal and crepuscular pollinators. Quart. Rev. Biol. 36:64-73.

————. 1963. Evolutionary mechanisms in pollination biology. Science 139:877-883.

————. 1965. The evolution of the cultivated kapok tree: A probable West African product. *In* D. Brokensha (Editor). Ecology and economic development in tropical Africa. Inst. Internat. Studs., Univ. California, Berkeley, Research Series 9:185-216.

————. 1970a. Evolution in the tropics. Biotropica 2:101-111.

————. 1970b. Two cases of bat pollination in Central America. Rev. Biol. Trop. 17:187-197.

————. 1972a. Seed weight in relation to environmental conditions in California. Ecology (in press).

————. 1972b. Migrations of weeds. *In* D. H. Valentine (Editor). Taxonomy and phytogeography of higher plants in relation to evolution. Academic Press. New York [In press.]

Baker, H. G., R. W. Cruden and Irene Baker. 1971. Minor parasitism in pollination biology and its community function: The case of *Ceiba acuminata*. BioScience 21:1127-1129.

Baker, H. G., G. W. Frankie, and P. A. Opler. 1972. Seed characteristics and seed dispersal systems in relation to forest vegetation types in Costa Rica. [In manuscript.]

Baker, H. G., and B. J. Harris. 1957. The pollination of *Parkia* by bats and its attendant evolutionary problems. Evolution 11:449-460.

————. 1959. Bat-pollination of the silk-cotton tree, *Ceiba pentandra* (L.) Gaertn. (sensu lato) in Ghana. J. West Afr. Sci. Ass. 5:1-9.

Baker, H. G., and W. M. Hatheway. 1970. Reproductive strategies in *Pithecellobium* and *Enterolobium*—further information. Evolution 24:253-254.

Baker, H. G., and P. D. Hurd. 1968. Intrafloral ecology. Ann. Rev. Entom. 13:385-414.

Balle, S. 1960. Contribution à l'étude des *Viscum* de Madagascar. Lejeunia, Mém. 11:1-151.

Bridwell, J. C. 1918. Notes on the Bruchidae and their parasites in the Hawaiian Islands. Proc. Haw. Ent. Soc. 3:465-505.

Brown, J. H. 1968. Activity patterns of some neotropical bats. J. Mammal. 49:754-757.

Burtt, B. L. 1970. Studies in the Gesneriaceae of the Old World, XXXI: Some aspects of functional evolution. Notes Roy. Bot. Gard. Edin. 30:1-42.

Carvalho, C. T. de. 1960. Das visitas de morcegos às flores. Anais Acad. Bras. Ciências 32:359-377.

Corner, E. J. H. 1949. The durian theory or the origin of the modern tree. Ann. Bot., n.s. 13:367-414.

————. 1964. The life of plants. Weidenfeld. London.

Croat, T. B. 1969. Seasonal flowering behavior in central Panama. Ann. Miss. Bot. Gard. 56:295-307.

Danser, B. H. 1929. Bestuiving van *Parkia* door vleermuizen. De Trop. Natuur 18:118-119.

Docters van Leeuwen, W. M. 1933. Bestuiving van *Parkia* door vleermuizen en bijen. De Trop. Natuur 22:119.

Duke, J. A. 1969. On tropical tree seedlings, I: Seeds, seedlings, systems and systematics. Ann. Miss. Bot. Gard. 56:125-161.

Eames, A. J. 1961. Morphology of the angiosperms. McGraw-Hill Book Co. New York.

Eisentraut, M. 1945. Biologie der Flederhunde (Megachiroptera). Biologia Generalis 18:327-435.

Frankie, G. W., H. G. Baker, and P. A. Opler. 1972. Phenological studies of woody plants in two lowland tropical forest ecosystems in Costa Rica. [In manuscript.]

Glass, B. P. 1970. Feeding mechanisms of bats. Pp. 84-92, *in* B. H. Slaughter and D. W. Walton (Editors). About bats. Southern Methodist Univ. Press. Dallas.

Harris, B. J., and H. G. Baker. 1958. Pollination in *Kigelia africana* Benth. J. West Afr. Sci. Ass. 4:25-30.

————. 1959. Pollination of flowers by bats in Ghana. Nigerian Field 24:151-159.

Hart, H. J. 1897. Bats fertilizing the flowers of *Bauhinia megalandra* Griseb. (n. sp.). Bull. Misc. Info. Trinidad Roy. Bot. Gard. 3(2):30-31.

Hopkins, B. 1969. Seasonal changes in the vegetation of West African savanna. Invited symposium paper at XI International Botanical Congress, Seattle, 1969.

Heirtzler, J. R. 1969. The theory of sea-floor spreading. Die Naturwissenschaften 56:342-347.

Jaeger, P. 1954a. Note sur l'anatomie florale, l'anthocinétique et les modes de pollinisation du Fromager (*Ceiba pentandra* Gaertn.). Bull. I.F.A.N. 16, Sér. A:370-378.

————. 1954b. Les aspects actuals du problème de la cheiropterogamie. Bull. I.F.A.N. 16, Sér. A:796-821.

Janzen, D. H. 1967. Synchronization of sexual reproduction of trees within the dry season in Central America. Evolution 21:620-637.

————. 1969. Seed-eaters versus seed size, number and dispersal. Evolution 23:1-27.

————. 1970. Herbivores and the number of tree species in tropical forests. Amer. Nat. 104:501-528.

————. 1971a. Escape of juvenile *Dioclea megacarpa* (Leguminosae) vines from predators in a deciduous tropical forest. Amer. Nat. 105:97-102.

————. 1971b. Euglossine bees as long-distance pollinators of tropical plants. Science 171:203-205.

————. 1971c. The fate of *Scheelea rostrata* fruits beneath the parent tree: Predispersal attack by bruchids. Principes 15:89-101.

————. 1972. Escape in space by *Sterculia apetala* seeds from the bug *Dysdercus fasciatus* in a Costa Rican deciduous forest. Ecology 53:350-361.

Jepsen, G. L. 1966. Early Eocene bat from Wyoming. Science 154:1333-1339.

————. 1970. Bat origins and evolution. Pp. 1-64, *in* W. A. Wimsatt (Editor). Biology of bats. vol. 1. Academic Press. New York.

Jones, E. W. 1956. Ecological studies on the rain-forest of southern Nigeria, IV. J. Ecol. 44:83-117.

Keay, R. W. J. 1957. Wind-dispersed species in a Nigerian forest. J. Ecol. 45:471-478.

Koopman, K. F. 1970. Zoogeography of bats. Pp. 29-50, *in* B. H. Slaughter and D. W. Walton (Editors) . About bats. Southern Methodist Univ. Press. Dallas.

Koopman, K. F., and J. K. Jones, Jr. 1970. Classification of bats. Pp. 22-27, *in* B. H. Slaughter and D. W. Walton (Editors). About bats. Southern Methodist Univ. Press. Dallas.

Kuijt, J. 1969. The biology of parasitic flowering plants. Univ. of California Press. Berkeley.

Matthews, L. H. 1971. The life of mammals. Vol. 2. Weidenfeld and Nicolson. London.

Maxwell, A. E., R. P. von Herzen, K. J. Hsü, J. E. Andrews, T. Saito, S. F. Percival, Jr., E. D. Milow, and R. E. Boyce. 1970. Deep sea drilling in the South Atlantic. Science 168:1047-1059.

Meschinelli, L. 1903. Un nuovo chirottero fossile (*Archaeopteropus transiens* Mesch.) Atti R. Inst. Veneto Sci., Lett., Arti 62:1330-1344.

Odum, E. P. 1959. Fundamentals of ecology (with H. T. Odum) 2nd edition. W. B. Saunders. Philadelphia.

Rick, C. M., and R. I. Bowman. 1961. Galápagos tomatoes and tortoises. Evolution 15:407-417.

Romer, A. S. 1966. Vertebrate paleontology. 3rd edition. Univ. of Chicago Press. Chicago.

Savage, D. E. 1951. A Miocene phyllostomatid bat from Colombia. Univ. Calif. Publ., Bull., Dept. Geol. Sci. 28: 357-365.

Silva Tabaoda, G., and R. H. Pine. 1969. Morphological and behavioral evidence for the relationship between the bat genus *Brachyphylla* and the Phyllonycterinae. Biotropica 1:10-19.

Slaughter, B. H. 1970. Evolutionary trends of chiropteran dentitions. Pp. 51-83, *in* B. H. Slaughter and D. W. Walton (Editors). About bats. Southern Methodist Univ. Press. Dallas.

Smith, A. C. 1970. The Pacific as a key to flowering plant history. H. L. Lyon Arb. Lecture 1:1-27. Univ. Hawaii. Honolulu.

Smythe, N. 1970. Relationships between fruiting seasons and seed dispersal methods in a neotropical forest. Amer. Nat. 104:25-35.

Snow, D. W. 1965. A possible selective factor in the evolution of fruiting seasons in tropical forests. Oikos 15:274-281.

Tarling, D., and M. Tarling. 1971. Continental drift. Anchor Books. Garden City, New York.

Tuttle, M. D. 1968. Feeding habits of *Artibeus jamaicensis.* J. Mammal. 49:787.

van der Pijl, L. 1936. Fledermäuse und Blumen. Flora 131: 1-40.

————. 1937. Disharmony between Asiatic flower-birds and American bird-flowers. Ann. Jard. Bot. Buitenz. 48: 17-26.

————. 1941. Flagelliflory and cauliflory as adaptations to bats in *Mucuna* and other plants. Ann. Jard. Bot. Buitenz. 51:83-93.

————. 1956. Remarks on pollination by bats in the genera *Freycinetia, Duabanga* and *Haplophragma*, and on chiropterophily in general. Acta Bot. Néerl. 5:135-144.

————. 1957. The dispersal of plants by bats (chiropterochory). Acta Bot. Néerl. 6:291-315.

————. 1969. Principles of dispersal in higher plants. Springer-Verlag, Berlin.

van der Pijl, L., and C. H. Dodson. 1967. Orchids and their pollinators. Univ. Miami Press. Miami.

van Heurn, W. C. 1929. Bestuiving van *Parkia* door vleermuizen. De Trop. Natuur 18:140.

Vogel, S. 1958. Fledermaüsblumen in Südamerika. Österr. Bot. Zeitschr. 104:491-530.

————. 1968-1969. Chiropterophilie in der neotropischen Flora. Neue Mitteilungen, I, II and III. Flora, Abt B 157:562-602; 158:185-222; 158:289-323.

Voorhoeve, A. G. 1967. Liberian high forest trees. Belmontia, Agric. Res. Reports 652:1-416. Wageningen.

Wolf, L. L. 1969. Female territoriality in a tropical hummingbird. Auk 86:490-504.

Wolf, L. L., and F. G. Stiles. 1970. Evolution of pair cooperation in a tropical hummingbird. Evolution 24:759-773.

A Comparison of the Hylean and Congo-West African Rain Forest Ant Faunas

WILLIAM L. BROWN, JR.

Introduction

The ants—Formicidae—are one of the world's truly spectacular animal families. They make up in numbers what they lack in individual size, and their activity rates and ubiquity in tropical countries bespeak their powerful role as transformers of energy in warm-country ecosystems. Unfortunately, we still lack quantitative data on biomass and energy transfer for tropical ants, and even in the temperate zone such information is exceedingly scarce and local, and remains largely unchecked. This is a pity, because collectors' impressions of ant distribution and abundance over the earth indicate a very interesting state of affairs. There is no doubt that temperature is a very important factor controlling the occurrence of ants. But the temperature limitation does not work in a straightforward way. Ants occur to and above the treeline in the Arctic and on most of the world's highest mountain chains, but they are absent above about 2300 meters in closed-canopy broadleaf forest everywhere, and specifically in the tropics. This fact never ceases to surprise me as I start hopefully into lush forest (after considerable trouble to get there!) at say, 2300 to 2500 meters in the Colombian Andes, in the Nilgiri Hills of southern India, or in the Ankaratra of central Madagascar. Even at 2100 meters in most tropical mountain forest, ants are exceedingly scarce, and in any one locality are represented by very few species. Yet at much higher altitudes (of 3500 or even 4000 meters) on treeless slopes of the Andes or the Himalayas, ants may be locally abundant. This suggests that radiant heat controls distribution; that is, cool shady mountain forests just don't provide enough warmth to allow ants to forage efficiently, or their larvae to develop fast enough, or both, while high-altitude or high-latitude open situations may offer sufficient radiant energy, even if in a brief seasonal burst, to do the trick (see review by Brian, 1965:68–69). All this suggests controlled-environment experiments that are beyond the scope of this paper.

We may say that the great lowland rain forests of Africa and South America appear to demonstrate by their fulsome and diverse ant faunae that temperature is probably not a serious limiting factor there. Yet we should not be too sure of this, because some sun-baked thorn-scrub, savanna, and even shrub semidesert environments, both in and out of the tropics, have astonishingly rich ant complements. This leads again to an appeal for quantitative data on colony density, species diversity, and eventually, biomass, for the different major habitat types over the world. All caveats considered, though, it is clear that the tropical forests have at least the greatest *diversity* at both the genus and species levels.

WILLIAM L. BROWN, JR., Department of Entomology, Cornell University, Ithaca, New York 14850. ACKNOWLEDGMENTS: My thanks are due to Paul D. Hurd, Jr., for the opportunity to study the University of California's collection of Chiapas Amber ants, and I am grateful to C. Baroni Urbani, B. Bolton, W. W. Kempf, R. W. Taylor, and E. O. Wilson for checking and adding names to the list of genera. For samples and information enlarging my understanding of the African ant genera, I am indebted to Barry Bolton and Jean Lévieux, and also to Jean Lévieux and Ivan Lieberburg for their help with the preparation of the manuscript and computer lists.
Behind this zoogeographical summary lies many years of work in field and museum, much of it made possible by successive grants (G-23680, GB-2175, GB-5574 and GB-24822) from the U. S. National Science Foundation.

It seems likely that evolution of most genera has prevailingly proceeded from a forest base into more open (xeric) habitats, some older opinions to the contrary notwithstanding. We should be cautious about this, however, because for ants wet forest areas are also in some degree "species sinks," as Jago here finds for his grasshoppers (p. 191 herein). We should also remember Wilson's (1959) findings for Melanesian ponerine ants, where many stocks apparently have moved from continental or large-island bases outward via "marginal" (more open, hence drier) habitats, and eventually occupied peripheral forested island areas.

The main rain forests of South America and Africa today are, except for interruptions caused by human activity, distributed in relatively continuous blocs. These blocs may have been somewhat subdivided during dry periods in the past (Moreau 1966, Vanzolini and Williams 1970), but their respective ant faunae are remarkably widespread within them as now known. Thus, the fauna of Guyana predicts to a remarkable degree that of the Rio Beni in lowland Bolivia, and the species list of south Cameroon is similar to that of the gallery forests of northern Angola. This is not to say that all species occur everywhere; actually, many of them have a very spotty distribution within this broad area. Of course many also extend far beyond the forest proper.

In order to consider the geography of ants, certain characteristics of the family need to be understood. All true ants are social and live in groups, normally representing the offspring of one or more queens, or sexually developed females. Workers are genetically females, but arrive at the adult stage as neuters after differential influences, probably mainly in the kind and quality of food they receive, affect them during development. In most ant species, the queens are produced in winged form together with winged males; copulation usually takes place during or after a nuptial flight of one or both sexes at a distance from the nest. This nuptial flight may offer great opportunity for dispersal by wind of fertile winged queens, yet the very nature of most flights as far as they are known also seems to militate against wide dispersal because of the rapidity with which the queens, once fecundated, drop to the substrate and actively divest themselves of their wings. Probably a certain amount of habitat selection is practiced by most newly fertilized queens, but it is hard to know how much of this occurs before the queen alights. Flying queens have been taken at above 900 meters in airplane traps, and also in shipboard traps at sea, but these queens have not been tested for nest-founding ability. A large minority of ant species have wingless queens (e.g., all army ants), or the queen is lacking and is replaced by worker reproductives. In still other species (e.g., at least some *Monomorium*, *Crematogaster*, *Rhoptromyrmex*, and known parasitic forms), the queen may have wings at some stage, yet may be unable to found a colony without the assistance of workers of her own or another species. Such species of course have reduced ability to colonize new areas across sea barriers.

Another obvious possibility for long-distance oversea transport is rafting. A number of kinds of ants, especially in the tropics, make their nests in preformed plant cavities, such as hollow twigs, burrows of wood-boring insects, hollow nuts or capsules, leaf-bases and the like. Often these nest cavities can be closed off to the outside by carton or sawdust plugs, or even by the plug-shaped heads of specialized soldiers. No doubt nests so barricaded can last for long periods of inclemency, for the brood can be eaten when starvation threatens, and adult workers and queens can often live a long time without food or free water. One can imagine a tree floating in the ocean with branches on the light side held well out of the water and bearing twigs containing numerous ant colonies. Such trees might make the voyage across the Atlantic in a matter of a few months in the Equatorial Current if the winds held fair (see Darlington 1957:17). I have no doubt that the voyage has been made safely on numerous occasions. Establishment of the voyagers on a foreign shore is a more difficult matter, in the face of potential competitors and predators, but the possibility still cannot be eliminated.

The third important way that ants are transferred between continents is of course through human commerce. In the tables below, tramp-species distributions are discounted for obvious reasons, but one thing about such distributions should be mentioned. If we consider the known or probable origin of tramp ant species, and the new places to which they go, a trend is obvious. Most of them originate in the Old World tropics or subtropics and establish themselves in the New World, Australia, or assorted oceanic islands. The reverse tendency, from the Americas to the Old World, does show itself (with species such as *Iridomyrmex humilis* and *Solenopsis geminata*), but this is obviously a weaker counter-current. The position of Australia, New Zealand, and smaller Oceanic islands in the cross-exchange hierarchy is even lower than that of the Americas, and in fact the "emigration potency" of each of the areas conforms to a Darlingtonian scheme, whereby those lands with a combination of large area and favorable (i.e., warm) climate produce animal species that tend to dominate and spread into lands increasingly smaller in area and less favorably endowed with climate.

The possibility exists that the current of tramps outward from tropical Africa-Asia is merely a reflection of (1) shipping practices, with ant-laden ballast going mainly from Europe to the New World, and (2) the longer time that Old World ants have had to adapt to humanly disturbed environments, which are mostly the kinds of places in which tramp species are found as immigrants. However, arguments can be made for the *reverse* movement of materials likely to carry ants, and some of the effective tramp species do not seem to be particularly anthropophilic in their home countries. As we shall see, the ant distribution data tend to support the conclusion that the prevailing *direction* of tramp flow is largely independent of human influence. This whole question is open to experimental study that has never been properly started.

Noting that ants had apparently arrived in central Polynesia before man, but had not reached eastern Polynesia until carried there by him (Wilson and Taylor 1967), we may characterize the ants' colonizing capability over the oceans as only slightly poorer than that of the bats, and better than that of the Trichoptera, Isoptera, and some other insect orders (see Zimmerman 1948).

World Distribution of the Ant Genera

Table 1 shows the distributions of the genera of living ants, taking into account published revisions and some of my own projections, many of them from my manuscripts toward a reclassification of the Formicidae. Naturally, those subfamilies and tribes that have been most studied in recent years, whether or not they have been published on, are those in which I consider the genera here listed to be relatively firmly established. In those groups of genera not yet closely studied, particularly among the Myrmicinae, there are a number of amalgamations yet to come. I might for example point to *Lordomyrma* and *Rogeria* as genera doubtfully distinct from one another and from *Stenamma*, and the separation of *Tetramorium* from *Xiphomyrmex* and *Triglyphothrix* comes perennially into question. Small genera such as *Promeranoplus, Prodicroaspis, Romblonella, Willowsiella,* and *Tetramyrma* remain unsatisfactorily defined; the relationships of *Leptothorax* to *Podomyrma, Atopula* and relatives in the Oriental-Australian and Ethiopian regions on the one hand, and to the *Macromischa* in the Caribbean area on the other, are still unclear. *Leptothorax* and *Mychothorax* may be species-groups within one genus, or they may be two genera. The New World *Iridomyrmex* almost certainly does not belong to that genus, but it is still not clear how it fits in with the other New World Tapinomini. The neotropical *Tapinoma* apparently provide a similar case. The *glaber* group of *Iridomyrmex* may really belong to *Turneria*. The Attini are probably over-split, and 5–6 or even fewer genera could well distill out from the current 10. In the Cephalotini, I may be a bit impulsive in recognizing only *Procryptocerus* and a much-enlarged *Cephalotes*, but it does seem a less tortured arrangement than the one we have inherited, with latter-day nomenclatural juggling, from Emery and Wheeler. A number of genera of Ponerini subside back into *Pachycondyla*, whence they originally came. The slowly multiplying genera of Leptanillinae are all listed with doubt; so little is known about them that their taxonomy must be arbitrary. The genera and subgenera of Dacetini that I introduced in 1948 and 1953 have suffered casualties, mainly because the flood of species of this tribe found since 1950 have included connecting forms, particularly among the short-mandibulate genera. It seems that even more genera in this group will sink as Berlese funnel collecting increases around the world, even though some bizarre new genera are still turning up from time to time.

I should acknowledge the obvious; my list excludes all subgenera on principle. I will also agree that after closer study a few of these subgenera might possibly be worth resurrecting as good genera, but I have found the subgeneric category so weak and so inconsistent in application, and so productive of taxonomic confusion and wasted effort, that for me the burden of proof is on those who would use it.

Many social-parasitic "satellite" genera of important taxa such as *Myrmica, Leptothorax, Monomorium,* and *Pheidole* are not listed in the table, either because their taxonomic distinctness is prima facie questionable, or because despite great modification they seem to represent relatively sporadic and evanescent offshoots of their host genera. Such parasites apparently often undergo a particularly rapid kind of "degenerative" evolution leading to confusing "reduction convergences" that arise at places remote from one another on the globe, so their zoogeographical significance is more than ordinarily doubtful.

The disposition of genera and subgenera that may be familiar to the myrmecological reader, but not named in the zoogeographic tables, can be traced through the alphabetical list in the Appendix. Names, even obvious ones, probably have been omitted by inadvertance despite the protracted effort I have made to include all of those current within recent decades. I would appreciate hearing of omissions so that I can correct my list for a new distribution to taxonomists.

The columns in the tables give first the "conventional" zoogeographical regions, subdivided into appropriate compass directions (N, E, S, W). The arrangement is intended to show as nearly as is possible in a linear array, the principal Tertiary-Quaternary faunal-exchange connections among the regions, according to the conception of P. J. Darlington (1957). It should be emphasized, though, that the entries in the tables are based on the available data, including many unpublished records, and not a priori on Darlington's or any other zoogeographical theory.

Neotropical, S: Chile, Argentina, southeastern Brazil, Bolivian highlands, coastal and Andean Peru.

Neotropical, N: Central and northern Brazil, Amazonian lowlands of Bolivia and Peru, northern South America, Central America north into the Mexican states of Tamaulipas and Michoacan, West Indies.

Nearctic, S: North America south of about 40° N, except for the higher mountains of the United States; Mexican Plateau and the high mountains south to about 19°N.

Nearctic, N: North America north of about 40° N, and higher elevations in the United States south of 40°.

Palearctic, N: Europe north of the Pyrenees, Alps and Black Sea; highlands of Central Asia and Tibet south to Burma; Mongolia and eastern USSR; central and eastern China and Japan north of about 34°.

Palearctic, S: Mediterranean lands, including the Sahara; Asia Minor and northern Arabia, Iran and Afghanistan; lowland deserts of Central Asia south of the Aral Sea and the Tien Shan. In the Far East, typical southern Palearctic elements, such as *Messor* and *Cataglyphis,* are present, but here the northern Palearctic is customarily considered to merge directly into the Oriental Region.

Ethiopian, N: Africa from the southern Sahara (about 15° N) to about the Zambezi River (15° S); extreme southwestern Arabia.

Ethiopian, S: Africa from the Zambezi southward.

Malagasy: Madagascar with the Comoro Islands, Mauritius, and Reunion.

Oriental, W: India, Ceylon, and Pakistan south of the Pamir and Himalayas.

Oriental, E: China from the Tsinling Mountains and the Tibetan Scarp southward and eastward; Burma through Southeast Asia and the East Indies to Timor and Celebes; southern Japan, Formosa, Philippines.

Australian, N: New Guinea and nearby islands, including the Moluccas, Bismarck Archipelago, and Solomons; rest of Melanesia, Micronesia, and Central Polynesia (Samoa, etc.); northern half of the Northern Territory, Cape York Peninsula, and the Gulf of Carpentaria in Australia; New Caledonia. The ant fauna of Farther Polynesia is entirely man-introduced.

Australian, S: Continental Australia south of about 20° S; Tasmania, New Zealand, and nearby islands.

The entries themselves are intended to convey by code two biotic modes: "M" or "m" (for *mesic*) signifies the moister kinds of forest habitats, those with closed canopies. "X" or "x" refers to more xeric or open habitats—open woodland, savanna, thorn forest, semidesert, desert, and the like. The emphasis, as befits the title of this paper and the general topic of the symposium, is deliberately placed on the comparison between African and South American rain forests. It should be borne in mind that the biotic formations around the world between these two provinces are predominantly graded or clinal with reference to almost any single diagnostic character, be it plant or animal taxonomic, plant-physiognomic, or whatever. Furthermore, the "diagnostic" characteristics and "indicator" taxa tend to be distributed discordantly one from another, each according to its own ecological valency. For these reasons, any partitioning of the earth's surface into biogeographical regions is bound to be arbitrary and misleading to some degree. If, in spite of these serious difficulties, the typological mind persists in drawing zoogeographical lines "for convenience," then it is clear that the finer the subdivisions used, the more will be the information contained in the scheme. Fineness of subdivision unfortunately soon runs into practical difficulty in a table, and the resulting compromise usually ends up something like what I offer below.

One more characteristic that I have tried to show through the entries is the "importance," for the genus concerned, of its occurrence in a particular habitat in a particular zoogeographical region. This importance is indicated by either an upper-case "M" or "X," or lower-case "m" or "x." If a genus barely enters a region outside its main range, a letter is entered in the appropriate column in lower case. The same holds for a genus that is very rare and sporadic (i.e., relict) in one region as compared to another. An example is *Prionopelta,* (see Table 1) which is very common in rain forest in parts of tropical America and Melanesia, has extra-limital occurrences in northern Florida and southeastern Australia, and is rarely collected in different parts of tropical Africa. This convention is *not* intended for important comparisons *between* genera; the closely related genera *Prionopelta* and *Amblyopone,* for example, are both rated "M" in the northern neotropical region, but as far as we now know, *Prionopelta* is 2–3 orders of magnitude more abundant and much more continuously distributed than is *Amblyopone* in this same region.

Table 1. World Distribution of the Ant Genera

Genus	*Neotropical*		*Nearctic*		*Palearctic*		*Ethiopian*		*Malagasy*	*Oriental*		*Australian*	
	S	N	S	N	N	S	N	S		W	E	N	S
MYRMECIINAE													
Nothomyrmecia	–	–	–	–	–	–	–	–	–	–	–	–	X
Myrmecia	–	–	–	–	–	–	–	–	–	–	–	mx	MX
PONERINAE													
Amblyopone	M	M	Mx	Mx	–	MX	MX	–	–	Mx	M	Mx	MX
Mystrium	–	–	–	–	–	–	m	–	M	–	m	mx	–
Myopopone	–	–	–	–	–	–	–	–	–	–	M	M	–
Prionopelta	M	M	m	–	–	–	m	m	m	–	M	M	m
Onychomyrmex	–	–	–	–	–	–	–	–	–	–	–	M	m
Apomyrma	–	–	–	–	–	–	mX	–	–	–	–	–	–
New Genus A.	–	–	–	–	–	–	M	–	–	–	–	–	–
Paraponera	–	M	–	–	–	–	–	–	–	–	–	–	–
Acanthoponera	M	M	–	–	–	–	–	–	–	–	–	–	–
Heteroponera	M	M	–	–	–	–	–	–	–	–	–	M	Mx
Rhytidoponera	–	–	–	–	–	–	–	–	–	–	m	MX	MX
Ectatomma	X	MX	–	–	–	–	–	–	–	–	–	–	–
Aulacopone	–	–	–	–	–	M	–	–	–	–	–	–	–
Gnamptogenys	Mx	MX	x	–	–	–	–	–	–	m	M	M	–
Proceratium	m	Mx	M	M	m	M	m	–	m	–	M	M	m
Discothyrea	M	M	m	–	–	–	Mx	M	–	–	M	M	M
Typhlomyrmex	M	M	–	–	–	–	–	–	–	–	–	–	–
Platythyrea	m	M	m	–	–	–	Mx	mX	–	Mx	Mx	MX	MX
Probolomyrmex	–	M	–	–	–	–	M	M	–	–	M	–	x
Sphinctomyrmex	M	–	–	–	–	–	MX	–	–	mx	m	MX	MX
Cerapachys	m	M	x	–	–	x	M	M	Mx	Mx	M	MX	MX
Simopone	–	–	–	–	–	–	M	M	M	–	M	M	–
Cylindromyrmex	M	Mx	–	–	–	–	–	–	–	–	–	–	–
Acanthostichus	MX	M	X	–	–	–	–	–	–	–	–	–	–
Thaumatomyrmex	M	M	–	–	–	–	–	–	–	–	–	–	–
Harpegnathos	–	–	–	–	–	–	–	–	–	MX	M	–	–

TABLE 1. World Distribution of the Ant Genera—*Continued*

Genus	*Neotropical* S	*Neotropical* N	*Nearctic* S	*Nearctic* N	*Palearctic* N	*Palearctic* S	*Ethiopian* N	*Ethiopian* S	*Malagasy*	*Oriental* W	*Oriental* E	*Australian* N	*Australian* S
Centromyrmex	Mx	Mx	–	–	–	–	MX	M	–	MX	M	–	–
Dinoponera	m	M	–	–	–	–	–	–	–	–	–	–	–
Streblognathus	–	–	–	–	–	–	–	X	–	–	–	–	–
Paltothyreus	–	–	–	–	–	–	MX	MX	–	–	–	–	–
Megaponera	–	–	–	–	–	–	X	X	–	–	–	–	–
Odontoponera	–	–	–	–	–	–	–	–	–	M	M		
Pachycondyla	Mx	MX	mx	–	–	–	Mx	MX	MX	MX	MX	MX	MX
New Genus B	–	–	–	–	–	–	M	–	–	–	–	–	–
New Genus C	–	–	–	–	–	–	M	–	–	–	–	–	–
New Genus D	–	–	–	–	–	–	M	–	–	–	–	–	–
Ophthalmopone	–	–	–	–	–	–	X	X	–	–	–	–	–
Hagensia	–	–	–	–	–	–	–	X	–	–	–	–	–
Euponera	–	–	–	–	–	–	–	–	M?			–	–
Brachyponera	–	–	–	–	–	–	MX	MX	–	MX	MX	MX	X
Cryptopone	–	m	M	–	–	M	–	m	–	–	M	M	M
Simopelta	m	M	–	–	–	–	–	–	–	–	–	–	–
Belonopelta	–	M	–	–	–	–	–	–	–	–	–	–	–
Emeryopone	–	–	–	–	–	–	–	–	–	–	M	–	–
Ponera	–	m	M	M	m	Mx	–	–	–	–	M	M	m
Hypoponera	M	Mx	MX	–	–	M	Mx	Mx	Mx	Mx	M	M	MX
Plectroctena	–	–	–	–	–	–	Mx	MX	–	–	–	–	–
Psalidomyrmex	–	–	–	–	–	–	M	–	–	–	–	–	–
Asphinctopone	–	–	–	–	–	–	M	–	–	–	–	–	–
Leptogenys	Mx	MX	MX	–	–	–	MX	MX	MX	MX	MX	MX	MX
Prionogenys	–	–	–	–	–	–	–	–	–	–	–	M	–
Odontomachus	Mx	MX	MX	–	–	Mx	MX	MX	MX	MX	MX	MX	mX
ECITONINAE													
Eciton	M	M	–	–	–	–	–	–	–	–	–	–	–
Labidus	mx	MX	x	–	–	–	–	–	–	–	–	–	–
Nomamyrmex	Mx	Mx	x	–	–	–	–	–	–	–	–	–	–
Neivamyrmex	MX	MX	MX	–	–	–	–	–	–	–	–	–	–
Cheliomyrmex	–	M	–	–	–	–	–	–	–	–	–	–	–
Leptanilloides	–	M	–	–	–	–	–	–	–	–	–	–	–
LEPTANILLINAE													
Leptanilla	–	–	–	–	–	X	–	–	–	–	X	M?	X
Leptomesites	–	–	–	–	–	–	–	–	–	M?	–	–	–
Phaulomyrma	–	–	–	–	–	–	–	–	–	–	M?	–	–
Scyphodon	–	–	–	–	–	–	–	–	–	–	M?	–	–
Noonilla	–	–	–	–	–	–	–	–	–	–	M	–	–
DORYLINAE													
Dorylus	–	–	–	–	–	X	MX	MX	–	MX	MX	–	–
Aenictus	–	–	–	–	–	mx	Mx	Mx	–	MX	M	Mx	m
Aenictogiton	–	–	–	–	–	–	M	–	–	–	–	–	–
PSEUDOMYRMECINAE													
Pseudomyrmex	Mx	MX	MX	–	–	–	–	–	–	–	–	–	–
Tetraponera	–	–	–	–	–	X	MX	MX	MX	MX	MX	MX	mx
MYRMICINAE													
Myrmica	–	–	Mx	MX	MX	MX	–	–	–	m	–	–	–
Manica	–	–	–	MX	MX	MX	–	–	–	–	–	–	–

TABLE 1. World Distribution of the Ant Genera—*Continued*

Genus	*Neotropical* S	*Neotropical* N	*Nearctic* S	*Nearctic* N	*Palearctic* N	*Palearctic* S	*Ethiopian* N	*Ethiopian* S	*Malagasy*	*Oriental* W	*Oriental* E	*Australian* N	*Australian* S
Hylomyrma	M	M	–	–	–	–	–	–	–	–	–	–	–
Pogonomyrmex	X	X	X	X	–	–	–	–	–	–	–	–	–
Ephebomyrmex	MX	X	X	–	–	–	–	–	–	–	–	–	–
Aphaenogaster	–	MX	MX	MX	mx	MX	–	–	MX	MX	M	MX	MX
Messor	–	–	–	–	x	X	X	X	–	X	X	–	–
Veromessor	–	–	X	x	–	–	–	–	–	–	–	–	–
Goniomma	–	–	–	–	–	X	–	–	–	–	–	–	–
Oxyopomyrmex	–	–	–	–	–	X	–	–	–	–	–	–	–
Pheidole	MX	MX	MX	X	x	MX	MX	MX	MX	MX	MX	MX	MX
Proatta	–	–	–	–	–	–	–	–	–	–	M	–	–
Stenamma	–	M	Mx	Mx	Mx	Mx	–	–	–	–	m	–	–
Rogeria	Mx	Mx	–	–	–	–	–	–	–	–	–	MX	–
Lordomyrma	–	–	–	–	–	–	M	–	–	M	M	M	m
Lachnomyrmex	m	M	–	–	–	–	–	–	–	–	–	–	–
Geognomicus	–	–	–	–	–	–	M	–	–	–	–	–	–
Dacetinops	–	–	–	–	–	–	–	–	–	–	M	M	–
Adelomyrmex	–	M	–	–	–	–	M	–	–	–	–	M	–
Prodicroaspis	–	–	–	–	–	–	–	–	–	–	–	M	–
Promeranoplus	–	–	–	–	–	–	–	–	–	–	–	M	–
Calyptomyrmex	–	–	–	–	–	–	Mx	M	–	–	M	M	–
Mayriella	–	–	–	–	–	–	–	–	–	–	M	M	Mx
Meranoplus	–	–	–	–	–	–	X	X	X	MX	M	MX	MX
Podomyrma	–	–	–	–	–	–	–	–	–	–	–	MX	MX
Dilobocondyla	–	–	–	–	–	–	–	–	–	–	M	M	–
Terataner	–	–	–	–	–	–	M	–	M	–	–	–	–
Atopomyrmex	–	–	–	–	–	–	X	X	–	–	–	–	–
Poecilomyrma	–	–	–	–	–	–	–	–	–	–	–	M	–
Atopula	–	–	–	–	–	x	M	–	–	Mx	x	–	–
Brunella	–	–	–	–	–	–	–	–	M?	–	–	–	–
Ireneopone	–	–	–	–	–	–	–	–	m	–	–	–	–
Peronomyrmex	–	–	–	–	–	–	–	–	–	–	–	M?	M?
Vollenhovia	–	–	–	–	–	–	–	–	–	M	Mx	M	–
Rhopalomastix	–	–	–	–	–	–	–	–	–	M	M	–	–
Metapone	–	–	–	–	–	–	–	–	MX	M	M	M	MX
Melissotarsus	–	–	–	–	–	–	MX	–	m	–	–	–	–
Liomyrmex	–	–	–	–	–	–	–	–	–	–	M	M	–
Leptothorax	MX	MX	MX	MX	MX	MX	Mx	Mx	M	Mx	–	M	–
Harpagoxenus	–	–	–	Mx	Mx	–	–	–	–	–	–	–	–
Tetramorium	–	–	–	–	X	X	MX	MX	MX	MX	MX	MX	–
Xiphomyrmex	–	–	X	–	–	–	M	MX	MX	M	M	MX	X
Decamorium	–	–	–	–	–	–	–	X	–	–	–	–	–
Rhoptromyrmex	–	–	–	–	–	–	M	X	–	MX	MX	M	–
Triglyphothrix	–	–	–	–	–	–	MX	MX	–	MX	MX	M	–
Eutetramorium	–	–	–	–	–	–	–	–	M	–	–	–	–
Teleutomyrmex	–	–	–	–	X	–	–	–	–	–	–	–	–
Anergates	–	–	–	–	X	–	–	–	–	–	–	–	–
Strongylognathus	–	–	–	–	X	X	–	–	–	–	–	–	–
Macromischoides	–	–	–	–	–	–	M	–	–	–	–	–	–
Tetramyrma	–	–	–	–	–	–	–	X	–	–	–	–	–
Monomorium	–	MX	MX	x	–	X	MX	MX	MX	MX	MX	mX	mX
Huberia	–	–	–	–	–	–	–	–	–	–	–	–	MX
Chelaner	–	–	–	–	–	–	–	–	–	–	–	MX	MX
Hagioxenus	–	–	–	–	–	X	–	–	–	X	–	–	–
Syllophopsis	–	–	–	–	–	–	M	MX	–	–	–	–	–

TABLE 1. World Distribution of the Ant Genera—*Continued*

Genus	*Neotropical*		*Nearctic*		*Palearctic*		*Ethiopian*		*Malagasy*	*Oriental*		*Australian*	
	S	N	S	N	N	S	N	S		W	E	N	S
Anillomyrma	–	–	–	–	–	–	–	–	–	M?	m?	–	–
Diplomorium	–	–	–	–	–	–	MX	MX	–	–	–	–	–
Paedalgus	–	–	–	–	–	–	MX	–	–	M?	–	–	–
Allomerus	M	M	–	–	–	–	–	–	–	–	–	–	–
Megalomyrmex	MX	MX	–	–	–	–	–	–	–	–	–	–	–
Nothidris	MX	–	–	–	–	–	–	–	–	–	–	–	–
Oxyepoecus	MX	M	–	–	–	–	–	–	–	–	–	–	–
Solenopsis	MX	MX	MX	MX	MX	MX	X	X	–	MX	MX	M	MX
Carebara	MX	M	–	–	–	–	X	X	–	MX	MX	–	–
Carebarella	X?	m	–	–	–	–	–	–	–	–	–	–	–
Pheidologeton	–	–	–	–	–	–	MX	MX	–	MX	MX	M	–
Oligomyrmex	MX	MX	X	–	–	X	MX	MX	M	MX	MX	MX	MX
Tranopelta	MX	M	–	–	–	–	–	–	–	–	–	–	–
Brownidris	MX	–	–	–	–	–	–	–	–	–	–	–	–
Adlerzia	–	–	–	–	–	–	–	–	–	–	–	–	X
Machomyrma	–	–	–	–	–	–	–	–	–	–	–	–	X
Anisopheidole	–	–	–	–	–	–	–	–	–	–	–	–	X
Trigonogaster	–	–	–	–	–	–	–	–	–	X	X	–	–
Lophomyrmex	–	–	–	–	–	–	–	–	–	MX	MX	–	–
Stereomyrmex	–	–	–	–	–	–	–	–	–	X?	–	–	–
Xenomyrmex	–	X	X	–	–	–	–	–	–	–	–	–	–
Myrmecina	–	m	Mx	–	M	Mx	–	–	–	M	M	M	–
Pristomyrmex	–	–	–	–	–	–	M	m	m	MX	M	M	M
Acanthomyrmex	–	–	–	–	–	–	–	–	–	M	M	M	–
Perissomyrmex	–	M	–	–	–	–	–	–	–	–	–	–	–
Ocymyrmex	–	–	–	–	–	–	x	X	–	–	–	–	–
Myrmicaria	–	–	–	–	–	–	mX	mX	–	MX	MX	–	–
Cardiocondyla	–	–	–	–	–	X	mX	X	X	MX	MX	MX	X
Ochetomyrmex	MX	MX	mx	–	–	–	–	–	–	–	–	–	–
Romblonella	–	–	–	–	–	–	–	–	–	–	MX	MX	–
Willowsiella	–	–	–	–	–	–	–	–	–	–	–	M?	–
Crematogaster	MX	MX	MX	MX	–	MX	MX	MX	MX	MX	MX	MX	MX
Stegomyrmex	M	M	–	–	–	–	–	–	–	–	–	–	–
Phalacromyrmex	M	–	–	–	–	–	–	–	–	–	–	–	–
Tatuidris	–	M	–	–	–	–	–	–	–	–	–	–	–
Basiceros	m	M	–	–	–	–	–	–	–	–	–	–	–
Aspididris	m	M	–	–	–	–	–	–	–	–	–	–	–
Creightonidris	–	M	–	–	–	–	–	–	–	–	–	–	–
Octostruma	m	Mx	–	–	–	–	–	–	–	–	–	–	–
Rhopalothrix	M	M	–	–	–	–	–	–	–	–	–	m	M
Eurhopalothrix	M	Mx	x	–	–	–	–	–	–	–	M	M	m
Cataulacus	–	–	–	–	–	–	MX	MX	MX	MX	MX	–	–
Daceton	–	M	–	–	–	–	–	–	–	–	–	–	–
Acanthognathus	M	M	–	–	–	–	–	–	–	–	–	–	–
Orectognathus	–	–	–	–	–	–	–	–	–	–	–	M	M
Epopostruma	–	–	–	–	–	–	–	–	–	–	–	–	MX
Mesostruma	–	–	–	–	–	–	–	–	–	–	–	–	X
Trichoscapa	–	–	–	–	–	X	X	–	–	X	X	–	–
Colobostruma	–	–	–	–	–	–	–	–	–	–	–	M	MX
Microdaceton	–	–	–	–	–	–	Mx	MX	–	–	–	–	–
Strumigenys	MX	MX	Mx	–	–	–	MX	MX	M	MX	M	M	MX
Neostruma	M	Mx	–	–	–	–	–	–	–	–	–	–	–
Smithistruma	Mx	Mx	MX	MX	–	MX	Mx	MX	–	–	M	–	–
Kyidris	–	–	–	–	–	–	–	–	M	–	M	M	–

TABLE 1. World Distribution of the Ant Genera—*Continued*

Genus	*Neotropical*		*Nearctic*		*Palearctic*		*Ethiopian*		*Malagasy*	*Oriental*		*Australian*	
	S	N	S	N	N	S	N	S		W	E	N	S
Serrastruma	–	–	–	–	–	–	Mx	Mx	M	–	–	–	–
Glamyromyrmex	M	M	–	–	–	–	M	–	–	–	–	M	–
Dorisidris	–	X	–	–	–	–	–	–	–	–	–	–	–
Dysedrognathus	–	–	–	–	–	–	–	–	–	–	M	–	–
Epitritus	–	–	–	–	–	MX	M	–	–	M	M	–	–
Pentastruma	–	–	–	–	–	–	–	–	–	–	M?	–	–
Miccostruma	–	–	–	–	–	–	M	M	–	–	–	–	–
Quadristruma	–	–	–	–	–	–	X?	–	–	–	–	M	–
Tingimyrmex	–	M	–	–	–	–	–	–	–	–	–	–	–
Procryptocerus	MX	MX	–	–	–	–	–	–	–	–	–	–	–
Cephalotes	MX	MX	X	–	–	–	–	–	–	–	–	–	–
Apterostigma	m	M	–	–	–	–	–	–	–	–	–	–	–
Cyphomyrmex	MX	MX	X	–	–	–	–	–	–	–	–	–	–
Mycocepurus	MX	MX	–	–	–	–	–	–	–	–	–	–	–
Myrmicocrypta	MX	MX	–	–	–	–	–	–	–	–	–	–	–
Mycetarotes	MX	–	–	–	–	–	–	–	–	–	–	–	–
Trachymyrmex	MX	MX	MX	X	–	–	–	–	–	–	–	–	–
Sericomyrmex	MX	MX	–	–	–	–	–	–	–	–	–	–	–
Acromyrmex	MX	MX	X	–	–	–	–	–	–	–	–	–	–
Atta	MX	MX	X	–	–	–	–	–	–	–	–	–	–
DOLICHODERINAE													
Aneuretus	–	–	–	–	–	–	–	–	–	M	–	–	–
Leptomyrmex	–	–	–	–	–	–	–	–	–	–	–	Mx	Mx
Dolichoderus	Mx	Mx	MX	MX	MX	MX	–	–	–	MX	MX	M	MX
Monoceratoclinea	–	–	–	–	–	–	–	–	–	–	–	M	–
Linepithema	–	M?	–	–	–	–	–	–	–	–	–	–	–
Semonius	–	–	–	–	–	–	X?	mX	–	–	M?	–	–
Axinidris	–	–	–	–	–	–	X	–	–	–	–	–	–
Liometopum	–	–	X	X	–	X	–	–	–	–	mX	–	–
Turneria	–	–	–	–	–	–	–	–	–	–	–	X	MX
Froggattella	–	–	–	–	–	–	–	–	–	–	–	–	X
Iridomyrmex	MX	M	X	X	–	–	–	–	–	–	MX	MX	MX
Dorymyrmex	X	X	X	X	–	–	–	–	–	–	–	–	–
Forelius	X	x	X	x	–	–	–	–	–	–	–	–	–
Neoforelius	X	–	–	–	–	–	–	–	–	–	–	–	–
Bothriomyrmex	–	–	–	–	–	X	–	–	–	X	X	X	X
Azteca	M	MX	–	–	–	–	–	–	–	–	–	–	–
Engramma	–	–	–	–	–	–	M	–	–	–	–	–	–
Tapinoma	MX	MX	MX	MX	–	mX	mX	mX	X?	mX	MX	X	–
Ecphorella	–	–	–	–	–	–	–	X	–	–	–	–	–
Technomyrmex	–	–	–	–	–	–	Mx	MX	M	MX	M	M	MX
Anillidris	X?	–	–	–	–	–	–	–	–	–	–	–	–
Zatapinoma	–	–	–	–	–	–	–	–	–	X?	–	X	–
FORMICINAE													
Myrmoteras	–	–	–	–	–	–	–	–	–	M	Mx	–	–
Oecophylla	–	–	–	–	–	–	MX	–	–	MX	MX	MX	–
Gesomyrmex	–	–	–	–	–	–	–	–	–	mx	Mx	–	–
Myrmecorhynchus	–	–	–	–	–	–	–	–	–	–	–	–	Mx
Melophorus	–	–	–	–	–	–	–	–	–	–	–	–	X
Notoncus	–	–	–	–	–	–	–	–	–	–	–	–	MX
Pseudonotoncus	–	–	–	–	–	–	–	–	–	–	–	–	MX
Prolasius	–	–	–	–	–	–	–	–	–	–	–	–	Mx

TABLE 1. World Distribution of the Ant Genera—*Continued*

Genus	*Neotropical* S	*Neotropical* N	*Nearctic* S	*Nearctic* N	*Palearctic* N	*Palearctic* S	*Ethiopian* N	*Ethiopian* S	*Malagasy*	*Oriental* W	*Oriental* E	*Australian* N	*Australian* S
Lasiophanes	MX	–	–	–	–	–	–	–	–	–	–	–	–
Acropyga	m	M	–	–	–	m	M	–	–	MX	MX	MX	MX
Aphomomyrmex	–	–	–	–	–	–	M	M	–	–	–	–	–
Cladomyrma		–	–	–	–	–	–	–	–	M	M	–	–
Brachymyrmex	MX	MX	MX	MX	–	–	–	–	–	–	–	–	–
Myrmelachista	MX	MX	–	–	–	–	–	–	–	–	–	–	–
Pseudaphomomyrmex	–	–	–	–	–	–	–	–	–	–	M	–	–
Plagiolepis	–	–	–	–	mx	MX	MX	MX	MX	MX	MX	X	X
Anoplolepis	–	–	–	–	–	–	MX	MX	–	MX	X	X	–
Acantholepis	–	–	–	–	–	X	MX	MX	–	MX	MX	M?	–
Stigmacros	–	–	–	–	–	–	–	–	–	–	–	–	MX
Prenolepis	–	MX	MX	MX	–	MX	–	–	–	MX	mx	–	–
Euprenolepis	–	–	–	–	–	–	–	–	–	–	M	M	–
Paratrechina	MX	MX	MX	MX	–	MX	MX	MX	MX	MX	MX	MX	MX
Pseudolasius	–	–	–	–	–	–	MX	MX	–	Mx	Mx	M	–
Lasius	–	–	MX	MX	MX	MX	–	–	–	–	MX	–	–
Acanthomyops	–	–	MX	MX	–	–	–	–	–	–	–	–	–
Myrmecocystus	–	–	X	X	–	–	–	–	–	–	–	–	–
Teratomyrmex	–	–	–	–	–	–	–	–	–	–	–	–	M
Cataglyphis	–	–	–	–	–	X	x	–	–	–	–	–	–
Proformica	–	–	–	–	–	MX	–	–	–	–	–	–	–
Formica	–	–	MX	MX	MX	MX	–	–	–	–	mx	–	–
Polyergus	–	–	X	X	X	X	–	–	–	–	–	–	–
Rossomyrmex	–	–	–	–	–	X	–	–	–	–	–	–	–
Gigantiops	–	M	–	–	–	–	–	–	–	–	–	–	–
Santschiella	–	–	–	–	–	–	M	–	–	–	–	–	–
Opisthopsis	–	–	–	–	–	–	–	–	–	–	–	MX	x
Notostigma	–	–	–	–	–	–	–	–	–	–	–	M	MX
Camponotus	MX	MX	MX	MX	MX	MX	MX	MX	MX	MX	MX	MX	MX
Phasmomyrmex	–	–	–	–	–	–	M	–	–	–	–	–	–
Overbeckia	–	–	–	–	–	–	–	–	–	–	M	–	–
Dendromyrmex	–	M	–	–	–	–	–	–	–	–	–	–	–
Calomyrmex	–	–	–	–	–	–	–	–	–	–	–	MX	X
Echinopla	–	–	–	–	–	–	–	–	–	M	M	M	–
Polyrhachis	–	–	–	–	–	X	MX	MX	–	MX	MX	MX	MX
Forelophilus	–	–	–	–	–	–	–	–	–	–	M	–	–

Faunal Relationships Among the Main Tropical Areas

Given the data array in the table above, our problem is to extract from it the faunal differences and similarities among the three major regions of the earth that contain most of the mesic tropical forest. In the New World, most of such forest is found in the northern half of South America, with extensions north and south. In Africa, the main mesic tropical forests are in the Congo and along the underside of the West African bulge. A fragmented "Oriental-Australian" belt stretches from southwestern India to northeastern Australia and the Melanesian chains. In addition, Madagascar has a strip of wet tropical forest in the east that is nearly a thousand miles long from north to south. I shall not discuss Madagascar further, except to say that its incompletely known ant fauna, while rich in endemic species, contains relatively few genera, most of them occurring in both tropical Africa and in the Oriental region. In the lists given below, tramp species spread by human commerce are listed only for their assumed native homelands.

The first thing we count are the genera that occur in all three regions: Neotropical, Ethiopian, and Oriental-Australian (i.e., "tropicopolitan" genera). For the moment, we shall accept the fact that the Oriental merges gradually into the Australian region within the tropics; in other words, they are not separated by a cold-temperature barrier. (The genera indicated by an asterisk (*) occur now outside the tropics in the Northern Hemisphere, or are represented there by Tertiary fossils.) Then we count those genera shared by each pair of major tropical regions, but not occurring in the third region (Table 2).

"Tropicopolitan" Genera
(total: 29)

*Amblyopone	*Hypoponera	*Strumigenys
*Prionopelta	*Leptogenys	*Smithistruma
*Proceratium	*Odontomachus	Glamyromyrmex (including Codiomyrmex)
*Discothyrea	*Pheidole	*Tapinoma
*Platythyrea	Adelomyrmex	Acropyga
Probolomyrmex	*Leptothorax	*Paratrechina
Sphinctomyrmex	*Monomorium	*Camponotus
*Cerapachys	*Solenopsis	
Centromyrmex	Carebara	
*Pachycondyla	*Oligomyrmex	
*Cryptopone	*Crematogaster	

Ethiopian and Oriental-Australian Genera
(total: 30)

Mystrium	*Tetramorium	*Epitritus
Simopone	*Xiphomyrmex	Quadristruma
*Brachyponera	*Rhoptromyrmex	Semonius
*Dorylus	Triglyphothrix	Technomyrmex
*Aenictus	Paedalgus	*Oecophylla
*Tetraponera	Pheidologeton	*Plagiolepis
Lordomyrma	*Pristomyrmex	Anoplolepis
Calyptomyrmex	Myrmicaria	*Acantholepis
Meranoplus	*Cardiocondyla	Pseudolasius
Atopula	*Cataulacus	*Polyrhachis

Neotropical and Oriental-Australian Genera
(total: 12)

*Heteroponera	*Stenamma	Rhopalothrix
*Gnamptogenys	Rogeria	*Dolichoderus
*Ponera	*Myrmecina	Iridomyrmex
*Aphaenogaster	*Eurhopalothrix	*Prenolepis

Ethiopian and Neotropical Genera
(none)

TABLE 2. Genera of Ants Known to be Shared between Different Pairs of the Main Tropical Mesic Forest Areas of the Earth (The totals entered above the diagonal include the 29 genera shared by all three areas; those below the diagonal exclude the widespread 29.)

Region	*Ethiopian*	*Neotropical*	*Oriental-Australian*
Ethiopian	—	29	59
Neotropical	0	—	41
Oriental-Australian	30	12	—

As we move from Africa toward Australia, the number of genera shared by subregions decreases with increasing distance. Thus, if we compare Africa with just the Australian region, we find that they have only 22 genera in common. And if we exclude the 10 African genera that enter only the northern part of the Australian region, the number of genera shared by sub-Saharan Africa and the main part of the Australian continent drops to 12. So we see that the Oriental-Australian region and South America compared on the one hand, and Africa with Australia on the other, have a similar level of generic sharing.

ENDEMIC GENERA

The genera endemic to the four main regions make up an interesting category. In counting, I include as "endemic" to the neotropical region those genera centered in the American tropics, but with a few species entering the southern nearctic region. Also, those genera that extend from Africa into the Malagasy region, but not beyond, are considered Ethiopian endemics in these counts.

Neotropical	65	Oriental	22
Ethiopian	31	Australian	32

About 10 more genera occur only in the Australian region and the eastern part of the Oriental region.

As one might expect, the neotropical region shows the highest endemism, and the Oriental the lowest, in consonance with their differing degrees of geographical isolation.

The Ethiopian and Australian regions have about the same number of endemic genera, but in

Africa more of the genera seem to be tied to the closed-canopy forest, they tend to have fewer species, and they seem more often than not to have evolved from derivative rather than primitive relatives. In Australia, more of the endemic genera seem to have radiated into drier vegetational zones, and radiation has been more extensive; also, they have in general a more primitive complexion than do Ethiopian-region endemics. If these impressions can be trusted, I think they add to the general picture of the Ethiopian fauna being relatively younger than the Australian.

SPECIES-GROUP RELATIONSHIPS

A brief look at distribution of species and species-groups shows us quickly that the Africa-South America distributional gap is as clearcut at these levels as it is at genus level. Even tropicopolitan genera tend to be represented by different species groups in Africa and South America, and, except for obvious tramps, I do not know of a single *species* that is shared by these two continents. The few shared species groups that might be cited are mostly those with a high proportion of twig- or other plant cavity-inhabiting species, such as the *"Nesomyrmex"* group of *Leptothorax*, and doubtfully some groups in *Camponotus, Crematogaster,* and *Monomorium*; relationships in these last three genera are still very uncertain. Still other genera with a strong twig- or tree trunk-inhabiting component (e.g., *Pseudomyrmex, Tetraponera, Cataulacus, Procryptocerus, Cephalotes, Azteca, Cylindromyrmex, Simopone, Myrmelachista*) occur on one side of the Atlantic or the other, but not on both sides, showing that this ocean has been a formidable barrier even to the most likely rafting taxa.

In contrast, there is a great deal of sharing of species groups, and even of species, between the Ethiopian and Oriental regions. As taxonomic revision proceeds, these ties are certain to get stronger, because quite a number of species in several genera (e.g., *Pachycondyla, Brachyponera, Odontomachus, Tetramorium, Monomorium, Camponotus*) appear to be distinguishable only by their distribution east or west of the Indian Ocean. The Oriental-Australian and neotropical regions also share some species groups, and even one species (*Pachycondyla stigma*), if it is not a tramp.

Fossil Ant Faunas Compared with Modern Ones

The fossil remains of ants so far have come almost entirely from the Northern Hemisphere, and there chiefly from localities now within the Temperate Zone. The earliest known ant is *Sphecomyrma freyi* from the Cretaceous of New Jersey, placed in a distinct subfamily, and quite different from any living member of the Formicidae. Eocene ant fossils are few and incomplete or poorly preserved, and tell us little. By Oligocene-Miocene times, though, fossilization overtook two major and several minor assemblages of ants that we can recognize as related to genera and species alive today.

The Oligocene Baltic Amber contains insect remains trapped in the transformed resin of pine-like trees that formed mild climate forests in what is now north-central Europe (Wheeler 1914). The Florissant Shales of central Colorado and similar deposits scattered through the western United States entombed their rich insect remains in the sediments of shallow lakes (Carpenter 1930). In addition, we have fragmentary faunules in amber (Sicily, Chiapas, Burma, etc.) and in shales in southeastern Europe and elsewhere. From these Tertiary beds we have samples totaling over 20,000 specimens worth studying, and these represent nearly 200 species, including a few that have been examined but not yet described.

The best and most informative sample (Wheeler 1914) is that from the Baltic Amber (Oligocene). From the biogeographical point of view, Baltic Amber ants sort into four groups:

1. Extant genera still most prominent in the temperate Northern Hemisphere, and nearly limited to it, chiefly *Myrmica, Stenamma, Liometopum, Lasius* and *Formica.*

2. Living genera now widespread in the North Temperate Zone and more or less so in the tropics: *Ponera, Aphaenogaster, Leptothorax, Dolichoderus, Prenolepis, Camponotus.*

3. Living genera now found chiefly in the tropics and South Temperate regions (in some cases in the warmer parts of North Temperate Zone): *Platythyrea, Gnamptogenys, Pachycondyla, Tetraponera, Vollenhovia, Oligomyrmex, Monomorium, Iridomyrmex, Oecophylla, Gesomyrmex, Pseudolasius, Plagiolepis.*

The affinities of category 3 are predominantly Old World. Seven of these 12 genera (including true *Iri-*

domyrmex) are today restricted to the Old World, except for tramps. The remaining five are found in both Old World and New. All 12 now occur in the Oriental-Australian regions, but only eight are in the Ethiopian region.

4. Extinct genera (19), most of them allied to genera now living in the Oriental and Australian regions. Again here the relationships to the Ethiopian region are not quite as strong as they are to the Oriental-Australian, and links to the neotropical region are much weaker.

The Sicilian Amber ants, a small middle Miocene assemblage studied by Emery (1891), comprise only a few genera, such as *Cataulacus, Oecophylla,* and *Sicilomyrmex,* predominantly of Old World tropical affinities. The report of male *Crematogaster* in the Sicilian Amber by Emery was later retracted by him.

The Miocene Florissant Shale (Carpenter 1930), and similar but less productive western North American beds, such as Ruby Basin (Montana) and Latah (Washington State) have yielded thousands of specimens, mostly winged queens and males, from shallow-water lacustrine deposits. These are predominantly genera such as *Lasius, Formica, Liometopum,* and *Protazteca* (the last related to *Iridomyrmex*), with a sprinkling of other ponerine, pseudomyrmecine, myrmicine, dolichoderine, and formicine genera, some of them now extinct. Affinities are on the whole with genera occurring in the same region today, as well as with some now found only farther south in the American tropics. *Crematogaster* is absent from these sedimentary beds, and *Pheidole* is represented only by two doubtful winged queen specimens from Florissant.

The Chiapas Amber of southern Mexico is supposed to be Miocene in age. Though it contains a few ants, none has yet been formally described. I have examined most of the available samples (about 110 specimens), and of these about half are fragmentary, badly decomposed or otherwise in such poor condition that their genus, and often even their subfamily, cannot be determined with certainty. Fairly common among identifiable remains are males of three or more species of ectatommine Ponerinae, and light-colored workers of what seems to be one species of *Azteca*. A few workers represent the *pyramicus* group of *Dorymyrmex*, and there are a few poor examples of *Camponotus* and possibly of *Lasius*. A winged queen lacking a head and pronotum is *Pachycondyla* (=*Trachymesopus*) *stigma* or something very close to it, and some winged myrmicines appear to be attine males resembling those of *Mycetosoritis*. Two shrunken and distorted workers resemble *Stenamma,* and three workers in fair condition could well be minors of *Pheidole*; this last identification is fairly firm, but the specimens need to be recut and studied in detail. There is also a winged male, unfortunately with the dorsal side largely obscured by bubbles, that is almost certainly a *Crematogaster*; the wing venation agrees very well with that of many recent species of that genus, and the petiole, postpetiole and gastric base, while obscured and twisted, are apparently crematogastrine in form. The antennal pedicel is short and subglobular.

The Chiapas Amber faunule is not too different from what we might expect of a small resin-trapped sample of ants found in southern Mexico today. The specimens of greatest interest are the possible *Crematogaster* and *Pheidole*. The finding of *Crematogaster* workers and *Pheidole* soldiers or winged forms is needed to confirm these identifications. Meanwhile, for me at least, the presence of these two genera in the Miocene of tropical Mexico must be considered likely, unless the amber dating is questioned.

Taken all together, Tertiary fossil faunas of the Northern Hemisphere are an interesting mixture. Certain species of *Ponera, Dolichoderus, Liometopum, Formica, Lasius,* and *Prenolepis* of Amber times (Oligocene-Miocene) can often be matched rather closely to species of these genera existing today in the North Temperate Zone. At least, it can be claimed that they often represent the same species groups. This fact has been used to call ant evolution "stagnant since the Miocene" (Mayr 1942:140), an opinion that ignores important extinctions and geographical contractions of many ant genera since mid-Tertiary times, and also overlooks the world-wide expansion since then of now-dominant genera such as *Pheidole, Crematogaster, Tetramorium,* and *Camponotus,* whose combined species certainly number in the thousands. These expansions are worth outlining here.

Pheidole, with hundreds of described and undescribed species, is a dominant genus in tropical rain forest, warm semidesert, and some mild temperate areas in most parts of the earth. It does not seem to be represented in any pre-Miocene deposits, but has been reported from the Florissant Shale (Carpenter 1930) based on two winged females. Now I have found three (minor?) workers in the Chiapas Amber that may well be a species of *Pheidole*. Due to the indifferent preservation of both amber and shale specimens, especially the latter, these identifications should be regarded with reserve, as has already been stated above.

Pheidole today has many species that forage intensively on the trunks, branches, and foliage of herbs, shrubs, and trees; and some of these in all main distribution areas live under tree bark or in the epiphytes growing on the trees. One would expect it to have been caught in the resin had it been present in the Baltic Amber forests. In some rain forests (e.g., lowland Costa Rica) I have found *Pheidole* to be far and away the dominant ant genus collected by beating understorey foliage. I think we can assume that *Pheidole* was absent, or at least very rare, in the Northern Hemisphere through middle Miocene times. If this genus arose in Africa or South America during the Tertiary, it has had a spectacularly explosive evolutionary history since the Miocene.

Crematogaster, another dominant and widespread myrmicine genus, also has a blank fossil history up to the Miocene. Like *Pheidole,* it must have spread mainly since the mid-Tertiary, since the only known fossil is a single male in the Chiapas Amber. The distribution of *Crematogaster* is especially interesting when compared with that of the dolichoderine genus *Iridomyrmex*. Although these two genera are of course not at all closely related, they have entered a very similar adaptive zone. Both have many species that form powerful, populous colonies. The colonies often attend Homoptera on plant stems and foliage, and they form long, often dense columns from nest to food source. According to species, the nests may be situated in the ground, in termite nests, in natural plant cavities, in rotting logs or tree trunks, or in epiphytes. *Crematogaster* tends to build primary or auxiliary nests of carton in shrubs or trees, whereas this tendency is weak or absent in most *Iridomyrmex*. Both genera have defensive secretions emitted from the gastric apex. These secretions repel other arthropods and also become gummy on exposure to air, so that they can glue an arthropod enemy's antennae and limbs together if it gets smeared with the stuff. Both *Crematogaster* and *Iridomyrmex* have the waist and gaster so constructed that the latter can be raised vertically, and even thrust forward above the head, in order to direct the tip of the gaster against a potential foe. The structural modifications that allow this acrobatic defense system are quite different in detail, and the method of application is also very divergent: *Crematogaster* holds its drop of viscous poison on the end of its flexible (and often spatulate) sting, while *Iridomyrmex*, effectively lacking a sting, simply extrudes its gluey poison through an orifice under the tip of the gaster. Thus, while the two systems of defense are obviously very different in evolutionary origin, they apparently have converged to do much the same kind of job.

In view of their adaptive convergence, it is easy to see why *Crematogaster* and *Iridomyrmex* are distributed over the earth in such a complementary pattern. Unfortunately, the taxonomy of both these genera is at present chaotic. The *Iridomyrmex* of the New World differs from that of the Old World in both internal and external gastric structure of the worker caste, and they cannot be considered as congeneric. In my opinion, the New World *Iridomyrmex* species are very close to *Forelius,* and through *Forelius* they apparently connect with the *Dorymyrmex* complex of species, which is also confined to the New World. The situation requires a thorough revision utilizing karyotypic and other cryptic characters. Work is in progress, but for now I have no definitive arrangement to offer. At any rate, the true *Iridomyrmex* of the Old World and the *"Iridomyrmex"-Forelius-Dorymyrmex* complex of the Americas appear to be cognate lineages within tribe Tapinomini, and it makes little difference if we consider them together as one taxon for purposes of contrast to *Crematogaster*.

The genus *Crematogaster* forms a tribe, isolated and distinct among the Myrmicinae, and up to now with no identifiable relatives in that subfamily. The genus has been split into subgenera, but these are apparently only species groups, and some of them weak at that. The species-level taxonomy is difficult, due to the great number of species, the close relationships among them, and their considerable variability, which is often allometric. Unfortunately, some of the most irresponsible and profligate descriptive publication ever visited upon the ants has left *Crematogaster* a taxonomic shambles. Buren, Kempf, and one or two other myrmecologists have begun to sift through this trash-heap in an effort to sort out the names—one to a species; but it will be a long time until these workers can bring enough order for us to make fairly accurate species lists. More than 900 names (species, subspecies, and varieties) have been proposed, but I doubt that more than half of these will prove to be valid species. The nearctic fauna, which through the revisionary efforts of Buren (1968) is probably somewhere near being worked out, numbers about 25 species. Tropical America should have at least another 50, possibly more. From the existing literature, I would *guess* the real numbers of species in Africa with Madagascar to be about 175, in the southern palearctic about 15, in the Oriental region about 125, in Melanesia about 50, and in Australia about 30. Interestingly, *Crematogaster* fails to reach Fiji, New Zealand, or Polynesia; it is not yet known from Chile, and in Argentina it is sparsely represented only in the north.

The weight of distribution of *Iridomyrmex* in the Old World regions is just the opposite: Africa and palearctic, 0; Oriental, about 5, in the eastern part only; Melanesia, 25–30; Australia, about 80. *Iridomyrmex* does not reach far into the Pacific, although a few tramp species have been carried into New Zealand and Polynesia by man. It is interesting to note that as one goes from north to south on the Australian continent, *Crematogaster* seems to become less abundant generally and less varied, while *Iridomyrmex* species tend to become more dominant and diverse; it seems fair to say that the latter is the overall dominant ant genus in southern Australia. In New Guinea, the two genera are much more evenly balanced. The New World *Iridomyrmex* counterparts are most abundant and diverse, with perhaps some 40 species, in southern South America, mostly in Argentina, Chile, and southeastern Brazil. They are modestly represented in the Andes and on the dry west coast of South America, and three or four species reach northward into the drier and warmer sections of the nearctic region. Notably, members of this complex are very poorly represented in the Amazonian and Central American rain forest, even if we allow that the handful of "*Tapinoma*" species known from these areas may really belong to the "*Iridomyrmex*"-*Forelius-Dorymyrmex* complex.

Going back to the Tertiary, it is clear from the Baltic Amber that *Iridomyrmex* was a dominant genus in the Oligocene. The five *Iridomyrmex* species recognized by Wheeler (1914) comprised well over half of the nearly 12,000 specimens that he and other specialists determined from that formation. In the Florissant Shale, *Iridomyrmex* occurs in much reduced numbers, but the related genus *Protazteca* was dominant (more than a quarter of the approximately 5600 specimens), and the subfamily Dolichoderinae still represented 63 percent of the total specimens, a proportion not significantly changed from the 64 percent of the Baltic Amber. In both the Baltic Amber and the Florissant Shales, the Formicinae make up about one-third (32%) of the individuals, and the Myrmicinae 5 percent or less, though the latter subfamily shows a considerable diversity of genera and species in both formations. This pattern compared with the modern distribution implies the massive replacement of the Dolichoderinae, primarily by genera of the Myrmicinae, in the Northern Hemisphere since the Miocene. The myrmicine taxa most widely involved in this replacement were *Pheidole, Crematogaster,* and *Tetramorium*, with *Myrmica* prominent in the cooler regions.

This hypothesis runs directly counter to that put forward by Haskins (1939:158–162), which has the Dolichoderinae and Formicinae, with their thin, flexible integuments, replacing the relatively heavily armored Myrmicinae and Ponerinae. Haskins' evolutionary scheme rests on two questionable assumptions, the first of which is that such species as *Iridomyrmex humilis* (the "Argentine Ant") are "world-conquerors." He believes that *I. humilis* "undertook a campaign of expansion which has left almost no part of the tropical world which is inhabited by humans unknown to it." Apparent victories in the struggle for territory have been won by the Argentine Ant, but only in lands with a more or less warm-temperate, especially a winter-rainfall or Mediterranean-type climate. In the last 20 years, in fact, *I. humilis* has arrived at an apparently rather stable distribution in its adopted countries, and this distribution describes a well-defined double belt around the earth, lying mainly outside the tropics. The chief mortal enemy that *I. humilis* is supposed to be vanquishing is *Pheidole megacephala*, a myrmicine which, however, now appears to be holding the line at the midlength of the Florida peninsula, in midcoastal Queensland, and elsewhere at the outer boundaries of the tropics.

Within the tropics, *P. megacephala* excludes *I. humilis* everywhere in culture areas except in the South American uplands in which *I. humilis* and its relatives are endemic. *I. humilis* has done no better against the imported Fire Ant (*Solenopsis invicta*, a myrmicine), which has successfully invaded Argentine Ant strongholds in the southern United States and flourished there (see also Fluker and Beardsley 1970).

Haskins' second assumption (1939:45,159) is that the dolichoderines and formicines, because of their often thinner and more flexible integument, have greater sensory contact with the environment, and somehow, in connection with this, are more adaptable in their relationships with the outside world. Inasmuch as the relationship between integumental thickness and actual density of sensory receptors remains to be established, this assumption is unwarranted. Furthermore, it is not a foregone conclusion that myrmicine integuments are prevailingly thicker than those of dolichoderines and formicines; in fact, the situation has never been properly surveyed.

Regardless of these considerations, the evidence of zoogeography and paleontology rather conclusively reverses the hypothesis that the Dolichoderinae are today evolutionary winners and the Myrmicinae evolutionary losers.

Tetramorium is the central genus of a complex also including the extant genera *Xiphomyrmex*, *Triglyphothrix*, *Macromischoides*, *Rhoptromyrmex*, and *Strongylognathus*. Of these genera, *Xiphomyrmex* (worker-queen antennae 11-merous and sting with a rounded-spatulate appendage) is only weakly differentiated from *Tetramorium* and *Triglyphothrix* (worker antennae 12-merous, sting appendage perpendicular to shaft and sharply dentiform or pennant-shaped). *Triglyphothrix* is distinguished mainly by its branched pilosity. All of these characters may show exceptions or intergradient conditions, and only a careful revision will tell whether the three genera deserve separate status. At any rate, no member of the tetramoriine complex is native to the New World except *Xiphomyrmex spinosus*, a perfectly typical member of this genus that occurs widely in the Sonoran arid lands of North America (Brown 1957, 1964). How this single tetramoriine reached its present range, so far from any of its congeners, we can only guess. The tetramoriines are apparently unrepresented in Tertiary deposits, but they are very common now (in places co-dominant) throughout the southern palearctic and Africa, and are common and diverse over much of the Oriental, Malagasy, and northern Australian regions, though poorly represented in the southern Australian region. The present and fossil distributions suggest that *Tetramorium* and its offshoots represent another group that has radiated and spread mainly or entirely since the middle of the Tertiary.

Camponotus, the largest and certainly one of the most important living ant genera, probably contains over a thousand valid species. It is also the most widespread and ecologically tolerant genus, reaching as it does Chile and central Argentina, the Arctic Circle, Mauritius, central Polynesia, New Zealand, and Tasmania. A single species is present in the Baltic Amber, where it constitutes only about 1 percent of the identified formicid specimens. In Miocene formations, *Camponotus* is still not abundant in individuals, but several species are present. Since species of this genus usually attend Homoptera and are very frequent foragers on trees and shrubs, one would expect good representation in the amber deposits. The pattern of evolution suggested for *Camponotus* by the fossil record and present distribution is one of a slow but continuous radiation and expansion from a modest beginning made during or just before the Oligocene.

Conclusions and Summary

1. The ant faunas of sub-Saharan Africa and the neotropical region, including those of their rain forests, are very different from one another at both the species group and generic levels. They share only 29 genera, all are widespread in the tropics, and most of them are also in the Northern Hemisphere now or in Tertiary deposits.

2. Analysis of distributions of ant genera suggests the hypothesis that at least from mid-Tertiary times, evolution of world-dominating new taxa has proceeded mainly from combined tropical Africa-southern Asia. Warm-country dominant taxa, such as *Pheidole* and *Crematogaster*, probably originated in this area, and have spread explosively over the rest of the earth from about the Miocene. *Camponotus*, which arose earlier in the Tertiary, may also fit this pattern in a general way.

3. Dominant Old World genera, such as *Dorylus, Tetramorium, Acantholepis, Anoplolepis* and *Polyrhachis*, are in earlier stages of the same kind of spread. *Monomorium* is distributed in a pattern intermediate between these genera and *Camponotus*.

4. Genera or genus groups well represented in the neotropical and Indo-Australian regions, and absent or very rare in Africa, are the peripheral relicts of older taxal waves that are now being replaced from the central Old World tropics.

5. Contrary to an earlier hypothesis, it appears that genera of Myrmicinae, especially *Pheidole* and *Crematogaster*, now have the upper hand as expanding world-dominant taxa at the expense mainly of the Dolichoderinae, which are contracting toward the periphery of the ant-inhabited world. The Formicinae appear to be holding their own.

6. Generic distributions offer no encouragement to the hypothesis of extensive direct exchanges among the southern landmasses, except that the impoverished New Zealand ant fauna shows clear signs of having been derived from Australia by transoceanic immigration.

7. The total evidence for the ants tends to support the proposition that all of the living genera could have evolved and reached their present distributions since the beginning of the Tertiary and within a geographical frame of reference substantially like that of today. Fossil and present distributions of ant taxa fit well the Darlingtonian zoogeographical model based on vertebrate and coleopteran patterns.

References

Brian, M. V. 1965. Social insect populations. vii + 135 pp. Academic Press. London and New York.

Brown, W. L., Jr. 1957. Is the ant genus *Tetramorium* native in North America? Brev. Mus. Comp. Zool. Harv. 72:1-8.

———. 1964. Solution to the problem of *Tetramorium lucayanum* (Hymenoptera: Formicidae). Entomol. News 75:130-132.

Buren, W. F. 1968. A review of the species of *Crematogaster*, sensu stricto, in North America (Hymenoptera, Formicidae), Part II: Descriptions of new species. J. Georgia Entomol. Soc. 3:91-121.

Carpenter, F. M. 1930. The fossil ants of North America. Bull. Mus. Comp. Zool. Harv. 70:1-66, 11 pl.

Darlington, P. J., Jr. 1957. Zoogeography: The geographical distribution of animals. xiv + 675 pp. John Wiley and Sons. New York.

Emery, C. 1891. Le formiche dell'ambra siciliana nel museo mineralogico dell'Università di Bologna. Mem. R. Accad. Sci. Inst. Bologna (5)1:141-165, pl. 1-3.

Fluker, S. S., and J. W. Beardsley. 1970. Sympatric associations of three ants: *Iridomyrmex humilis, Pheidole megacephala*, and *Anoplolepis longipes* in Hawaii. Ann. Entomol. Soc. Amer. 63:1290-1296.

Haskins, C. P. 1939. Of ants and men. vii + 244 pp. Prentice-Hall, Inc. New York.

Mayr, E. 1942. Systematics and the origin of species. 334 pp. Columbia Univ. Press. New York.

Moreau, R. E. 1966. The bird faunas of Africa and its islands. ix + 424 pp. Academic Press. New York and London.

Vanzolini, P. E., and E. E. Williams. 1970. South American anoles: The geographic differentiation and evolution of the *Anolis chrysolepis* species group (Sauria, Iguanidae). Arq. Zool., São Paulo 19:1-298.

Wheeler, W. M. 1914. The ants of the Baltic Amber. Schrift. Phys.-ökon Ges Königsberg 55[1915]:1-142.

Wilson, E. O. 1959. Adaptive shift and dispersal in a tropical ant fauna. Evolution 13:122-144.

Wilson, E. O., and R. W. Taylor. 1967. The ants of Polynesia (Hymenoptera: Formicidae). Pacific Insects Monogr. 14: 1-109.

Zimmerman, E. C. 1948. Insects of Hawaii, I: Introduction. xii + 206 pp. Univ. Hawaii Press. Honolulu.

Appendix: Generic and Subgeneric Names Proposed in the Family Formicidae

Some readers will find that familiar names have not been included in the zoogeographical table. The omissions are accounted for in the list below, in which I have tried to set down every available generic and subgeneric name that has ever been proposed for the Formicidae. The equality sign (=) indicates synonymies, both those long recognized by myrmecologists and a good many more "projected synonymies" that have never been proposed anywhere formally in print. It should be understood that almost all of the synonyms listed, whether widely accepted or here projected, are subjective ones. For this reason, I make no distinction between them, except that I have placed a question mark (?) after especially controversial cases. As already explained in the body of the text preceding Table 1, more projected synonyms could easily be added to this list. I have not made such additions because the information available now does not allow a reasonable guess as to how these cases will be settled.

Certain of the projected synonyms will doubtless be unacceptable to someone, and some of them will probably be rejected when more evidence is in. But

I do not think that any of them are completely unreasonable in the light of our present information. The main reason why these provisional decisions have been made here is that they greatly shorten and simplify the zoogeographical tables and the conclusions based on the tables. It does not make a great deal of difference whether, for example, the Cephalotini contain two genera or four; this tribe is a tightly knit, obviously monophyletic New World lineage. On the other hand, it is important that the diverse elements I here include in *Pachycondyla* (e.g., *Mesoponera, Trachymesopus, Myopias, Trapeziopelta, Neoponera, Bothroponera, Pseudoponera, Wadeura, Ectomomyrmex*) should be recognized as very closely related, even if characters are eventually found to split this group into two or more formal genera.

Acalama M. R. Smith 1948 = Vollenhovia
Acamatus Emery 1894 = Neivamyrmex
Acanthidris Weber 1941 = Rhopalothrix
Acanthoclinea Wheeler 1935 = Dolichoderus?
Acanthognathus Mayr 1887
Acantholepis Mayr 1861 [preoccupied]
Acanthomyops Mayr 1862
Acanthomyrmex Emery 1892
Acanthoponera Mayr 1862
Acanthostichus Mayr 1887
Acidomyrmex Emery 1915 = Rhoptromyrmex
Acrocoelia Mayr 1852 = Crematogaster
Acromyrmex Mayr 1865
Acropyga Roger 1862
Acrostigma Emery 1891 = Podomyrma
Acrostigma Forel 1902 = Stigmacros
Adelomyrmex Emery 1897
Adformica Lomnicki 1925 = Formica
Adlerzia Forel 1902
Aenictogiton Emery 1901
Aenictus Shuckard 1840
Aeromyrma Forel 1891 = Oligomyrmex
Aethiopopone Santschi 1930 = Sphinctomyrmex
Agroecomyrmex Wheeler 1910 [fossil only]
Alaopone Emery 1881 = Dorylus?
Alfaria Emery 1896 = Gnamptogenys
Alistruma Brown 1948 = Colobostruma
Alloformica Dlussky 1969 = Proformica
Allomerus Mayr 1877
Allopheidole Forel 1912 = Pheidole
Amauromyrmex Wheeler 1929 = Pheidologeton
Amblyopone Erichson 1842
Amblyopopone Dalla Torre 1893 [emendation] = Amblyopone
Ameghinoa Viana and Haedo Rossi 1957 [fossil only]
Ammomyrma Santschi 1922 = Dorymyrmex?
Amyrmex Kusnezov 1953
Anacantholepis Santschi 1914 = Plagiolepis
Anacanthoponera Wheeler 1923 = Heteroponera
Ancylognathus Lund 1831 = Eciton
Ancyridris Wheeler 1935 = Lordomyrma?
Andragnathus Emery 1922
Aneleus Emery 1900 = Oligomyrmex
Anergates Forel 1874
Anergatides Wasmann 1915 = Pheidole?
Aneuretus Emery 1892
Anillidris Santschi 1936
Anillomyrma Emery 1913
Anisopheidole Forel 1914
Anochetus Mayr 1861 = Odontomachus
Anomma Shuckard 1840
Anonychomyrma Donisthorpe 1947 = Iridomyrmex?
Anoplolepis Santschi 1914
Anoplomyrma Chapman 1963 = Polyrhachis
Antillaemyrmex Mann 1920 = Leptothorax?
Aphaenogaster Mayr 1853
Aphantolepis Wheeler 1930 = Technomyrmex
Aphomomyrmex Emery 1899
Aphomyrmex Ashmead 1905 = Pseudaphomomyrmex
Apomyrma Brown, Gotwald and Lévieux 1971
Aporomyrmex Faber 1969 = Plagiolepis
Apsychomyrmex Wheeler 1910 = Adelomyrmex
Apterocrema Wheeler 1936 = Crematogaster
Apterostigma Mayr 1865
Aratromyrmex Stitz 1938 = Liomyrmex
Araucomyrmex Gallardo 1919 = Dorymyrmex?
Archaeomyrmex Mann 1921 = Myrmecina
Archaeatta Gonçalves 1942 = Atta
Archimyrmex Cockerell 1923 [fossil only]
Archiponera Carpenter 1930 [fossil only]
Arctomyrmex Mann 1921 = Adelomyrmex
Arnoldidris Brown 1950 = Orectognathus
Arotropus Provancher 1881 = Amblyopone
Asemorhoptrum Mayr 1861 = Stenamma
Asphinctopone Santschi 1914
Aspididris Weber 1950
Asymphylomyrmex Wheeler 1915 [fossil only]
Atopodon Forel 1912 = Acropyga
Atopogyne Forel 1911 = Crematogaster
Atopomyrmex Ern. André 1889
Atopula Emery 1912
Atta Fabricius 1804
Attomyrma Emery 1915 = Aphaenogaster
Attopsis Heer 1850 [fossil only]
Aulacomyrma Emery 1921 = Polyrhachis
Aulacopone Arnoldi 1930

Australasius Faber 1969 = Lasius
Axinidris Weber 1941
Azteca Forel 1878
Aztecum Bertkau 1879 [emendation] = Azteca

Barbourella Wheeler 1930 = Gnamptogenys
Basiceros Schulz 1906
Belonopelta Mayr 1870
Biconomyrma Kusnezov 1952 = Dorymyrmex?
Bisolenopsis Kusnezov 1953 = Solenopsis
Blepharidatta Wheeler 1915 = Ochetomyrmex
Bondroitia Forel 1911 = Diplomorium
Borgmeierita Brown 1953 = Glamyromyrmex
Bothriomyrmex Emery 1869
Bothroponera Mayr 1862 = Pachycondyla
Brachymyrmex Mayr 1868
Brachyponera Emery 1901
Bradoponera Mayr 1868 [fossil only]
Bradyponera Mayr 1886 = Pachycondyla
Bregmatomyrma Wheeler 1929
Brownidris Kusnezov 1957
Bruchomyrma Santschi 1922 = Pheidole?
Brunella Forel 1917 [preoccupied]
Bryscha Santschi 1925 = Brachymyrmex

Cacopone Santschi 1914 = Plectroctena
Calomyrmex Emery 1895
Calyptites Scudder 1878 [fossil only]
Calyptomyrmex Emery 1887
Campomyrma Wheeler 1911 = Polyrhachis
Camponotus Mayr 1861
Campostigmacros McAreavey 1957 = Stigmacros
Camptognatha Gray 1832 = Eciton
Cardiocondyla Emery 1869
Cardiopheidole Wheeler 1914 = Pheidole
Carebara Westwood 1840
Carebarella Emery 1905
Carebarelloides Borgmeier 1937 = Carebarella
Cataglyphis Foerster 1850
Cataulacus Fr. Smith 1853
Caulomyrma Forel 1915 = Leptothorax?
Cautolasius Wilson 1955 = Lasius
Centromyrmex Mayr 1866
Cephalomorium Forel 1922 = Pheidole
Cephalomyrma Karavaiev 1935 = Polyrhachis
Cephalomyrmex Carpenter 1930 [fossil only]
Cephalotes Latreille 1802
Cephaloxys Fr. Smith 1864 = Smithistruma
Cepobroticus Wheeler 1925 = Megalomyrmex
Cerapachys Fr. Smith 1857
Ceratopachys Schulz 1906 [emendation] = Cerapachys
Ceratobasis Fr. Smith 1860 = Basiceros
Ceratopheidole Pergande 1895 = Pheidole
Chalcoponera Emery 1897 = Rhytidoponera
Chalepoxenus Menozzi 1923 = Leptothorax?
Champsomyrmex Emery 1891 = Odontomachus
Chapmanella Wheeler 1930 = Euprenolepis
Chariomyrma Forel 1915 = Polyrhachis
Chariostigmacros McAreavey 1957 = Stigmacros
Chelaner Emery 1914
Cheliomyrmex Mayr 1870
Chelystruma Brown 1950 = Glamyromyrmex?
Chronoxenus Santschi 1920 = Bothriomyrmex?
Chrysapace Crawley 1924 = Cerapachys
Chthonolasius Ruzsky 1912 = Lasius
Cladomyrma Wheeler 1920
Clarkistruma Brown 1948 = Colobostruma
Codiomyrmex Wheeler 1916 = Glamyromyrmex?
Codioxenus Santschi 1931
Colobocrema Wheeler 1927 = Crematogaster
Colobopsis Mayr 1861 = Camponotus
Colobostruma Wheeler 1927
Commateta Santschi 1929 = Gnamptogenys
Condylodon Lund 1831 = Pseudomyrmex
Condylomyrma Santschi 1928 = Camponotus
Conomyrma Forel 1913 = Dorymyrmex?
Conothoracoides Strand 1935 = Pheidole
Conothorax Karavaiev 1935 = Pheidole
Coptoformica Mueller 1933 = Formica
Corynomyrmex Viehmeyer 1916 = Monomorium
Cosmaecetes Spinola 1853 = Dorylus
Cosmaegetes Dalla Torre 1893 [variant spelling of Cosmaecetes] = Dorylus
Crateropsis Patrizi 1948 = Oligomyrmex?
Cratomyrmex Emery 1892 = Messor
Creightonidris Brown 1949
Cremastogaster Mayr 1861 [emendation] = Crematogaster
Crematogaster Lund 1831
Croesomyrmex Mann 1920 = Leptothorax?
Cryptocephalus Lowne 1865 = Meranoplus
Cryptocerus Latreille 1804 = Cephalotes?
Cryptopone Emery 1892
Ctenopyga Ashmead 1905 = Acanthostichus
Cyathocephalus Emery 1915 = Cephalotes?
Cyathomyrmex Creighton 1933 = Cephalotes?
Cylindromyrmex Mayr 1870
Cyphoidris Weber 1952 = Lordomyrma?
Cyphomannia Weber 1938 = Cyphomyrmex
Cyphomyrmex Mayr 1862
Cyrtomyrma Forel 1915 = Polyrhachis
Cyrtostigmacros McAreavey 1957 = Stigmacros
Cysias Emery 1902 = Cerapachys

Dacetinops Brown and Wilson 1957
Daceton Perty 1833
Dacetum Agassiz 1846 [emendation] = Daceton
Dacryon Forel 1895 = Podomyrma?
Decacrema Forel 1910 = Crematogaster
Decamera Roger 1863 = Myrmelachista
Decamorium Forel 1913

Decapheidole Forel 1912 = Pheidole
Dendrolasius Ruzsky 1912 = Lasius
Dendromyrmex Emery 1895
Deromyrma Forel 1913 = Aphaenogaster
Descolemyrma Kusnezov 1951 = Mycocepurus
Diabolus Karavaiev 1926 = Dolichoderus
Diacamma Mayr 1862
Diagyne Santschi 1923 = Solenopsis
Diceratoclinea Wheeler 1935 = Dolichoderus
Dichothorax Emery 1895 = Leptothorax
Dichthadia Gerstaecker 1863 = Dorylus?
Dicroaspis Emery 1908 = Calyptomyrmex
Dilobocondyla Santschi 1910
Dimorphomyrmex Ern. André 1892 = Gesomyrmex
Dinomyrmex Ashmead 1905 = Camponotus
Dinoponera Roger 1861
Diodontolepis Wheeler 1920 = Notoncus
Diplomorium Mayr 1901
Diplorhoptrum Mayr 1855 = Solenopsis
Discothyrea Roger 1863
Dodecamyrmica Arnoldi 1968 = Myrmica
Dodous Donisthorpe 1946 = Pristomyrmex
Doleromyrma Forel 1907 = Iridomyrmex
Dolichoderus Lund 1831
Dolichorhachis Mann 1919 = Polyrhachis
Donisthorpea Morice and Durrant 1914 = Lasius
Dorisidris Brown 1948
Doronomyrmex Kutter 1945 = Leptothorax?
Dorothea Donisthorpe 1948 = Vollenhovia
Dorylozelus Forel 1915 = Leptogenys
Dorylus Fabricius 1793
Dorymyrmex Mayr 1866
Drepanognathus Fr. Smith 1858 = Harpegnathos
Drymomyrmex Wheeler 1915 [fossil only]
Dyclona Santschi 1930 = Cardiocondyla?
Dyomorium Donisthorpe 1947 = Vollenhovia
Dysedrognathus Taylor 1968

Echinopla Fr. Smith 1857
Eciton Latreille 1804
Ecphorella Forel 1909
Ectatomma Fr. Smith 1858
Ectomomyrmex Mayr 1867 = Pachycondyla
Elaeomyrmex Carpenter 1930 [fossil only]
Elasmopheidole Forel 1913 = Pheidole
Electromyrmex Wheeler 1910 [fossil only]
Electropheidole Mann 1921 = Pheidole
Electroponera Wheeler 1915 [fossil only]
Emeryella Forel 1901 = Gnamptogenys
Emeryia Forel 1890 = Cardiocondyla
Emeryopone Forel 1912
Emplastus Donisthorpe 1920 [fossil only]
Eneria Donisthorpe 1948 = Strumigenys
Engramma Forel 1905
Enneamerus Mayr 1868 [fossil only]
Eoformica Cockerell 1921 [fossil only]
Eomonocombus Arnoldi 1968 = Cataglyphis
Eoponera Carpenter 1929 [fossil only]
Ephebomyrmex Wheeler 1902
Epiatta Borgmeier 1950 = Atta
Epimyrma Emery 1915 = Leptothorax?
Epipheidole Wheeler 1904 = Pheidole
Epitritus Emery 1869
Epixenus Emery 1908 = Monomorium
Epoecus Emery 1892 = Monomorium
Epopostruma Forel 1895
Equessimessor Santschi 1936
[emendation of Equestrimessor] = Monomorium
Equestrimessor Santschi 1919 = Monomorium
Erebomyrma Wheeler 1903 = Oligomyrmex
Ericapelta Kusnezov 1955 = Amblyopone
Erimelophorus Wheeler 1935 = Melophorus
Eriopheidole Kusnezov 1952 = Pheidole
Escherichia Forel 1910 = Probolomyrmex
Eubothroponera Clark 1930 = Platythyrea
Eucrema Santschi 1918 = Crematogaster
Eucryptocerus Kempf 1951 = Cephalotes?
Eulithomyrmex Carpenter 1935 [fossil only]
Eumecopone Forel 1901 = Pachycondyla
Euophthalma Creighton 1930 = Solenopsis
Euponera Forel 1891
Euprenolepis Emery 1906
Eurhopalothrix Brown and Kempf 1960
Eusphinctus Emery 1893 = Sphinctomyrmex
Eutetramorium Emery 1900
Evelyna Donisthorpe 1937 = Polyrhachis
Examblyopone Donisthorpe 1949 = Prionopelta

Florencea Donisthorpe 1937 = Polyrhachis
Forelifidis M. R. Smith 1954 = Oxyepoecus
Forelius Emery 1888
Forelomyrmex Wheeler 1913 = Pogonomyrmex
Forelophilus Kutter 1931
Formica Linnaeus 1758
Formicina Shuckard 1840 = Lasius
Formicium Westwood 1854 [to Siricoidea; fossil only]
Formicoxenus Mayr 1855 = Leptothorax?
Froggattella Forel 1902
Fulakora Mann 1919 = Amblyopone

Gallardomyrma Bruch 1932 = Pheidole
Gauromyrmex Menozzi 1933 = Vollenhovia
Geognomicus Menozzi 1924
Gesomyrmex Mayr 1868
Gigantiops Roger 1862
Glamyromyrmex Wheeler 1915
Glaphyromyrmex Wheeler 1915 [fossil only]
Glyphopone Forel 1913 = Centromyrmex
Glyptomyrmex Forel 1885 = Myrmicocrypta
Gnamptogenys Roger 1863

Gonepimyrma Bernard 1948 = Leptothorax?
Goniomma Emery 1895
Goniothorax Emery 1896 = Leptothorax?
Granisolenopsis Kusnezov 1957 = Solenopsis
Gymnomyrmex Borgmeier 1954

Hagensia Forel 1901
Hagiomyrma Wheeler 1911 = Polyrhachis
Hagiostigmacros McAreavey 1957 = Stigmacros
Hagioxenus Forel 1910
Halmamyrmecia Wheeler 1922 = Myrmecia
Harnedia M. R. Smith 1949 = Cephalotes?
Harpagoxenus Forel 1893
Harpegnathos Jerdon 1851
Hedomyrma Forel 1915 = Polyrhachis
Hemioptica Roger 1862 = Polyrhachis
Hendecapheidole Wheeler 1922 = Pheidole
Hendecatella Wheeler 1927 = Oligomyrmex
Heptacondylus Fr. Smith 1857 = Myrmicaria
Heptastruma Weber 1934 = Rhopalothrix
Hercynia Enzmann 1947 = Ochetomyrmex
Heteromyrmex Wheeler 1920 = Vollenhovia
Heteroponera Mayr 1887
Hexadaceton Brown 1948 = Epopostruma
Hincksidris Donisthorpe 1944 = Myrmelachista
Hiphopelta Forel 1913 = Pachycondyla
Holcomyrmex Mayr 1878 = Monomorium
Holcoponera Mayr 1887 = Gnamptogenys
Holcoponera Cameron 1891 = Cylindromyrmex
Holopone Santschi 1924 = Eciton
Hoplomymrus Gerstaecker 1858 = Polyrhachis
Huberia Forel 1890
Hylidris Weber 1941 = Pristomyrmex
Hylomyrma Forel 1912
Hypercolobopsis Emery 1920 = Camponotus
Hypochira Buckley 1866 = Formica?
Hypoclinea Mayr 1855 = Dolichoderus?
Hypocryptocerus Wheeler 1920 = Cephalotes?
Hypocylindromyrmex Wheeler 1924 = Cylindromyrmex
Hypopomyrmex Emery 1891 [fossil only]
Hypoponera Santschi 1938

Icothorax Hamann and Klemm 1967 = Leptothorax?
Idrisella Santschi 1937 = Pheidologeton
Imhoffia Heer 1849 [Formicinae incertae sedis; fossil only]
Irenea Donisthorpe 1938 = Polyrhachis
Ireneella Donisthorpe 1941 [Myrmicinae incertae sedis]
Ireneidris Donisthorpe 1943 = Monomorium
Ireneopone Donisthorpe 1946
Iridomyrmex Mayr 1862
Irogera Emery 1915 = Rogeria
Ischnomyrmex Mayr 1862 = Pheidole
Isholcomyrmex Santschi 1936 [variant spelling of Isolcomyrmex] = Monomorium
Isolcomyrmex Santschi 1917 = Monomorium
Isopheidole Forel 1912 = Pheidole

Janetia Forel 1899 = Pogonomyrmex
Johnia Karavaiev 1927 = Polyrhachis

Karavaievia Emery 1925 = Camponotus
Karawajewella Donisthorpe 1944 = Dolichoderus
Kyidris Brown 1949

Labauchena Santschi 1930 = Solenopsis
Labidogenys Roger 1862 = Strumigenys
Labidus Jurine 1807
Lachnomyrmex Wheeler 1910
Lampromyrmex Mayr 1868 = Monomorium
Laparomyrmex Emery 1887 = Liomyrmex
Lasiophanes Emery 1895
Lasius Fabricius 1804
Lecanomyrma Forel 1913 = Oligomyrmex
Leonomyrma Arnoldi 1968 = Leptothorax?
Lepidopone Bernard 1952 = Asphinctopone
Lepisiota Santschi 1926 = Acanthomyrmex
Leptalaea Spinola 1851 [variant spelling of Leptalea] = Pseudomyrmex
Leptalea Erichson 1839 = Pseudomyrmex
Leptanilla Emery 1870
Leptanilloides Mann 1923
Leptogenys Roger 1861
Leptomesites Kutter 1948
Leptomyrma Motschulsky 1863 = Pheidole
Leptomyrmex Mayr 1862
Leptomyrmula Emery 1912 [fossil only]
Leptopone Arnold 1916 = Centromyrmex
Leptothorax Mayr 1855
Leucotaphus Donisthorpe 1920 [fossil only]
Lilidris Kusnezov 1957 = Solenopsis?
Limnomyrmex Arnold 1948 = Leptothorax?
Linepithema Mayr 1866
Liometopum Mayr 1861
Liomyrmex Mayr 1865
Lioponera Mayr 1878 = Cerapachys
Lithomyrmex Carpenter 1930 = Eulithomyrmex
Lithomyrmex Clark 1928 = Amblyopone
Lobognathus Enzmann 1947 = Veromessor
Lobomyrmex Kratochvil 1941 = Tetramorium
Lobopelta Mayr 1862 = Leptogenys
Lonchomyrmex Mayr 1867 [fossil only]
Loncyda Santschi 1930 = Cardiocondyla?
Lophomyrmex Emery 1892
Lordomyrma Emery 1897
Lundella Emery 1915 = Hylomyrma

Machaerogenys Emery 1911 = Leptogenys
Machaeromyrma Forel 1916 = Cataglyphis
Machomyrma Forel 1895
Macromischa Roger 1863 = Leptothorax?
Macromischoides Wheeler 1920
Macropheidole Emery 1915 = Pheidole

Malacomyrma Emery 1922 = Acropyga
Manica Jurine 1807
Manniella Wheeler 1921 = Camponotus
Martia Forel 1907 = Oxyepoecus
Mayria Forel 1878 = Camponotus
Mayriella Forel 1902
Mayromyrmex Ashmead 1905 = Eciton
Megalomyrmex Forel 1885
Megaloponera Emery 1877 = Megaponera
Megaponera Mayr 1862
Melissotarsus Emery 1877
Melophorus Lubbock 1883
Meranoplus Fr. Smith 1853
Mesanoplolepis Santschi 1926 = Anoplolepis?
Mesocrema Santschi 1928 = Crematogaster
Mesomyrma Stitz 1911 = Dilobocondyla
Mesoponera Emery 1901 = Pachycondyla
Mesostruma Brown 1948
Mesoxena Fr. Smith 1860 = Echinopla
Messor Forel 1890
Metacylindromyrmex Wheeler 1924 = Cylindromyrmex
Metapone Forel 1911
Mianeuretus Carpenter 1930 [fossil only]
Miccostruma Brown 1948
Microbolbos Donisthorpe 1948 = Leptogenys
Microdaceton Santschi 1913
Micromyrma Dufour 1857 = Tapinoma
Mictoponera Forel 1901 = Gnamptogenys
Miomyrmex Carpenter 1930 [fossil only]
Mitara Emery 1913 = Monomorium
Moellerius Forel 1893 = Acromyrmex
Monacis Roger 1862 = Dolichoderus
Monoceratoclinea Wheeler 1935
Monocombus Mayr 1855 = Cataglyphis
Monomarium Fr. Smith 1859 = Aphaenogaster
Monomorium Mayr 1855
Morleyidris Donisthorpe 1944 = Polyrhachis
Mycetarotes Emery 1913
Mycetophylax Emery 1913
Mycetosoritis Wheeler 1907
Mychothorax Ruzsky 1904 = Leptothorax?
Mycocepurus Forel 1893
Myopias Roger 1861 = Pachycondyla
Myopopone Roger 1861
Myrafant M. R. Smith 1950 = Leptothorax
Myrma Billberg 1820 = Polyrhachis
Myrmacantha Emery 1920 = Phasmomyrmex?
Myrmacrhaphe Santschi 1926 = Camponotus
Myrmamblys Forel 1912 = Camponotus
Myrmammophilus Menozzi 1924 = Leptothorax
Myrmapatetes Wheeler 1929 = Odontomachus
Myrmaphaenus Emery 1920 = Camponotus
Myrmatopa Forel 1915 = Polyrhachis
Myrmecia Fabricius 1804
Myrmecina Curtis 1829
Myrmecinella Wheeler 1922 = Xenomyrmex
Myrmecocystus Wesmael 1838
Myrmecopsis Fr. Smith 1865 = Opisthopsis
Myrmecorhynchus Ern. André 1896
Myrmegis Rafinesque 1815 = Atta
Myrmelachista Roger 1863
Myrmentoma Forel 1912 = Camponotus
Myrmepinotus Santschi 1921 = Camponotus
Myrmepomis Forel 1912 = Camponotus
Myrmespera Santschi 1926 = Camponotus
Myrmetaerus Soudek 1925 = Leptothorax?
Myrmeurynota Forel 1912 = Camponotus
Myrmex Guérin 1845 = Pseudomyrmex
Myrmhopla Forel 1915 = Polyrhachis
Myrmica Latreille 1804
Myrmicaria Saunders 1841
Myrmicites Foerster 1891 [fossil only]
Myrmicium Heer 1870 [fossil only] = Promyrmicium
Myrmicocrypta Fr. Smith 1860
Myrmisolepis Santschi 1921 = Camponotus
Myrmobrachys Forel 1912 = Camponotus
Myrmocamelus Forel 1914 = Camponotus
Myrmocladoecus Wheeler 1921 = Camponotus
Myrmodirhachis Emery 1925 = Camponotus
Myrmogigas Forel 1912 = Camponotus
Myrmogonia Forel 1912 = Camponotus
Myrmolophus Emery 1920 = Camponotus
Myrmomalis Forel 1914 = Camponotus
Myrmonesites Emery 1920 = Camponotus
Myrmopalpella Staercke 1934 = Camponotus
Myrmopelta Santschi 1921 = Camponotus
Myrmophyma Forel 1912 = Camponotus
Myrmopiromis Wheeler 1921 = Camponotus
Myrmoplatypus Santschi 1921 = Camponotus
Myrmoplatys Forel 1916 = Camponotus
Myrmopsamma Forel 1914 = Camponotus
Myrmopytia Emery 1920 = Camponotus
Myrmorhachis Forel 1912 = Phasmomyrmex?
Myrmosaga Forel 1912 = Camponotus
Myrmosaulus Wheeler 1921 = Camponotus
Myrmosericus Forel 1912 = Camponotus
Myrmosphincta Forel 1912 = Camponotus
Myrmostenus Emery 1920 = Camponotus
Myrmotarsus Forel 1912 = Camponotus
Myrmotemnus Emery 1920 = Camponotus
Myrmoteras Forel 1893
Myrmothrinax Forel 1915 = Polyrhachis
Myrmothrix Forel 1912 = Camponotus
Myrmotrema Forel 1912 = Camponotus
Myrmoturba Forel 1912 = Camponotus
Myrmoxenus Ruzsky 1902 = Leptothorax?
Myrmoxygenys Emery 1925 = Camponotus
Myrmus Schenck 1853 = Strongylognathus
Mystrium Roger 1862
Myrtoteras Matsumura 1912 = Odontomachus

Neaphomus Menozzi 1935 = Myrmelachista?
Neivamyrmex Borgmeier 1940
Nematocrema Santschi 1918 = Crematogaster
Neoamblyopone Clark 1927 = Amblyopone
Neoatta Gonçalves 1942 = Atta
Neocolobopsis Borgmeier 1928 = Camponotus
Neocrema Santschi 1918 = Crematogaster
Neoforelius Kusnezov 1953
Neoformica Wheeler 1913 = Formica
Neomyrma Forel 1914 = Manica
Neomyrmamblys Wheeler 1921 = Camponotus
Neophyracaces Clark 1941 = Cerapachys
Neoponera Emery 1901 = Pachycondyla
Neostruma Brown 1948
Nesolasius Wheeler 1935 = Pseudolasius
Nesomyrmex Wheeler 1910 = Leptothorax?
Nimbamyrma Bernard 1952 = Oligomyrmex?
Nomamyrmex Borgmeier 1936
Noonilla Petersen 1968
Nothidris Ettershank 1966
Nothomyrmecia Clark 1934
Nothomyrmica Wheeler 1910 [fossil only]
Nothosphinctus Wheeler 1918 = Sphinctomyrmex
Notomyrmex Emery 1915 = Chelaner
Notoncus Emery 1895
Notostigma Emery 1920
Novomessor Emery 1915 = Aphaenogaster
Nycteresia Roger 1861 = Labidus
Nylanderia Emery 1906 = Paratrechina
Nystalomyrma Wheeler 1916 = Aphaenogaster

Ochetomyrmex Mayr 1877
Octella Forel 1915 = Oligomyrmex
Octostruma Forel 1912
Ocymyrmex Emery 1886
Odontomachus Latreille 1804
Odontomyrmex Ern. André 1905 = Pristomyrmex
Odontopelta Emery 1911 = Leptogenys
Odontoponera Mayr 1862
Oecodoma Latreille 1818 = Atta
Oecophthora Heer 1852 = Pheidole
Oecophylla Fr. Smith 1860
Oedaleocerus Creighton 1930 = Solenopsis
Oligomyrmex Mayr 1867
Onychomyrmex Emery 1895
Ooceraea Roger 1862 = Cerapachys
Ophthalmopone Forel 1890
Opisthopsis Emery 1893
Opisthoscyphus Mann 1922 = Gnamptogenys
Orectognathus Fr. Smith 1853
Oreomyrma Wheeler 1914 = Manica
Orthocrema Santschi 1918 = Crematogaster
Orthonotomyrmex Ashmead 1906 = Camponotus
Orthonotus Ashmead 1905 = Camponotus
Otomyrmex Forel 1891 = Cataulacus

Overbeckia Viehmeyer 1915
Oxyepoecus Santschi 1926
Oxygyne Forel 1901 = Crematogaster
Oxyopomyrmex Ern. André 1881

Pachycondyla Fr. Smith 1858
Pachysima Emery 1912 = Tetraponera
Paedalgus Forel 1911
Palaeatta Borgmeier 1950 = Atta
Paltothyreus Mayr 1862
Paracolobopsis Emery 1920 = Camponotus
Paracrema Santschi 1918 = Crematogaster
Paracryptocerus Emery 1915 = Cephalotes?
Paraenictus Wheeler 1929 = Aenictus
Paraformica Forel 1915 = Cataglyphis
Paraholcomyrmex Emery 1915 = Monomorium
Parameranoplus Wheeler 1915 [fossil only]
Paramycetophylax Kusnezov 1956 = Mycetophylax
Paramyrmamblys Santschi 1926 = Camponotus
Paramyrmica Cole 1957 = Myrmica
Paranamyrma Kusnezov 1954 = Solenopsis
Paraneuretus Wheeler 1915 [fossil only]
Paranomopone Wheeler 1915 = Heteroponera
Paraparatrechina Donisthorpe 1947 = Paratrechina
Paraphacota Santschi 1919 = Monomorium
Parapheidole Emery 1915 = Pheidole?
Paraplagiolepis Faber 1969 = Plagiolepis
Paraponera Fr. Smith 1859
Paraprionopelta Kusnezov 1955 = Amblyopone?
Parasima Donisthorpe 1948 = Tetraponera
Parasyscia Emery 1882 = Cerapachys
Paratopula Wheeler 1919 = Atopula
Paratrechina Motschulsky 1863
Parectatomma Emery 1911 = Gnamptogenys
Parholcomyrmex Emery 1915 [emendation of Paraholcomyrmex]
Pedetes Bernstein 1861 = Odontomachus
Pentastruma Forel 1912
Perissomyrmex M. R. Smith 1947
Peronomyrmex Viehmeyer 1922
Petraeomyrmex Carpenter 1930 [fossil only]
Phacota Roger 1862 = Monomorium?
Phalacromyrmex Kempf 1960
Pharaophanes Bernard 1952 [nomen nudum] = Monomorium
Phasmomyrmex Stitz 1910
Phaulomyrma G. C. and E. W. Wheeler 1930
Pheidolacanthinus Fr. Smith 1864 = Pheidole
Pheidole Westwood 1840
Pheidologeton Mayr 1862
Phidole Bingham 1903 [variant spelling of Pheidole]
Phrynoponera Wheeler 1920 = Pachycondyla
Phyracaces Emery 1902 = Cerapachys
Physatta Fr. Smith 1857 = Myrmicaria
Physocrema Forel 1912 = Crematogaster

Pityomyrmex Wheeler 1915 [fossil only]
Plagiolepis Mayr 1861
Planimyrma Viehmeyer 1914 = Aphaenogaster
Platystruma Brown 1953 = Smithistruma
Platythyrea Roger 1863
Plectroctena Fr. Smith 1858
Podomyrma Fr. Smith 1859
Poecilomyrma Mann 1921
Pogonomyrmex Mayr 1868
Polyergus Latreille 1804
Polyhomoa Azuma 1950 = Kyidris
Polyrhachis Fr. Smith 1857
Polyrhachis Shuckard 1840 [nomen nudum]
Ponera Latreille 1804
Poneracantha Emery 1897 = Gnamptogenys
Poneropsis Heer 1867 [fossil only]
Prenolepis Mayr 1861
Prionogenys Emery 1895
Prionomyrmex Mayr 1868 [fossil only]
Prionopelta Mayr 1866
Pristomyrmecia Emery 1911 = Myrmecia
Pristomyrmex Mayr 1866
Proatta Forel 1912
Probolomyrmex Mayr 1901
Procerapachys Wheeler 1915 [fossil only]
Proceratium Roger 1863
Procryptocerus Emery 1887
Prodicroaspis Emery 1914
Prodimorphomyrmex Wheeler 1915 [fossil only]
Prodiscothyrea Wheeler 1916 = Discothyrea
Proformica Ruzsky 1903
Prolasius Forel 1892
Promeranoplus Emery 1914
Promyopias Santschi 1914 = Centromyrmex
Promyrma Forel 1912 = Liomyrmex
Promyrmecia Emery 1911 = Myrmecia
Promyrmicium Baroni Urbani 1971
[fossil only; incertae sedis]
Propodomyrma Wheeler 1910 = Vollenhovia
Proscopomyrmex Patrizi 1946 = Strumigenys
Prosopidris Wheeler 1935 = Cardiocondyla
Protamblyopone Clark 1927 = Amblyopone
Protaneuretus Wheeler 1915 [fossil only]
Protazteca Carpenter 1930 [fossil only]
Protholcomyrmex Wheeler 1922 = Chelaner
Protoformica Dlussky 1967 [fossil only] = Formica?
Protomognathus Wheeler 1905 = Harpagoxenus
Psalidomyrmex Ern. André 1890
Psammomyrma Forel 1912 = Dorymyrmex
Pseudaphomomyrmex Wheeler 1920
Pseudoatta Gallardo 1916 = Acromyrmex?
Pseudocamponotus Carpenter 1930 [fossil only]
Pseudocolobopsis Emery 1920 = Camponotus
Pseudocryptopone Wheeler 1933 = Ponera
Pseudocyrtomyrma Emery 1921 = Polyrhachis
Pseudodichthadia Ern. André 1885 = Labidus
Pseudolasius Emery 1886
Pseudomyrma Guérin 1844 = Pseudomyrmex
Pseudomyrmex Lund 1831
Pseudoneoponera Donisthorpe 1943 = Pachycondyla
Pseudonotoncus Clark 1934
Pseudopodomyrma Crawley 1925 = Podomyrma?
Pseudoponera Emery 1901 = Pachycondyla
Pseudosphincta Wheeler 1922
[variant spelling of Pseudosysphincta]
Pseudostigmacros McAreavey 1957 = Stigmacros
Pseudosysphincta Arnold 1916 = Discothyrea
Pteroponera Bernard 1949 = Ponera
Pyramica Roger 1862 = Strumigenys

Quadristruma Brown 1949

Raptiformica Forel 1913 = Formica
Renea Donisthorpe 1947 = Prionopelta
Rhachiocrema Mann 1919 = Crematogaster
Rhinomyrmex Forel 1886 = Camponotus
Rhizomyrma Forel 1893 = Acropyga
Rhogmus Shuckard 1840 = Dorylus?
Rhopalomastix Forel 1900
Rhopalomyrmex Mayr 1868 [fossil only]
Rhopalopone Emery 1897 = Gnamptogenys
Rhopalothrix Mayr 1870
Rhoptromyrmex Mayr 1901
Rhytidoponera Mayr 1862
Rogeria Emery 1894
Romblonella Wheeler 1935
Rossomyrmex Arnoldi 1928

Santschiella Forel 1916
Schizopelta McAreavey 1949 = Chelaner
Scrobopheidole Emery 1915 = Pheidole
Scyphodon Brues 1925
Selenopone Wheeler 1933 = Ponera
Semonius Forel 1910
Sericomyrmex Mayr 1865
Serrastruma Brown 1948
Serviformica Forel 1913 = Formica
Shuckardia Emery 1895 = Dorylus
Sicelomyrmex Wheeler 1915 [fossil only]
Sicilomyrmex Wheeler 1926 [fossil only;
emendation] = Sicelomyrmex
Sifolinia Emery 1907 = Myrmica
Sima Roger 1863 = Tetraponera
Simopelta Mann 1922
Simopone Forel 1891
Smithistruma Brown 1948
Solenomyrma Karavaiev 1935 = Vollenhovia
Solenops Karavaiev 1930 = Oligomyrmex
Solenopsis Westwood 1840
Sommimyrma Menozzi 1925 = Myrmica?
Spalacomyrmex Emery 1889 = Centromyrmex

Spaniopone Wheeler and Mann 1914 = Gnamptogenys
Spelaeomyrmex Wheeler 1922 = Oligomyrmex
Sphaerocrema Santschi 1918 = Crematogaster
Sphecomyrma Wilson, Carpenter and Brown 1967 [fossil only]
Sphegomyrmex Imhoff 1852 = Dorylus
Sphinctomyrmex Mayr 1866
Spinomyrma Kusnezov 1952 = Dorymyrmex?
Sporocleptes Arnold 1948 = Oligomyrmex
Stegomyrmex Emery 1912
Stegopheidole Emery 1915 = Pheidole
Stenamma Westwood 1840
Stenomyrmex Mayr 1862 = Odontomachus
Stenothorax McAreavey 1949 = Adlerzia
Stereomyrmex Emery 1901
Stictoponera Mayr 1887 = Gnamptogenys
Stigmacros Forel 1905
Stigmatomma Roger 1859 = Amblyopone
Stigmomyrmex Mayr 1868
Stiphromyrmex Wheeler 1915 [fossil only]
Streblognathus Mayr 1862
Strongylognathus Mayr 1853
Strumigenys Fr. Smith 1860
Sulcomyrmex Kratochvil 1941 = Tetramorium
Syllophopsis Santschi 1915
Symbiomyrma Arnoldi 1930 = Myrmica
Symmyrmica Wheeler 1904 = Leptothorax?
Sympheidole Wheeler 1904 = Pheidole
Synsolenopsis Forel 1918 = Solenopsis
Syntaphus Donisthorpe 1920 [fossil only]
Syntermitopone Wheeler 1936 = Pachycondyla
Syscia Roger 1861 = Cerapachys
Sysphincta Mayr 1865 [emendation of Sysphingta] = Proceratium
Sysphingta Roger 1863 = Proceratium

Talaridris Weber 1941 = Rhopalothrix
Tammoteca Santschi 1929 = Gnamptogenys
Tanaemyrmex Ashmead 1905 = Camponotus
Tapinolepis Emery 1925 = Anoplolepis
Tapinoma Foerster 1850
Tapinoptera Santschi 1925 = Tapinoma
Tatuidris Brown and Kempf 1968
Technomyrmex Mayr 1872
Teleutomyrmex Kutter 1950
Temnothorax Mayr 1861 = Leptothorax
Terataner Emery 1912
Teratomyrmex McAreavey 1957
Termitopone Wheeler 1936 = Pachycondyla
Tetramorium Mayr 1855
Tetramyrma Forel 1912
Tetraponera Fr. Smith 1852
Tetrogmus Roger 1857 = Tetramorium
Thaumatomyrmex Mayr 1887
Theryella Santschi 1921 = Stenamma
Thlipsepinotus Santschi 1928 = Camponotus
Tingimyrmex Mann 1926
Tomognathus Mayr 1861 = Harpagoxenus
Trachymesopus Emery 1911 = Pachycondyla
Trachymyrmex Forel 1893
Trachypheidole Emery 1915 = Pheidole
Trachyponera Santschi 1928 [lapsus calami for Trachymesopus] = Pachycondyla
Tranetera Arnold 1952 = Terataner?
Tranopelta Mayr 1866
Tranopeltoides Wheeler 1922 = Crematogaster
Trapeziopelta Mayr 1862 = Pachycondyla
Trichomelophorus Wheeler 1935 = Melophorus
Trichomyrmex Mayr 1865 = Monomorium
Trichoscapa Emery 1869
Tricytarus Donisthorpe 1947 [Myrmicinae incertae sedis]
Triglyphothrix Forel 1890
Trigonogaster Forel 1890 [preoccupied]
Turneria Forel 1895
Typhlatta Fr. Smith 1857 = Aenictus
Typhlomyrmex Mayr 1862
Typhlopone Westwood 1839 = Dorylus?
Typhloteras Karavaiev 1925 = Centromyrmex

Veromessor Forel 1917
Viticicola Wheeler 1920 = Tetraponera
Vollenhovenia Dalla Torre 1893 [emendation] = Vollenhovia
Vollenhovia Mayr 1865

Wadeura Weber 1939 = Pachycondyla
Wasmannia Forel 1893 = Ochetomyrmex
Weberidris Donisthorpe 1948 = Calyptomyrmex
Weberistruma Brown 1948 = Smithistruma
Wessonistruma Brown 1948 = Smithistruma
Wheeleria Forel 1905 = Monomorium
Wheeleriella Forel 1907 = Monomorium
Wheelerimyrmex Mann 1922 = Megalomyrmex
Wheeleripone Mann 1919 = Gnamptogenys
Willowsiella Wheeler 1934
Woitkowskia Enzmann 1952 = Neivamyrmex

Xenhyboma Santschi 1919 = Monomorium?
Xenoaphaenogaster Baroni Urbani 1964 = Monomorium?
Xenometra Emery 1917 = Cardiocondyla
Xenomyrmex Forel 1884
Xeromyrmex Emery 1915 = Monomorium
Xiphocrema Forel 1913 = Crematogaster
Xiphomyrmex Forel 1887
Xiphopelta Forel 1913 = Pachycondyla
Xymmer Santschi 1914 = Amblyopone

Zacryptocerus Ashmead 1905 = Cephalotes?
Zasphinctus Wheeler 1918 = Sphinctomyrmex
Zatapinoma Wheeler 1928
Zealleyella Arnold 1922 = Anoplolepis

The Genesis and Nature of Tropical Forest and Savanna Grasshopper Faunas, with Special Reference to Africa

NICHOLAS D. JAGO

The following discussion will contrast the life styles, behavior, and functional morphology of forest and savanna grasshoppers. This having been done, two interesting and complementary concepts arise: first, that grasshopper speciation is likely to be most active under dry transequatorial conditions (here called the "equatorial species dynamo") and, secondly, that forest speciation and adaptation are to be considered an evolutionary dead end and the forest ecosystem a "species sink."

Fossil evidence of the origins of the Acridoidea is scanty. Many genera seem to be of great antiquity and can be recognized among fossil material of Miocene age (Leakey, in litt.) from East Africa. While recognizing that rates of speciation will differ, it is to the modern faunas that we turn for clues to the origins of species and genera based mainly on analysis of morphological and ethological divergence. It is also clear that many of the larger taxonomic groups are artificial aggregations and that we find ourselves in difficulties when discussing the evolution of such groups. Good examples of such nonphylogenetic groupings are the subfamilies Acridinae and Catantopinae (Dirsh 1961).

It is evident that forests have offered a major problem to exploitation by grasshoppers. Mixed and forb feeders have had a clear advantage over grass feeders, because forest grasses are either unpalatable or not extensive enough in the forest ecosystem to provide an adequate niche. Apart from the Eumastacidae, Trigonopterygidae, and Pyrgomorphidae, forests provide a home for only a small proportion of each family or subfamily that embraces forest species. Grass feeding groups, e.g., Gomphocerinae and Acridinae, have relatively few forest members (Jago 1971). In Africa, the Acridinae are richer in forest species than in South America, but in Southeast Asia, India, and Africa, their forest radiation has been limited to very few genera (e.g., *Odontomelus* and *Holopercna* in Africa and India, *Phloeoba* and its allies in southeastern Asia). The Gomphocerinae have adapted even less to forest niches, *Pseudoutanacris* and *Silvitettix* in South America being two of the few examples. Apart from species associated with agricultural practice (Jago 1968) the groups best represented in tropical forests are the Pyrgomorphidae and Eumastacidae (both considered to be rather ancient groups, the latter being in the main highly specialized to arboreal life among bushes and trees), Lentulidae (many of whose members also live in dry savanna woodlands), and, among the Acrididae, elements of the subfamilies Hemiacridinae (e.g., *Glauningia* and *Kassongia*), Oxyinae, Coptacridinae, Eyprepocnemidinae, and Catantopinae. Although the last subfamily contains the greatest number of forest forms, it is also a nonphylogenetic group of great complexity. These species all have in common the tendency to live among forest trees, understorey herbs or saplings, on the ground, or in the field layer. In Africa, there does not seem to be a rich canopy fauna

NICHOLAS D. JAGO, Centre for Overseas Pest Research, London, England.

equivalent to that among the Catantopinae of South America and southeastern Asia, but this may merely reflect lack of research.

Forest Acridoidea display seven major specialized trends in their general morphology (Figures 1–9).

1. In general, they are brightly colored, especially around the apex of the abdomen, on the inner side of the posterior femora, on the posterior tibiae, on the ends of the antennae, and on the sides of the thorax. These conspicuous markings may be toned down in species living on the forest floor; in genera such as *Apoboleus, Pseudophialsphera, Serpusia,* and *Auloserpusia,* for example, there is a general tendency to divide between bright coloration in species inhabiting the vegetation and somber pigmentation in those living among the litter.

2. The compound eyes are abnormally large and give overlapping fields of vision anteriorly.

3. The antennae are proportionally much longer in forest forms than in the nearest savanna relatives of a given genus.

4. Auditory organs, though sometimes normal in size, are liable to be reduced, or even absent. The Eumastacidae and Lentulidae as a whole lack tympanic membranes at the base of the abdomen. In other groups, a trend in this direction is evident in members of the same genus, e.g., in *Auloserpusia,* some of which have a tympanum while others show various degrees of reduction. Reduction or absence correlates closely with the degree of reduction of the tegmina (Jago 1970). Reduction clearly nullifies the utility of the tegmina for flight and reduces their stridulatory function. It can be inferred that loss of the auditory organs is correlated with loss of stridulatory ability. The tegmina may persist as lateral scales covering the tympanic membranes, but in species with greatly reduced hearing organs, the wings and tegmina are reduced to vestigial folds in the cuticle. When present, the tegminal scales presumably serve a protective function; they can be actively raised in many species even when greatly reduced in size (e.g., in *Aresceutica*).

5. There is a tendency for males to search for females when they are sexually mature. In this process, sight, ability to judge distance, and possibly scent (hence the long antennae and the bright colors of the males) play an important role. In many genera, some species may lack sexual dichromatism, e.g., *Auloserpusia impennis* (Rehn), while it is conspicuous in others, e.g., *Auloserpusia potamites* (Jago). Where it occurs, females are cryptically colored and match their food plants. This may well be a compromise situation, in which the need for visual recognition competes with predation pressure on the females. There is also a tendency for suppression of the feature seen in the Gomphocerinae for males to aggregate, sing, and cause assembly of the females, although in the genera *Apoboleus* and *Pseudophialspherera,* the rather large dorsally overlapping wing scales can be used to produce rasping notes.

6. There is a tendency for reduction of the organs of flight. This may be a primary specialization connected with the requirements of the forest ecosystem. As is the case in species living in deserts, marshy savannas, and mosaic subalpine habitats, the forest grasshopper is generally severely circumscribed by its specialized niche requirements. Few of these are known in detail. These limitations may be minimal in far-ranging and rather exceptional forest species like *Holopercna* (Acridinae), which rove freely through the forest understorey and are omnivorous plant feeders. Most species, however, are confined to clearings in normally closed canopies of the forest. In areas unaffected by man these niches are found, for example, where a large tree has fallen causing a clearing, or in steep topographies where rainy season torrents scour open spaces beside streams. Such areas will be occupied by rich fern associations and by plants such as *Afromomum, Marantochloa, Pollia, Brillantaisia* (Ahn 1958). They offer a microclimate with lower humidity, sunshine for basking, and warmer oviposition sites. Such clearings offer the foodplants of forest grasshoppers enough sunlight for growth and seed formation until the succession eliminates them once again.

We do not know just how much population movement there is in flightless forest acridoids, but presumably it is sufficient to permit access to new clearings as they form. Regarding more circumscribed movements, *Aresceutica subnuda* (Karsch) in East Africa follows a very tight but clear diurnal movement pattern exactly like that of extra-forest species, with descent from night roosts after an early morning sunning period, followed by little activity until the dropping of temperatures in the mid-afternoon, at which time limited feeding oc-

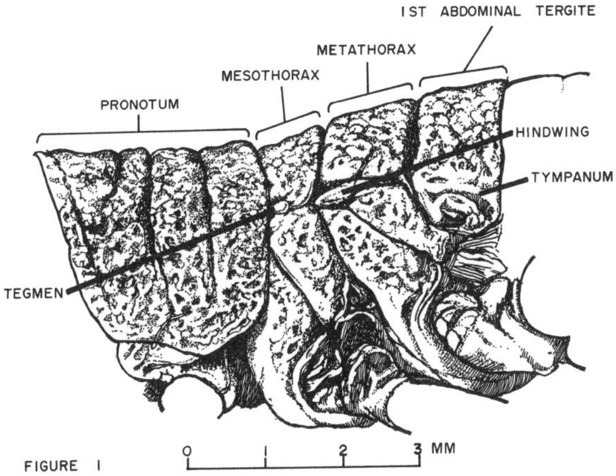

FIGURE 1

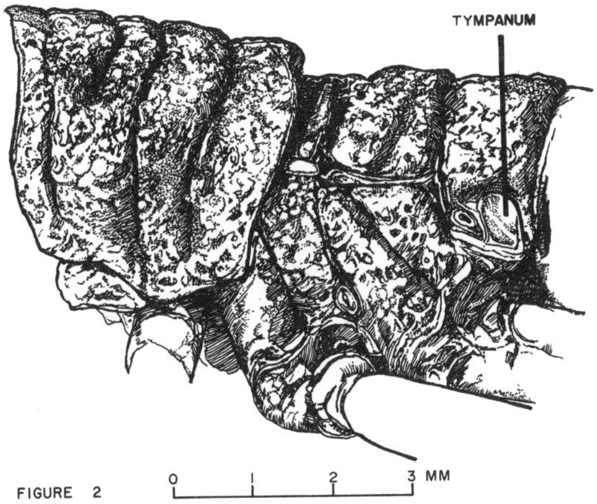

FIGURE 2

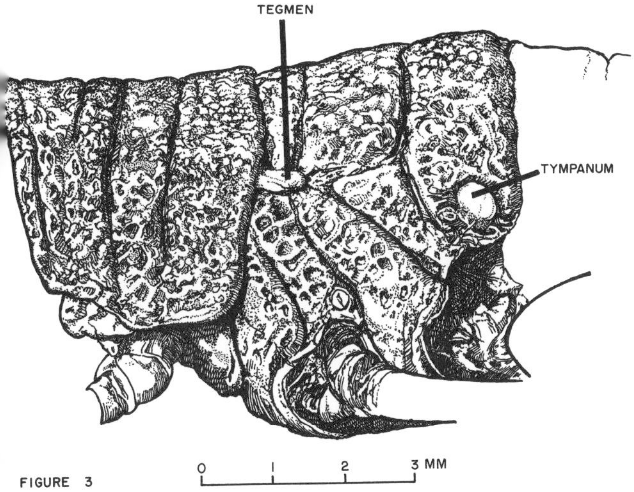

FIGURE 3

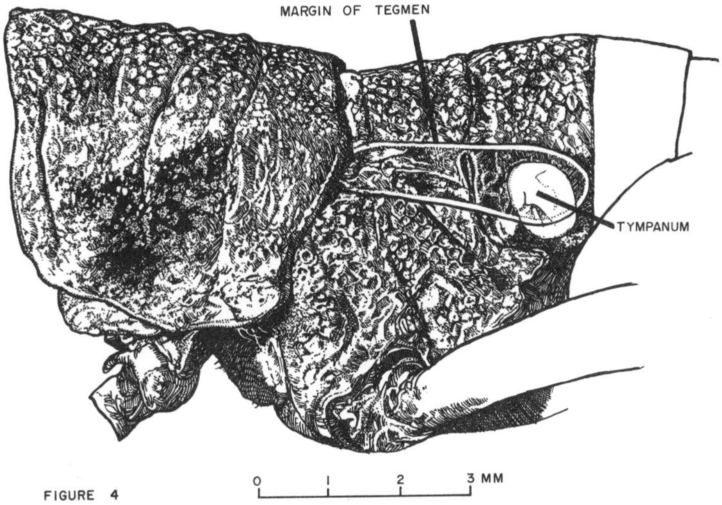

FIGURE 4

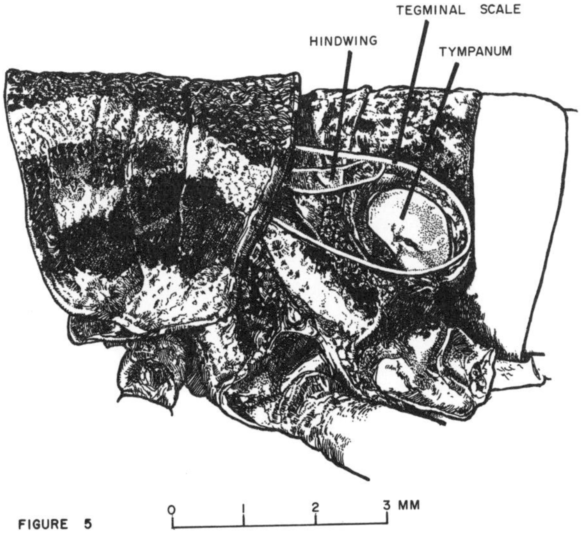

FIGURE 5

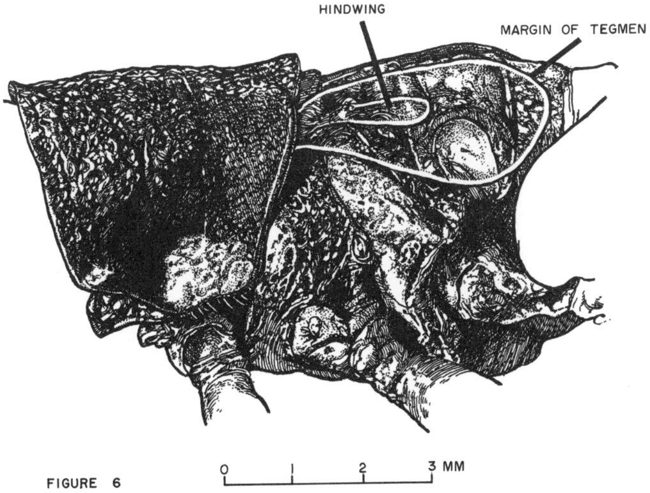

FIGURE 6

FIGURE 1. Lateral aspect pronotum, mesothorax, metathorax, and first abdominal tergite, showing extent of tegminal, hindwing and tympanal reduction in adult forest grasshoppers of *Auloserpusia picta* Ramme. Note the extreme reduction of the male tegmen (tiny lobe), hindwing (fold in cuticle), and tympanum (closed arcuate groove). Head to left. FIGURE 2. *A. poecila* Jago, male, showing slightly less extreme specialization with heavily sclerotised and reduced tympanum. Other details as in Figure 1. Head to left. FIGURE 3. *Auloserpusia lacustris* Rehn, female, showing especially heavy cuticle of dorsal plates on mesothorax, metathorax, and first abdominal tergite, making them similar to each other in appearance. Reduction of flight organs and tympanum are similar to those of *A. poecila* Jago (Figure 2). Head to left. FIGURE 4. *Serpusia opacula* Karsch, male, showing absence of membranous hindwing, presence of enlarged and possibly protective tegmen, and large hyaline tympanic membrane (probably functional). Head to left. FIGURE 5. *Veseyacris ufipae* Dirsh, male, a species very closely related to *Aresceutica morogorica* Dirsh, showing enlarged lobate tegminal scale, lanceolate hindwing fragment, and large hyaline tympanum on the first abdominal tergite. Head to left. FIGURE 6. *Aresceutica vansomerini* Kevan, male, with fore and hindwing shown in transparency; greater proportional development of tegmen and fragmentary hindwing. Head to left.

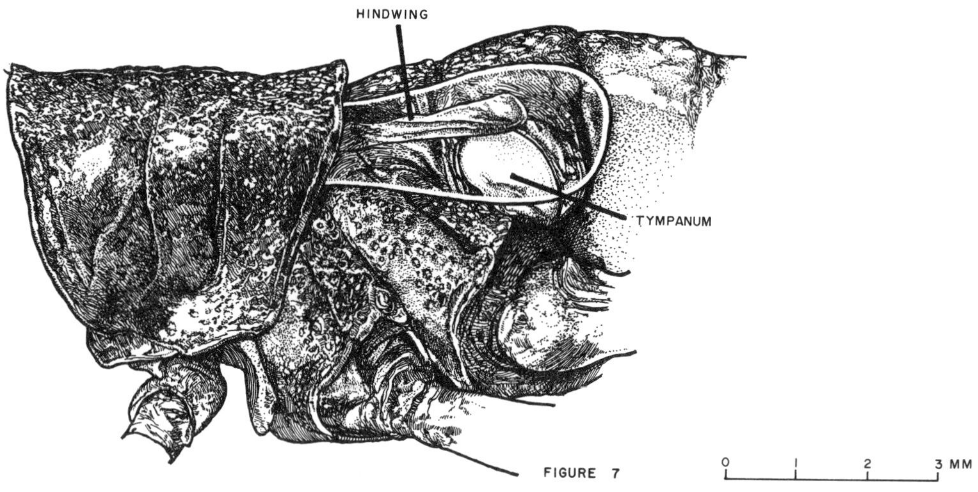

FIGURE 7

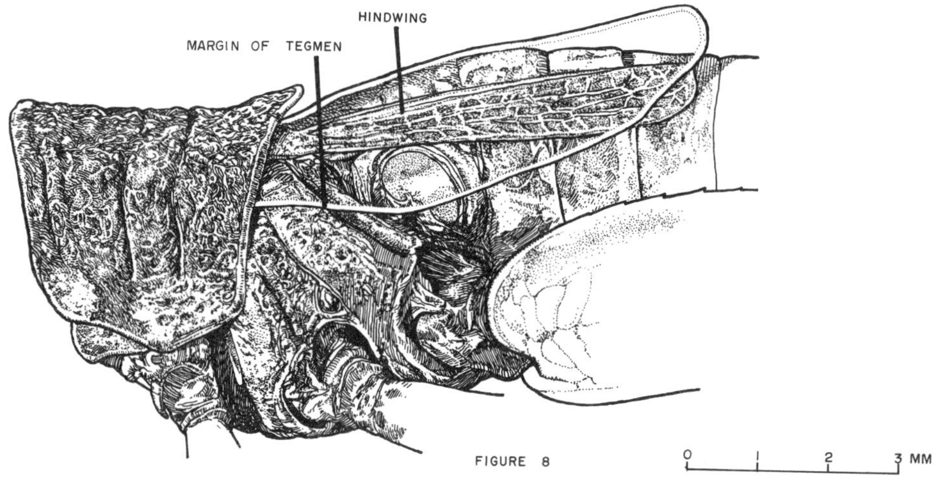

FIGURE 8

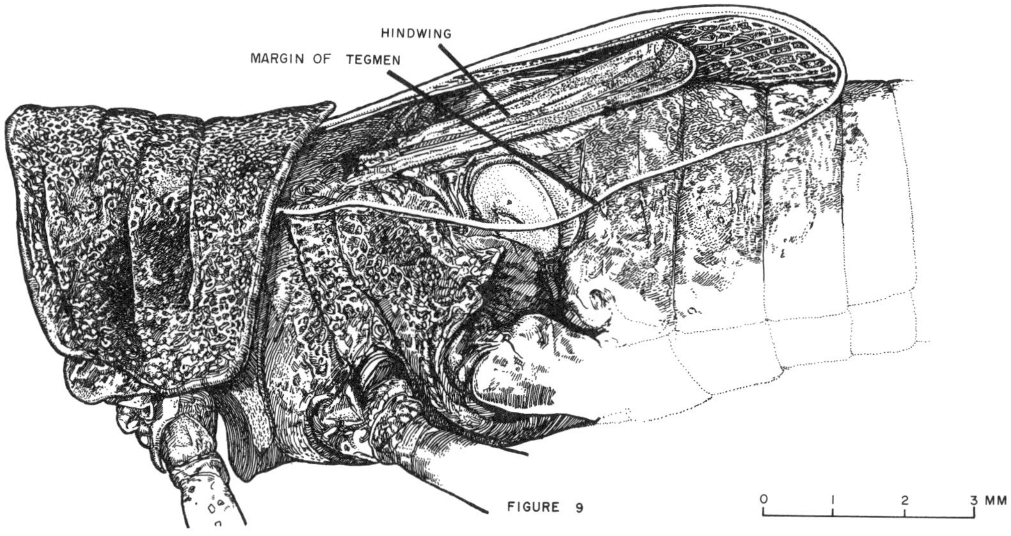

FIGURE 9

FIGURE 7. *Aresceutica morogorica* Dirsh, male, with hindwing enlarged sufficiently to show anal fan and some venation. Note the especially large tympanum. Head to left. FIGURE 8. *Apoboleus sudanensis* Dirsh, male, with hindwings showing almost complete venational sequence, though not used for flight. Tegmen is shown in transparency. Head to left. FIGURE 9. *Pseudophialosphera severini* (Ramme), male, showing same features as Figure 8. Head to left.

curs; there is then ascent into saplings followed by nocturnal feeding. This last is a feature that can also be observed in *Serpusia*. Nocturnal activity may be a special feature of forest grasshoppers, but it is not yet known just how widespread this habit is or what effect it has had in selection pressures for reduced flight.

7. The slender ovipositor valves of many forest species, e.g., especially in the Oxyinae (Hollis 1973), are indicative of scarcity of terrestrial egg-laying sites.

On the assumption that decreased flight movement is of supreme importance to these grasshoppers, it would appear that this feature alone could have been the trigger for generating other characteristics such as inability to stridulate and bright colors associated with species recognition. In forest habitats, the surrounding closed canopy vegetation for all its apparent lushness is effectively an ecological desert in which species have become extremely specialized. This specialization is apparent not only in morphological characteristics, but in their requirements for high humidity, relatively low temperature (hence the difficulty of transporting them back to laboratories and rearing them there without air conditioning), and features like light-adapted eyes (upper and lower parts of the compound eye showing differential pigment movement in *Aresceutica*).

Some forest genera show signs of fairly recent ecological adaptability, e.g., *Usambilla* (Lentulidae) on the Ufipa Plateau which displays two species, one of which has become adapted to life in low growing perennial herbs and bushes outside the declining plateau forests, the rest of the genus being denizens of wet forest. Forest specialization, however, has typically produced species so conservative that they are obligate forest dwellers. Their speciation in a habitat where breeding is possible throughout the year is probably generated in the main by the fragmentation of forest blocks. That seems to have been a feature of African forests. Much of the semideciduous forest in West Africa periodically has been reduced to almost nothing in the not too distant geological past (Moreau 1966: 1–60). It is suggestive that certain forest nuclei, especially those at slightly higher elevations, survived in a relatively intact state for a considerable period of time, while the faunas of other surrounding forest areas have been eliminated. The Sierra Leone-Guinea-Liberia highlands, the Atewa hills of central Ghana, the Togo forests, the Benin forests of Nigeria, the forests of eastern Congo and Ruanda and the Kibale forest (Langdale-Brown et al. 1964: 46) of Uganda, are particularly rich sanctuaries for forest species.

If we look at dry season savanna populations of the Catantopinae in rugged topographies where montane grasslands interdigitate with forests, we can see a likely route by which species originally adapted to savanna habitats gained access to the forest. During the dry season, such forests tend to shed their leaves and the forest floor, though dry, is more moist than that of the surrounding grassland. Many savanna species are at this time found in the forest. *Exopropacris modica mellita* in Ghana and Togoland may be an example of just such a series of populations, derived from *E. m. modica*, but now living in woodland and upland forest. Thus, wooded and forested mountainous areas probably favor induction of forest species by reason of the greater permanence of their forests (the rainfall being higher among such hills) and by the possibility they provide for the intimate juxtaposition of widely differing ecological niches. Only later would forest adaptation become more profound (and indeed less likely to be reversible in a behavioral sense) and deep penetration into more specialized forest niches become possible.

When the process of speciation in forest grasshoppers is viewed against the geological time scale, one is led irresistibly to the conclusion that tropical forest is not a great generator of new species. Admittedly it induces oddity and specialization, but in the long term such species are (if we are to judge by the experience of recent climatic fluctuations) doomed to quicker extinction than their more adaptable brethren in the more continuous savanna and savanna-woodland zones. Hence, as far as grasshoppers are concerned the tropical forest is a *species sink*, continuously seducing species into its confines and then eliminating them by fading away during interpluvial periods.

This hypothetical model of the tropical forest as a species sink immediately poses the question of where one might expect the greatest generation of grasshopper species to be taking place at the present time. Indeed, the site of such activity might be termed a species dynamo on a global scale and might be a model for speciation applicable to many other organisms in similar circumstances to those of grasshoppers.

Laboratory studies by Norris (1959, 1965a, 1965b) have shown conclusively that at least two acridid species are highly sensitive to changes in day-length. In the case of *Nomadacris septemfasciata* (Serville), nymphal and adult development during increasing day-length causes sexual diapause in thc adults. In the case of *Anacridium aegyptium* (Walker), female adults emerging in the spring north of the equator undergo immediate maturation, while under decreasing day-lengths they enter diapause. In *A. aegyptium*, males are nondiapausing under all day-lengths conditions. In nature this would insure that females of *A. aegyptium* remained immature until the spring, thus keying their life cycle into the climatic regime of the Mediterranean region. In *Nomadacris septemfasciata* in its outbreak center south of the equator, the adults pass through the driest part of the year (June-October) in an inmature state, responding only by a slight increase in weight. After the height of the southern winter, the adults mature under increasing day-length. This also coincides with the major rains, thus insuring that hatchlings have an abundant food supply. Individuals developing under decreasing day-length both as nymphs and adults have a longer adult diapause. If the nymphal period is spent in static or increasing day-length and the adult stage in decreasing day-length, they have a shorter adult diapause. Adult diapause is induced, therefore, if the decreasing day-length is experienced only by the adult. Increasing day-length experienced in both nymphal and adult development allows for continuous development of the adult. It will be noted here that maturation is affected by increasing day-length in both cases, but that geographical position north or south of the equator ensures that the two species are temporarily out of phase as sexually mature adults. The critical day-length period for diapause or nondiapause lies between 12 and 13 hours, exactly the order of magnitude that *N. septemfasciata* would experience some 10° north or south of the equator (Figure 10) (Woolley and Gurnette 1960).

The adaptation of savanna and savanna-woodland species to day-length has far-reaching implications. The species are rendered virtually incapable of premature egg-laying, which adherence to a mechanism tied to the caprice of rainfall would entail. The danger involved in the latter type of mechanism would be greatest in the equatorial region itself, where the rains are particularly unpredictable. It should, however, not be overlooked that some species, such as *Locusta migratoria migratorioides* (Reiche and Fairmaire 1850, Albrecht 1971, Farrow 1972) survive the dry season as adults and mature in specialized flooded areas. Food shortage and lack of breeding sites are overcome by development of great mobility. Photoperiodic effects in *Locusta* have not yet been totally elucidated but its southward invasions across the equator have seldom shown any ability to survive for long periods at high population densities.

It is clear that, since the populations on one side of the equator are tied to a day-length regime which is the complement of that on the opposite side (Figure 10, middle area) and since the rainfall patterns are also complementary (Figure 11, compare rainfall graph for Gulu, northern Uganda, with that for Tabora, Tanzania), breeding populations of any particular species straddling the equator are always out of phase. This is so even if they are mobile. This discontinuity in relatively mobile species is reflected in the development of distinct subspecies of *Schistocerca gregaria* (Forskal) north and south of the equator. Moreover, it may not be just the whim of the taxonomist that very few species extend their range uninterruptedly across the equator. In 99 percent of the cases, there is a different subspecies, or a related species, on the other side of this line. Where apparently uniform populations span the equator, the species concerned are either highly mobile or are extremely tolerant of very dry conditions and consequently able to continue breeding, e.g., *Morphacris fasciata* (Thunberg). Even here, however, the apparent morphological uniformity may mask unrecognized genetic differences (see John and Lewis 1965, in the case of *Eyprepocnemis plorans meridionalis* (Uvarov), an eastern and south equatorial subspecies, and *E.p. ornatipes* (Walker), a north equatorial race).

Where a species does cross the equator (Jago 1967), it is only for a short distance, e.g., *Acorypha glaucopsis* (Walker), indicating that intrusive movements are held in check. Species keyed to decreasing day-length north of the equator experience difficulty in adjusting maturation and timing of hatchling emergence to fit rains that are 6 months out of phase across the equator. Consider a chance

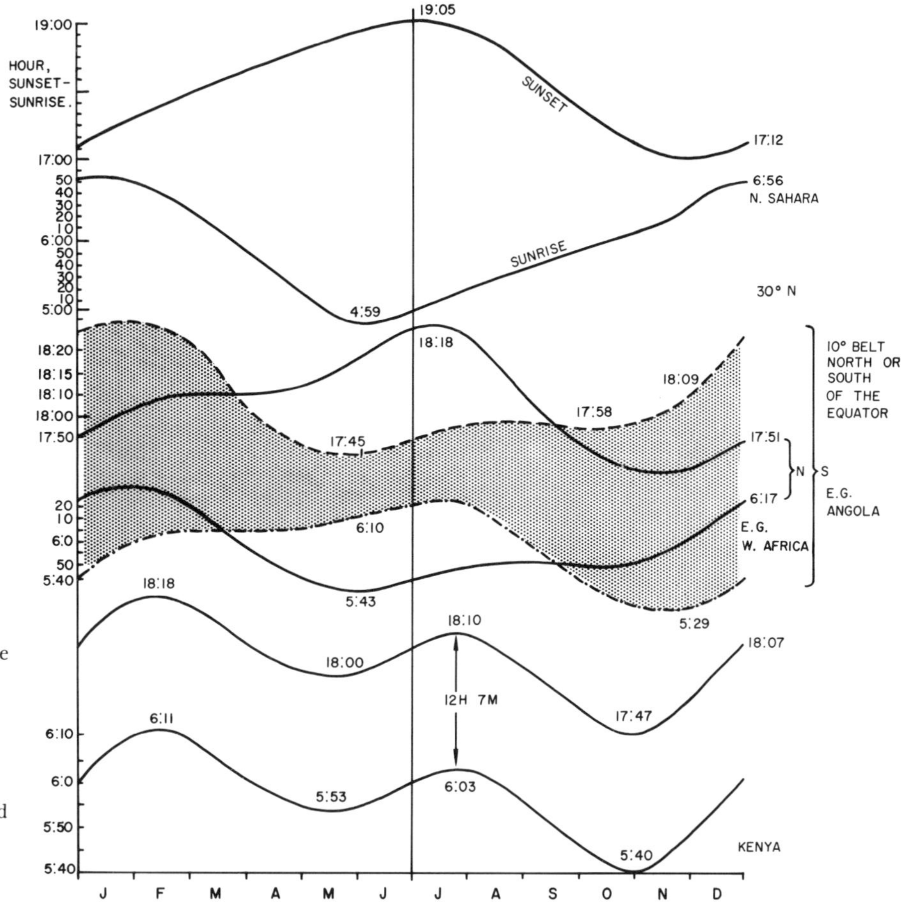

FIGURE 10. Graphical representation of sunrise and sunset throughtout the year for observers in four positions on the earth's surface. *Upper:* Northern edge of the Sahara; *Middle:* 10 degrees north (unshaded) and 10 degrees south (shaded); *Lower:* On the equator. The vertical axis represents the time of sunrise and sunset, with the middle of the day omitted to show more clearly the interrelationship of sunrise and sunset in the different months.

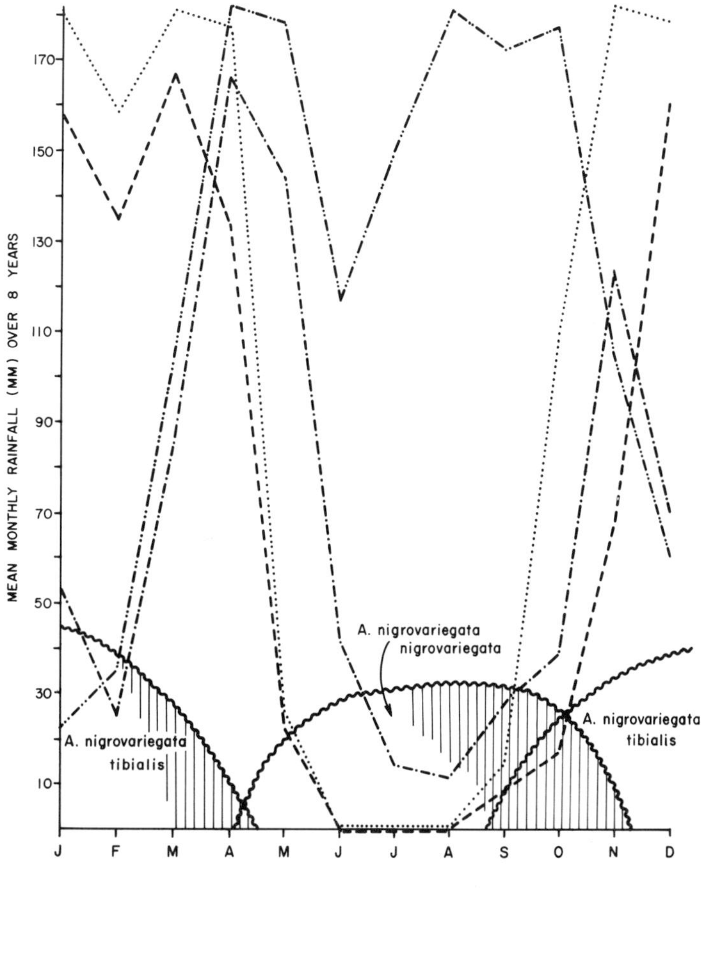

FIGURE 11. Graphical representation of rainfall at chosen localities just north and south of the equator. Two dots and dash, Gulu, northern Uganda; one dot and dash, Kenya, Nairobi; dashes only, Tabora, Tanzania; dots only, Nova Lisboa, Angola. Note the change in rainfall pattern from the West African conformation in Uganda to the south equatorial one at Nairobi. The schematic curves show the months when adults of *Acorypha nigrovariegata nigrovariegata* and *A. n. tibialis* (Kirby) occur.

incursion of adult *A. glaucopsis* south of the equator from a northern habitat. This species undergoes dry season egg diapause north of the equator, and matures at the end of the rains in August and September, in shortening day-lengths. Adults crossing the equator would experience increasing day-lengths at the beginning of the southern rains. Assuming they were not inhibited by the day-length regime, they would lay eggs and these would presumably hatch in an attempt to coincide with the onset of the northern rains in April and May. This would be disastrous south of the equator, since the nymphs would emerge at the beginning of the dry season, for which they are ill-adapted.

This argument begs the question of how universal the adherence to maturation ordered by the day-length really is among the Acridoidea. I think that examination of more species in the laboratory will reveal that this phenomenon is indeed widespread. It will not be developed in forest species, however, but will be delicately and incisively ingrained in transequatorial savanna and savanna-woodland species. Just south of the equator, however, (Figure 10, lower area), we meet another problem. Here day-length does not alter, but the timing of the sunrise is advanced and retarded in an interestingly asymmetrical way twice a year. Even 10° north or south of the equator, the increase and decrease of day-length is not symmetrical (as is implied in Danilevskii 1965:33, fig. 5). North of the equator, for example, (Figure 10, unshaded middle area), increase of day-length in April-May is achieved by greater emphasis on the advance of sunrise than on the retreat of sunset, while in the period of September-October shortening of day-length stems from the rapid change of sunset to an increasingly earlier hour. It may be that Norris' results (1965b), which are interpreted as purely due to changes in total day-length, would be greatly sharpened by testing for the effect of changing the timing of sim-

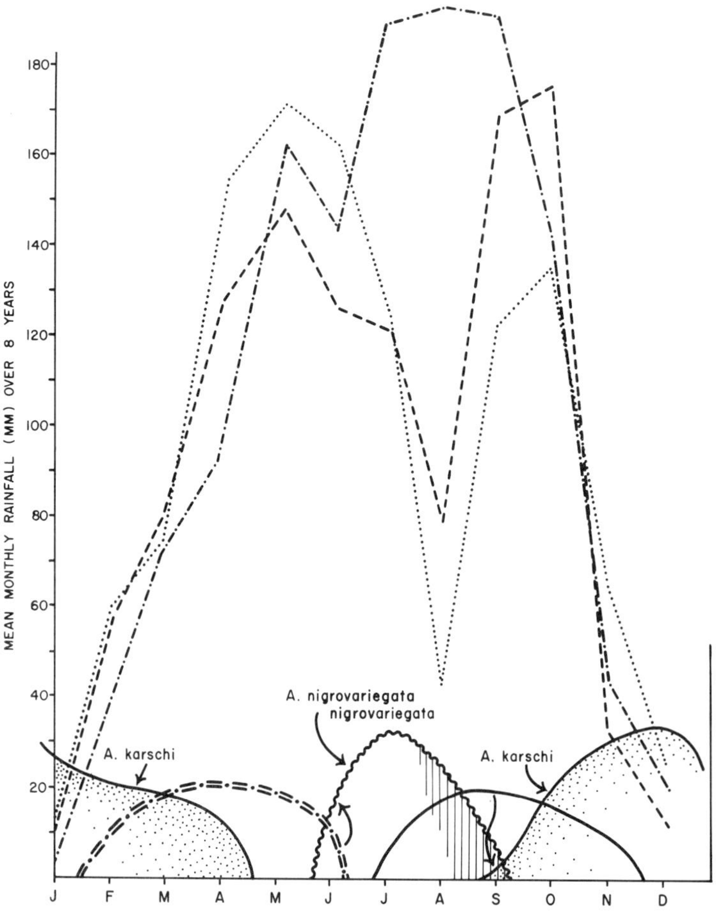

FIGURE 12. Graphical representation of rainfall in West Africa. Dotted line, Dimbokro; dashed line, Bouake; dash and dot, Ferkessedougou. The curves below represent the occurrence of nymphs and adults of *Acorypha n. nigrovariegata* and *A. karschi* depicted in schematic fashion from museum records.

ulated sunrise and sunset. With relatively little change in day-length, the "disturbance" due to the movement of sunrise relative to the period of darkness might have profound effects, and indeed might explain why apparent day-length effects still operate below the equator, where this is the only system to which the insects would be responding. Moreover, the double oscillation of the day at the equator would allow subequatorial populations to react to, say, advance and retreat of sunrise twice in the year, and hence allow better coordination with the complex rainfall system in this region. Much of this is speculation, but evidence from seasonal records is strongly suggestive of some such mechanism. Laboratory experimentation is required for confirmation, using an extension of Norris' techniques.

As an example of the suggestive evidence supporting the hypothesis of an *equatorial species generator,* the genus *Acorypha* (Acrididae, Calliptaminae) give us some good examples. This genus has become divided broadly into two groups of species, one typified by *A. glaucopsis* (Walker) and *A. nigrovariegata* (I. Bolivar) with egg diapause for dry season survival, and the other typified by *A. karschi* (Martinez), which has adult diapause akin to that of *Nomadacris septemfasciata* (Serville). The latter life style excludes the species from the extremely arid zones occupied by *A. glaucopsis.* Such species tend to live in regions of slightly higher elevation or adjacent to hills where some rain is experienced even during the dry season. Although closely related and sympatric in West Africa, *A. karschi* and *A. nigrovariegata* never interbreed. Thus, as seen in Figure 12 the rains in Africa north of the equator follow the pattern indicated with a single peak system as one reaches the Sudan savanna belt. *A. nigrovariegata* hatches at the beginning of the rains (dot and double dash line) with nymphal instars up to the beginning of June; from then on (wavy line), these give rise to adults that mature in August and September (vertical shading). *A. karschi* hatches in July (smooth line), becomes adult in the late part of the rains, and only matures in March and April (shaded dotting). These two closely related sibling species thus never overlap in the time of occurrence of sexually mature adults (according to data provided by Gillon, in litt.).

A. karschi is replaced south of the equator by *A. johnstoni* (Kirby), but *A. n. nigrovariegata* occurs in East Africa as a different south-equatorial subspecies, *A.n. tibialis* (Kirby) (Figure 11). North of the equator, *A.n. nigrovariegata* matures under decreasing day-length at the end of the rains. South of the equator, but east of the western rift valley system of Lake Tanganyika, it does precisely the same and consequently is adult and mature in March and April (wavy line and vertical shading). Transequatorially, genetic isolation is thus complete in this region. In the western half of southern Africa, however, the situation is different. The species occurs as subspecies *A.n. nigrovariegata* (Figure 11, wavy line center and shading), which has a radically different life style. It lives in the Angola highlands and thus becomes capable of surviving under essentially the same pattern of rainfall (see that for Nova Lisboa, dotted line), but with adult diapause. Consequently its adults mature at the beginning of the rains in September to November and are out of phase with the eastern subspecies. A fairly narrow area of overlap is seen in northern Zambia. It is of interest that the nominate subspecies is morphologically much closer to the West African populations of the species than to subspecies *A.n. tibialis.* The adult populations of both mature at the same time of year and the similarity may be due either to ancestral genetic traits held in common or to the presence of occasional population interchange down that side of Africa. The latter possibility cannot be ruled out since many grasshoppers are known to make extensive migratory movements, especially at the end of the rains. These movements present yet another problem, which is in urgent need of investigation.

If this hypothesis of the transequatorial generation of species—the equator as a *species dynamo*—is shown to be true, it would seem that we have the possibility of a much faster species generating system than is demonstrated among forest forms. Savanna species concerned in this process have retained great powers of dispersal and are much more generalized. They give the impression of great evolutionary flexibility. Adaptation to the rigors of the seasonally dry niche into which they fit demands this. Few transequatorial excursions seem to be successful in establishing viable populations. Once established, however, against all the odds set against them, such populations would be quite cut off from the parent populations. Forest speciation patterns seem to depend much more on the long slow processes of forest block isolation and changes in community structure due to the balance of climate between pluvials and interpluvials, than on the kind of relatively rapid evolutionary processes envisaged in the emergence of these new arid-zone species. Moreover, forest speciation may stem from a set of chromosomal phenomena different from those found in more mobile savanna species populations. It is perhaps to be predicted that such phenomena as the well known "neo-Y, X plus autosome" system found in temperate grassland and boreal alpine species will recur in forest grasshoppers, but be lacking in savanna forms. The underlying effects of mobility on population genetics have been little studied in grasshoppers and fundamental differences will probably be revealed when enough comparative data have been amassed.

References

Ahn, P. 1958. Regrowth and swamp vegetation in the western forest areas of Ghana. J. West Afr. Sci. Ass. 4:163-173.

Albrecht, F. O. 1971. The regulation of numbers in locust populations: Laboratory studies in reproduction and survival. *In* Hemming, C. F., and Taylor, T. H. C. (Editors). Proc. Int. Study Conf. Current and Future Problems Acridology London, 1970. Centre for Overseas Pest Control, London.

Danilevskii, A. S. 1965. Photoperiodism and seasonal development in insects. 282 pp. Oliver and Boyd. Edinburgh and London.

Dirsh, V. M. 1961. A preliminary revision of the families and subfamilies of the Acridoidea (Orthoptera, Insecta). Bull. Brit. Mus. (Nat. Hist.) (Ent.) 10:351-419, 34 figs.

Farrow. 1972. The African migratory locust in its main outbreak area on the Middle Niger: Quantitative studies of solitary population in relation to environmental factors. Thesis, University of Reading, England.

Hollis, D. 1973. Review on the Oxyinae. [In manuscript.]

Jago, N. D. 1967. A key, checklist and synonymy to the species formerly included in the genera *Caloptenopsis* I. Bolivar, 1889 and *Acorypha* Krauss, 1877. Eos, Madrid 42:397-462.

————. 1968. A checklist of grasshoppers (Orthoptera, Acrididae) recorded from Ghana, with biological notes and extracts from the recent literature. Trans. Amer. Entomol. Soc., 94:209-353.

————. 1970. A review of the genus *Auloserpusia* Rehn, 1914 (Orthoptera, Acrididae, Catantopinae) and its evolution in the eastern Congo and western Uganda. Trans. Roy. Entomol. Soc. London 122:145-183.

————. 1971. A review of the Gomphocerinae of the world with a key to the genera (Orthoptera, Acrididae). Proc. Acad. Nat. Sci. Philad. 123(8):205-343.

John, B., and K. R. Lewis. 1965. Genetic speciation in the grasshopper *Eyprepocnemis plorans*. Chromosoma 16:308-344.

Langdale-Brown, I., H. A. Osmaston, and J. G. Wilson. 1964. The vegetation of Uganda and its bearing on land use. 159 pp., maps. Uganda Government Publishers. Entebbe.

Moreau, P. 1966. The bird fauna of Africa and its islands. 424 pp. Academic Press. New York.

Norris, M. J. 1959. The influence of daylength on imaginal diapause in the red locust *Nomadacris septemfasciata* (Serville). Entomol. exp. appl. 2:154-168.

————. 1965a. Reproduction of the grasshopper *Anacridium aegyptium* L. in the laboratory. Proc. Roy. Entomol. Soc. London (A)40:19-29, 1 fig.

————. 1965b. The influence of constant and changing photoperiods on imaginal diapause in the red locust (*Nomadacris septemfasciata* (Serville)). J. Insect Physiol. 11:1105-1119.

Woolley, R. van der R., and B. L. Gurnette. 1960. Nautical almanac for 1962. H. M. Stationery Office. London.

Diptera Parasitic on Vertebrates in Africa South of the Sahara and in South America, and their Medical Significance

F. ZUMPT

Diptera or two-winged flies parasitic on vertebrates and having a proven or potential medical and veterinary importance are found in the following families occurring in South America and/or Africa south of the Sahara: Suborder Brachycera: Calliphoridae (blowflies), Cuterebridae (rodent bot flies), Gasterophilidae (stomach flies), Glossinidae (tsetse flies), Hippoboscidae (louse flies), Muscidae (house, horn, and stable flies), Oestridae (nasal and warble flies), Sarcophagidae (flesh flies), Tabanidae (horse flies); Suborder Nematocera: Ceratopogonidae (biting midges), Culicidae (mosquitoes), Psychodidae (moth and sand flies), Simuliidae (black flies).

Suborder Brachycera

CALLIPHORIDAE

Calliphoridae or blow flies are only parasitic in the larval stage, during which they may cause myiasis syndromes in humans and animals. But in this respect, some of them have an enormous economic importance in Africa, as well as in America. There are facultative as well as obligatory myiasis producers in this family.

Two of these facultative myiasis producers, namely *Lucilia sericata* (Meigen) and *L. cuprina* (Wiedemann), have a world-wide distribution, but originated in the Old World. The former is evidently a palearctic element, and as a typical synanthropic fly may have reached the Americas in very ancient times. *L cuprina* is restricted to the tropical and subtropical zones and has probably evolved in Africa. It is the principal blowfly causing sheep strike in South Africa and Australia. In South America, it has been named *L. pallescens* Shannon, until Waterhouse and Paramonov (1950) stated its synonym with *L. cuprina*. It is not known to me whether this fly has any economic importance in South America.

Two calliphorid genera which replace each other are *Cochliomyia* in the New World and *Chrysomya* in the Old World. These genera contain several species, but two of each genus are of special interest, namely *Cochliomyia macellaria* (Fabricius) and *C. hominovorax* (Coquerel) in Central and South America, and *Chrysomya chloropyga* (Wiedemann) and *C. bezziana* Villeneuve in Africa south of the Sahara. *C. macellaria* and *C. chloropyga* develop normally like other members of these genera as saprobionts in decaying organic matter and appear as facultative myiasis producers under certain circumstances. *C. hominivorax* and *C. bezziana,* however, have lost the ability to live as saprobionts and have become obligatory parasites, completely dependant on living vertebrates. Both species play important economic roles, but they have remained restricted to their respective continents.

Calliphorid flies causing a sanguinivorous myiasis in mammals are restricted to Africa south of the

F. Zumpt, South African Institute for Medical Research, Johannesburg, South Africa.

Sahara and belong to the genera *Pachychoeromyia* and *Auchmeromyia*. The larvae are found in the burrows of ant bears, warthogs, and carnivores, and one species has become adapted to man, namely *Auchmeromyia luteola* (Fabricius). Curiously enough, the primary host of the fly has not yet been detected. From these genera, a subcutaneous mode of parasitism has developed (as in *Protocalliphora* parasitizing birds) and is represented by species placed into the genera *Cordylobia* and *Neocordylobia*. *Neocordylobia roubaudi* (Villeneuve) is host-specific to the African ant bear (*Orycteropus afer*), and *Cordylobia ruandae* Fain to the forest mouse (*Grammomys dolichurus*), but *Cordylobia rodhaini* Gedoelst and *C. anthropophaga* (Blanchard) parasitize a great number of wild and domestic animals, and also humans. The larvae develop in skin boils.

In recent times, visitors to Africa from Europe have repeatedly acquired infections with the larvae of *C. anthropophaga* and taken them back to their home country. While there is apparently no danger that this fly could establish itself in Europe, an introduction to South America by humans, or more probably by dogs, could result in a permanent establishment.

CUTEREBRIDAE

This family contains only myiasis-producing species and is restricted to the New World. The morphology of the adult stages shows features suggesting that this group of Diptera represents a line of Calliphoridae that arose in the Paleocene with the rodents, of which they became specific parasites. From their basic stock, the Oestridae and Gasterophilidae of the Old World may have developed.

J. H. Guimarães (1967a) lists the following 5 genera with 53 species and subspecies which parasitize Rodentia and Lagomorpha: *Alouattamyia* (1), *Cuterebra* (43), *Rogenhofera* (6), *Pseudogametes* (2), and *Montemyia* (1). Only very occasionally, larvae of these genera may be found also in other mammals, including dogs, cats, monkeys, and man. *Dermatobia hominis* (Linnaeus), however, the single representative of the genus, parasitizes a large range of mammals and even birds. In America, it is called the "human bot fly," although the most commonly attacked vertebrates are cattle, and the damage done by this fly runs into millions of dollars annually. The clinical symptoms caused by the larvae are similar to those of the African tumbu fly larvae (*Cordylobia anthropophaga* and *C. rodhaini*). However, the economic significance of the latter flies is much smaller, because they mainly infest rodents and carnivores, whereas *D. hominis* prefers larger domestic animals. It also shows a unique mode of life in respect to the transportation of its eggs to the host. The females capture other Diptera which normally visit the hosts, and glue the eggs to their abdomens. The other rodent infesting Cuterebridae, however, do not behave in this way, but deposit their eggs on the hosts or at the resting places as *Cordylobia* flies do. None of the Cuterebridae have succeeded in reaching the Old World.

GASTEROPHILIDAE

The Gasterophilidae are myiasis producers of the Old World, and the larval stages of members of the genera *Gasterophilus, Gyrostigma,* and *Cobboldia* parasitize the alimentary canals of equids and rhinoceroses (order Perissodactyla) and elephants (order Proboscidae). The genus *Gasterophilus* contains nine species, seven of which occur in Africa south of the Sahara and two species have reached South America with horses, namely *Gasterophilus intestinalis* (De Geer) and *G. nasalis* (Linnaeus). The larvae of two species of *Gyrostigma* are found in the stomachs of rhinoceroses in the Ethiopian and Oriental regions; of the genus *Cobboldia* four species are known, of which three are found in the stomach of the African elephant, the other in the Indian elephant (Wetzel 1969).

Two other species, the larvae of which develop in the skin of the African elephant, are *Ruttenia loxodontis* Rodhain and *Neocuterebra squamosa* Grünberg. Their relationship within the higher Diptera is not clear, but Zumpt (1965) places them tentatively in the Gasterophilidae.

GLOSSINIDAE

The family Glossinidae or tsetse flies comprises 20 species (Zumpt 1966) restricted to the Ethiopian zoogeographical region, comprising Africa south of the Sahara and the southern part of Arabia. They are bloodsucking as adults, and the larvae are deposited by the female in their third stage, just before pupating.

These flies have an enormous economic importance as vectors of human and animal trypanosomiasis (sleeping sickness and nagana), but because of the bionomics of tsetse flies, there should be no danger that they might be accidentally introduced into South America, although their existence in the New World in earlier geological times is indicated by four species of fossil *Glossina* described from the Oligocene shales of Colorado.

As is mentioned under Tabanidae (p. 201), however, there is a hypothesis concerning surra, which proposes that this animal disease has evolved from a nagana syndrome caused by *Trypanosoma brucei*. This disease (surra) has spread to many other tropical and subtropical parts of the world including South America.

The so-called American trypanosomiasis or Chagas' disease, caused by *Trypanosoma cruzi* and transmitted by bugs of the subfamily Triatominae, has certainly no relation to African trypanosomiasis and has evolved independently in South America.

HIPPOBOSCIDAE

The Hippoboscidae or louse flies are ectoparasitic bloodsuckers in the adult stage and, as in the tsetse flies, the larvae are deposited just before pupating. In contrast to the bloodsucking Nematocera and the Tabanidae and most Stomoxyinae of the Brachycera, their association with the host is very close and many species show a more or less pronounced host-specificity. In some genera, the wings are rudimentary or they break off after a host has been found. The majority of Hippoboscidae parasitize birds, relatively few being found on mammals.

The following genera have been recorded from America south of the United States (L. R. Guimarães 1968) and from Africa south of the Sahara (Zumpt 1966), with the number of species given in parentheses—the first figures from America (totaling 45), the second from Africa (totaling 44): *Crataerina* (1, 2); *Hippobosca* (0, 7); *Lipoptena* (2, 4); *Lynchia* (9, 13); *Melophagus* (1, 1); *Microlynchia* (4, 0); *Myriophthiria* (1, 0); *Neolipoptena* (1, 0); *Olfersia* (7, 3); *Ornithoica* (2, 2); *Ornithomyia* (5, 7); *Pseudolynchia* (2, 2); *Stilbometopa* (5, 0); *Struthiobosca* (0, 1).

Only five species are found on both sides of the Atlantic. These are *Pseudolynchia canariensis* (Macquart), a common parasite of pigeons, *Melophagus ovinus* (Linnaeus), the well-known sheep ked, and three species of *Olfersia* which infest sea birds and have a more or less pantropic distribution. The first two species are of Old World origin and came, like their main hosts, from the Temperate Zone. *Olfersia* is to be regarded as a genus of the New World that spread with some host-species to other regions.

An interesting fact is that none of the mammal-infesting *Hippobosca* species have been imported with domestic animals either to North or to South America. The well-protected endoparasitic *Gasterophilus* species by contrast have been introduced with their equine hosts.

The hippoboscids have no medical importance, and the veterinary one is slight.

MUSCIDAE

This large family is distributed over the whole world and may contain about 4000 to 5000 species. Most of them are free-living in all stages, and only a relatively small number is parasitic in the adult or larval stages. With respect to the vetebrates, two groups of parasitic Muscidae may be recognized. The one consists of the Stomoxyinae which are bloodsucking in the adult stage; the other is composed of species belonging to several nonrelated genera which are parasitic in the larval stage and cause myiasis in humans and other vertebrates.

The Stomoxyinae represent a very interesting group of monophyletic origin which contains 8 genera with 43 species. In Africa south of the Sahara, 6 genera containing some 33 species are found. (These figures are preliminary, because a monograph of the Stomoxyinae is in preparation.) These are: *Rhinomusca* (2), *Bruceomyia* (2), *Stygeromyia* (2), *Stomoxys* (14), *Haematobia* (=*Lyperosia* auct.) (5), and *Haematobosca* (=*Haematobia* auct.) (8). Only the genus *Neivamyia* with four species is restricted and indigenous to South America. On the other hand, two species with presently an almost cosmopolitan distribution have America, respectively. These are *Stomoxys calcitrans* (Linnaeus) and *Haematofia irritans* (Linnaeus).

The *Neivamyia* species of South America are poorly known and have been collected on domestic animals which have been introduced in historic times. Their original (primary) hosts are unidentified, but should be found among indigenous vertebrates. *Neivamyia* species have not succeeded in filling the niche for bloodsucking Muscidae created by the introduced domestic animals, but this has been done by *Stomoxys calcitrans* which is now the most common representative of the subfamily in South America.

Stomoxyinae have no medical importance and the veterinary one is slight, apart from the fact that in certain areas these flies represent a serious pest by virtue of the numbers in which they attack, and their blood-sucking habit which causes unrest and loss of weight in the animals.

Myiasis-producing Muscidae are few in number and consist (with the exception of two genera parasitic on birds) of facultative parasites only (cf., Zumpt 1965, and James 1947). Facultative myiasis producers are mainly found in the genera *Musca* and *Fannia* which are occasionally involved in cases of dermal, intestinal, and urogenital myiasis. These flies (*Musca domestica* Linnaeus, *Fannia canicularis* (Linnaeus), and *F. scalaris* (Fabricius)) are now of almost cosmopolitan distribution and represent in South America foreign elements from the Old World or North America, respectively.

Of no practical importance, but of great academic interest, are species that are obligatory parasites of birds. They belong to *Passeromyia* in the Old World and to *Philornis* in the New World. *Passeromyia heterochaeta* (Villeneuve) is apparently distributed over the whole of Africa south of the Sahara; in the Oriental region it is known from India, Java, southern China, and Taiwan. The larvae are temporary bloodsuckers and live in the nests of passeriform birds. A second species, *P. longicornis* (Macquart), known from eastern Australia and Tasmania, has larvae that have become subcutaneous parasites. According to Hindwood (1930), the eggs are usually deposited under the wings of the nestlings, and the hatched larvae disperse over the body, pierce the skin and commence to feed on the blood of the young birds. As the larvae grow, they move under the skin, leaving the posterior segment protruding slightly.

In South America, the genus *Passeromyia* is replaced by *Philornis*, of which seven species are listed by Séguy (1937). The larvae are, however, all subcutaneous parasites like *P. longicornis*.

In the Holarctic region, the niche is filled by members of the genus *Protocalliphora*, belonging to the Calliphoridae. There are 19 species known (Stone et al. 1965, and Zumpt 1965), most of them bloodsucking in the larval stages, but a few have also evolved the subcutaneous manner of development.

OESTRIDAE

This family has been subdivided by Zumpt (1965) into two subfamilies, the Oestrinae or nasal flies and the Hypodermatinae or warble flies. All species are obligatory parasites in the larval stages, and due to their rudimentary mouth parts, they are nonfeeding in the adult stages, like the Gasterophilidae.

In Central and South America no indigenous species exist, and only the sheep nasal fly (*Oestrus ovis* Linnaeus) has been introduced from the Old World and become established. It infests sheep and goats, and the first larval instars are sometimes erroneously dropped by the female fly into the orbits of humans, where they cause a short and normally benign ophthalmomyiasis.

In North America, several indigenous species of the deer-infesting *Cephenemyia* exist, as well as *Hypoderma bovis* (Linnaeus) and *H. lineatum* (Villers), the warble flies of cattle, and *Oedemagena tarandi* (Linnaeus), the reindeer warble fly. All three species have been introduced with their imported hosts to South America, but apparently have never become established.

The oestrid fauna of Africa south of the Sahara is very rich. The following list gives the recorded parasitic species with their regular hosts (in brackets): Subfamily Hypodermatinae: *Przhevalskiana corinnae* (Crivelli) and *P. silenus* (Brauer) [Grant's gazelle (*Gazella granti*)]; *Strobiloestrus clarkii* (Clark) [Neotragine antelopes (*Raphicerus, Oreotragus*), other records doubtful]; *S. ericksoni* (Poppius) and *S. vanzyli* Zumpt [Lechwe (*Kobus leche*]. Subfamily Oestrinae: *Kirkioestrus blanchardi* (Gedoelst) and *K. minutus* (Rodhain and Bequaert) [Alcelaphine antelopes (*Connochaetes, Alcelaphus, Damaliscus*)]; *Gedoelstia cristata* Rodhain and Bequaert and *G. haessleri* Gedoelst [Alcelaphine antelopes (*Connochaetes, Alcelaphus, Damaliscus*)]; *Oestrus aureoargentatus* Rodhain and Bequaert and *O. variolosus* (Loew) [Alcelaphine and Hippotragine antelopes (*Connochaetes, Alcelaphus, Damaliscus, Hippotragus*)]; *O. dubitatus* Basson and Zumpt [Wildebeest (*Connochaetes taurinus*)]; *O. macdonaldi* Gedoelst [Alcelaphine antelopes (*Alcelaphus* and *Damaliscus*)]; *O. ovis* Linnaeus [Sheep, goat]; *Pharyngobolus africanus* Brauer [Elephant (*Loxodonta africana*)]; *Rhinoestrus antidorcitis* Zumpt and Bauristhene and *R. vanzyli* Zumpt and Bauristhene [Springbuck (*Antidorcas marsupialis*)]; *R. giraffae* Zumpt [Giraffe (*Giraffa camelopardalis*)]; *R. hippopotami* Grünberg [Hippopotamus (*Hippopotamus amphibius*)]; *R. nirvaleti* Rodhain and Bequaert [Bushpig (*Potamochoerus porcus*)]; *R. phacochoeri* Rodhain and Bequaert [Warthog (*Phacochoerus aethiopicus*)]; *R. purpureus* (Brauer) [Horse, donkey, and zebra (*Equus burchelli*)]; *R. steyni* Zumpt [Zebras (*Equus zebra* and *E. burchelli*)]; *R. usbekistanicus* Gan [Zebra (*Equus burchelli*) and probably horse and donkey].

The Hypodermatinae develop in the dermal and subdermal tissues and the females are oviparous; the Oestrinae develop in head cavities and the female flies are larviparous. Only three of the species listed above are introductions to Africa south of the Sahara, namely *Rhinoestrus purpureus* and *R. usbekistanicus* with horses or donkeys, and *Oestrus ovis* with sheep or goats. They may have reached tropical Africa in very early times and became fully established. Other species that have been imported many times with cattle are *Hypoderma lineatum* and *H. bovis*, but as in South America, they have not adapted to the new environments for reasons which are not yet understood. None of the species indigenous to Africa south of the Sahara have spread to South America, and this is also not to be expected for biological and other reasons.

SARCOPHAGIDAE

This family should actually be united with the Calliphoridae, as has been done by several former and contemporary authors. As in the Calliphoridae there are no adult biting representatives, but the larval stages of some species cause facultative or obligatory myiasis syndromes in vertebrates.

One species of the large genus *Sarcophaga* has an almost cosmopolitan distribution and is commonly found in Africa, as well as in America. This is *S. haemorrhoidalis* (Fallén) which has been recorded as acting occasionally as a facultative parasite, and may be involved in cases of wound and intestinal myiasis.

In the genus *Wohlfahrtia* there are facultative as well as obligatory myiasis producers. However, the obligatory *W. vigil* (Walker) and *W. magnifica* (Schiner) are restricted to the holarctic region, and *W. nuba* (Wiedemann) of the Sudan lives like other African species as a saprobiont and its larvae have only occasionally been found as facultative parasites in wounds of man and domestic animals. No species of *Wohlfahrtia* have been recorded from South America.

TABANIDAE

The large and cosmopolitan family Tabanidae is represented with many genera and species in South America as well as Africa south of the Sahara, but no species occur simultaneously in both parts. However, at least one disease transmitted by tabanids found its way from the Old World to the Americas, namely surra, a highly fatal disease to horses and camels. Other domestic and wild animals also are more or less seriously affected. The causal agent, *Trypanosoma evansi*, is mechanically transmitted, and not only may tabanids of several genera and species act as vectors, but also other biting flies.

Most probably, this disease evolved from *Trypanosoma brucei*, one of the causal agents of animal trypanosomiasis (nagana) in tropical Africa and cyclically transmitted by tsetse flies (*Glossina*). In northern Africa, it became adapted to a purely mechanical transmission by tabanids and spread to many other parts of the world, including South America. Also, *Trypanosoma venezuelense*, *T. hippicum*, and *T. equinum*, described from South America and pathogenic for horses, are evidently nothing else than surra strains.

A human filarial disease of tropical Africa, cyclically transmitted by tabanids is loaiasis (fugitive swellings). The causal agent (*Loa loa*) is spread by *Chrysops silacea* Austen, *C. dimidiate* Wulp, and *C. distinctipennis* Austen, all species which are restricted to Africa. Monkeys evidently act as reservoirs. A spread to other parts of the world has not taken place.

Suborder Nematocera

CERATOPOGONIDAE

The Ceratopogonidae are mostly very small biting Diptera with fifty or more genera, of which *Culicoides*, *Forcipomyia*, and *Leptoconops* attack humans and other warmblooded vertebrates. Their medical and veterinary importance is probably greater than known so far. Many species occur mutually in the Old and the New World, and a spreading of ceratopogonid-borne diseases from one continent to the other appears easily possible.

One of these diseases which only recently found its way from Africa to the United States is bluetongue of sheep, a viral infection (Kettle 1965). It has also been recorded from Spain and Portugal as well as from Japan, but not yet from South America. As transmitter in South Africa, *Culicoides pallidipennis* Carter et al. has been detected, and in North America, *C. variipennis* (Coquillet) acts as a vector. An introduction of this disease to South America should not be unexpected. Another viral disease, African horsesickness, is apparently also transmitted by Ceratopogonidae, but has not yet found its way to the New World.

Culicoides species also transmit several filarial infections. The non- or slightly pathogenic *Dipetalonema perstans,* found in Africa in humans and apes, is transmitted by *C. austeni* Carter et al. and *C. grahami* Austen. This infection has now also been recorded from Guyana and New Guinea (James and Harwood 1969). *Dipetalonema streptocerca* is so far only known from tropical Africa, where it is transmitted by *C. grahami* to humans and apes. It is more pathogenic than *D. perstans.* In Central and South America, the related *Mansonella ozzardi* is spread among humans by *Culicoides furens* (Poey).

CULICIDAE

Human malaria is probably the best known mosquito-borne disease. It is caused by the Protozoa *Plasmodium falciparum* (tropical malaria), *P. vivax* and *P. ovale* (tertian fever) and *P. malariae* (quartan fever). Transmitters are exclusively mosquitoes of the genus *Anopheles,* the most important ones being the following: AFRICA: *A. dureni* (Edwards), *A. funestus* Giles, *A. gambiae* Giles, *A. hancocki* Edwards, *A. melas* Theobald, *A. moucheti* Evans, *A. nili* Theobald, *A. pharoensis* Theobald, *A. pretoriensis* Theobald, *A. rufipes* (Gough); SOUTH AMERICA: *A. albimanus* Wiedemann, *A. albitarsus* Lynch-Arribalzaga, *A. aquasalis* Curry, *A. argyritarsis* Rob.-Desvoidy, *A. bellator* Dyar and Knab, *A. cruzi* Dyar and Knab, *A. darlingi* Root, *A. nuneztovari* Gabaldon, *A. pseudopunctipennis* Theobald, *A. punctimacula* Dyar and Knab.

There is not one species in common between South America and Africa. However, in 1930, *A. gambiae* was imported to Brazil, invading an area of about 12,000 square miles and causing severe epidemics. In the town of Natal, there were for instance 10,000 cases of malaria out of a population of 12,000, with a fatality rate of up to 15 percent. *A. gambiae* was eventually eradicated from Brazil in 1940. The incident shows how great the danger is of highly suitable vectors being introduced to areas previously free of malaria or inhabited by *Anopheles* species that are only poor transmitters.

Another mosquito that found its way from tropical Africa to the Americas is *Aedes aegypti* Linnaeus, the vector of urban yellow fever. The first recognizable epidemic of this viral disease occurred probably about 1648 in Yucatan, Mexico, followed in 1649 by another one in Havana. Since then, numerous epidemics of yellow fever have ravaged the Americas, especially the sea ports. As late as 1905, southern ports of the United States had 5000 cases with 20 percent mortality.

The home of *Aedes aegypti,* as well as yellow fever, is certainly tropical Africa where the disease has existed since prehistoric times as a zoonosis between monkeys and mosquitoes of the *Aedes aegypti* group (subgenus *Stegomyia*), which includes *A. africanus* (Theobald), *A. metallicus* (Edwards), *A. simpsoni* (Theobald), and *A. vittatus* (Bigot). This endemic sylvatic yellow fever gave rise to epidemic urban yellow fever transmitted by *Aedes aegypti,* which became a synanthropic species in Africa. After its introduction to Middle and South America, yellow fever again found its way to the forests and became once more sylvatic by adaptation to indigenous primates and mosquitoes, such as species of *Haemagogus,* particularly *H. spegazzinii* Brèthes, and *Aedes leucocelaenus* Dyar and Shannon. Soper (1955) suggested that the jungle type of yellow fever may have existed in the New World in pre-Columbian days; however the evidence, based on some Mayan chronicles, is very poor.

Another viral disease exclusively associated with *Aedes* mosquitoes of the subgenus *Stegomyia,* and particularly with *A. aegypti* L., is dengue. It is comprised of four or even six distinct strains and found in tropical and subtropical parts of both the New and Old World. Because the subgenus *Stegomyia* was originally restricted to the Old World, dengue must also have been imported in historic times to the Americas. The role of animals as reservoirs is not clear.

Yellow fever and dengue both belong to group B arbo-viruses. Another pathogenic agent of this group is SLE-virus (St. Louis encephalitis), which was isolated from four species of *Culex* and from *Psorophora ferox* (Humboldt) in Trinidad, from *Sabethes belisarioi* Neiva in Brazil, and serological investigations suggest its presence in Colombia and Argentina. It is also known from several places in Central America. It caused epidemics with thousands of cases and hundreds of deaths only in the United States, where it is mainly transmitted by members of the *Culex pipiens* group. This suggests, I think, that the virus has been introduced in historic times from South or Central America to North America. It has, however, not found its way to the Old World, where the SLE-virus is replaced by other viruses causing encephalitis. In Africa, these viruses are not yet known. West Nile, Spondweni, and Wesselsbron viruses isolated from various Culicinae in Africa south of the Sahara, are still incompletely studied, as well as the Ilheus virus in Central and South America.

Arbo-viruses of group A, pathogenic for humans and transmitted by mosquitoes, are represented in South America as well as in Africa south of the Sahara, but none occur in both areas: AFRICA: Chikungunya, O'Nyong-Nyong, and Sindbis; SOUTH AMERICA: Mayaro, Mucambo, Pixuna, Venezuelan equine encephalitis, Western encephalitis, and Eastern encephalitis.

Viruses belonging to antigenic groups other than A and B have been isolated from mosquitoes, man, and domestic animals in South America and Africa. Many of them are only imperfectly studied. Their identity is not fully clear and the epidemiology of the diseases they cause is still the subject of investigations (see Zumpt 1968, and James and Harwood 1969).

One of the most important pathogenic agents to be mentioned in this context is the Rift Valley fever virus of East and South Africa causing enzootic hepatitis in domestic animals, but also involving man. An epizootic in South Africa in 1950/51 caused the death of about 100,000 sheep and cattle. Approximately 20,000 human infections were detected, the course of which was, however, much milder than in animals and no fatal cases occurred. As transmitters, mosquitoes of the genera *Eretmapodites, Aedes,* and *Culex* were disclosed, and wild animals, especially rodents, may act as natural reservoirs. This disease is certainly a potential danger to other parts of the world and especially to South America, as well as certain other arbo-viruses.

Another group of diseases transmitted by mosquitoes is that caused by filarial nematodes. Bancroftian filariasis occurs in many tropical and subtropical regions of the world (Edeson and Wilson 1964) and is caused by *Wuchereria bancrofti*; the only known vertebrate host is man. Identified as transmitters are 11 species of *Anopheles,* 1 *Mansonia,* 6 *Aedes,* and 3 or 4 *Culex*. The primary vectors are apparently *Anopheles* species, and in tropical Africa *A. gambiae* and *A. funestus* are most important. The cosmotropical *Culex pipiens quinquefasciatus* Say (=*fatigans* (Wiedemann)) should be regarded as a secondary, but extremely well adapted transmitter, which acts as a nocturnal as well as a diurnal vector. It is interesting that in South America only this species and the indigenous *Anopheles darlingi* Root have been recorded so far as transmitters. This may support the theory that this infection has also been introduced from the Old World with *Culex pipiens,* originating from *Anopheles* species and re-infecting *A. darlingi* (and perhaps also other *Anopheles* species) in South America.

A major veterinary problem in much of the world is infection with *Dirofilaria immitis,* the heart worm, in dogs, cats, and wild carnivores. It is transmitted mainly by *Aedes aegypti* and members of the *Culex pipiens* complex, but also by other species of mosquitoes. Another filarial infection of dogs and cats with reservoirs in wild carnivores is caused by *Dirofilaria repens.* Hawking and Worms (1961) list a great number of filaroid nematodes that infect domestic and wild animals and are transmitted by mosquitoes. Occasional infections with these helminths occur in humans and may constitute a health problem (Beaver and Orihel 1965). All these infections may easily be interchanged between South America and Africa with the transport of dogs and cats.

PSYCHODIDAE

Only one genus, *Phlebotomus,* of the subfamily

Phlebotominae, or sand flies, has a pronounced medical importance. Many species are recorded from South America as well as Africa, but none is present in both geographical areas.

Phlebotomus species are vectors of sand fly fever, a viral disease, kala-azar (caused by *Leishmania donovani*), oriental sore (*Leishmania tropica*), espundia (*Leishmania brasilinensis*), leishmaniasis mexicana (*Leishmania mexicana*), and Carrion's disease, a bacterial infection caused by *Bartonella bacilliformis*. Oriental sore is restricted to the Old World. Kala-azar occurs in the Old World as well as in South America, and Carrion's disease, espundia, and leishmaniasis mexicana are endemic diseases of South and Central America. The following *Phlebotomus* species are the most important vectors known so far (their concommitant diseases indicated in parentheses): AFRICA SOUTH OF THE SAHARA: *P. orientalis* Parrot and *P. martini* Parrot (kala-azar); *P. papatasi*(?) (oriental sore). SOUTH AMERICA: *P. longipalpis* Lutz and Neiva (kala-azar); *P. intermedius* Lutz and Neiva, *P. longipalpis* Lutz and Neiva, *P. renei* Martins et al. (espundia); *P. flaviscutellatus* Mangabeira (leishmaniasis mexicana); *P. columbianus* Ristorcelli and Van Tyl and *P. verrucarum* Townsend (Carrion's disease).

Our knowledge about the distribution of leishmaniasis in Africa south of the Sahara is still quite scanty. Only recently (Grové 1970) have cases of oriental sore been discovered in Southwest Africa, but as yet the transmitter has not been detected. In the Mediterranean, *P. papatasi* Scopoli and *P. sergenti* Parrot are proved vectors of cutaneous leishmaniasis, but they invade the Ethiopian region only in the extreme northern parts and have not been found in central or southern Africa.

In this connection, it should be mentioned that Barnett (1962) suggested that only one species of *Leishmania* may be involved, and that the ability of leishmaniasis to cause different clinical syndromes may be a consequence of the different *Phlebotomus* species involved in the transmission.

Whether an interchange of *Leishmania*-infections between South America and Africa has taken place in historic times is difficult to say, but theoretically it is quite possible that diseased Negro slaves from West Africa reached South America, where indigenous *Phlebotomus* became infected.

SIMULIIDAE

In certain parts of Africa and Central and South America, *Simulium* species are responsible for the transmission of a dreadful disease to humans, Onchocerciasis, also known as Robles' disease or blinding filarial disease. In Guatemala, for instance, 35 percent of the population is infected and blindness occurs in about 5 percent. The most important transmitters in Africa are *Simulium damnosum* Theobald and *S. neavei* Roubaud; in America *S. callidum* (Dyar and Shannon), *S. exiguum* Roubaud, *S. metallicum* Bellardi and *S. ochraceum* Walker. It is apparent that this disease has also been introduced to America with the slave trade, but indigenous *Simulium* species became adapted to the transmission of the causal agent, the filarial worm *Onchocerca volvulus*.

There are no other diseases of medical or veterinary importance in South America and Africa south of the Sahara, in which transmission Simuliidae are involved.

Discussion and Summary

Parasitic Diptera act as causal agents or transmitters of human and animal diseases. It has been shown that the dipterous fauna on the species level in South America is different from that in Africa south of the Sahara. There are relatively few species occurring both in South America and in the Old World, and in many cases it can be proved that this duplication is of recent origin and traceable to man's interference. It is interesting that although South America received species from Africa or Europe, no neotropic parasitic flies of economic importance have reached the Old World. The former African slave trade played an important role in this respect.

Some of the parasitic flies that have reached the New World during historic times are the following: Culicidae: *Anopheles gambiae* Giles, *Aedes aegypti* Linnaeus; Muscidae: *Stomoxys calcitrans* (Linnaeus), *Haematobia irritans* (Linnaeus); Calliphoridae: *Lucilia sericata* (Meigen), *Lucilia cuprina* (Wiedemann); Sarcophagidae: *Sarcophaga haemorrhoidalis* (Fallén); Gasterophilidae: *Gasterophilus intestinalis* (De Geer), *Gasterophilus nasalis* (Linnaeus); Oestridae: *Oestrus ovis* Linnaeus; and Hippoboscidae: *Melophagus ovinus* (Linnaeus).

The economic importance of the introduction of these Diptera to South America lies in the establishment of yellow fever and dengue by *Aedes aegypti,* and the former great epidemics of malaria in Brazil caused by *Anopheles gambiae.* As bloodsuckers, *Stomoxys calcitrans* and *Haematobia irritans* are troublesome to domestic animals, and *Lucilia* and *Sarcophaga* species are involved in occasional cases of myiasis. The two *Gasterophilus* species are parasites of horses, and *Oestrus ovis,* as well as *Melophagus ovinus,* pose veterinary problems to sheep husbandry.

Some diseases of the Old World have become established in South America by adaptation to indigenous species of Diptera. These are kala-azar and bancroftian filariasis in humans, and some filarial infections of dogs and cats. Other diseases pose a danger to South America by accidental introduction, namely bluetongue of sheep, Rift Valley fever of sheep and cattle, horsesickness, and the myiasis producing *Cordylobia* and *Hypoderma* species in man and cattle, respectively.

References

Barnett, H. C. 1962. Sandflies and sandfly-borne diseases. Pp. 83-91, *in* K. Maramarosch. Biological transmission of disease agents. Academic Press. New York and London.

Beaver, P. C., and T. C. Orihel. 1965. Human infection with filariae of animals in the United States. Amer. J. Trop. Med. Hyg. 14:1010-1029.

Dalmat, H. T. 1955. The black flies (Diptera, Simuliidae) of Guatemala and their role as vectors of Onchocerciasis. Smithsonian Misc. Coll. 125 (1):1-425.

Edeson, J. F. B., and T. Wilson. 1964. The epidemiology of filariasis due to *Wuchereria bancrofti* and *Brugia malayi.* Ann. Rev. Ent. 9:245-268.

Grové, S. S. 1970. Cutaneous leishmaniasis in South West Africa. S. Afr. Med. J. 44:206-207.

Guimarães, J. H. 1967a. A catalogue of the Diptera of the Americas south of the United States, 105: Family Cuterebridae. 11 pp. Dep. Zool., Secr. Agricultura, São Paulo.

————. 1967b. A catalogue of the Diptera of the Americas south of the United States, 106: Family Oestridae (including Hypodermatidae). 4 pp. Dep. Zool., Secr. Agricultura, São Paulo.

Guimarães, L. R. 1968. A catalogue of the Diptera of the Americas south of the United States, 99: Family Hippoboscidae. 17 pp. Dep. Zool., Secr. Agricultura, São Paulo.

Hawking, F., and M. Worms. 1961. Transmission of filariod nematodes. Ann. Rev. Ent. 6:413-432.

Hindwood, K. A. 1930. A sub-cutaneous avian parasite. Emu 30:131-137.

James, M. T. 1947. The flies that cause myiasis in man. U.S. Dept. Agric., Misc. Publ. 631:1-175.

James, M. T., and R. F. Harwood. 1969. Herms's medical Entomology. 6th edition. 484 pp. Macmillan Co. London.

Kettle, D. S. 1965. Biting ceratopogonids as vectors of human and animal diseases. Acta Tropica 22:356-362.

Séguy, E. 1937. Diptera, Fam. Muscidae. Gen. Insct. 205, 604 pp.

Soper, F. L. 1955. Yellow fever conference. Amer. J. Trop. Med, 4:571-661.

Stone, A., C. W. Sabrosky, W. W. Wirth, R. H. Foote, and J. R. Coulson. 1965. A catalog of the Diptera of America north of Mexico. 1696 pp. Agr. Res. Serv., U.S. Dept. Agric., Washington, D.C.

Waterhouse, D. F., and S. J. Paramonov. 1950. The status of the two species of *Lucilia* (Diptera, Calliphoridae) attacking sheep in Australia. Austral. J. Sci., Res. B 3:310-336.

Wetzel, H. 1969. Myiasis-Fliegen (Diptera: Calliphoridae, Oestridae) beim Afrikanischen und Indischen Elefanten (*Loxodonta africana* (Blumenbach) und *Elephas maximus* Linnaeus). Z. angew. Ent. 56:489-502.

Zumpt, F. 1965. Myiasis in man and animals in the old world. xv + 267 pp. Butterworths. London.

————. 1968. Human- und veterinärmedizinische Entomologie. Handbuch der Zoologie, IV. Band. 2. Hälfte: Insecta (2. Aufl.) 1. Teil: Allgemeines. 49 pp. Walter de Gruyter & Co. Berlin.

Zumpt, F., (Editor). 1966. The arthropod parasites of vertebrates in Africa south of the Sahara (Ethiopian region). Publ. S. Afr. Inst. Med. Res. 13(52):1-283.

Problems Related to the Transoceanic Transport of Insects, Especially Between the Amazon and Congo Areas

C. G. JOHNSON AND J. BOWDEN

Introduction

This paper deals with insects, a phylum scarcely considered by Darlington (1957) and which, because of the extreme mobility of its members, may not fit the current views of some zoogeographers. Ideas about insect dispersal and migration have changed greatly in the last 25 years. The dispersal of insects by wind cannot be regarded merely as accidental and occasional, in an otherwise localized life, but rather as an adaptive and widespread feature of the phylum. The former idea that long-distance migrants, such as locusts and some Lepidoptera, control the main direction of their major movements has been disproved and it is now known that the direction of displacement of many strongly flying migrants is controlled by the wind within large-scale weather systems. Geographical isolation is not as absolute, with many species, as was formerly supposed, at least as far as immigration in contrast to establishment is concerned; this applies particularly to pests that colonize temporary habitats such as annual crops and which must migrate to survive. Large populations of many such insects regularly travel immense distances on the wind, with great rapidity, as part of their way of life.

We shall therefore discuss insect pests particularly (as well as some other insects) because many migrate, because of their importance in the development of the two areas drained by the Amazon and Congo rivers, and because they are relatively well documented. We shall first consider some examples of long-distance flights and then the wind fields over the tropical Atlantic Ocean in relation to the possible transport of pests by natural wind-aided flight between Africa and South America. We shall consider also records of insects over the Atlantic Ocean and the occurrence of limited groups of pests in Africa and South America in relation to natural flight and to transport by human agencies.

The information on many of these topics is rather scattered and difficult to obtain. The introduction of insects through ports and airfields involves quarantine and since analysis of the records of quarantine stations was beyond our means, this special problem, though relevant to the present symposium, is outside the scope of this paper. Moreover there are many unscheduled aircraft flights, and it is difficult or impossible to obtain information about the passages and cargoes of many ships in which pests could be carried. This paper, therefore, is not comprehensive but merely a review of present perspectives.

We are not zoogeographers discussing the origins of faunas. We discuss in a preliminary way only those events, and possible events, affecting transfer

C. G. Johnson and J. Bowden, Department of Entomology, Rothamsted Experimental Station, Harpenden, Hertfordshire, England. Acknowledgments: Miss Jennifer Sherrard has helped throughout, especially with the bibliography and illustrations, and has our special thanks. We are grateful also to the late Mr. G. W. Hurst for much help on meteorological matters; to Dr. V. F. Eastop, Dr. L. A. Mound, Dr. D. J. Williams, Mr. B. Bolton, and Mr. C. A. Collingwood for information on aphids, thrips, whitefly, coccids, and ants; to Dr. C. B. Williams, Mr. R. A. French, Dr. D. R. Ragge, and Dr. Z. Waloff who allowed us to publish their records of insects at sea.

of insects at the present time in relation to the future development of both areas. Each species of insect has its own peculiarities and needs special consideration. As Wilson (1959:122) wrote "There is a need for a 'biogeography of the species' orientated with respect to the broad background of biogeographic theory but drawn at the species level and correlated with studies on ecology, speciation and genetics."

Long-distance Displacement of Insects

To show nonentomologists the capabilities of flight possessed by some insects we give a few well-documented examples of adaptive migration on the wind and of long-distance flight over the northern Atlantic Ocean; apparently there are no records for the tropical Atlantic. Many insects produce migratory forms more or less regularly, whose physiology enables them to fly for longer distances than nonmigratory, or less migratory, individuals of the same species. For example, migratory forms of the desert locust make long-distance flights during periods of several weeks. Other species, notably moths that normally fly at night, also produce individuals that can flap their wings continuously, day and night, for 1–2 days at a stretch in the laboratory; for example *Chorizagrotis auxilliaris* (Grote), *Pseudaletia separata* (Walker). Others can also travel across the sea in single flights of 3–5 days duration (*Spodoptera exigua* Hübner, *Plutella xylostella* (L.)). The air speed of most insects is small and contributes relatively little of direction or the long distances flown, and this is particularly so with small insects such as aphids. In all long-distance displacements by flying insects the direction of wind has a major effect on the direction of travel, and the duration of active wing-flapping (the length of time the insect remains actively airborne), combined with wind speed, largely determines the distance flown. It should be stressed that active wing-flapping seems essential for long-distance displacement; for though it may contribute relatively little to the speed and direction of displacement, without it insects soon fall out of the air (Johnson 1969).

The date of arrival of many migrants, particularly locusts, leafhoppers, and moths, at particular places is accurately known. From synoptic weather charts the trajectory of the air-mass in which they were believed to have flown can be back-tracked to the possible sources of the insects. A few examples show that some long-distance migrants can travel far over the northern Atlantic Ocean and even traverse it.

Plutella xylostella (L.) (=*maculipennis* Curtis), a small moth about as big as a clothes moth and a pest of brassicas, invaded the northeast areas of England and Scotland in millions on 28–30 June 1958; moths reached Ireland on 2 July; on 4 July thousands were reported on a weather ship 300 miles south of Iceland, 500 miles from the northern coast of Scotland and about 900 miles from the coast of Norway. Back-tracks of the wind associated with the moths ended in the Baltic area and in western parts of the USSR, which was almost certainly where the moths came from and where the thundery, convective weather was suitable for the start of the journey at a high altitude. It is doubtful if the weather ship was the limit of their journey, and a flight of many millions nonstop across the seas (a descent onto the sea would have been fatal) for at least a thousand miles, and probably 2000 miles, undoubtedly occurred. Some of the insects were, therefore, continuously airborne for four days, or more (French and White 1960). Similar, though smaller invasions probably occur frequently (Shaw and Hurst 1969).

The lesser army worm moth, *Spodoptera exigua*, invades Britain (where there is no permanent population) from February to May probably in most years. Many back-tracks show that the moths come from North Africa, over the sea past Portugal and over the Bay of Biscay, making nonstop flights of 4–5 days duration and at least 1500 miles (Hurst 1965, French 1969) (Figure 1).

The moth *Phytometra biloba* Stephens, a native of North America, has twice been recorded in Britain where it does not breed (once arriving at the same time as North American birds), evidently borne on winds originating in Virginia and Newfoundland. The journeys of 2500 and perhaps 3500 miles were estimated to take 3–4 days. There are many other similarly documented examples in other parts of the world (Figure 2) and they are discussed fully by Johnson (1969).

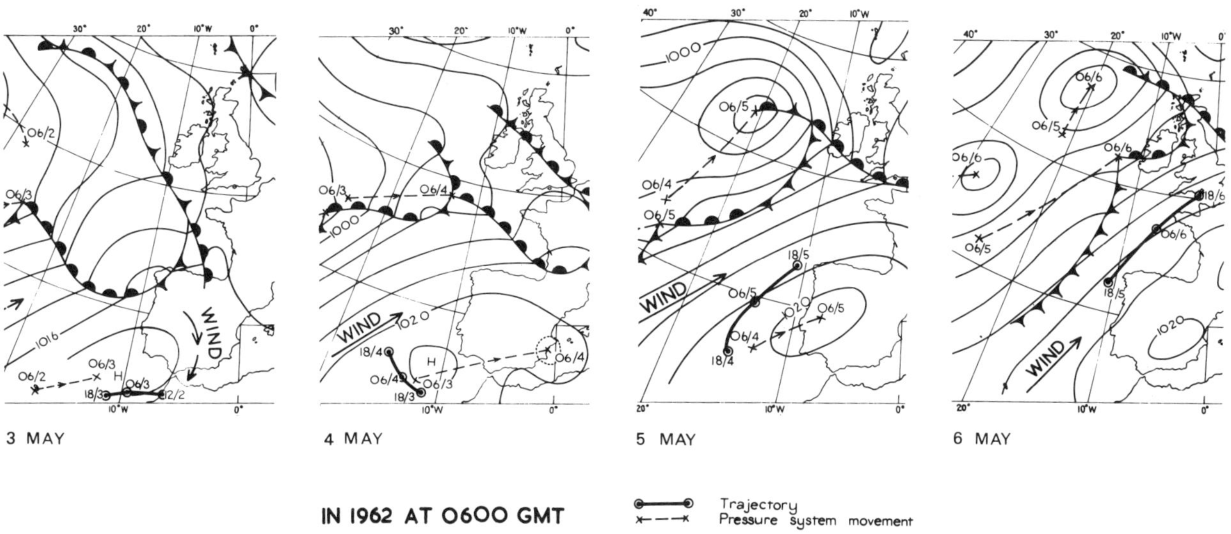

FIGURE 1. An example of a back-track, with the aid of synoptic charts, of *Spodoptera exigua* arriving in south England from Morocco. [From Hurst 1965.]

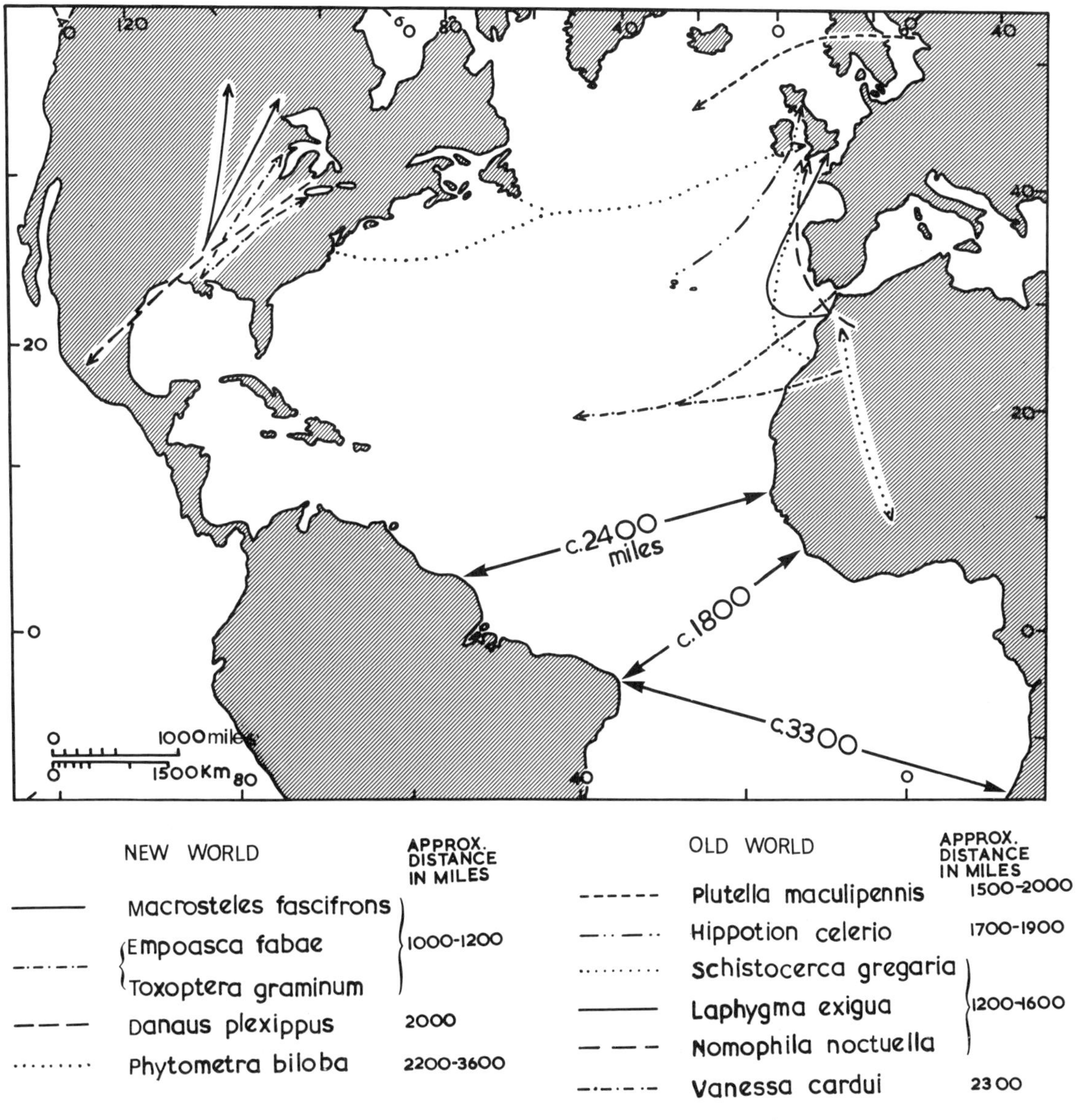

FIGURE 2. Some examples of documental trajectories of insects on both sides of the Atlantic Ocean together with approximate straight-line distances in miles between tropical West Africa and South America.

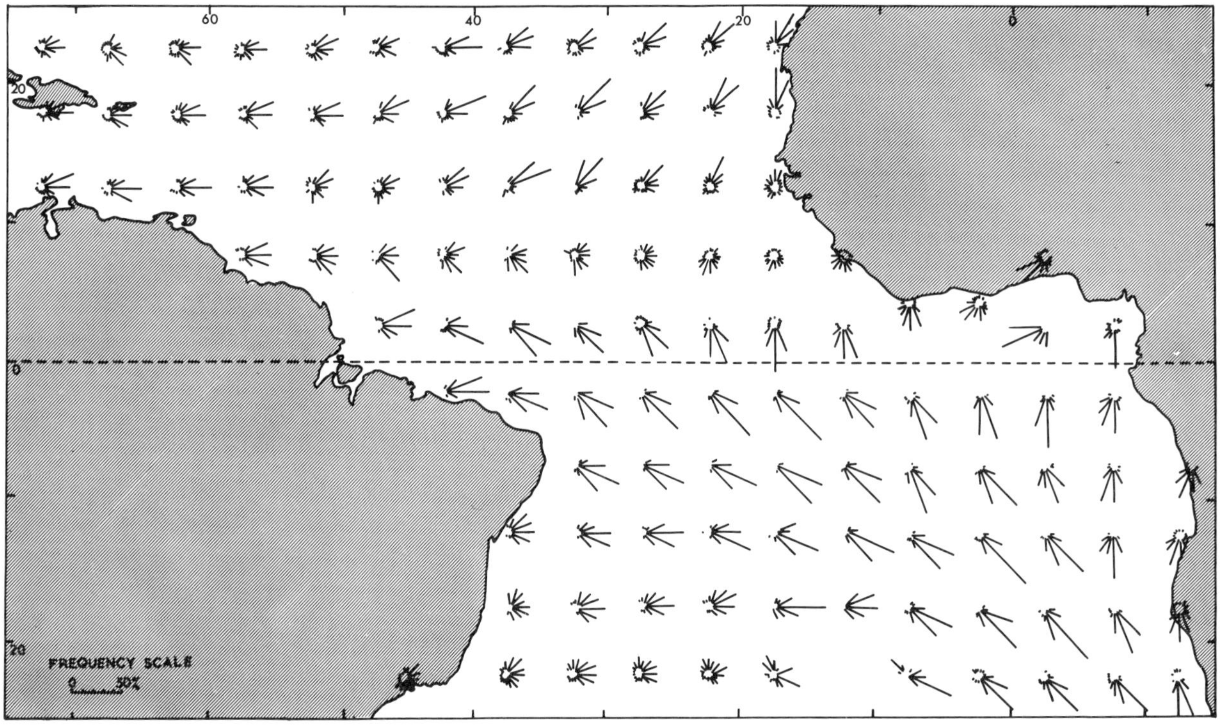

FIGURE 3. Percentage frequencies of winds between 25°N and 25°S in October, for 67 years to the south and 29 years to the north of the equator. [From Monthly Meteorological Charts of the Atlantic Ocean.]

Possibilities of Trans-Atlantic Flight in the Tropics

The frequency, duration, and strength of the prevailing surface winds over the Atlantic Ocean in the tropics are given in *Monthly Meteorological Charts of the Atlantic Ocean* and are based on 61 years of records for the South Atlantic and 29 years for the North Atlantic. Figure 3 shows records for October; wind fields at other times of the year differ mainly in the position of the intertropical convergence zone, where the Northeast and Southeast Trades meet, and little in frequency and direction (Figure 4). Winds up to an altitude of about 1500 meters, the zone where most transport of insects occurs, do not differ greatly from the surface winds (U.S. Navy Department 1958). High altitude jet streams are of no significance in insect transport because their air is too cold for active flight; whether warm low level jet streams exist over the Atlantic Ocean is not known. Thus the following conclusions seem justified.

1. A continuous trajectory from South America to any part of western Africa is not likely, perhaps not possible, at any time of year, within 25° north or south of the equator. Apart from local off-shore winds in South America (too transient to be shown in Figure 3) and winds in various directions along the coasts of western Africa, all winds flow across the ocean from an easterly sector.

2. The shortest distance between the two continents, from near Freetown to Natal (Brazil), is about 1800–2000 miles; this is across the region of the Doldrums where winds are either light or towards the land on the African side. Insects are unlikely to be carried far enough from the land of Africa in this area to join up with a persistent wind to South America, although occasional disturbances moving off-shore could carry insects far out to sea. However, such long trajectories (Figure 2) are extremely unlikely in the light winds on the Doldrums, and a trans-Atlantic flight unaided by wind across the narrowest part of the ocean is beyond the known capability of any insect.

3. To the south the ocean widens. Winds come generally from savanna areas of Africa (which are a better source of migrants than forested regions) or even from deserts. The shortest southerly distance, from the mouth of the Congo River to Natal (Brazil), is more than 3000 miles and a crossing in the fastest trade winds (about 25–27 mph; US Navy 1958) would take about five days, which, as far as

FIGURE 4. Mean directions, mean strengths, and frequencies of winds for February, July, and October for 54 years, showing also the position of the Intertropical Convergence Zone. [From U. S. Marine Climatic Atlas.]

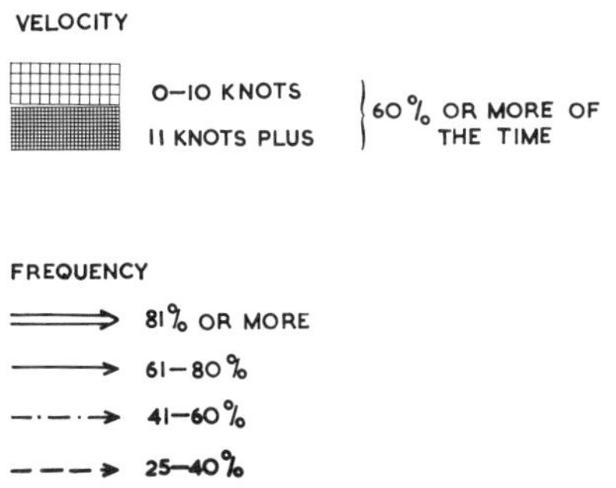

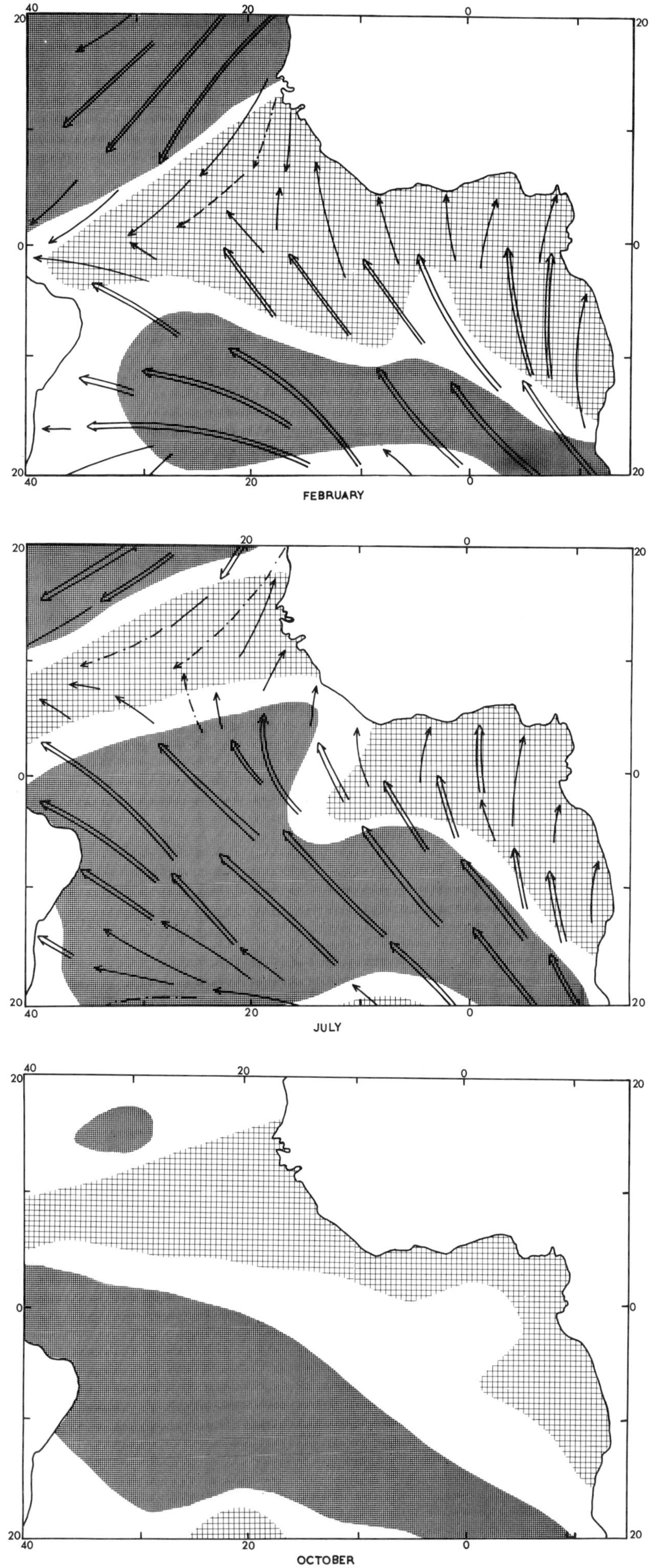

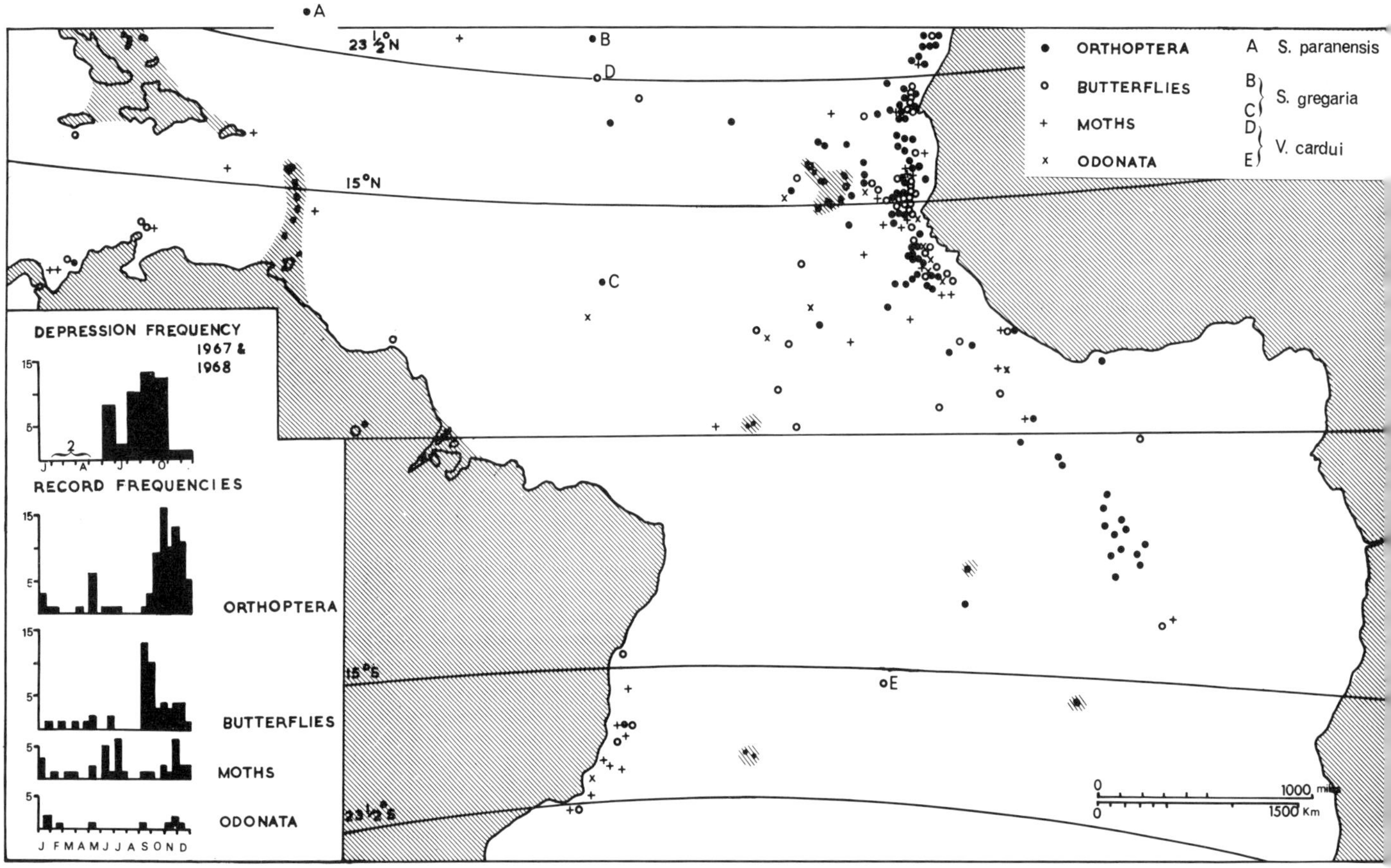

FIGURE 5. Records of insects at sea. Frequency of depressions originating in West Africa. [From Simpson et al., 1969.]

one can judge from published evidence, is probably close to the limit of a long-flying insect's endurance. Some species might cross in very favorable conditions.

4. Crossing from east to west is more likely north of the equator, but would probably still be near or beyond the limits of endurance for most insects. Winds from Africa are generally from savanna rather than from forest areas, and in January and April flow directly into the Amazon Delta, 2000–2500 miles away. In the autumn, however, the winds are weaker and inconstant; yet most migration occurs then (Figure 5). However, from Cape Verde Islands, which could be a source (or, more doubtfully, a refueling point for some insects) as well as from Dakar, the distance is only about 2000 miles. Winds are often of force 5–6 (19–31 mph), which would allow some larger insects with 8–10 mph air speed to cross in about 3–5 days.

Thus a trans-Atlantic crossing by flight seems possible from areas of Senegal, Gambia, and Guinea, less likely from Angola, and unlikely from the region of Freetown around the Gulf of Guinea to the Congo Delta.

There remains the possiblity that transport from Africa to South America is helped by line squalls and tropical disturbances. There are many kinds of tropical disturbances that originate over the land, well to the east in the continent of Africa, and move toward the western seaboards and thence out over the Atlantic Ocean. These systems are usually turbulent or convective and almost certainly, on occasion, lift insects, even heavy ones, to altitudes of several hundred meters. They therefore might provide the initial component for transporting insects from the land out over the sea and thence into wind systems that might carry them westward. The two general systems that we shall consider are tropical disturbances and depressions and line squalls.

TROPICAL DISTURBANCES. A tropical disturbance says Simpson et al. (1968:252–253) is "a discrete system of apparently organised convection, generally 100–300 mi. in diameter, originating in the tropics and subtropics, having a non-frontal, migratory character and having maintained its identity for 24 hours or more. It may or may not be associated with a detectable disturbance in the wind field." These disturbances can intensify and become tropical depressions with a definite closed circulation with maximum wind speeds less than 39 mph. These depressions weaken as they move into the Atlantic Ocean but occasionally develop into named hurricanes that affect the Caribbean area (see also Simpson et al. 1969).

Such systems are ideal for transporting insects into the Atlantic Ocean but particles carried with them tend to make looped trajectories because the winds circulate as the depression moves westwards. These looped trajectories extend northwards before turning southwards to continue the westerly track, and the extent of the deflection depends on the relative speeds of the circulating winds and of the depression itself (Johnson 1969). The effects of a depression might thus be two-fold; namely, carrying the insects out to sea and into the trade winds, but also creating a looped and therefore longer trajectory. The insect may thus fly a long time before moving far to the west and so exhaust its fuel by the time it arrives in mid-Atlantic. In such depressions the chances of crossing might be less than on normal Trade winds.

By contrast tropical "disturbances" move out across the sea, perhaps without a change of wind direction, and might take insects into the Trades without hindrance.

LINE SQUALLS. A line squall is a series of tropical storms that sometimes extends for hundreds of miles overland with very strong surface winds and convection to high altitudes. Line squalls are usually orientated north-south and travel east to west (Eldridge 1957). They may be expected to carry insects out to sea but not so far as the systems described above. Line squalls occur to the south of the intertropical convergence zone and, as this zone moves, the north to south incidence of line squalls changes accordingly. The maps of wind frequencies (Figures 3,4) show that, at any season of the year, insects carried out to sea from the west coast of Africa from the mouth of the Congo to Dakar would probably encounter winds that would carry them back to Africa. Line squalls are unlikely to be effective agents in trans-Atlantic transport.

The above descriptions are of general systems and we have not considered variations on this general pattern. Exceptional circumstances may occur but their significance cannot be appreciated until there are more regular and frequent observations of insects over the Atlantic Ocean.

Insects Over the Intertropical Atlantic Ocean

We have obtained as many records as possible of insects recorded by ships at sea. Space does not permit a full list of the records, but Figure 5 indicates their pattern between the tropics. The records are necessarily of large strong-flying insects that are easily noticed and identified. Each large insect was probably accompanied by many small ones in the same airstream; there seem to be no records over the Atlantic Ocean of such small insects, but some over the Pacific Ocean show that Empididae (Diptera) can fly more than 1000 miles, and other small species for at least hundreds of miles (summarized in Johnson 1969).

Coastal records predominate, reflecting the greater number of insects near land and also the regular shipping routes to West and South Africa (Figure 5). Near the coast of South America, the records are of South American species. A few records far out to sea need special mention in relation to the wind fields and to possible sources and distances traveled.

Schistocerca gregaria Forskål, a very strong flier, apparently can remain airborne for 3–4 days (Rainey 1963). It has been recorded several times more than 1000 miles from the nearest land, possibly having flown much farther than the direct distance indicated. There is no record of this insect crossing the Atlantic Ocean from one side to the other, but two females of *S. paranensis* (Burmeister) were seen over the Sargasso Sea (25°20′N, 62°30′W) on 24 May 1906 (see Figure 5). It is tempting to speculate about these insects. If the identification is correct they could have come from the Western Hemisphere, but Dirsch of the Centre for Overseas Pest Research, London (in litt.) now considers *S. gregaria* and *S. paranensis* to be members of the same species. It seems likely, therefore, that the insects identified as *S. paranensis* could have come from Africa on depressions that might have deflected them northwards.

The record of the Lygaeid bug *Pachybrachius vinctus* (Say) needs comment. One male was trapped in April 1964 in mid-Atlantic (22°01′–23° 20′N, 39°15′–40°12′W) (Scudder 1968). This species is widely distributed in South America, the West Indies, and in southern parts of the United States (with no record of it having been found anywhere in the Eastern Hemisphere; W. J. Knight, British Mus. (Nat. Hist.), pers. comm.). This specimen was caught, however, in the region of the Southwest Trades (Figures 3, 4), and it seems impossible that it could have come on the wind from the Western Hemisphere. Scudder (pers. comm.), therefore, kindly re-examined the specimen and confirmed that it was wrongly identified; it is now determined as the African species *Pachybrachius dubius* Reuter, which is known from Ghana, Republic of Guinea, Cote d'Ivoire, and Senegal, and probably occurs elsewhere in Africa.

A record of *Vanessa cardui* L. at 23°N 41°W (Figure 3) was shown by Williams (1958) to be associated with a wind from North Africa or Spain, implying a journey of half way across the Atlantic Ocean, 1300–1400 miles from North Africa or 2000 miles from Spain.

A swarm of the butterfly *V. cardui* (identified by C. B. Williams and J. A. Downes of Dept. of Agriculture, Ottawa) was also recorded at 16°S 20°W on 12 October 1939, 1000 miles west of St. Helena, 1100 miles east of Bahia (Figure 5). This butterfly is unknown in South America but is common in Africa. Wind field charts for October (Figure 3) suggest that the most likely source was the southwestern African coast (2000–2500 miles distant), but St. Helena is also a possible source. The insect is unknown in South America, so in spite of such migrational capabilities successful crossings (if any) have not resulted in establishment.

By contrast, an individual of the related species *V. virginensis* Drury form *brasiliensis*, which occurs in South America, was found on the island of Tristan da Cunha in April 1944 (Williams 1958). The nearest point in South America is Cabo Frio in Brazil, about 2000 miles away, from which strong winds flow frequently in April in these southerly latitudes.

Insects, Especially Pests, Common to South America and Africa

We must now consider the smaller less conspicuous insects not recorded at sea, many of which are pests. Many species of insect pests are common to Africa and South America, although some are present on only one of the continents.

It is not appropriate to consider all the 265 important pests listed in the *Distribution Maps of Insect Pests* published by the Commonwealth Institute of Entomology. Sixty-nine are found in both Africa and South America within 15°N and 15°S. Thirty-three South American species are not recorded from Africa; 31 of them occur in the north but only 20 to the south of the equator. Forty-eight found in Africa are not recorded from South America, and 43 of these occur both north and south of the equator (Table 1).

TABLE 1. Analysis of Distribution of Pests 15°N–15°S from Data in *Commonwealth Institute of Entomology; Distribution Maps of Insect Pests*, Nos. 1-275.

	South America		Africa		Both areas	
	No. of species	%	*No. of species*	%	*No. of species*	%
Coccids	4	12	9	19	28	41
Mites	1	3	1	2	4	5
Aphids	3	9	4	9	14	21
White flies	2	6	1	2	0	0
Thrips	0	0	2	4	4	5
Others	23	70	31	64	19	28
Totals	33	100	48	100	69	100

Of the 69 species common to both continents between 15°N and 15°S, 32 are wingless forms (28 coccids and 4 mites); 56 of these 69 occur in or around the Amazon and 50 in or around the Congo; 66 occur to 15°N in both continents; to the south, however, only 50 occur in South America and 63 in Africa. This disparity in South America south of the equator probably reflects an absence of economic entomologists.

In cases of adaptation for dispersal by wind, it might be thought that African species would be better represented in South America than South American species in Africa. However, such a simple effect is blurred by other factors; for example, transport by means other than flight can carry insects against the direction of prevailing winds. Many species have now become widely distributed in this way, though their manner of traveling is speculative. Gaps in distribution records may indicate either a lack of suitable breeding places or of entomologists. Some species have been named several times and their identity in different areas is sometimes questionable. There is also a striking dearth of records from northeastern Brazil, perhaps because the area is arid.

We deal below both with pests that are naturally adapted for transport by wind (many of which are known, or suspected, of being able to fly at least some hundreds of miles) as well as with wingless forms. Many of both kinds are known to be transported intercontinentally by human agencies.

APHIDS. A few aphids are restricted to Africa and are evidently of African origin, but there is only one certain record of such an African aphid in America, namely of a single, winged individual of the *Paoliella* group, trapped in Venezuela in 1968. The manner of its traveling is unknown (Eastop, British Museum (Nat. Hist.) in litt.), but if it flies across the Atlantic Ocean frequently, which seems doubtful, its establishment in America is evidently infrequent.

By contrast, other aphids have crossed the Atlantic Ocean from America to Africa in the opposite direction to the trade winds, by human agency. *Aphis spiraecola* Patch, a pest of citrus, was introduced into Morocco (Chapot and Delucci 1964) and Europe, presumably from North America and also into South Africa and Rhodesia, within the last 20 years. In 1967 it was recorded in Nigeria and the Ivory Coast, but how it got there is unknown (Commonwealth Institute of Entomology, map 256). *Hysteroneura setariae* (Thomas), a North American aphid that feeds on graminaceous plants (and whose primary host is *Prunus* subspecies), is now in the northern part of South America. It also crossed to West Africa, possibly on aircraft during World War II. Since the early 1960s it has spread to southeast Asia, to the Pacific Islands, and to Australasia, possibly by military aircraft operating from grassy airfields (Eastop, in litt., Commonwealth Institute of Entomology, map 255). Very small numbers of aphids brought into a country by accident can become serious pests; the spotted alfalfa aphid *Therioaphis maculata* (Buckton) introduced into Texas, probably by aircraft from India in 1953 as a single individual or small colony, is now a serious pest over tens of thousands of square miles in the United States of America (Dickson 1959, Hille Ris Lambers and van den Bosch 1964).

Many species of aphids generally distributed around the tropics have uncertain origins and unknown histories of transport. Few have been recorded in the Amazon area. Such a common worldwide pest as *Myzus persicae* Sulzer, occurs in Nigeria, the Congo, Tanzania, and in many countries in South America, but has not been recorded in the Amazon Basin. Surely such gaps in recorded distribution result from the absence of aphidologists.

By contrast, the absence of records of some species reflects lack of suitable habitats or hosts; for example, *Myzus circumflexus* (Buckton) is absent from both Amazon and Congo areas, no doubt because there are no glasshouses or bulb crops; and woolly aphis, *Eriosoma lanigerum* (Hausman), is not to be expected where apple trees are not grown.

THYSANOPTERA. Thrips are among the smallest flying insects, some being only 1-2 millimeters long. They are very light, with a slow falling speed, and therefore have a good chance of being transported by air currents for long distances even when flight ceases or becomes inefficient. They are also transported by human agencies. Nevertheless, some groups of thrips found in South and Central America and the West Indies are, apparently, absent from Africa. For example, *Heterothrips*, with about 50 species, is unknown in Africa; and *Frankliniella* with perhaps more than 100 species in South America is represented by only one in Africa and that species, *F. schultzei* Trybom, is cosmopolitan. It is not surprising that these thrips have not flown or been blown from America to Africa, because the winds are generally in the opposite direction. But they are flower-dwelling species and should be able to reach Africa through commerce, though apparently they have not yet done so.

Some thrips, like some aphids, are now cosmopolitan, and their country of origin is uncertain. Thrips are bisexual, but some are also parthenogenetic, and the males tend to disappear from populations established far from the original source. For example, *Haplothrips gowdeyi* (Franklin) is cosmopolitan but males are known only from Africa; *Haplothrips robustus* Bagnall, a minor pest in the United States, has males only in Australia. The males of *Thrips tabaci* Lindeman, a cosmopolitan species present in tropical Africa and South America, are common in Iraq and Saudi Arabia; it probably originated in western Asia where its primary host, *Allium cepa*, also originated. By contrast the cacao thrips, *Selenothrips rubrocinctus* (Giard), is probably African in origin, but both sexes occur in South America and Africa (Mound, in litt.).

Natural flight cannot wholly account for the international spread of these insects, though it must have been responsible for much dissemination at least over land. Insects may be transported for long distances by commerce or related means. Probably *Nesothrips propinquus* (Bagnall) was carried round the Southern Hemisphere in hay on board ships, and *Hoplandrothrips flavipes* Bagnall was similarly distributed in firewood. Many species dwell in leaf litter, and the sand and soil used for ballast in sailing ships, which was often unloaded on to the shore, was probably a common vehicle by which some thrips became widespread (Mound 1970) (as with some carabid beetles; Lindroth 1957).

ALEYRODIDAE OR WHITE FLIES. White flies are small delicate insects able to fly at least a few miles. They are not known to migrate long distances but might conceivably do so on the wind. The subfamily Aleurodicinae has many species in South America, but is not known in western Africa in spite of its conspicuous appearance and extensive searches in Nigeria and the Congo, where suitable hosts exist. As with other insects, uncertainties of nomenclature make the problems of deciding about distribution more difficult. For example, the tobacco white fly, *Bemisia tabaci* (Gennadius) is common in Africa and is also found in Brazil under other names (*bahiana*; *costa-limai*, and *signata*) and *B. tuberculata* Bondar in Brazil is possibly the same as *B. hancocki* in Africa. These trans-Atlantic distributions are probably attributable to the slave trade (Mound, British Museum (Nat. Hist.), in litt.).

OTHER PEST SPECIES. Insects that normally fly only short distances can nevertheless be rapidly transported long distances by ships or aircraft. *Anopheles gambiae* Giles, the main vector of malaria in tropical Africa was absent from South America until it was introduced late in 1939 or early in 1940 almost certainly as the adult, transported in less than 100 hours either by French mail ships, or destroyers used by the French government in a survey of a trans-Atlantic air route, or conceivably by one of the early trans-Atlantic experimental flights (Soper and Watson 1943, Shannon 1930, Whitfield 1939, Smart 1944).

Another species that has become distributed widely from a known area of origin by unknown means is *Nezara viridula* (L.) (Heteroptera; Pentatomidae), native to the southern United States but not tropicopolitan (Commonwealth Institute of Entomology, map 27). It had spread to the Mediterranean by 1884 and was also found in Asia before the end of the last century. It is a migratory species, and there is some evidence that in West Africa it regularly migrates, possibly for long distances; but its migratory powers apparently were not a major factor in its cosmopolitan, in contrast to its regional, distribution. It can certainly survive short periods of transport through commerce, and presumably in this way has been slowly distributed to many countries during the last three or four hundred years.

By contrast, there are other well-known pest species on both sides of the Atlantic that regularly migrate long distances in a north-to-south (meridional) direction and some examples are given below, but those in Africa evidently have not crossed to South America by natural flight.

In Africa long journeys are made by the army worms, *Spodoptera exempta* (Walker) and *S. exigua* (Hübner), and the desert locust, the last two ranging into Europe (Figure 2). *Empoasca lybica* (de Berg) in Africa probably migrates as other potato leafhoppers do in the United States and much evidence suggests that large numbers of many species are regularly wind-borne throughout Africa for distances of hundreds and perhaps thousands of miles. However, long-distance migration on the wind in Africa is not well documented.

More species are known to travel long distances in America and though they would not be expected to cross to Africa they might, with favorable winds, invade the Amazon area. Most records are from North America; for example the locust, *Schistocerca paranensis*, and the cotton leafworm, *Alabama argillacea* (Hübner), are notable migrants, the latter ranging, as adults, as far north as Canada. In North America, the potato leafhopper, *Empoasca fabae* Harris moves seasonally on the warm southerly winds from the Mississippi Delta up through Ohio, Illinois, Missouri, Iowa, and over the Canadian border where cold fronts terminate its progress after a journey of about 1000 miles in several days. Another leafhopper, *Macrosteles fascifrons* (Stål), makes similar migrations (Figure 2). Marked individuals of the monarch butterfly, *Danaus plexippus* (L.) have been recaptured in Mexico 1870 miles in a straight line from a northern release-point in Canada and there are many such records of journeys of 1000-1400 miles. *Toxoptera graminum* (Rondani) and other cereal aphids and the cotton bollworm beetle, *Anthonomus grandis* Boheman, are all adapted to travel, not necessarily nonstop, hundreds of miles (Figure 2; see also other examples in Johnson 1969). It is certain that similar, long-distance migrations occur in South America and probably involve the Amazon Basin. We have, however, no documented examples.

THE TRANSPORT OF ANTS. So far we have dealt only with insects inimical to man. Though this includes some ants, they are, in general, a dominant part of the insect ecosystem in the tropics, either as fierce predators or as protectors of particular insects and so have major economic and ecological importance.

As their powers of dispersal by natural flight are limited, ants are particularly suitable for zoogeographical study in Africa and America (Brown 1971). The genera and species on both sides of the Atlantic in the tropics are distinct, reflecting the geological past. But there has also been much transport by human agency, which is likely to continue with consequences that whether good or bad, can sometimes be great. Although it is very difficult to study "tramping" in ants because of the great proportion of unrecognized synonyms among tramp species (Brown 1954 and pp. 162–163 herein), there are some well-established instances.

Iridomyrmex humilis Mayr, an Argentine and Brazilian species, becoming cosmopolitan, was probably introduced into South Africa in freight during the Boer War (Wheeler 1910). The two Ethiopian species *Strumigenys rogeri* Emery and *S. scotti* Roger have become established as "tramps" outside Africa, and the other dacetines that have become widespread, e.g., *Trichoscapa membranifera* Emery and *Quadristruma emmae* (Emery), are almost certainly African in origin. For example, *S. rogeri* workers outside Africa are uniform, as would be expected if the population started from a small original "tramp" stock, even perhaps from a single female. Studies on Dacetini show that world-wide "tramp habitats" in the tropics are occupied by species and genera of African origin (Brown 1954).

Means by which females become distributed are various. For example, a colony of workers of *Serrastruma lujae* (Forel) was intercepted in Honolulu by the U.S. Plant Quarantine on plants arriving from the Belgian Congo (Brown 1952). *Tetramorium lucayanum* Wheeler, widespread in Central and West Africa, arrived in the West Indies (and is now in Cuba, Puerto Rico, Jamaica, and Virgin Islands) probably in ballast, or timber, or perhaps with the slave trade in the early days of American colonization (Brown 1964a). Numerous species, e.g., *T. guineense* Fabricius, *T. simillimum* F. Smith, and *T. pacificum* Mayr are "tramp" species of Old World origin. The first two, almost certainly African, became well established in tropical South America; *T. guineense, Monomorium pharaonis* L., and *Paratrechina longicornis* Latreille were found in Manaos, Brazil; *Tapinoma melanocephalum* Fabricius in São Luis, Maranhão, Brazil, and *M. floricola* Jerdon far up the Amazon; all were domestic, mostly in new buildings. (The curious ways in which transport is possible are illustrated by *T. melanocephalum*, which was carried into Massachusetts and became established there, in a comic carved head made from a coconut still in its husk [Brown 1964].)

Transport in the opposite direction is exemplified by *Odontomachus haematodes* L., which reached Africa from South America, occupying all islands between, long before the advent of modern ships (Bolton, British Museum (Nat. Hist.), in litt.). More recently *Wasmannia auropunctata* Roger was introduced accidentally into Cameroun from America, and is now used deliberately by cocoa farmers to protect their trees from attack by mirids which (as well as *Crematogaster* ants) *Wasmannia* eliminates (de Mire 1969). This biological control of a pest by an ant involves the danger that the ant might destroy local fauna, or alternatively increase mealybugs or psyllids, which it attends and encourages on cocoa, or become a domestic nuisance. Another ant introduced into West Africa (Nigeria) from South America is *Solenopsis geminata* Fabricius. One wonders if or when the American leaf cutting ants of the genus *Atta*, which devastate pasture in Texas and South America and

Citrus in the Caribbean Islands, will reach Africa, and what the effect might be.

Stages in tramping are listed by Brown (1954), who considers that the critical step involves chance transport of a continental propagule to favorable off-shore islands with a limited native fauna. If not already saturated with competing tramps, a large dense population is likely to develop quickly, being free from the continental predation, parasitization, and competition. Other contacts will spread this concentrated source to many other places. However, whether a tramp species becomes eliminated or established when introduced into another continental area depends on the species and local conditions. Other examples of transport of ants are quoted by Donisthorpe (1927).

AERIAL TRANSPORT OF INERT FORMS

It is commonly believed that insects borne on the wind are carried "passively" like dust particles. No doubt some insects, wafted to heights where it is too cold for flight or rendered inert by exhaustion, carried some distance before they reach the surface of the earth. However, the usual mode of long-distance displacement almost certainly depends on the insects flapping their wings to remain airborne, and thus to be blown on the wind.

The distance insects travel without supporting themselves by active flight is conjectural and depends on the falling speed of the particular insects in relation to the strength and frequency of up-currents. The falling speed of an aphid is approximately 1-2 meters per second, and less for smaller insects. The frequency and strength of up-currents are variable and, over the tropical Atlantic Ocean, unlikely to be strong or frequent enough to support inert insects over a 2000-mile journey. The wingless insects most likely to be transported inertly by wind are the crawlers of scale insects and mealybugs (the matter is discussed more fully in Johnson 1969).

SCALE INSECTS AND MEALY BUGS (COCCOIDEA). Females of coccids, or scale insects, are flightless, but the small wingless first-instar larvae are distributed in air currents, probably only for short distances. Nevertheless, many species, especially those that are pests, become widely distributed by commercial traffic. The *Distribution Maps of Insect Pests* show that coccids are among the most cosmopolitan in spite of their flightlessness. Brazil has its own coccid fauna of 363 species, (d'Araujo e Silva et al. 1968), of which 258 are not known outside the Western Hemisphere, but 73 (i.e., about one-fifth) occur in West and East Africa. However, no true African species has yet been found in Brazil (Williams, British Museum (Nat. Hist.), in litt.). Of the 69 species of insect pests mapped as common to both Africa and South America within 15°N, 15°S of the equator, 28 or 41 percent are coccids. As with so many other species, they have almost certainly been carried commercially on plant material.

FUNGUS SPORES. Although it is not an entomological topic, brief mention must be made of the transport of pathogenic fungi as spores. Some spores can remain airborne for long periods and are not limited in flight by temperature or fuel. Their terminal velocity is much less than that of inert insects and they remain airborne much longer. Dust particles of similar size-range are known to cross the Atlantic in about five days (Prospero and Carlson 1970) and there is little doubt that many fungus spores also make the crossing. Whether they remain viable is another matter. The recent discovery of coffee rust in Brazil (Anon. 1970), following its discovery in Angola in 1966, indicates the possibility of its trans-Atlantic transport in the Trade winds that blow from Angola to Brazil (Figure 5; Bowden, Gregory, and Johnson 1971).

An example of transport of a fungus across the Atlantic in the opposite direction, from America to Africa, is that of *Puccinia polysara* Underwood, the pathogen of maize rust. There is evidence that the fungus has existed in the southeastern United States since at least 1879, though it causes little damage. It appeared in West Africa (Sierra Leone) only in 1949, where it severely damaged maize because the African varieties are susceptible. It was introduced into Africa probably with seed corn or corn-on-the-cob flown into Africa from the United States as food for the military forces there. It then spread, evidently by wind, and reached other parts of West Africa by 1951, the Congo and East Africa by 1952 and the islands in the Indian Ocean by 1955 (Gregory 1966, Cammack 1958, 1959). These authors also discuss the general problem of long-distance transport of fungus spores by wind currents. Long-distance trajectories of pathogenic spores are also considered by Hirst, Stedman, and Hogg (1967), Hirst, Stedman, and Hurst (1967); Hirst and Hurst (1967); Ogilvie and Thorpe (1961), and Zadoks (1967).

The main factor limiting long-distance transport of fungi by air currents is the viability of spores; many can cross oceans and deserts but die in transit. As with insects, it seems that extension of the geographical range often awaits transport by man, which may cause far-reaching and often devastating effects (Gregory 1966).

INTERCONTINENTAL DISTRIBUTION OF WINGED AND WINGLESS FORMS

Only a rough idea can be formed of the proportions and significance of different higher insect taxa represented by pests on either or on both sides of the Atlantic Ocean. Such taxa may be present in different proportions in either continent because of differences in the original composition of the fauna and in agricultural practices. A study of particular groups, for example aphids or coccids, is stimulated by the presence of a few important pests in that group and may produce many more records than of taxa with less economic importance. The effects of such imponderable factors cannot be gauged accurately. In spite of these and other uncertainties, the estimated distributions of flying and flightless forms are significant.

Table 1 compares the numbers of different species of coccids and mites (which have no wings), with those of aphids, thrips, white flies, and all the other larger winged pests.

Most of the groups common to both Africa and South America are wingless (Coccidae and mites), although some are small insects adapted for aerial spread (aphids and thrips) which nevertheless, like coccids, have apterous forms that live all their lives on plants. If ants were included (few are given in the Commonwealth Institute maps) this proportion would be increased.

By contrast, those pests present on one or other side of the Atlantic Ocean, but not on both sides, are predominantly large winged forms that spend at least part of their life history away from the host plant (e.g., moths).

It seems paradoxical that the groups that have most successfully crossed the ocean are those most adapted to stay in a restricted habitat and least capable of aerial migration for long distances. This is because the habitat itself, namely the plant or soil, is moved and because many of the apterous forms such as coccids and aphids are parthenogenetic and not faced with the problem of mating before or after a long flight in which contact between the sexes may be lost. Those species seemingly most adapted for long-distance travel have either not cross the ocean or, if so, have not established themselves successfully afterwards.

We have dealt only with plant pests; other forms whose habitat itself is transported, for example, parasitic arthropods and other contaminants of vertebrates, especially of birds, are probably as ubiquitous as many coccids.

Comparison of Meridional and Transoceanic Migration

There is an apparent anomaly in the existence of successful and habitual long-range movements within the continents but not between them. The wind fields over the ocean in the tropics are such that it is unlikely that any of the American species would be transported by wind into the tropical eastern Atlantic areas; this is not so in more northerly latitudes where much stronger winds blow and where trans-Atlantic crossings from west to east have been recorded. It is perhaps surprising, however, that the African-European species have not been recorded as migrating into the Americas. This suggests that, although some species have evolved long distance dispersal mechanisms associated with wind systems over a continuous habitat range and thus are able to land periodically and to refuel or to mate (as over landmasses), long-range transoceanic dispersal, even with a continuously favorable wind system and with no possibility of periodic landing, does not often occur on a scale large enough either to establish new or to influence existing populations. In this context an important limiting factor in establishment is the time of mating. Many migrant species mate only at the end of migration (*S. gregaria*) or during it (*D. plexippus*). Those species that begin to migrate already fertilized or that reproduce parthenogenetically stand the maximum chance of establishing themselves after a transoceanic journey. This limits the number of species able to make a viable crossing and is perhaps why *S. gregaria* and *V. cardui* are not established in South America. Such species are adapted for long-distance aerial migration between habitats on the same or contiguous continental masses, and the practical limit for transoceanic transport by flight (irrespective of the problem of mating) seems to be about 2000–2500 miles. Sometimes this distance may be exceeded for reasons that depend on the particular species. Nevertheless, only one certain record exists of a specifically African migrant insect in Central and South America (*Paoliella*), although the distribution maps of insects recorded over the tropical Atlantic show *S. gregaria* more than half way across.

The absence of the African pest species in South America may be more apparent than real, as closely similar species, whose taxonomic distinction is sometimes doubtful, occur on each side, e.g., *S. gregaria*, in Africa, *S. paranensis* in South America, *Heliothis armigera* (Hübner) in Africa, *H. zea* (Boddie) in America. The evidence for *Heliothis* (Hardwick 1965) suggests that the taxa on each side of the Atlantic are genetically distinct, but the distinction of *S. gregaria* and *S. paranensis* is doubtful.

The distribution of pests on crops common to both sides of the Atlantic, e.g., cocoa, coffee, bananas, sugar cane, rice, maize, and cotton, suggests that there is no continuing and effective contact between Africa and South America. Pests such as *Pectinophora gossypiella* (Saunders) on cotton, *Cosmopolites sordidus* (Germar) on bananas, *Hypothenemus hampei* (Ferrari) on coffee are found in both Africa and South America, but human activity is certainly responsible for these distributions. Apart from these and a few more widespread pests, each crop has developed its own pest complex with equivalent habits, e.g., cocoa mirids: *Monalonion* spp. in America and *Distantiella* and *Sahlbergella* in Africa; stem borers: *Diatraea* spp. in America and *Sesamia* spp. in Africa; coffee leaf miners: *Leucoptera coffeella* Guérin-Méneville in America and *L. meyricki* Ghesquière in Africa.

Man has accidentally introduced important pests from one continent to the other, but there is no evidence yet that unaided aerial transport by winds has introduced, with subsequent establishment, a major pest on either continent. Most risk to either a developing Congo or Amazon region comes from local sources rather than from trans-Atlantic windborne contamination. Development of each area will change habitats so that existing species may become major pests in new niches (as for instance *Sahlbergella* now on cocoa in West Africa originally had *Cola* and *Ceiba* as its forest hosts) or will encourage movement of savanna species into formerly forest areas. The alteration of habitats or the provision of new ones may affect the balance of species within each area. Long-distance meridional transport over land, where extraordinary long nonstop flights are not necessary for the ultimate invasion of new territories, may introduce pests into developing areas from other parts of the same continent.

Thus the Congo and Amazon forest basins cannot be treated as naturally linked entities for interchange of insects, and to suppose that a higher rate of exchange of insect faunas exists between Ethiopian and neotropical regions merely because they are closer than lands on the Pacific side of South America (Kuschel 1969) is not supported by our analysis. The distribution of vegetation around each basin shows a progression from rain forest through various associations of wetter and drier deciduous or semideciduous forest and forest-savanna mosaics to savanna vegetation, the transition to savanna being generally more abrupt in Africa than in South America. There are concomitant changes in animal communities. The range 15°N to 15°S of the equator embraces both forest and contiguous savanna areas. Development of each area is likely to follow the same pattern, but independently as regards pests.

Forest insect species are less mobile than savanna species because the habitat is more permanent and there is little evidence of long-distance migration in forest species. Savanna habitats are less permanent and subject to more marked seasonal change, so savanna insects are usually the more migratory, e.g., the subspecies of *Schistocerca*, *Spodoptera*, and *Heliothis* on either side of the Atlantic.

Agricultural development of either forest area will cause a shift towards a savanna-type ecosystem to an extent depending on the type of crops grown. Tree crops with a long rotation, such as rubber or cocoa, eventually produce a fairly stable single-dominant secondary forest, with minimum change, but annual crops inevitably introduce a savanna-type seasonal alteration in habitats, even when they form a stable, nondestructive agricultural system in an otherwise aforested area. In such conditions, savanna insects become established. Examples in West Africa are the graminaceous stem borer *Busseola fusca* (Fuller), the sorghum midge *Contarinia sorghicola* (Coquillet), and the cocoa mirid *Distantiella theobroma* (Distant), the latter being almost certainly originally a species of savanna-forest mosaic. By contrast, there is no evidence that forest species move into savanna habitats, even when cultivation extends continuously from forest to savanna; for example *Sesamia botanephaga* (Tams and Bowden) attacks maize in the forest but not in the savanna though *B. fusca* is invading forest areas.

Discussion

This paper, though limited to present-day pests, is the first to give a general perspective to the problem of trans-Atlantic transfer of insects. Several aspects that we have not had time or space to consider, or on which information is scanty or unsystematized, should therefore be mentioned.

In the future development of both the Congo and Amazon areas, it seems likely that the transfer of pests from contiguous areas on the same continent will be of prime importance, and trans-Atlantic transfer possibly of secondary importance. Anyone searching the records for insects, especially economically important ones, is immediately struck by the lack of records in the Amazon area even of common pests that might be expected to occur there. Some systematic recording of insect pests in the area, even in its relatively undeveloped state, is a prime need.

The insects most likely to be transferred across the Atlantic are apterous forms, carried with the habitat (such as plant material or soil). Many of these are also parthenogenetic and thus relieved of the limiting factor of mating to establish themselves successfully. The same applies to some insects of medical and veterinary importance. A study of the parasites of birds known to cross the Atlantic Ocean and of other vertebrates carried over by man, would be interesting. Insects that attack man and animals are not usually long-distance migrants; however, many can be carried in aircraft. Military aircraft are a particular source of danger because many if not all escape standard quarantine procedures. The dangers have become more acute in recent years because the speed and numbers of aircraft have greatly increased. This subject was treated voluminously by Whitfield (1939) but a modern review now seems desirable.

Another method of trans-Atlantic transfer results from the biological control of insect pests and weeds by parasites and predators. This too is a subject beyond our scope for it involves a consideration of successful establishment of particular species in defined ecological niches.

Transfer of pests by natural flight is limited to the one-way passage from Africa to South America. Although at present this possibility seems slight, it could increase in the future. As agriculture develops and a greater acreage of annual crops is grown, much larger populations of migrant pests than exist at present might develop. Such an increase in size of populations at the source would enhance the chances of a few successfully crossing the ocean, especially if the females mate before migrating.

Chances of crossing may have been better in past ages, when the continents were closer together than they are now. This also is a subject beyond our present scope. Its proper treatment would involve the taxonomy of different taxa in both continents in relation to past climates and particularly to modern views on wind-aided flight, flight potentials, and evolutionary tendencies for dispersal. This is an almost untouched subject.

References

Bowden, J., P. H. Gregory and C. G. Johnson. 1971. Possible wind transport of coffee leaf rust across the Atlantic Ocean. Nature, London 229:500-501.

Brown, W. L., Jr. 1952. Revision of the ant genus *Serrastruma*. Bull. Mus. Comp. Zool., Harvard 107:67-86.

———. 1954. The ant genus *Strumigenys* Fred Smith in the Ethiopian and Malagasy regions. Bull. Mus. Comp. Zool., Harvard 112:1-34.

———. 1964a. Some tramp ants of Old World origin collected in tropical Brazil. Bull. Mus. Comp. Zool., Harvard 75:14-15.

———. 1964b. Solution to the problem of *Tetramorium lucayanum* (Hymenoptera: Formicidae). Ent. News 75: 130-132.

———. 1971. A comparison of the Hylaean and Congo: West African rain forest ant faunas. (In press.)

Cammack, R. H. 1958. Studies on *Puccinia polysora* Underw., I: World distribution of forms of *P. polysora*. Trans. Brit. Mycol. Soc. 41:89-94.

———. 1959. Studies on *Puccinia polysora* Underw., II: Consideration of the method of introduction of *Puccinia polysora* into Africa. Trans. Brit. Mycol. Soc. 42:1-6.

Chapot, H., and V. L. Delucchi. 1964. Maladies, troubles et ravageurs des agrumes un Maroc. 339 pp. Inst. Nat. Recherche Agron. Rabat.

Commonwealth Institute of Entomology. Distribution maps of insect pests, Nos. 1-275. London.

d'Araujo e Silva, A. G. et al. 1968. Quarto catálogo dos insetos que vivem nas plantas do Brasil seus parasitos e predadores, Parte II. 1 Tomo. 622 pp. Rio de Janeiro.

Darlington, P. J., Jr. 1957. Zoogeography: The geographical distribution of animals. 675 pp. John Wiley & Sons, Inc. London and New York.

de Mire, Pl. B. 1969. Une fourmi utilisée au Cameroun dans la lutte contre les mirides du Cacaoyer *Wasmannia auropuncta* Roger. Café, caco, thé. 13:209-212.

Dickson, R. C. 1959. On the identity of the spotted alfalfa aphid in North America. Ann. Ent. Soc. Amer. 52:63-68.

Donisthorpe, H. St. J. K. 1927. British ants: Their life history and classification. 2nd edition. London.

Eldridge, R. N. 1957. A synoptic study of West-African disturbance lines. Quart. J. Roy. Met. Soc. 83:303-314.

French, R. A. 1969. Migration of *Laphygma exigua* Hubner (Lepidoptera: Noctuidae) to the British Isles in relation to large-scale weather systems. J. Anim. Ecol. 38:199-210.

French, R. A., and J. N. White. 1960. The diamond-back moth outbreak of 1958. Pl. Path. 9:77-84.

Gregory, P. H. 1966. Dispersal. Volume 2:709-732, in G. C. Ainsworth and A. S. Sussman, (Editors). The fungi. Academic Press. New York and London.

Hardwick, D. F. 1965. The corn earworm complex. Mem. Ent. Soc. Canada 40:1-247.

Hille Ris Lambers, D., and R. van den Bosch. 1964. On the genus *Therioaphis* Walker 1870, with descriptions of new species (Homoptera, Aphididae). Zool. Verh., Leiden 68: 1-47.

Hirst, J. M., and G. W. Hurst. 1967. Long distance spore transport. Pp. 307-344, *in* P. N. Gregory and J. L. Monteith (Editors). Airborne microbes. 17th Symp. Soc. Gen. Microbiol. Cambridge Press.

Hirst, J. M., O. J. Stedman, and W. N. Hogg. 1967. Long distance spore transport: Methods of measurement, vertical spore profiles and the detection of immigrant spores. J. Gen. Microbiol. 48:329-355.

Hirst, J. M., O. J. Stedman, and G. W. Hurst. 1967. Long distance spore transport: Vertical sections of spore clouds over the sea. J. Gen. Microbiol. 48:357-377.

Hurst, G. W. 1965. A note on patterns for air trajectories to the British Isles from Biscay and beyond. Agric. Mem. 117. [Unpublished paper available. Meteorological Office Library, Blacknell, England. (See also French 1969, Johnson 1969).]

Johnson, C. G. 1969. Migration and dispersal of insects by flight. 763 pp. Methuen. London.

Kuschel, G. 1969. Biogeography and ecology of South American Coleoptera. Biogeography and ecology in South America. 946 pp. [see pp. 709-22]. W. Junk. The Hague.

Lindroth, C. H. 1957. The faunal connections between Europe and North America. 344 pp. Almquist and Wiksell. Stockholm.

Meteorological Office. 1959. Monthly meteorological charts of the Atlantic Ocean. M.O. 483. London.

Mound, L. A. 1965. An introduction to the Aleyrodidae of western Africa (Homoptera). Bull. Brit. Mus. (Nat. Hist.), Entom. 17(3):113-160.

————. 1970. Thysanoptera from the Solomon Islands. Bull. Brit. Mus. (Nat. Hist.), Entom. 24(4):85-126.

Ogilvie, L., and J. G. Thorpe. 1961. New light on epidemics of black stem rust of wheat. Sci. Progs. (London) 49:209-227.

Prospero, J. M., and T. N. Carlson. 1970. Radon 222 in the North Atlantic Trade Winds: Its relationship to dust transport from Africa. Science 167:974-977.

Rainey, R. C. 1963. Meteorology and the migration of desert locusts. Tech. Notes. World Met Org. 54:1-115.

Scudder, G. G. E. 1968. Air-borne Lygaeidae (Hemiptera) trapped over the Atlantic, Indian and Pacific oceans, with the description of a new species of *Appolonius* Distant (*A. errabundus* Spin.) Pacif. Insects 10(1):155-160.

Shannon, R. C. 1930. O aparecimento de una espécie africana de Anopheles no Brasil. Brasil Medico 44:515-516.

Shaw, M. W., and G. W. Hurst. 1969. A minor immigration of the diamond-back moth *Plutella xylostella* (L.) (*maculipennis* Curtis). Agric. Meteorol. 6:125-132.

Simpson, R. N., N. Frank, D. Shideler, and H. M. Johnson. 1968. Atlantic tropical disturbances 1967. Monthly Weather Review 9:251-259.

————. 1969. Atlantic tropical disturbances of 1968. Monthly Weather Review 97:240-255.

Smart, J. 1944. Invasion of the New World by *Anopheles gambiae*. Nature, London 153:765-766.

Soper, F. L., and D. B. Watson. 1943. *Anopheles gambiae* in Brazil 1930 to 1940. 262 pp. Rockefeller Foundation. New York.

U. S. Navy Department. 1958. Marine climatic atlas of the world, IV: South Atlantic Ocean. Navaer 50-1C-531 Office, Chief of Naval Operations, Washington, D.C.

Wheeler, W. M. 1910. Ants: Their structure, development and behaviour. Columbia Univ. Press, Biol. 9. New York.

Whitfield, F. G. S. 1939. Air transport, insects and disease. Bull. Ent. Res. 30:365-442.

Williams, C. B. 1958. Insect migration. 235 pp. Collins. London.

Wilson, E. O. 1959. Adaptive shift and dispersal in a tropical ant fauna. Evolution 13:122-144.

Zadoks, J. C. 1967. International dispersal of fungi. Neth. J. Pl. Path 73 (Suppl. 1):61-80.

Limnology of the Congo and Amazon Rivers

G. MARLIER

Introduction

Notwithstanding their location at approximately the same latitude, near the equator, and their huge size, the Congo and Amazon rivers contrast in a number of features. Primary differences exist in altitude and relief, which have strong bearing on limnological aspects. Further differences are mainly geological. The chemical content of the waters is greatly influenced by the nature of both the geological strata from which the rivers originate and subsequent geological formations through which they flow. Chemical content is also affected by plant cover and by conditions of insolation. Lastly, some of the differences may have historical explanations. In spite of these factors, the lowland equatorial environments have given the two rivers a similar general aspect and many common features.

Physical Features of the Congo Basin

The Congo Basin occupies about 68 percent of the area of the Amazon Basin, or 3,822,000 square kilometers. The Congo River is 4650 kilometers long, 83 percent of the length of the Amazon. This means that the watershed area per kilometer of river is about 822 square kilometers for the Congo and about 1000 square kilometers for the Amazon.

The Congo River lies entirely within the intertropical zone. Its upper course, which bears the name of Lualaba (Figure 1), flows northward toward the equator. From there (Kisangani), the Congo turns west, crosses the equator and advances to the second degree north (Lisala). Then it bends again toward the southwest, recrossing the equator and continuing southwestward to the Atlantic Ocean, which it enters at Banana (6°S). The major part of the watershed is situated in the southern hemisphere, but the river itself is nearly equatorial; some important tributaries are even north equatorial. The basin extends between 13 degrees south and 8 degrees north latitude.

The Congo has several distinctive features. The headwaters, known as the Lualaba, flow from what may loosely be termed the 1500 meter contour line near the northern border of Zambia, which is the dividing line between the Congo and Zambesi watersheds. The great southern tributaries of the Congo arise at a similar elevation, flowing northward through the central basin.

The Congo watershed, excepting its lower course, has the shape of a shallow saucer, the rims of which are at 1500 meters on the south, 1000 meters on the north and west, and nearly 3000 meters on the east (see Figure 1). The lowest point of the saucer is near Lake Tumba in the central Congo with an elevation of 375 meters. The upper course of the Congo receives an important contribution from the waters of the western Rift Valley. Waters from Lake Kivu in the Kivu Province flow southward into Lake Tanganyika, which also receives waters from the East African plateaus south of Lake Victoria. All of these waters collect in Lake Tanganyika the outlet of which is the Lukuga, a tributary of the Lualaba. The upper affluents of these great lakes originate at elevations of about 2500 meters, whereas the Malagarasi, the East African tributary of Lake Tanganyika, is below 2000 meters. Other large lakes that drain into the Lualaba system are Lake Bangweulu and Lake Mweru, which feed the Luvua, a tributary that discharges into the Congo at Ankoro.

G. Marlier, Institut Royal des Sciences Naturelles, Brussels, Belgium.

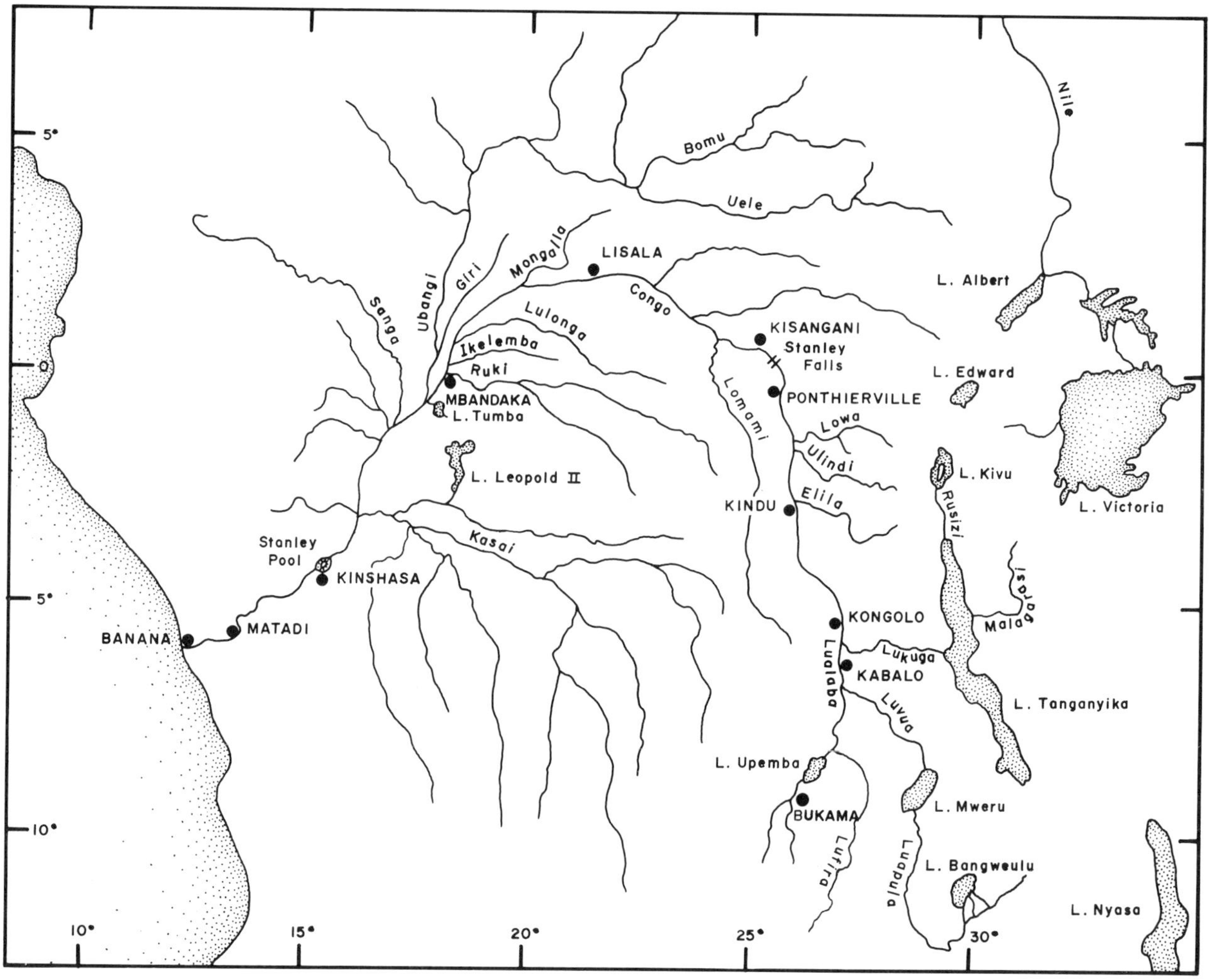

FIGURE 1. Congo watershed, showing the principal tributaries and lakes composing the drainage system.

Another peculiarity of the Congo is the existence of extensive depressions giving rise to swampy areas in some portions of the basin. Some of these depressions may be considered ancient lakes that have evolved from the endorheic into the exorheic type, later becoming swamps after changes in the climate from dry to wet and the acquisition of an outlet to the central basin. An example is the Kamolondo depression in northern Katanga, a swampy area of about 10,000 square kilometers inundated during the rainy season by the overflow of various affluents. These swamps include several areas of free water that are called lakes; namely, Lakes Kabwe, Kabale, Upemba, and Kisale, all linked to the Lualaba by narrow channels. They are invaded by water plants and surrounded by thick papyrus.

Another depression of this type, to which further reference will be made, is the Bangweulu Lake area, which drains into the Lualaba-Congo system through the Luapula, Lake Mweru, and the Luvua. The Bangweulu depression lies at an elevation of 1148 meters and is 4150 square kilometers in area. It receives water from Zambia, south of Lake Tanganyika. Its tributaries might easily be considered the true sources of the Lualaba, owing to their long courses. The Bangweulu-Luvua watershed has an area of over 218,000 square kilometers. Its output into the Lualaba fluctuates from 300 to over 950 cubic meters per second. Farther downstream, the Lualaba receives the Lukuga, the outlet of the great lakes watershed (244,500 km^2), but its flow is only 100 to 300 cubic meters per second.

Thereafter, the Lualaba receives several minor affluents on its right side that originate in the highlands forming the eastern rim of the basin. These tributaries supply a significant amount of water. The combined contribution of the Ulindi, Elila, and Lowa is estimated at between about 2000 and 5000 cubic meters per second. This rate of flow is of the same order of magnitude as that of the Lualaba proper at Kindu (near the Elila mouth), which means that between Kindu and Ponthierville the volume of the Lualaba has doubled. At this point, the Congo is 625 meters wide, nearly 15 meters deep, and has a rate of flow of about 9000 cubic meters per second.

Henceforth, it receives additional affluents from the central plain, the left ones originating at a little above 500 meters and the right ones coming from the northeastern and northern divides at elevations of about 1000 meters. The latter affluents generally have a rather lengthy course interrupted by many falls and rapids. They run parallel to the Congo for a considerable distance and only in their lower portion bend southward to join the main river. The left bank tributaries drain the central plain, some rising from the plain itself, others rising on the slopes of the southern plateau, and in the case of the Lomami even as far south as Katanga.

It is only after having crossed the equator for the second time that the Congo receives the second mightiest of its tributaries, the Ubangi, which now separates the Democratic Republic of Zaïre from the Congo Republic and from the Central African Republic. This important river, with a watershed of about 777,000 square kilometers, is formed by the confluence of the Bomu and the Uele. The combined length of the latter with the Ubangi is 2300 kilometers. In its lower course, the Ubangi drains a swampy forest area of about 45,000 square kilometers, the most important part of which is known as the Giri swamps.

A little downstream from the mouth of the Ubangi, the Congo enters what is called the "channel," a narrow passage 800 to 1000 meters wide and 210 kilometers long, between the high sandstone cliffs of the Bateke plateau. Here, the water is from 23 to 30 meters deep and the current accelerates accordingly. At this point, the Congo River receives its principal left bank tributary, the Kasai, which originates in Angola and is fed by a rich network of affluents in the central and western sections of the Congo Basin. The watershed of the Kasai is about 900,000 square kilometers. Its average discharge is about 1200 cubic meters per second, with very large seasonal variation. The upper courses of the Kasai and its affluents are interrupted by falls and rapids along the edge of the Lunda plateau. Their sources are between 1000 and 1500 meters in elevation.

After having received the Kasai, the Congo rushes on through the "channel" and then expands into the "Stanley Pool" near Kinshasa-Brazzaville. There, its discharge fluctuates between 23,000 and 50,000 cubic meters per second. The bed widens into a lake behind a barrier formed by a rocky sill. The pool has an area of about 500 square kilometers and a large island in its center. The outlet of the pool is the beginning of the cataracts that break the course of the river for more than 300 kilometers, with intermittent stretches of calm water. The drop is 270 meters over the distance between Kinshasa and Matadi, where the maritime course begins. The Congo is now a deep river (12 to 75 m) with a current of 4 to 6 kilometers per hour. Its width is 1800 meters and its discharge between 30,000 and 60,000 cubic meters per second.

Another limnological feature of this huge river is its slope, which is highly variable. From the sources to the rapids above the Kamolondo depression, the inclination is 0.59 meters per kilometer. In the rapid section, it is 2.37 meters per kilometer. From there to Kinshasa, the last point before the cataracts, the slope is never steeper than 8 centimeters per kilometer except at the Stanley Falls, where the drop over 150 kilometers is around 42 meters (or about 28cm/km). With the exception of the maritime section, the declivity of the Congo always exceeds 5 centimeters per kilometer. This illustrates a peculiar aspect of the Congo Basin and of many other African rivers; their watersheds are often a succession of flat plateaus separated by rather abrupt changes of elevation. This characteristic, which results in the formation of rapids, is pronounced in several affluents, among them the Ubangi and the Kasai. Such stretches of rapids occur throughout the basin and are especially common on its eastern and southern margins. Some are so steep that they have caused isolation and consequent differentiation of the aquatic fauna.

The steep sections of the Lualaba from Kongolo to Kindu and from Ponthierville to Kisangani are probably to be explained by the phenomenon of captures between the upper Lualaba and the Congo during the recent erosion cycle. This process of capture is of course responsible for the peculiar nature of the course below Stanley Pool, where the old lacustrine expansion of the Congo at 275 meters elevation was captured by a coastal river that cut its way through the Crystal Mounts by regressive erosion. The navigable portions of the Congo, except the maritime section and the channel, are rather shallow. They rarely exceed 15 meters and are often less than 2.5 meters in depth at low water. Most of the riverbed is obstructed by islands and sandbars, as are the courses of the Ubangi and the Kasai. This phenomenon is due to the strong erosion of the upper stretches of these rivers and to the slow current and flat morphology of the basin.

The fluctuation of the Congo River level is small, in part because of the presence on the upper course of lakes and swampy depressions that absorb the effects of heavy rains and also in part because of the very flat structure of the river banks. In the middle course, some are so low that the divides between neighboring watersheds disappear at high water, making it possible to navigate from one to the other in a flat-bottomed boat. This explains why a considerable increase in rate of flow may produce only moderate changes in water level.

The regime of the floods is unimodal on the upper course of the river and rather regular in regions away from the equator. The amplitude of annual fluctuations is rarely in excess of 3 meters (Luapula River, 4 m; Lake Mweru, 1.28 m; Luvua River, 2.65 m; Lake Tanganyika, 4 m; Lukuga River, 1.14 m; Lualaba, upper course, 2.38 m; Bukama upstream from the Kamolondo depression, 2.28 m; Lualaba at Kindu, 2.39 m). The unimodal tropical regime begins shifting toward bimodality approaching the equator and from the second parallel northward two maxima and two minima are perceptible, with an annual amplitude of about 2 meters. The tributaries from the Kivu highlands on the east have a very irregular discharge and also tend toward bimodality when approaching the equator. The Elila (3°S) is unimodal with an amplitude of 2.24 meters. The Ulindi (1° 40′S) is also unimodal, but very irregular (amplitude 1.64 m). The Lowa (1° 20′S) is irregular but shows a tendency toward bimodality with pronounced oscillations and an average amplitude of 2.78 meters. There are two maxima, one in April-May and the other in November-December. From Kisangani downstream, the regime is definitely bimodal and the annual amplitude decreases from 2.68 at Kisangani to 1.81 at Mbandaka. In the "channel," the annual amplitude rises to 3.5 meters, but decreases in Stanley Pool to a little over 2 meters. The central affluents of the Congo (e.g., the Ruki) also have a bimodal regime with an amplitude of 2.00 to 2.31 meters. The Kasai has a generally unimodal regime, but tends to become bimodal near its junction with the Congo. Its annual amplitude is around 3 meters. The Ubangi has a unimodal regime with an amplitude of 2.14 meters. The bimodal regime continues to the Congo mouth.

Physical Features of the Amazon Basin

We know less about the physical characteristics of the Amazon. The river is about 5500 kilometers long and drains an area of some 5.6 million square kilometers, to which should be added the accessory basin of the Tocantins with about 845,000 square kilometers. The watershed extends between 50° and 82° W longitude and between 5° N and 17° S latitude, placing it somewhat more to the south than the Congo Basin. Nowhere does the main river cross the equator and very few of its principal tributaries do so in their lower portion. The upper course of the Amazon, called the Marañon (Figure 2), originates in Peru, in the Laguna Lauricocha, at an elevation above 4300 meters. We may divide the watershed (according to Gibbs 1967b) into three parts: mountainous, mixed, and tropical. The mountainous portion contains the upper Marañon itself, the Huallaga, and the Ucayali, all situated mainly in the Andean highlands. The upper course of the Marañon is 725 kilometers long and runs generally from south to north in the central Andean depression. Its watershed area is 407,000 square kilometers. It enters the lowland at the rapids of Menseriche and from there runs eastward for more than 4500 kilometers (changing names en route, first to Solimões and then to Amazon). Parallel and to the east of the Marañon is the Huallaga; it originates south of Cerro de Pasco (Lago Punrun) at an altitude of 4000 meters and flows 820 kilometers before entering the Amazon (Marañon) 290 kilometers downstream from the rapids of Menseriche. The watershed of this tributary is probably similar to that of the Marañon. The Ucayali is longer than the Huallaga and its upper

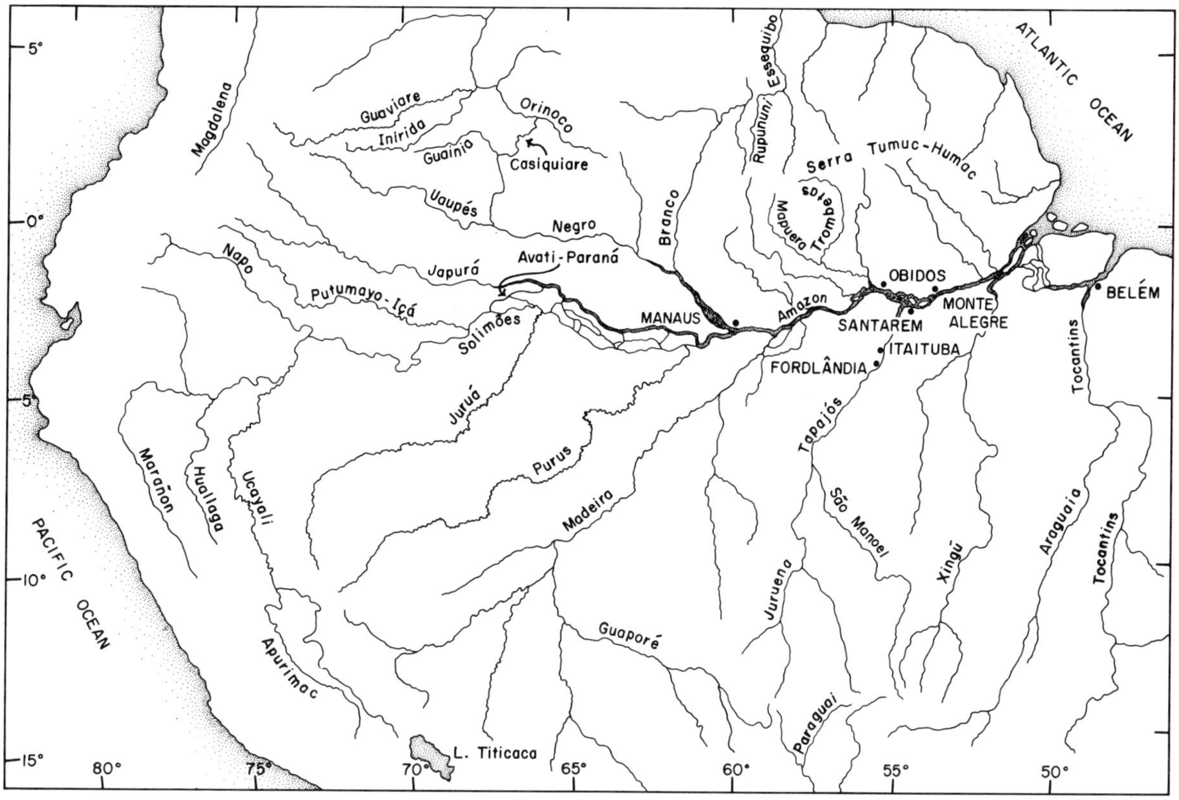

FIGURE 2. Amazon watershed, showing the principal tributaries. Noteworthy is the proximity of the headwaters in several parts of the basin to affluents of other drainages, in particular those of the Orinoco and Essequibo on the north and the Paraguai on the south.

affluent, the Apurimac, has its sources near 4000 meters also. The watershed of the Ucayali has been estimated at 406,000 square kilometers. The discharges of the Marañon and the Ucayali are .343 and .301 $\times$ 10^{12} cubic meters per year respectively, or 10,870 and 9540 cubic meters per second.

The mixed part of the Amazonian watershed is composed of tributaries that rise in the Andean highlands but have a rather long course over the lowland plain. Of these, the Napo (watershed area 122,000 km^2) the Putumayo-Içá (148,000 km^2), and the Japurá (289,000 km^2) are of moderate importance. The Madeira, however, is the most important of all the Amazon tributaries. It is 3381 kilometers long and the watershed has an area of 1,380,000 square kilometers. It rises in the highlands of Bolivia near Lake Titicaca and in the Sierra de Cochabamba. The total average discharge has been computed by Gibbs (1967b) to be .992 $\times$ 10^{12} cubic meters per year. The greater part of its course lies in lowland forest areas.

The third category of affluents of the Amazon forms the tropical watershed. These tributaries rise in the lowlands at elevations between 500 and 1000 meters or on the slopes of the eroded remnants of the Guayana and Central Brazilian shields. The slope of these rivers is very gentle, as is that of the Amazon itself. Their courses are sinuous as a rule and navigable in their lower reaches. Among the numerous tributaries forming this group is the Purus, with a length of about 3000 kilometers, a rather deep channel (between 5 and 30 m), and what is probably the most meandering course on earth. This important affluent supplies .341 $\times$ 10^{12} cubic meters of water per year to the Amazon, entering from the right side at no less than five different places. It has this peculiarity in common with the Madeira, which joins the Amazon well below the Purus by three mouths. Other important affluents are the Juruá (watershed area 217,000 km^2), the Negro (650,000 km^2), the Tapajós (500,000 km^2), the Xingú (540,000 km^2), and the Tocantins-

Araguaia. Of these, only the Negro emerges from the left side and originates north of the equator. It has a length of 3000 kilometers and its sources are at the base of the Colombian Andes at an elevation of about 1000 meters. According to Gibbs (1967b), it has a discharge of 1.407×10^{12} cubic meters per year, while Ungemach (1967) places the discharge at 0.851×10^{12} cubic meters per year. The Negro is thus the second largest tributary of the Amazon, after Madeira. A well-known peculiarity of this river is its link with the middle course of the Orinoco, which drains north into the Caribbean. This link is not a very surprising feature considering the slight relief of the region. The slope of the Negro is gentle (4.2 cm/km). Its principal affluent, the Rio Branco, drains the western slopes of the Guayana shield, a watershed approximately 190,000 square kilometers in area.

Continuing downstream, the next important affluent of the Amazon is the Tapajós, which enters from the right side. It drains from the Central Brazilian plateau and is formed by the confluence of the Juruena and the São Manoel. The Tapajós is nearly 2000 kilometers long and has an estimated average discharge of $.224 \times 10^{12}$ cubic meters per year (Gibbs 1967b) or of .392 (Le Cointe 1922). Its slope is gentle and its middle and upper courses are obstructed by numerous falls and rapids. The last important tributary of the lower Amazon is the Xingú, which rises in the same general region and at the same elevation as the Tapajós, in the Serra do Roncador. Its watershed is of the same order of magnitude (540,000 km^2) and its discharge is comparable ($.243 \times 10^{12}$ m^3/year). The length of the Xingú is almost 2000 kilometers. As with the Tapajós, the course is obstructed by many rapids.

We have not included the Tocantins and its principal affluent, the Araguaia, in the watershed of the Amazon, as this stream is gradually becoming independent of the main river. It communicates with the latter by a network of narrow channels (the Furos of Breves), through which navigation upstream from Belém takes place.

The Amazon and some of its tributaries carry a heavy load of loamy sediment. Owing to the slight declivity and moderate flow of these rivers, this silt settles on the bottom and along the banks creating a huge sedimentary territory. This area is nearly completely inundated annually and the riverbed and margins are raised by new alluvium with each period of high water. Erosion occurs simultaneously and the result is a continuous redistribution, which creates new sandbars and islands and eliminates others. This inundation and alluvial zone, called the "várzea," is widespread along the Amazon proper and occurs to a smaller extent along some white water affluents, such as the Purus, the Madeira, and the Juruá. The highland tributaries are also of this type, but their currents are so strong that no alluvium is deposited along their courses. Only the lowland portion of the Marañon and the upper Amazon (or Solimões) possess a significant várzea deposit. Other affluents of the Amazon have a very low content of suspended matter and are thus clear-water rivers, more or less colored by humic matter. Such rivers do not form a true várzea, but their upper courses may show a deposit of sandy sediment below major rapids. Such sedimentation occurs on the Tapajós between Itaituba (5°S) and Escrivão (3°S). The extensive sedimentation zone of the Rio Negro above Manaus and below the mouth of the Rio Branco is caused by suspended matter introduced by the latter river.

The depth of the Amazon is variable, but always considerable and much greater than that of the Congo. At the mouth of the Rio Negro, it is more than 60 meters deep. A maximum of 61 meters and a mean of 48 meters was recorded in July 1963 at Obidos, a narrow passage along the lower Amazon. Sioli (1965) has measured a depth of 47 meters above and 80 meters below the mouth of the Rio Negro. The depth of the Rio Negro at its mouth was found to be 93 meters. The affluents are also generally rather deep at their mouths, especially those with a small load of sediment. Their connection with the Amazon frequently takes the form of a so-called "mouth-lake" with a very weak current. It is now generally believed that these mouth lakes are drowned valleys, resulting from the general rise in sea level since the last glaciation. The beds of these mouth lakes, as well as that of the lower and middle Amazon, are well below sea level.

At its mouth, the Amazon is about 200 kilometers wide, excluding the mouth of the Tocantins. The volume of fresh water entering the ocean is estimated at 175×10^3 cubic meters per second or 5.5×10^{12} cubic meters per year. This huge quantity of water is pushed far out to sea by the pressure of the flow and is progressively bent toward the north under the influence of the Equatorial Current.

Resulting sedimentation has formed a true várzea along the Guiana coast, north of the principal outlet of the river. The tide may be felt very far inland; even at Santarem, 850 kilometers upstream, it is still 45 centimeters high. Sea water does not enter the river, however, even at the mouth. The regime of the river seems to be unimodal due to the great preponderance of southern tributaries over northern ones, and also because of the buffering effect of the huge várzea zones. Only in the Peruvian portion, where the várzea is insignificant and where northern affluents counterbalance southern ones, is the regime bimodal.

The height of the crest is considerable. It may reach 15 meters at Manaus and even more in the upper reaches of the western affluents (17 meters have been recorded on the upper Juruá). When the flood is at its maximum, water covers a great part of the sedimentation zone, including low islands, lakes, channels, and grasslands. Exceptionally, even the sedimentary dams built up by previous high inundations may be drowned, and buildings and fields destroyed. The invasion of tributary riverine lakes by Amazon water introduces a precious contribution of soluble salts and a less useful quantity of suspended matter, which tends to silt them up.

Water Quality

Consideration of the quality of the water in the two rivers must be prefaced by the observation that very little has been done in the way of analysis.

CONGO

Although fairly accurate information is available with regard to water chemistry in a number of watersheds, minor tributaries, sources, and lakes, the Congo itself has hardly been studied from this point of view. A very important paper by Symoens (1968) summarizes the existing information about several sections of the Congo Basin. Unfortunately, some very important tributaries remain unstudied and must consequently be left out of this summary.

The basin of the Luapula-Luvua introduces into the Lualaba about $17{,}800 \times 10^6$ cubic meters of water per year. As we have seen, this water comes from the Lake Bangweulu depression. Its origin in large swamps gives it a low mineral content and electrical conductivities from 10 to 20 microsiemens, characteristics that apply to nearly all tributaries of the Luapula. The water is enriched in salts (conductivity 70 to 125) while crossing Lake Mweru and about 350 kilometers farther on it enters the Lualaba without major change in its composition. Symoens has computed the mineral contribution of the Bangweulu-Luvua watershed at about 31.38 kilograms per second of solutes, equivalent to 989,600 tons per year. Of this total, silica amounts to nearly 6 kilograms per second, carbonates to more than 12 kilograms, calcium to 4 kilograms, and sulfates to 1 kilogram. Little information exists on other elements, but iron is always scarce and rarely amounts to more than .019 kilograms per second. The mineral contribution of this watershed is thus considerable, considering the volume of the river at the confluence. We have no information as to the quality of the water at the junction because the Lualaba crosses the Kamolondo depression after the last site of analysis. The observations of Magis (1967), however, indicate that the water of the upper Lualaba at the outlet of the Nzilo Dam is much more highly mineralized than that of the Luvua. The same is true of the Lufira, which enters the Lualaba at the northern end of the Kamolondo depression. We do not know anything about changes that the Lualaba waters may undergo within the depression.

The second important watershed feeding the Lualaba is the great lakes drainage, which enters via the Lukuga between Kabalo and Kongolo. This water is highly mineralized and the principal source of its mineral content must lie in the volcanic region around Lake Kivu. At the outlet of this lake, the water has an electrical conductivity of more than 1200 microsiemens. It is carried by the Rusizi into Lake Tanganyika, the deepest and second largest African lake. With few important affluents and a single outlet, it has a very low renewal rate. Computation has shown that the Lukuga carries away no more than one-third of the influx contributed by the Rusizi; water brought by the Malagarasi River, minor affluents, and rainfall is lost by evaporation. As Symoens (1968) has pointed out, the mineral contribution of the Lukuga to the Congo is reduced by the low discharge rate of this tributary, although its mineral content per liter is high (probably nearly 350 ppm of dry residue, of which 300 is due to salinity). This drainage contributes 0.04 kilogram per second silica, 15.0 kilograms carbonates, less than 1.0 kilogram calcium, 4.0 kilograms sulphates, but more than 7.0 kilograms metal alkalies.

Little is known about tributaries between the Lukuga and the central equatorial section of the basin, where several important affluents have been studied by Berg (1961). Whereas the Congo in this section has muddy water of a pH above neutrality and a composition rather rich in solutes (10 ppm Ca; conductivity 70 microsiemens), the aforementioned affluents have black humic water of low pH (3.6 to 5.2) and very low mineral content (hardness .0). All these rivers rise in the central plain under forest cover and are loaded with humic acids of very high oxidability. The largest are the Lulonga (draining 79,000 km^2), the Ruki (174,000 km^2), the Ikelemba (14,000 km^2), all on the left side, and the lower Ubangi tributaries and minor affluents on the right side between the lower Mongalla and the Sanga (total drainage area about 91,000 km^2). Some of these rivers are well known and the Ruki has been especially studied by Berg. On the basis of these studies, Symoens (1968) has computed that the Ruki contributes to the Congo 55.8 kilograms per second of dissolved solids, of which less than 3 kilograms are calcium, 8 kilograms carbonates and humic compounds, and 29 kilograms silica. The discharge of these forest affluents is important; that of the Ruki has been rated at 4450 cubic meters per second or $141{,}000 \times 10^6$ cubic meters per year. We know nothing of the mineral concentration and water quality of the Kasai, but from observation of its course and appearance we can infer that it is a "white" water river, poor in solutes but much richer than the preceding affluents.

Comparison of the waters of the Congo at its mouth with those of its principal affluents, when known, leads Symoens (1968) to the conclusion that the central plain affluents typified by the Ruki are the major contributors of solutes, the remaining ones being the upper affluents (the Bangweulu-Luvua system) and the Tanganyika watershed, in that order of importance. The discrepancy between this order of importance and the size of the respective areas (4.55%, 5.72%, and 6.40% of the basin) is worthy of note. The author attributes this to increased evaporation in the drier climate of the eastern part of the area.

Our survey of the streams and rivers of the African basin thus shows that the mineral content of the water is directly related to the geological composition of the sources and also to the ecology of the region through which a river flows. It is only under equatorial forest conditions that black humic water is formed, which is always very poorly mineralized. The high amount of silica contributed annually by the central Congo tributaries is certainly not derived from the lower forested section of these rivers, but comes from the upper savanna portions of their drainages. We have very few complete analyses of this type of water, but iron and manganese content may be rather high.

AMAZON

Much more is known about the water quality of the Amazon proper and its principal tributaries than about the principal rivers of the Congo Basin. Although few complete analyses exist, we have data on several important ions and also on suspended solids. Sioli has published analyses of the composition of small streams in a number of regions, showing the relations between pH, salinity, hardness, and calcium content. We are also indebted to him for the classification of Amazonian water according to color, hydrogen concentration, and calcium content into "white" or turbid waters, "clear" waters, and "black" waters. The Amazon and some of its larger affluents have white water; the latter are chiefly the upper mountainous tributaries issuing from the Andes. Gibbs (1967b) has showed that the salinity and turbidity of the Amazonian waters are chiefly determined by relief and weathering.

Clear waters are colorless and highly transparent. They rise on the central Brazilian or Guayana plateaus and include the Tapajós, the Xingú, and a large number of minor affluents flowing across Tertiary deposits of the "terra firme." Their slopes are moderate save where they cross the edges of the plateaus and enter the Amazon plain. There, the course is generally tumultuous and the water rushes through rocky channels. These stretches are always short, however, and the overall drop between the calm upper and lower courses is moderate. This fact, along with the hardness of the rocks over which they pass, explains why these rivers do not carry a heavy load of suspended matter. In any case, most of their load settles out soon after descent to the plain, so that the water entering the Amazon is nearly crystal clear.

As in the Congo drainage, black waters are common in the forested areas where the soil is always wet or inundated. These streams do not rise in the highlands, but on the plain. The outstanding example is the drainage of the upper and middle Rio Negro. Black waters resemble clear waters in their low content of suspended matter. Their color is due to humic substances from the forest soils where they originate. These substances are very oxidable and may reduce as much as 110 milligrams of potassium permanganate per liter. Having no suspensoids, they are transparent and the Secchi disk becomes invisible only below 2.9 meters depth.

The chemistry of these three types of water is very different. Small streams and creeks are likely to be more representative than large rivers because of the mixing of waters that often occurs in the latter. Sioli, who has analyzed a large number of streams and "igarapés" in various parts of Amazonia, has shown that the water quality depends upon the geological conditions at their sources and on conditions prevailing in their upper stretches. Clear waters originate in the igneous and metamorphic insoluble rocks of the Central Brazilian and Guayana shields. Thereafter, they flow through the Tertiary deposits of the "Barreira" series. The headwaters are located in brown loamy soils not subject to permanent flooding. The forest they drain is typical lowland tropical rain forest. Black waters, on the contrary, originate from flat land covered with poor and low forest called "caatinga" or with vegetation containing few if any trees ("campinas"). These localities are frequently inundated for many months each year and their soils are typical podsols with superficial bleached sands. The humic matter dissolved by rain and standing water percolates through the soil to the ground water. The water is thus colored with the humic compounds derived from decaying vegetation and soil humus. It does not contain dissolved mineral salts, which are scarce in such soils, and the pH is very low (generally below 4.5). These streams seem to be of the same nature as the black waters of the Congo and similar in origin. They predominate on the northern side of the middle Amazon, in the Rio Negro drainage, but the number of "rios pretos" or "rios negros" or igarapés with this name is evidence of the wide distribution of this type of water in the Amazon Basin. A good example is the Cururú, a right tributary of the lower Tapajós, well studied by Sioli (1967). Most Amazonian waters, excepting the rather rare streams originating in the Carboniferous zones north and south of the middle Amazon, have very low mineral content. Even the white waters have a salinity that varies from 120 to 140 parts per million at their sources to between 30 and 50 parts per million at their mouths.

It is as yet impossible to establish which watershed is the most important in determining the characteristics of Amazon water, since all the affluents are poor in most solutes. Gibbs (1967b) contends that 86 percent of the total salts are supplied by 12 percent of the Amazon Basin, where the mountainous affluents originate. There are few complete analyses of the Amazon itself or of its larger tributaries. We know much more about the small streams and creeks from the publications of Sioli (1954, 1955, 1963), Geisler (1967), and Braun (1952). These indicate that clear and black waters are extremely poor in electrolytes. The calcium concentration deduced from hardness measurements is around 1 part per million. The amount of dissolved silica depends on the origin of the water; black waters and clear waters always have a very low silica content, whereas white waters with a higher pH may show more dissolved silica. In this context, a "high" pH is rarely more than neutral (sometimes as high as 7.8) in richer waters. The white waters from the Amazon itself contain an appreciable quantity of solutes. Silica amounts to 9 parts per million, calcium to 10 parts per million, and bicarbonates, which control the buffering capacity of the water, reach 32 parts per million. In the black waters of the upper Rio Negro, bicarbonates are often absent, while in clear waters they do not exceed 2 or 3 parts per million. An interesting feature of the chemistry of the waters in the Amazon drainage, previously noted also in the Congo, is a reversal of the concentrations of iron and manganese and those of silica. Black waters lose their iron and aluminum ions as a consequence of chelation of these ions by humic compounds. But these waters do not carry dissolved silica in appreciable amounts. Clear and white waters contain no iron or aluminum, but are high in silica.

We do not have accurate measurements from black water rivers in the Congo Basin, making it difficult to compare the two watersheds from this point of view. The principal black water area in Amazonia is the right side of the upper and middle Rio Negro and the chief black water areas in the Congo are the Giri (lower Ubangi territory) and the right bank affluents of the middle Congo. These areas supply to their respective rivers a considerable amount of diluted water, which has lost its highly acidic properties. In the Amazon, the minerally rich streams are localized in the western basin, upstream from the Madeira. No white tributary introduces important quantities of salts below this point. In Africa, by contrast, the Kasai contributes waters to the lower Congo from the grasslands of the southern borders and from Katanga. We may suppose that these waters are more concentrated than the clear water tributaries of the lower Amazon. The latest and most complete analysis of the water of the Amazon proper (Oltman 1966) was made at Obidos on three different days. Comparison with the only analysis available for Congo waters, made at Matadi (and bearing in mind that below Obidos the Amazon receives two large clear water affluents) permits us to infer that the Congo may be a little more dilute, although slightly richer in silica (Table 1). From total salinity and suspended matter measurements, Gibbs (1967b) concluded that the Congo has a lower erosion rate than the Amazon (and the Mississippi) and a higher erosion percentage, due to chemical weathering.

TABLE 1. Simplified Comparison of the Water Composition of the Congo at Matadi (Symoens 1968) and the Amazon at Ôbidos (Oltman 1966). (All concentrations are in parts per million.)

Elements	*Congo*	*Amazon*		
Ca	2.71	4.3	10.0	3.0
Mg	1.81	1.1	0.4	0.6
Na	3.15	1.8	4.2	1.8
K	1.72	0.4	0.6	–
HCO_3	9.45	19.0	32.0	16.0
SO_4	6.48	3.0	6.4	1.0
Cl	5.32	1.9	4.5	1.6
SiO_2	10.74	7.0	9.0	9.0
Electrical Conductivity at 25°C	44.2	40.0	84.0	34.0
pH	7.32	6.5	7.1	6.5

The Lakes

Both the Congo and the Amazon drainages are very rich in lakes, but they are of different types and represent different ecological contexts.

LAKES IN THE AMAZON DRAINAGE

Amazonian lakes appear to be less diversified than African ones. This, at least, is the impression one gets from literature relative to lakes in Amazonia. If true, it may be due to the geographical, geological, and climatological uniformity of the huge basin, especially in its lower (eastern) part.

With the exception of those in the high Andean watershed, Amazonian lakes are generally closely related to rivers. In fact, most of them are generated by river activity. Many originate from former meanders in lowland stretches of white water affluents, such as the Purus. They are horseshoe-shaped, generally very narrow, and of moderate depth. They are evidence of the maturity of the river and their water is originally of the same type as the parent river.

Not very different are the lakes formed in the bed of the Amazon and separated from it by levees of sediment called "restingas," on which forest has progressively established itself. These lakes often retain a link with the river, called a "furo," through which they are fed at high water level. Their own level is generally higher than that of the river, however, and the shallow furo closes as the river recedes. These lakes are thus autonomous during periods of low and medium water level of the Amazon. With the passage of time, some have acquired nearly complete autonomy and their link with the river is operative only during years of unusually severe inundation. Some of these lakes, which are true "várzea lakes," are surrounded on all sides by river sediment and do not receive anything but river water; an example is Lago Redondo in the Várzea do Careiro near Manaus, which was studied by the author (Marlier 1967). Those located closer to the margins of the floodplain are bounded partly by sedimentary levees and partly by the terra firme. These lakes may receive river input at high water as well as runoff from local drainage. Consequently, they may be subject to considerable changes in chemistry from one season to another. An example of this "hybrid" type of várzea lake is the Lago Grande Curuay near Monte Alegre, described by Sioli (1951). Sioli has shown that this type of lake is alternately invaded by white water from the Amazon, bringing mineral salts and nutrients, and by black water from the igapós of the terra firme, which contains few dissolved nutrients but considerable organic matter and humic acids.

Another type of lake frequently encountered in Amazonia is formed by the damming of an affluent by sediment from the main river (type 52 lateral lake, Hutchinson 1957). This sediment increases year after year and may isolate the lake except for a narrow furo. If this happens in a region where the river is never high enough to overflow the dam, and the furo is long enough to prevent Amazon water from entering the lake, then the latter is chemically autonomous although its level is controlled by the regime of the principal river. When the river is able to enter the lake through the furo, however, it also controls the chemical quality of the water and thereby the lake's productivity. Lago Jurucui, on the lower Tapajós, which was thoroughly studied by Braun (1952) and visited twice by the author, is of the independent type. Its level fluctuates as much as 3.5 meters yearly, in part because of variations in the flow of the Tapajós. Another remarkable example exists on the lower Rio Preto da Eva. The terminal section of this Amazonian affluent is enormously lengthened by a levee of sediment through which the river winds for more than 30 kilometers. Above this furo, the river widens into a lake that never receives any influx from the Amazon and consequently is permanently composed of black water. The level changes considerably with the rise and fall of the Amazon, but the chemistry is not influenced by the white Amazonian water.

From the viewpoint of water quality, all these lakes are relatively poor in minerals, even the true várzea lakes, such as Lago Redondo. The salinity of the latter, which is highest at low water, does not exceed 53.9 parts per million total concentration and may fall as low as 27.2 parts per million at high water. These lakes are always very shallow and are strongly influenced in their composition and productivity by the invasion of Amazonian water. When the lakes experience an alternate influx of white and black water, their production and fertility are affected accordingly. During long periods of the year, a lake may contain water of both origins: white water on the side toward the Amazon and black water on the terra firme side. The true terra firme lakes always have transparent water of low pH; their salinity may be as low as 3 parts per million at high water. Since they receive no input of white water, fertility remains permanently low.

Another very common type of lake in the Amazon Basin is the true "mouth lake," which may be very extensive and of considerable depth. The most conspicuous examples occur on large tributaries, among them the Rio Negro, Rio Arapiuns, and Rio Tapajós, and are in fact freshwater "rias" or drowned valleys. Current is insignificant and depth is generally greater than that of the Amazon. The original chemical content is, of course, the same as that of the river, but nutrients are continually being supplied from the upper course. After the settling of suspended matter, when present, the water becomes very clear and a rich plankton can develop. This frequently happens in the Tapajós mouth lake, where a dense blue-green bloom may extend over wide areas. In black water mouth lakes, plankton production is always moderate because the influx of nutrients is meager and the color remains rather dark. Nevertheless, the author has observed that the black color of the water lightens with time under exposure to sunlight. This process never proceeds far enough, however, to transform waters as dark as those of the Rio Negro. Another difference between clear water and black water lakes is the permanence of acid reaction of the latter due to stable humic compounds. The pH of clear water mouth lakes seems to rise with progression downstream when the rate of flow is low, as in the Rio Arapiuns (Sioli 1954).

This classification of Amazonian lakes in terms of their relations with rivers is convenient but does not suffice to define the conditions in the lakes. We must now turn to their thermic features. We have long known that lakes in humid tropical countries must be of the so-called oligomictic type, but this merely signifies that their water circulates from the surface to the bottom at rare and irregular intervals. The lakes that have so far been studied in the Amazonian lowland are generally too shallow to develop oligomixis and, although they may show true stratification when the wind is temporarily slight, their water mixes very easily as a result of wind action or nightly cooling. Braun (1952) has showed that small shallow lakes along the Amazon and the Tapajós may develop periods of thermal stratification alternating with periods of continuous circulation. The same observation was made by the author in Lago Redondo.

In certain deeper lakes of the terra firme, conditions are less uniform. During most of the year, when the water level is low, periods of circulation alternate with periods of stratification. During high water, thermal stratification develops and may lead to exhaustion of the oxygen in the hypolimnion; the thermocline is then between 3 and 5 meters. This condition may persist for three months and the lake is thus to be classified among the monomictic lakes.

There are probably exceptions to this general summary of Amazonian lakes; for example, Braun (1952) has shown that the water of Lago Salgado is much more concentrated than normal. This is due to the existence of a thermal gypsitic source in its drainage. It should be pointed out that the geological strata where a lake is situated have a strong influence on the primary composition of its water and that all lakes lying in the narrow stretch of Carboniferous rocks north and south of the middle Amazon have more highly mineralized water than those so far investigated in other regions.

LAKES IN THE CONGO DRAINAGE

Some of the major lakes of Africa, and of the world, are part of the Congo drainage. Both Lake Kivu and Lake Tanganyika are of tectonic origin. Owing to their different histories, these two lakes have distinct shapes. Lake Kivu (alt. 1460 m), at least in its present state, was formed quite recently by the emergence of the Mufumbiro range of volcanoes in the valley between it and Lake Edward. This dammed valley, which then filled with water, developed an outlet to the south over a more ancient volcanic sill. This effluent eroded its way southward, creating a channel 500 meters deep in the Rusizi plain and terminating in Lake Tanganyika. With 2370 square kilometers of free water and 315 square kilometers of islands, Lake Kivu is probably the tenth largest lake in Africa. Its watershed is 7300 square kilometers. Its distinctive features are its great depth (475 m) and its very high mineral content, reaching 1430 parts per million, with a pH higher than 9. Its banks are very steep and the shore development is about 7. Volcanic activity has been strong since the formation of this lake and recent eruptions have contributed much mineral matter to the water.

From the limnological viewpoint, Lake Kivu has other interesting features. It is a typical meromictic lake; that is, its waters never circulate completely. The superior layers, about 60 meters thick, form the epilimnion. The 400-meter thick bottom layer is permanently stagnant, devoid of dissolved oxygen, and therefore completely azoic. The mineral content of this layer is higher than that of the upper layers and especially rich in methane and carbon dioxide, to the extent that these gases are now being exploited commercially. There are a few unimportant affluents and the sole outlet is the Rusizi, a torrential stream with a discharge of about 70 cubic meters per second. It drops to an altitude of 900 meters over a 30 kilometer distance interrupted by falls and rapids; this upper section has a slope of about 19 meters per kilometer. The remainder of the course is comparatively calm and crosses the "plain" with a slope of 2.3 meters per kilometer, reaching Lake Tanganyika at an elevation of 775 meters. The Rusizi is diluted to a final salinity of 600 parts per million after receiving a few small tributaries of low mineral content in this section. It makes a major contribution to the mineral content of Lake Tanganyika, being a primary affluent and certainly the most mineralized one.

Lake Tanganyika is the second largest African lake and the second deepest lake in the world, after Lake Baikal. It is larger than Belgium (34,000 km^2) and the shores are generally very steep. Maximum depth is 1470 meters, and mean depth is 572 meters (Hutchinson 1957). It too is a tectonic lake. Tanganyika has few large tributaries, the principal one being the Malagarasi, which drains the East African lowlands south of Lake Victoria. The Rusizi is second and the remainder are torrential streams flowing from the high ranges that line the Tanganyika trough. The latter are often small and have an irregular discharge; some cease to flow during the dry season. The lake water is rather warm (26° to 29°C) at the surface and most affluents are colder; their waters therefore sink to the bottom of the lake. The loss from evaporation is high (1.6 to 2.0 m according to Capart, 1952) and approximately balances gain from precipitation and runoff. Owing to the tremendous depth, the lake water does not mix completely and only the 100-meter upper layer in the south and the upper 20 to 80 meters in the north circulate continuously. Strong circulation of the upper layers (0 to 250 m) occurs during the dry and colder season, but the deeper water remains stagnant. These deeper layers should thus theoretically be much more concentrated in salts than the upper ones. Yet, except for the absence of oxygen below 200 meters, there is no marked difference between bottom and surface waters. Kufferath (1952) attributes this to slow movement of the deeper water, possibly induced by the wind, the stability of the Tanganyika being much inferior to that of Lake Kivu. Moreover, Tanganyika has a considerable north-south extension and thus is exposed to great differences in climate from one end to the other. As previously mentioned, Lake Tanganyika has an outlet toward the Lualaba via the Lukuga, a river of moderate importance whose flow has changed markedly during the present century. Lake Kivu with its headwaters and Lake Tanganyika, with the Malagarasi drainage and lesser tributaries, provide the East African contribution to the Congo Basin.

On its course to the Congo, the Luapula crosses Lake Mweru, which has an area of 5000 square kilometers. Unlike the preceding lakes, it is a shallow basin never exceeding 15 meters in depth; the mean depth is certainly under 5 meters. Lake Mweru does not develop true stratification because of the impact of its large affluent, the Luapula, and its equally large outlet, the Luvua, and also because of its high temperature (annual range 22.5° to 27.5°C). It is therefore polymictic. Annual variations in water level do not exceed 1.5 meters (average over 8 years, according to De Kimpe 1964). The influence of Lake Mweru on the Luapula-Luvua tributary of the Congo is small, save for its damping effect on fluctuations of the Luapula water level. In 1959, the fluctuations of the latter were higher than 3 meters, whereas those of the Luvua were only 0.96 meters.

The upper course of the Lualaba-Congo now contains several man-made lakes, which have the characteristics of this particular category of lake. All are located in Katanga, a little to the south of Lake Mweru. They are intended for energy production and are thus subject to major fluctuations in level. Their thermal behavior is that of monomictic lakes, with a circulation period during the dry season (May-July) and a stagnation period extending from September to April. Area is highly variable and depends on rate of consumption of water by the electric plants. Lake Mwadingusha may be as large as 400 square kilometers; Lake Koni does not exceed 4.5 square kilometers, and Lake Nzilo has about 200 square kilometers. Their elevations are respectively 1100, 985, and 1200 meters above sea level.

The Kamolondo depression, which lies across the course of the Lualaba at 536 meters above sea level, contains many lakes, some of which are merely widened sections of the river (as in the case of Lake Kisale). Other bodies of water are more independent of the river. We have little information about these lakes and only Lake Upemba has been the object of serious limnological study (Van Meel 1953). Lake Upemba resembles the other Kamolondo lakes in being a remnant of a large lacustrine expansion that filled the depression before the Lualaba was captured by the Congo. It is a shallow and swampy area, the freewater portion of which is highly variable; it covered about 400 square kilometers in June 1950. The depression is a labyrinth of true swamps, generally covered with dense stands of papyrus, and of grassland periodically flooded at high water, with thickets of the ambatch tree (*Aeschynomene*). The swamps also contain floating water plants, such as Nile lettuce (*Pistia*) and water chestnut (*Trapa*). Huge swampy areas of this kind may influence the regime and quality of the river in various ways. Evaporation of water has a concentration effect, the abundant humus deposited in the swamps absorbs much mineral matter, and the presence of papyrus and other weeds slows the rate of flow with consequent loss of suspended particles. Moreover, the temperature is raised and the productivity of the water is increased by the plankton produced under lacustrine conditions. These lakes are the first in our survey to be under the direct influence of the river and its floods. They are similar in this respect to the várzea lakes of the Amazon region.

Another type of lake in the Congo watershed is the forest lake, found in the lowlands of the central cuvette. An example is Lake Tumba, which occupies the lowest point in the central basin. It is located near the equator at an elevation of 350 meters and has an area of 765 square kilometers. It is very shallow, averaging only 3 to 5 meters deep, with occasional holes extending to depths of 8 meters. It receives several affluents from the south and flows into the Congo opposite the mouth of the Ubangi. It is a "lateral lake" (type 52, sensu Hutchinson 1957), caused by damming of an affluent by sediments deposited on the banks of the Congo. It is never invaded by Congo water due to the length of the Irebu channel and to the simultaneity of flooding in both river and lake. Variations in lake level are notable and may reach 3.73 meters. All the affluents of Lake Tumba are black water forest streams and drain a low swampy area that is completely flooded during a considerable part of the year. The water of Lake Tumba is thus very comparable to that of many Amazonian lakes, especially the Lago Preto da Eva. Being very shallow and traversed by important affluents, Lake Tumba never develops stratification. Temperature varies between 28° and 33°C. Plankton productivity is low.

Comparison of the Two Regions from the Biological Viewpoint

EXTENSION OF THE FOREST

In South America, about eighty percent of the lowlands and of the Andean portion of the Amazon drainage is covered with dense equatorial forest. We need not enter here into the classification and peculiarities of these forests, except to mention that the montane vegetation of the Marañon and Ucayali is different from that of the central Amazon or the State of Pará. Large sectors of the Rio Negro basin are covered with flooded forest of another type, sometimes called "caatinga," which grows on bleached sands; this type of soil and forest is associated with the formation of black waters. There are many other places in the Amazon valley where this sort of water and type of forest occur, and Sioli (1967) has studied such spots in the Tapajós basin intensively. In several black water areas, the vegetation consists of an impoverished savanna-like formation called "campina." The várzea bears another type of forest, which grows on sedimentary soils and is flooded for a variable duration each year. While the Amazon normally flows through forest, extensive grasslands or "campos" occur along both banks of the middle section. A large savanna also lines the course of the Rio Branco. Numerous minor unforested zones exist in the region between the left bank of the Amazon, the Rio Trombetas, and the Tumuc-Humac range on the Guiana border. In the south, campos intervene between the upper courses of the Madeira and occupy the Central Brazilian plateau and its northern extensions. The waters flowing from these campos are clear and the soils are brown loam.

In central Africa, by contrast, forest is less extensive and does not cover more than 34 percent of the Congo Basin. Here also, different types of forest are represented. Dense rain forest extends well beyond the limits of the basin, particularly in West Africa. In Katanga and Kasai on the south, the dominant forms of vegetation are deciduous forest or grassland. This is related to climate as well as to human activity (see Figure 1). The Congolese rain forest lies mainly in the area of Af type climate.

RIVER AND LAKE FAUNA

As is well known and will be discussed in detail later (see Roberts, pp. 239–252 herein), the fish fauna of the Amazon is much richer than that of the Congo, comprising more than 1300 species as against about 500 for the latter river. There is no doubt that this is partly attributable to the much larger area of the former basin. On the other hand, there is less difference between the fauna of Amazonian lakes and rivers than between African lakes and rivers. The cause lies in the different histories of the lakes in the two regions. In Africa, the lakes have not been formed by river action, except in rare cases. Tectonic lakes, such as Lake Tanganyika, have been isolated from the rivers for so long (since the Pliocene?) that a specialized fauna was able to evolve. The Tanganyikan fauna (as well as the Nyasan fauna) is incredibly rich in forms that occur nowhere else. These endemic forms are not confined to fish, but include groups as different as water snails and shrimp. This high level of endemicity has not been observed as yet anywhere in the Amazon Basin. A restricted distribution of certain species may also characterize small lakes separated by a geographic barrier from the river or its affluents. For instance, Lake Fwa in the Kasai watershed has a fish fauna differing generically from river fish of the same family. Endemicity is, of course, much less marked in lakes more accessible to the Congo and its major affluents.

MARINE GROUPS

Very few groups of marine animals thrive in Congo waters and the fauna is of pure freshwater origin. This is in opposition to the situation in the Amazon, where marine groups are well represented and where many freshwater fishes have a marine origin. The contrast is certainly due to the absence of rapids, permitting free access to the higher course of the Amazon by sea animals, which have adapted to life in fresh water in relatively recent times. This explanation is supported by the fact that some fish related to marine families inhabit the Niger and other African rivers that lack physical obstacles.

PRODUCTIVITY OF THE WATER

The productivity of Amazonian waters has been measured occasionally and has been found to be very low, even in the richer parts of the basin. No comparison can be made with the Congo, however, as no measurement has been made either in the river itself or in the lakes of the central plain, which are similar to Amazonian lakes and streams.

In the Amazon, the evidence indicates that the enormous biomass of fish and invertebrates depends for its subsistence on nutrients and organic material flushed down from the forest by the floods. The same conclusion will probably apply to the Congo when comparable data become available.

HUMAN DISEASES IN RELATION TO WATER QUALITY

The existence of human bilharziasis in the Amazon valley has been known for more than 20 years. As a matter of fact, Sioli (1953) has demonstrated that this parasitic disease is restricted to the region of Fordlândia on the middle Tapajós. Fordlândia is situated in the Carboniferous region, a narrow band that extends parallel to the lower Amazon less than 200 kilometers from the river. The geological composition of these Carboniferous rocks increases the pH and gives a higher calcium content to their drainage waters. The streams and brooks in this area are thus neutral and have better buffering properties, providing conditions favorable for the development of snails that serve as intermediate hosts to the worms responsible for the disease (*Schistosoma mansoni*). If, as in Fordlândia, an infected human has access to water where the mollusks thrive, the latter become infected.

The more acid clear and black waters occupying the greater part of the Amazon Basin are unsuitable habitats for snails and the disease cannot become established in them even if accidentally introduced. The white waters of the várzea are not completely immune, however, since their chemical quality is more tolerable, and the author has collected many snails of the genus *Tropicorbis* from várzea lakes. Nevertheless, bilharzia need not become a significant hazard in Amazonia except in the Carboniferous stretches, which are comparatively minor in extent.

The Congo situation is different. Large areas of the watershed are richly mineralized, especially the Kivu-Tanganyika area, the Katanga, and the highlands between the Congo and the Nile. These waters are well suited to the development of Planorbid snails. The human population is much denser than in Amazonia and the bilharzia disease is consequently widespread and a true hazard for human life. More than a quarter of the Congo Basin is pervaded by the disease, including the Katanga, part of the Kasai, the Uele and Kibali-Ituri, as well as the Kivu. The Pulmonate water snails are as rare in the acid regions of the central Congo as they are in the black and clear waters of the Amazon, making this region safe from contamination.

Influence of Human Settlement on the Limnology

A short statement must be devoted to the influence of human activities on the limnological features of the two watersheds. In the Amazon area, this problem is still rather simple. The human population is one of the sparsest in the world; the State of Amazonas, for example, has a density of .24 persons per square kilometer. Although this population is concentrated along the rivers and streams, its density is well below the point where it could affect the limnology. As a matter of fact, the huge Amazonian streams dominate human settlement, control its expansion, and dictate the conditions for agriculture and the food quest. Erosion is a threat only in small regions where "modern" forest exploitation and mechanical equipment have been introduced.

The problem in Africa is very different. Population density is 4.4 persons per square kilometer and the impact of human culture is pronounced, especially in the grassland regions around the central basin. The primitive agricultural methods of felling and burning the trees have led to erosion, which has strongly modified the quality of river water. Even when concentration of soluble salts is not significantly altered, turbidity is always increased with detrimental consequences in the form of increased sedimentation and silting, decrease of plankton and fish production, and disappearance of many bottom-dwelling animals unable to tolerate mud on their integuments and gills. In addition, the increased insolation resulting from deforestation seems to influence chemical composition, at least in the small water courses examined by the author in the eastern Congo. The waters from denuded areas have a higher pH and a higher concentration of salts than protected waters. Moreover, many tropical streams are rather low in nutrients and their faunal richness often depends on products introduced by runoff or litter falling from vegetation on the banks. Destruction of the forest thus results in a decline of aquatic productivity.

The comparatively dense population of the Congo Basin and the smaller size of the tributaries thus form a strong contrast to the conditions prevailing in Amazonia. The extensive flooding and huge inundation areas of the South American várzea and igapó have only minor counterparts in the Congo Basin, notwithstanding the size of the principal river. The many similarities between the limnology of these two drainages must not be allowed to obscure their significant differences.

References

Berg, A. 1961. Rôle écologique des eaux de la cuvette congolaise sur la croissance de la jacinthe d'eau. Acad. R. Sci. Outre-Mer. Classe Sci. Nat. et Médicales. Mém., n.s. T12, fasc. 3:1-120.

Braun, R. 1952. Limnologische Untersuchungen an einigen Seen im Amazonasgebiet. Schweiz. Z. Hydrol. 14:1-128.

Capart, A. 1952. Le milieu géographique et géophysique Expl. Hydrobiologique du lac Tanganika (1946-47), Résultats Scientifiques 1:3-28. Inst. R.S.N. Belg.

Davis, L. C. 1964. The Amazon's rate of flow. Nat. Hist. 73 (6):15-19.

De Kimpe, P. 1964. Contribution à l'étude hydrobiologique du Luapula-Moero. Ann. Mus. Roy. Afr. Centrale 8, Zool. 128.

Dubois, Th. 1959. Note sur la chimie des eaux du lac Tumba. Acad. R. Sci. Outre-Mer. Bulletin des Séances, n.s., 6:1321-1334.

Geisler, R. 1967. Zur Limnochemie des Igarapé Prêto (Oberes Amazonasgebiet). Amazoniana, Kiel, 1 (2):117-123.

Gessner, F. 1960. Untersuchungen über den Phosphathaushalt des Amazonas. Int. Rev. Gesamten Hydrobiol. 45(3):339-345.

Gibbs, R. J. 1967a. Amazon River: Environmental factors that control its dissolved and suspended load. Science 156:1734-1736.

————. 1967b. The factors that control the salinity and the composition and concentration of the suspended solids. Bull. Geol. Soc. Amer. 78:1203-1232.

Gresswell, R. K., and A. Huxley (Editors). 1964. Standard Encyclopedia of the world's rivers and lakes. Weidenfeld & Nicolson (Educational) Ltd. London.

Hutchinson, G. E. 1957. A treatise on limnology. Vol. 1, Geography, physics and chemistry. J. Wiley and Sons, New York.

Katzer, Fr. 1897. Das Wasser des unteren Amazonas. Sitzb. Gesell. Wissench. z. Prag. XVII:1-38.

————. 1900. Zur Geographie des Rio Tapajós. Globus 78 (18):281-284.

Kufferath, J. 1952. Le milieu biochimique. Expl. Hydrobiol. du Lac Tanganika (1946-47). Réultats Scient. 1:31-47. Inst. R.S.N. Belg.

Le Cointe, P. 1922. L'Amazonie brésilienne. Challamel Edit. Paris.

Livingstone, D. 1963. Chemical composition of rivers and lakes. Profess. Pap. U.S. Geol. Surv. 440G:1-63.

Magis, N. 1967. Le zooplancton des lacs artificiels du Haut Katanga méridional, étude faunistique et écologique. F.U.L.R.E.A.C. Univ. Liège, Inst. van Beneden, Travaux, fasc. 183.

Marlier, G. 1967. Ecological studies on some lakes of the Amazon valley. Amazoniana 1(2):91-115. Kiel.

Oltman, R. E. 1966. Reconnaissance investigations of the discharge and water quality of the Amazon. U.S. Geol. Surv. Circ. 552(1968):1-16.

Oltman, R. E., H. O'R. Sternberg, F. C. Ames, and L. C. Davis Jr. 1964. Amazon river investigations, reconnaissance measurements of July 1963. U.S. Geol. Surv. Circ. 486.

O.R.S.T.O.M. [Office de la Recherche Scientifique et Technique Outre Mer] 1958. Annuaire hydrologique de la France d'Outre-Mer. 1956: 324-327. Ministère de la France d'Outre-Mer.

Sioli, H. 1951. Sôbre a sedimentação na várzea do baixo Amazonas. Bol. Técn. Inst. Agron. N. No. 24:45-67.

————. 1953. Schistosomiasis and limnology in the Amazon region. Amer. J. Trop. Med. Hyg. 2(4):701-707.

————. 1954. Beiträge zur regionalen Limnologie des Amazonasgebietes. Arch. Hydrobiol. 49(4):441-447.

————. 1955. Beiträge zur regionales Limnologie des Amazonasgebietes III. Ueber einige Gewässer des oberen Rio Negro Gebietes. Arch. Hydrobiol. 50(1):1-32.

————. 1957. Sedimentation im Amazonasgebiet. Geologischen Rundschau 45(3):608-633.

————. 1963. Beiträge zur regionalen Limnologie des Brasilianischen Amazonasgebietes V. Die Gewässer der Karbonstreifen Unteramazoniens. Arch. Hydrobiol. 59(3): 311-350.

————. 1964. General features of the limnology of Amazonia. Verhandl. Intern. Verein. Limnol. 15:1053-1058.

————. 1965. Zur Morphologie des Flussbettes des unteren Amazonas. Die Naturwissenschaften 52(5):104-105.

————. 1966. General features of the limnology of Amazonia. Humid Tropical Research. 381-390. UNESCO.

————. 1967. The Cururú Region in Brazilian Amazonia, a transition zone between hylaea and cerrado. J. Indian Bot. Soc. 46(4):452-462.

Sioli, H., G. H. Schwabe, and H. Klinge. 1969. Limnological outlooks on landscape ecology in Latin America. Tropical Ecology 10:72-82.

Spronck, R. 1941. Mesures hydrographiques effectuées dans la région divagante du bief maritime du fleuve Congo. Mém. Inst. R. Col. Belge.

Symoens, J. J. 1968. La minéralisation des eaux naturelles. Résultats Scient. Cercle Hydrobiol. de Bruxelles. 1-199.

Ungemach, H. 1967. Sôbre o balanço metabólico de ionios inorgânicos da area do sistema do Rio Negro. Atas do Simpósio sôbre a Biota Amazônica 3:221-226. Belém.

Van der Ben, D. 1959. La végetation des rives des lacs Kivu, Edouard et Albert. Expl. Hydrobiol. des lacs Kivu, Edouard et Albert (1952-54), Résultats Scient. 4 (1). Inst. R.S.N. Belg.

Van Meel, L. 1953. Contribution à l'étude du lac Upemba. Expl. du Parc National de l'Upemba 9:1-190. IPNCB [Institut des Parcs Nationaux du Congo Belge].

Williams, P. M. 1968. Organic and inorganic constituents of the Amazon River. Nature 218:937-938.

Ecology of Fishes in the Amazon and Congo Basins

TYSON R. ROBERTS

Introduction

An overwhelming proportion of the species of fishes inhabiting continental fresh waters are unable to live in salt water, and have had a long history separate from that of marine fishes. The great majority of these "primary freshwater fishes" are members of a single order, the Ostariophysi. Whereas the present tropical marine shore fishes have had only one principal center of radiation—the Indo-Pacific—ostariophysans in Africa, South America, and Eurasia have radiated extensively in isolation from one another. Asia and Africa share several major ostariophysan and non-ostariophysan groups, indicating perhaps that the older South American and African ichthyofaunas had a common origin.

The formation of the Amazon Basin provided the opportunity for a remarkable radiation of Ostariophysi. The Amazon and Congo basins have more kinds of fishes than any other river basins in the world, and both exhibit a high degree of endemism. One of the principal reasons tropical fish faunas are richer than temperate ones is that they have not been as adversely affected by glaciation. Pleistocene glaciation probably caused extinction or withdrawal of marine shore fishes in the higher latitudes, especially in the North Atlantic (Briggs 1970). Western Europe now has a depauperate freshwater fish fauna of about 60 species, almost all derived from stocks that populated the area in postglacial times.

The richness of the Congo and Amazon fish faunas is not necessarily ancient. Fishes have undergone considerable diversification, with great increase in the number of species, in lakes considerably less than five million years old. It is conceivable that the present Amazonian fish fauna, with its large number of species, is the product of a million years of evolution from an original stock of two or three hundred founder species.

Draining two and a quarter million square miles, the Amazon Basin is the largest river basin in the world. Its mouth discharges an average of three to four million cubic feet of water per second (See Oltman 1968, table 1 for a selection of published estimates of flow of the lower Amazon River). The Congo, discharging 1.4 x 10^6 cubic feet per second from slightly over one and one-half million square miles, is the second largest. The vast area of these

TYSON R. ROBERTS, Museum of Comparative Zoology, Harvard University, Cambridge, Massachusetts 02138. This is an abridged and slightly amended version of a paper with the same title in *Bulletin of the Museum of Comparative Zoology* 143 (2). ACKNOWLEDGMENTS: Dr. P. E. Vanzolini, Director of the Museu de Zoologia of the University of São Paulo, enabled me to participate in field work of the Expedição Permanente de Amazônia on the Rio Solimões from 20 September to 5 November 1968. This provided an opportunity to see fishes in a great variety of habitats. EPA is a continuing joint effort of the Museu de Zoologia, Instituto Nacional de Pesquisas de Amazônia, and Museu Goeldi. It is financed by the Fundação de Amparo à Pesquisa of the state of São Paulo. The rich collection of fishes being assembled is kept at the Museu de Zoologia, under the curatorship of Sr. Heraldo A. Britski, to provide the basis for a systematic revision of the Amazon fish fauna.

I wish also to thank M. Pierre Brichard for helping me become acquainted with the fishes of Stanley Pool, and Sr. Willy Schwarz for providing an opportunity to visit the lower Rio Negro and Rio Jauaperi. Many colleagues have shared with me information about fishes from the Amazon and Congo. In this respect, I must particularly thank Prof. George S. Myers. Very helpful comments on drafts of this paper were provided by Dr. Betty J. Meggers, Dr. Gerald R. Smith, Dr. Robert R. Miller, Prof. Hilgard O'Reilly Sternberg, and Prof. Myers.

basins, with abundant water and varied habitats, undoubtedly contributes to the large number of fish species in them. Similar habitats, such as streams with high gradients or streams draining solid ground (*igarapés de terra firme*), are sometimes separated by hundreds of miles. Meandering creates a regular succession of habitats in the main course of the big rivers. The relatively stable and extensive aquatic habitat favors the existence of very large numbers of individuals, which in turn is conducive to the maintenance of large numbers of species (Preston 1962).

As of 1967 (the last year for which the Zoological Record has been issued) approximately 1300 species of fishes had been recorded from the Amazon and 560 from the Congo (including the Lualaba, but not Lakes Bangweulu and Mweru). The Mississippi Basin, in comparison, with an area (1,244,000 square miles) almost as large as that of the Congo Basin, has only 250 species. It is unlikely that many species remain unrecorded in the relatively well-studied Mississippi, but the number known from the Amazon and Congo will undoubtedly increase considerably as systematic studies continue.

The main part of this paper is divided into two sections. The first section deals with the interactions of fishes with physical aspects of environment in the Amazon and Congo basins, the second section with biological interactions. The rest of this introduction provides a brief sketch of the main groups of ostariophysans and other fishes under consideration.

OSTARIOPHYSI: THE PREDOMINANT FISHES IN BOTH BASINS

In the Amazon 43 percent of the fishes are characoids, 39 percent siluroids, and 3 percent gymnotoids. In the Congo 15 percent are characoids, 23 percent siluroids, and 16 percent cyprinoids. All belong to the order Ostariophysi, which thus comprises 85 percent of the Amazon and 54 percent of the Congo fish fauna. Ostariophysi differ from all other fishes in the modification of some of the neural arches and ribs of the first four vertebrae into an apparatus, the Weberian apparatus, which conducts vibrations from the swim bladder to the inner ear. There is no precise understanding of how the Weberian apparatus affects sound (and pressure?) perception, nor is much known about the effects of sound on the behavior of ostariophysans in nature. It is generally agreed, however, that they are "acoustic specialists," and that their world-wide predominance in fresh waters is somehow linked with the Weberian apparatus. Experimental work with various ostariophysans indicates that their auditory sensitivity and range are greater than in many other fishes.

GYMNOTOIDS. Gymotoids, the so-called electric eels, all have specialized electrogenic and electrosensory organs. The famous electric eel (*Electrophorus electricus* L.), studied by Alexander von Humboldt, Faraday, and others, has a very powerful discharge. The discharges of other gymnotoids are too weak to detect without the help of instruments and have only recently come to our attention. Many species have specialized trophic structures, and at least some are highly active at night, hiding or even burying in sand during the day. Their greatest diversity (18 genera and 35 species) is in the Amazon Basin. They are almost as well represented in the Guianas and Orinoco Basin, but outside these areas diversity declines markedly. A few species, belonging to wide-ranging genera, occur north as far as southern Central America (a single species reaching Guatemala) and south to the Plata Basin. They are almost absent west of the Andes (a single wide-ranging species reaching coastal Ecuador), and are absent in the numerous Atlantic coast drainages between the mouths of the Rio São Francisco and the Rio Paraíba. Gymnotoids evolved from characoids.

CHARACOIDS. Characoids or characins are mostly silvery or iridescent, laterally compressed, open-water fishes, active in the daytime. They usually have jaw teeth, often of a highly complex nature (Roberts 1967), and invariably lack barbels. With few exceptions, they are not known to produce biologically significant sounds (almost all fishes produce noises incidental to feeding and locomotion). Of the large groups of fishes inhabiting the earth's fresh waters, characoids (as a group) exhibit the least tolerance for salt or brackish water. They occur only in Africa and Central and South America, where the only fossils identified with certainty also occur (Weitzman 1960). In the light of evidence for continental drift and for characoid antiquity (Greenwood et al. 1966, Roberts 1969), it is reasonable to assume that characins were present in South America before it separated fully from Africa.

SILUROIDS. Siluroids or catfishes typically are denser-than-water, bottom-dwelling fishes with flattened bellies and nocturnal habits. The teeth usually are numerous, small, simple, conical elements set in bands of varying thickness. The barbels, typically two or three pairs, sometimes (as in Bagridae) four pairs, are almost invariably present (one exception in South America, none in Africa) and serve as tactile and gustatory organs. In African Mochokidae (represented by 37 species in the Congo), the barbels are highly branched. In contrast to characins, catfishes are noisy. They produce sound from various anatomical structures, most familiarly by stridulation between pectoral spine and girdle. Of the 31 families of catfishes recognized by Greenwood et al. (1966), 8 occur in Africa (3 of them endemic) and 14 in South and Central America (all but Ariidae endemic). Catfishes are more widely distributed geographically than any other ostariophysans, and their inter- and intrafamilial relationships are not well understood. Two of the living families (Ariidae and Plotosidae) are predominantly marine. Representatives of a South American endemic family, the Aspredinidae, occur along the Guiana coast (Myers 1960a). The startling diversity of endemic catfishes in South America and the presence in southern South America of the only two species in the family Diplomystidae, the most primitive of living catfishes, indicate that catfishes, like characoids, were in South America before the end of the Mesozoic.

CYPRINOIDS. Minnows, the only group of cyprinoids in the Congo Basin, more or less resemble characins except that they have protrusible jaws and frequently one or two pairs of small barbels, while jaw teeth and adipose fin invariably are absent (most characins and catfishes have a rayless adipose fin behind the dorsal fin). The lower pharyngeal teeth, however, are highly modified and minnows exhibit considerable diversification in feeding structures, as shown by Matthes (1963) for African forms. The cyprinoids are perhaps more diverse in North America and certainly much more diverse in Asia than they are in Africa. It is generally thought that they originated in Asia. However this may be, the number of minnow species in Africa is very high, especially in Ethiopia, southern Africa, and in the rockier headwaters of many of the larger tropical rivers, including the Congo.

NON-OSTARIOPHYSAN FRESHWATER FISHES

The only non-ostariophysan primary freshwater fishes in South America are a genus of Lepidosirenidae, two genera of Osteoglossidae, and two genera of Nandidae. The secondary freshwater fishes (sensu Myers 1949) consist of Cichlidae, Cyprinodontidae, Poeciliidae, Galaxiidae, and Percichthyidae. All of these families except Galaxiidae and Percichthyidae are present in the Amazon Basin, and all except Poeciliidae and Percichthyidae also occur in Africa.

The non-ostariophysan primary and secondary freshwater fish fauna of Africa is much more complex. It can be broken down into three main categories: (1) groups shared with South America—Lepidosirenidae, Osteoglossidae, Nandidae, Cichlidae, Cyprinodontidae, and Galaxiidae; (2) groups shared with Asia, and which probably originated in Asia—Notopteridae, Cyprinidae, Mastacembelidae, Anabantidae, and Ophiocephalidae or Channidae; and (3) an unparalleled assemblage of archaic primary freshwater forms known only from Africa—Polypteridae, Denticepitidae, Pantodontidae, Phractolaemidae, Kneriidae, Mormyridae, and Gymnarchidae. All of the families in these three groups, excepting Nandidae, Galaxiidae, Denticepitidae, and Gymnarchidae, are present in the Congo Basin, where most attain their greatest African diversity.

Interrelations between Fishes and the Physical Environment

Since the Amazon and Congo basins straddle the equator, they have two rainfall regimes, which come at somewhat different times of the year. As a consequence, the relative difference between maximum and minimum water levels in the lower courses of the Amazon and Congo rivers, although considerable, is less than in any other rivers in the world and periods of severe desiccation do not occur. These and other physical features have no doubt enhanced the evolution of faunal diversity.

SEASONAL FLUCTUATIONS IN WATER LEVEL

Seasonal fluctuations in water level have profound effects on feeding, reproduction, and dispersal of fishes (Matthes 1964, McConnell 1964). During high water many move into the flooded lands to feed and to reproduce. Growth is rapid and fishes are widely dispersed. As the waters recede, food becomes scarcer for most fishes except predators. Losses to predation are greatest during low water, when fishes are least dispersed. The seasonal fluctuations mean that certain habitats exist only part of the time. It should also be noted that these cyclic changes are more predictable than some other kinds of variability (especially in the Temperate Zones), and allow adaptive response to evolve, thus increasing the effective environmental heterogeneity—and increasing rather than decreasing the species diversity.

Forest streams overflowing in the wet season may be so reduced in the dry season that fishes must either leave or withstand extreme variations. Such constantly changing conditions must alter drastically the local faunal composition. The fact that many species are represented by widely separated populations over the entire basins probably contributes to evolution and maintenance of species diversity. Certainly it is difficult to conceive of the rapid extinction due to biotic factors of species dispersed in so broad a manner. Furthermore, the pattern suggests a favorable model for allopatric speciation.

PHYSICAL AND BIOLOGICAL NATURE OF WHITE, CLEAR, AND BLACK WATER RIVERS

The big rivers of the Amazon Basin are of three main types: white water, clear water, and black water. In black water and white water rivers, most of the food available for fishes must come from terrestrial sources (Marlier 1967) or floating vegetation. There is probably very little food for nonpredaceous species in the main Rio Negro, and most of its fishes must move during the wet season into temporary lakes and flooded forest to find sufficient food. In white water rivers, earth slides probably contribute considerable amounts of plant and lower animal life, and floating vegetation probably is also greater. The clear water rivers are more diverse in origin and possess a broader range of pH values—from 4.5 to 7.8—than black and white water rivers, indicating that they are chemically (and biologically) a heterogeneous assemblage. The only character all clear waters share is the relative lack of organic coloring material and suspended matter (Sioli 1967:33–34; cf. Marlier, herein).

In the Congo Basin, the rivers can be similarly classified according to their waters. The southern tributaries are mostly black water, and during part of the year the main Congo River is a deep-tinted brown approaching black water. The Ubangi River is the only major white water tributary, and its flood waters enter the Congo when contributions from the southern black water tributaries are at their lowest; the Congo never becomes sufficiently laden with silt, however, to be considered a white water river. As in the Amazon, Congo food chains must originate largely on the land.

There are many varieties of floating plants in the Amazon Basin, including *Eichhornia*, but these are little in evidence along the main river channels. The relatively recent introduction and widespread establishment of *Eichhornia* in the Congo Basin must have greatly increased the relative contribution of floating plants to food available in the main river courses. Its roots offer haven and presumably food to many small species of fishes and to the young of many larger species of catfishes, characins, and mormyroids in Stanley Pool (personal observation).

ACCESSIBILITY OF BASINS TO MARINE FISHES

Most of the African continent is relatively high above sea level and marine fishes ascending rivers usually do not get very far inland (Marlier 1967). Most fishes in the cuvette central of the Congo are primary freshwater fishes (sensu Myers 1949). Two families, Cyprinodontidae and Cichlidae, are secondary freshwater fishes and have undergone extensive radiation in fresh water. The remaining secondary freshwater fishes in the cuvette central belong to four families—Clupeidae, Eleotridae, Centropomidae and Tetraodontidae—all better represented in salt water than in fresh water. The particular groups involved, however, entered the fresh water of Africa a very long time ago, and must have invaded the Congo Basin via other river systems instead of directly from the sea. Not a single sporadic marine invader has been reliably recorded from the Congo River above the lower rapids.

In contrast to the Congo, the Amazon Basin is relatively open to invaders from the sea. Fourteen families of predominatly marine fishes, representing slightly more than 50 species, are widespread in the Amazon and about half are endemic. In most instances, the ancestral populations probably invaded directly from the sea. In addition to these 50, many more marine species have been recorded from the lowermost Amazon, especially from Pará. As noted by Marlier (1967), the great extension of brackish waters in the mouth of the river and along the coast has undoubtedly facilitated intrusion by marine forms. Such conditions must also have favored invasion of marine habits by catfishes. The three genera and four species of Aspredininae are restricted to the lowland, muddy coast of Guiana and Amazonia, occurring in the sea, in brackish water, and in the estuaries and tidal portions of rivers, including the Amazon delta (Myers 1960a).

TIDAL CONDITIONS IN THE LOWER AMAZON AND THEIR EFFECT ON FISH LIFE

Many marine and estuarine fishes ascend the lower courses of tropical rivers with the rising tide. Fishes apt to do this in the Amazon delta area include Centropomidae, Mugilidae, Belonidae, Sciaenidae, Pomadasyidae, Lutjanidae, Ariidae, Atherinidae, Carangidae, Clupeidae, Engraulidae, and Dasyatidae. Tidal bores ("pororoca") occur in much of the delta area (Branner 1884), and diminished tides are felt as far up river as Santarem and Obidos (Oltman 1968:9–10, fig. 5). The periodic rise in water level (occurring with great rapidity and force in certain areas) inundates vast areas, submerging terrestrial plants and insects upon which fishes then feed. This regular addition of terrestrial food, plus the higher titre of nutrients due to mixing with sea water, probably makes the mouth and delta the most productive area in the Amazon Basin. Variation in the seasonal availability of food is probably also least marked here.

SHORELINE AND ISLANDS

The amount of food available from terrestrial sources, of course, is proportional to the amount of shoreline. In addition, stabler bottom and clearer water resulting from slower currents along shore permit greater development of planktonic and benthic plants and animals. Higher plants fringing the banks provide shelter from currents and predators, as well as food for the adults and young of many species. Calm and deep places downstream from sand banks and islands, and backwaters in which mud and organic detritus accumulate, provide particularly suitable habitats for bottom-feeding fishes such as *Labeo*, *Citharinus*, and *Tilapia* (Gosse 1963). During high water, many islands are partially flooded, thus adding to the inundated areas accessible to fishes for feeding and reproduction. In a given section of the middle Congo River, fish productivity probably bears a strong relationship to the number and size of the islands.

The roles played by islands and shoreline in the middle Congo also apply to the Amazon. In the evening many small fish (especially characins) move close to the shoreline in order to feed, and larger fish, including predators and catfishes, come inshore from deeper waters. In the daytime, Curimatidae and Hemiodontidae are to be found feeding over sandy bottom near shore. Beginning about 40 miles above its confluence with the Solimões and continuing some 350 miles upstream, the Rio Negro incorporates a multitude of islands that no other river in the world can rival. The resulting high shoreline coefficient may partially compensate for the unproductivity of its black waters.

RIVER ANASTOMOSES AND STREAM CAPTURES

"River anastomoses" or interconnections occur both in the Congo Basin and the Amazon. Above Coquilhatville, the Congo River has a series of connections (Chenal de Bosesela, Chenal de Nyoi, etc.) with the Giri River (a large tributary of the Ubangi), some of which also intersect. What is probably the greatest complex of river anastomoses in the world occurs in Brazil between the Solimões and the Rio Japurá, in the vicinity of Fonte Boa (Furo Boia, Furo Auati-Paraná). The Auati-Paraná, a navigable link between the Japurá and Solimões, is 125 miles long. It seems likely that small connections between tributaries of the Rio Negro and Solimões exist some 500 miles above their confluence. Such features frequently do not show on maps, largely because they occur in relatively uninhabited and economically unimportant areas. In aerial surveys, small connections are concealed by vegetation. And of course maps cannot take into account all the minor changes in stream courses and the extent of flooding, which varies so much from year to year. During exceptionally wet years, the routes available for fish dispersal must be greatly augmented and the direction of flow depends on the relative water level in the rivers they connect. Thus if the Solimões is higher than the Japurá, the Auati-Paraná must flow towards the Japurá, and vice versa; if the water level is similar in both rivers, there is little or no current. If the level of both rivers should drop below that of the Auati-Paraná, it might be drained until only disconnected pools remained.

The Amazon Basin also has important connections with other basins. The largest of these, the Canal de Casiquiare, links the Rio Negro and Rio Orinoco. In Colombia, the Amazon Basin reputedly is linked to the Magdalena by the Japurá (although height of the Andean watershed makes this seem doubtful), to the Guaviare (which flows into the Orinoco) by the Uaupés, and to the Inírida (also flowing into the Orinoco) by the Guaniá (an affluent of the Rio Negro). The Mapuera reputedly links the Rio Trombetas and Essequibo (also doubtful). There is a link between the Rio Branco and the Essequibo via the Rupununi in the rainy season, however. In Mato Grosso, the Amazon Basin is reputedly linked to the Paraguai by the Tapajós and the Guaporé. The Tocantins is linked to the São Francisco. The Casiquiare is a good-sized waterway throughout the year. Some of the other connections probably are broken during particularly dry years. It seems likely, however, that within the recent past most of them permitted faunal exchanges. The Casiquiare undoubtedly has provided an easy route for exchange of fishes between the Orinoco and Amazon; many species have been recorded from it.

There also are some connections between the Congo Basin and adjacent river systems. Poll (1957) indicated "occasional hydrographic confluences between the Ogowe and Congo" (without stating where they are) as an explanation for the presence in the Ogowe of a number of characteristic Congo fish. Connections have been reported with the Nile in the region of Garamba; with Chad Basin by affluents of the Ubangi, Gribinqui, and Ouham; and with the Zambezi by affluents of the Lualaba (Bell-Cross 1965). As Grosse (1963:152) noted, some of these hydrographic connections occur in swampy areas, thereby limiting exchange to certain fishes.

The Congo and Amazon basins evidently have been growing by stream capture at the expense of adjacent basins. A very important capture was that of the Lualaba (probably once connected with the Upper Nile), which apparently occurred at "les portes de l'Enfer" (Poll 1957:60). There is a theory that the Tocantins was a separate basin before it was captured by the Amazon. To my knowledge, there are no reports of important Amazon or Congo tributaries being captured by adjacent basins. If this is true, it represents a mechanism by which the Amazon and Congo could have gained species from adjacent basins with little or no reciprocation.

SPECIAL BIOTOPES

The Congo and Amazon basins provide instances of endemic fishes that are restricted to the special biotopes in which they evidently originated. Several peculiar genera and species (most of them catfishes and cichlids) known only from the lower rapids of the Congo (Poll 1959, 1966, Roberts 1968) are almost certainly endemic there and unlikely to be found away from the rapids biotope. Two lakes in the central bowl of the Congo appear to have endemic fishes: Tumba Lake has an apparently endemic genus of Characidae (*Clupeopetersius*), an endemic species of catfish (*Eutropius tumbanus*), and an endemic subspecies of cichlid (*Tylochromis lateralis microdon*); Fwa Lake has two endemic genera of Cichlidae (*Cyclopharynx* and *Neopharynx*) and an endemic species of the cichlid genus *Haplochromis* (*H. rheophilus*) (Poll 1957: 60). The peculiar characoid *Paraphago*, known only from Lake Leopold II at Kutu, is probably a relict rather than an endemic form and may yet be found elsewhere.

As noted by Marlier (1967), extant Amazonian lakes are shallow, very young, and markedly dependent on the water level in the rivers with which they are linked. It may be well to consider the posisibility that many fishes of the Congo and Amazon are essential lacustrine forms. (This is not to deny the existence of many strictly riverine forms that seldom, if ever, enter lakes, such as cynodontids, various catfishes and gymnotoids.) There is evidence that during the Pleistocene much of the Amazon (Marlier 1967) and Congo basins were covered by a large lake or a series of lakes. One would expect such conditions to have had important consequences for fish evolution.

The Amazon has no rapids comparable to those in the Congo. Nevertheless, four strange genera of trichomycterid catfishes are known only from the São Gabriel rapids on the Rio Negro (Myers 1944). It seems likely that exploration of other rapids, such as those of the Araguaia, will divulge additional peculiar endemic forms.

EXTREME ENVIRONMENTAL FACTORS AND THEIR EFFECT ON FISH LIFE

Although McConnell (1969) has hypothesized that biotic pressures are of far greater importance than climatic or physical factors in the evolution of tropical freshwater fishes, major physical environmental factors have produced varied evolutionary responses. The blood of most temperate-water fishes is unable to exchange O_2 and CO_2 at pH's as low as those in which many tropical freshwater fishes live. Some Amazonian fishes apparently occur in white or clear waters, perhaps because they are unable to adapt physiologically to the acid conditions encountered in black waters. On the other hand, many Amazonian fishes are characteristic of black waters, and some of them seem able to reproduce only in water of such acidity that it is lethal to other kinds of fishes.

Attention should be called to the "friagens" or cold spells that cause mortalities in the Brazilian state of Acre and in other parts of western Amazonia. Bates (1892:289) attributed mortality of fry of different species of characins at Tefé to a very sudden and quite considerable drop in temperature caused by southerly winds. According to Geisler (1969), fish mortalities accompanying a friagem may be caused by uprising of water with little or no oxygen rather than a drop in temperature.

The nonbiotic factors that have had the most obvious, or at least best understood, effects on the evolution of tropical freshwater fishes are deoxygenation and drought. A variety of physiological, morphological, and behavioral adaptations of South American and African freshwater fishes permit them to survive such conditions (see Carter and Beadle 1931, McConnell 1964:132-134). Lewis (1970) has documented the morphological and behavioral adaptations that permit cyprinodontids to utilize the O_2-charged water of the first few millimeters immediately below the surface in habitats otherwise totally deficient in O_2.

Beebe (1945) reported an astonishing variety of fishes from a small "all but dried up mud-hole" in northeastern Venezuela. Some 34 species, comprising six catfishes (a trichomycterid, three callichtyids, and two loricariids), fifteen characins, a gymnotid (*Hypopomus*), four cyprinodontoids, six cichlids, *Polycentrus schomburgkii,* and *Synbranchus marmoratus,* were taken from malodorous mud and decayed vegetation covered by damp slime (but no free water) in what was left of a drying pool, which had been "almost unswimmable" for weeks. He estimated that in another week or ten days without water all would have perished. Of the 15 characins, two are air-breathers. *Copeina* and some of the others perhaps utilize oxygenated water immediately below the surface in a manner similar to that of cyprinodonts. *Copeina* is characteristic of swampy places and stagnant backwaters. But it is difficult to imagine how *Astyanax, Creagrutus, Moenkhausia, Paragoniates, Pristella,* and *Serrasalmus* could have survived in such a habitat for as long as they did. Most of the species found in this Venezuelan mud-hole also occur in the Amazon Basin.

PARENTAL CARE IN AMAZON AND CONGO FISHES

McConnell (1969) has observed that many tropical freshwater fishes have some form of parental care and implied that biotic factors such as predator pressure are of far greater importance than climatic factors in the evolution of such behavior. This seems to apply to cichlid fishes in Lakes Victoria, Tanganyika, and Nyasa, but not to the riverine fishes of tropical Africa and South America. Whereas an overwhelming proportion of the rich endemic cichlid fauna of these three lakes practices oral brooding, such behavior is far less common in riverine Cichlidae in Africa and in South America, where it has been reported only in some species of *Geophagus.* Several factors may contribute to this behavioral dichotomy. In the still, clear, littoral waters of the lakes, eggs and young are less readily dispersed and far more susceptible to visually oriented predators, whereas under riverine conditions, the eggs are likely to be separated from the parents by current. Moreover, in the black or white waters of many tropical rivers, the eggs and young would be extremely difficult to detect visually by predators, and, perhaps even more important, by the parents themselves.

In tropical rivers of Africa, South America, and Asia, parental care occurs mainly in fishes in which adults spend at least part of the time in swamps or other oxygen-deficient habitats; many are capable of air-breathing. In African and South American lungfishes, a nest is constructed and is guarded and aerated by the male. Notopteridae, osteoglossoids, and Pantodontidae are air-breathers and guard the young. *Gymnarchus*, which is probably capable of air-breathing, makes a floating nest of plants in dense swamps; it is the only mormyroid known to guard the young. Young *Gymnarchus* have external gills with numerous fine filaments and a highly vascularized, enlarged yolk sac, which play an important role in gas exchanges. No other mormyroids have such structures. Parental care might be expected of *Polypterus*; none has been recorded. Froth nests have been attributed to the African characoid *Hepsetus* (which is not known to be an air-breather, but may leave its young in habitats likely to be oxygen deficient). Nesting habits are also ascribed to characoids of the family Erythrinidae. Erythrinids tend to enter swampy regions, and *Hoplerythrinus* is evidently capable of air breathing. *Callichthys* and *Hoplosternum*, South American air-breathing catfishes, construct froth nests, whereas the related *Corydoras*, which probably are not air-breathers, generally scatter their eggs amidst plants.

In most characoids, although courtship and selection of spawning site may be highly complex, parental behavior probably ends once the eggs are deposited. There are no records of parental care in gymnotoids, excepting the unconfirmed report (Du Bois-Reymond 1882) that *Electrophorus* (the electric eel) practices oral brooding. In *Electrophorus*, a highly convoluted oral epithelium facilitates air-breathing (Carter 1935). (There is also reason to suspect that *Gymnotus*, which is related to *Electrophorus*, takes the young into its mouth.) It would appear that the great majority of fishes in the Amazon and Congo have no parental care. As pointed out by McConnell (1969:63-64), many Amazonian fishes engage in upstream spawning migration ("piracema"), producing extremely large numbers of eggs per female, all or most of these eggs being laid at one time at the start of, or early in, the rainy season. In all such fishes, parental care is probably nonexistent. None of the fishes with modifications for breathing air participate in piracema.

Biological Interactions of Fishes in the Amazon and Congo

Apart from predator-prey and host-parasite relationships, there is only one recorded instance of biological interaction between Amazon or Congo fishes and animals of other classes. The exception is commensalism of chironomid larvae attached to Loricariidae and Astroblepidae in the Amazon (Freihofer and Neil 1967). Insects aside, invertebrates apparently are of minor consequence in the Amazon and Congo river systems. Almost all of the marine groups with which tropical reef fishes display complex symbiotic and commensal relationships are absent, and nothing has taken their place. Most of the main rivers and streams in the Amazonian lowland are relatively poor in numbers of kinds and of individuals of mollusks and crustaceans. This is also true of the Congo. The paucity of these two groups is particularly striking when compared to their richness and abundance in lowland streams of portions of Southeast Asia. Aquatic leeches also seem to be more abundant in Southeast Asia. Perhaps mollusks, crustaceans, and leeches would be less abundant in Southeast Asia if the dominant ostariophysans there were characoids instead of cyprinoids. G. R. Smith of the University of Michigan Museum of Zoology informs me that shrimp are fairly abundant in some Amazonian headwaters. This may be related to the relative paucity of their fish fauna. Mollusks and crustaceans might find the black waters too acidic and poor in calcium, and the white waters generally too silty. Excepting parasites, then, almost the only animals with which the fishes can interact are other fishes.

CONSPICUOUS MARKINGS OF CONGO AND AMAZON FISHES

Although both Africa and South America are famous for brightly colored and strikingly patterned species, these are not evenly distributed among the river systems of the two continents. In general, brightly colored fishes are more common in forest rivers than in savanna rivers, and in black water or clear water rivers than in white water rivers. The highest proportion occurs in river systems with the highest numbers of species, and the trend toward distinctively marked representatives in the Congo and Amazon basins is evident in almost all groups. To cite just two examples from Africa: the catfish genus *Synodontis* is represented by 15 species in the Volta Basin and 37 species in the Congo. None of the Volta species can match the vivid colors or contrasting patterns of the Congo species *S. angelicus*,

S. ornatus, S. ornatipinnis, S. flavotaeniatus, S. decorus, S. nummifer, and *S. notatus.* Again, the Volta has three species of the characoid genus *Distichodus,* all with indistinct vertical bars and drab colors, while the Congo has eleven species, all but two of which are distinctively to strikingly marked. Equally good examples could be drawn from the Characidae, the cyprinid genera *Labeo* and *Barbus,* and the anabantoid genus *Ctenopoma.* The major exception to the trend in Africa is the uniformly drab or cryptically colored mormyrids, in which nonvisual sensory structures are highly specialized and the eyes are reduced. In South America, the gymnotoids constitute a similar exception.

Brilliant colors and striking markings are evidently meant to advertize the presence of their possessor. Unlike many gaudy tropical birds and insects, these characters are not confined to mature males, but generally appear at an early age in all individuals. If the fishes were distasteful, venomous, or harmful in some way, we could hypothesize that their function was to warn predators. At least for the majority of characoids, however (which provide most of the best examples both in the Amazon and in the Congo), there is no indication whatever that they are inedible or dangerous in any way to predators. In this respect, conspicuousness is probably a disadvantage. Many of the small, brightly colored Amazonian characins form schools, including the most brilliant characins of all, the neon tetras and cardinal tetras. Most of these live in black water or clear water igarapés, a habitat subject to fluctuation. Such populations may be frequently split up or dispersed and species recognition and schooling habits are probably important for their reconstitution.

ASSOCIATION OF AMAZONIAN CHARACOIDS IN MIXED SCHOOLS

Myers (1960b), Géry (1960), and McConnell (1969) have provided examples of similar appearing South American characoids that form mixed schools of two or more species. Myers (1960b:207) has reported that the small characoids, *Creagrudite maxillaris* and *Creagrutus phasma,* which look very similar, were collected together (it is unknown whether they were schooling together). He suggested that they form an instance of Batesian mimicry, with *Creagrudite* the model and *Creagrutus* the mimic. Reexamination of fishes identified as *Creagrudite* from the same area (upper Orinoco-upper Rio Negro) has revealed another species of *Creagrutus,* which appears to be *C. melanzonus* (Myers and Roberts 1967). All three species have a blackish crescentic humeral blotch and are closely similar in appearance. Géry (1960) suggested that cheirodontines and tetragonopterines that school together both benefit from the association, and thus the mimicry is Müllerian. Géry (1960:37) labeled the schools "protective associations," but did not identify the nature of the protection they supposedly provide. Perhaps mimicry permits small numbers of isolated individuals of two (or more) species to form a nucleus for aggregation with increased chances that breeding populations eventually will be reconstituted. Moynihan (1968) has discussed several instances of mimicry that seem to facilitate flocking in neotropical mountain birds.

AMAZON AND CONGO FISHES OF MINUTE SIZE

Minute body size in adult Amazon and Congo fishes is evidently a response to biotic pressures. In the Amazon, where biological interactions among fishes are probably greater than anywhere else in the world, we find the largest number of minute freshwater fishes, including the smallest oviparous and the smallest viviparous cyprinodonts (*Fluviphylax pygmaeus* and *Poecilia minor,* respectively); one of the smallest needlefishes (*Belonion apodion;* Collette 1966); two tiny species of Eleotridae (*Microphilypnus;* Myers 1927); and minute catfishes in several subfamilies of Trichomycteridae. Except the needlefish, which is very slender, everyone of these is less than 2.5 centimeters long when fully adult.

Excepting the catfishes, all diminutive species are secondary freshwater fishes and belong to groups whose presence can be thought of as "marginal." Large poeciliids are absent in the Amazon except for three good-sized species of *Poecilia,* which have penetrated no further inland than Pará. *Poecilia minor* seems confined to the lower Amazon; it is known from only two collections (separated by more than 100 years) within 100 miles or so upstream from Obidos. *Poecilia scalpridens,* the only other poeciliid in the interior of the Amazon, and not much larger than *P. minor,* is known from a few localities in the middle and lower Amazon. *Fluviphylax* is a phyletically isolated form, widespread in the Amazon Basin. Large gobioids occur in fresh water in many parts of the tropics where primary freshwater fishes are poorly represented. Although a number of large gobiids and eleotrids have been recorded from the mouth of the Amazon, the two minute species of *Microphilypnus* are the only gobioids known from its interior. A third species of *Microphilypnus*, perhaps the smallest one, occurs in the Orinoco Basin (Myers 1927). The Congo eleotrid, *Kribia nana*, while considerably larger than *Microphilypnus*, is nevertheless a very small fish. It is the only gobioid in the interior of the Congo. Until recently, the cyprinodont *Aplocheilichthys myersi* was the smallest fish known from the Congo Basin. We may note that at least some of these little fishes (*Fluviphylax*, *Poecilia minor*, and *Kribia*) apparently reproduce throughout the year; probably they all do. Their size may permit them to utilize food resources unavailable to larger adults and may place them below the threshold for attack by most predaceous fishes.

The best African example of a fish group with an essentially marginal distribution is the Kneriidae. The species of *Kneria* and *Parakneria*, some of which are 80-150 millimeters long, occur in high gradient streams around virtually the entire periphery of the Congo Basin (Poll 1966, 1969:360), but have yet to be found in the cuvette centrale or in the main rapids of the Congo River, where ecological conditions would seem to be suitable. The kneriid *Grasseichthys gabonensis* Géry, only 18-20 millimeters in standard length and very slender, was discovered in 1964 by Géry in forest streams in Gabon and by myself in forest streams in the western part of the cuvette centrale. This is now the smallest known species of fish in the Congo Basin.

ADAPTIVE RESPONSES TO PREDATION

Adaptations displayed by Amazon and Congo fishes to offset the toll of predation include the alarm substance and fright reaction in Ostariophysi, Kneriidae, and Phractolaemidae (Pfeiffer 1963, 1967); the ability of gymnotoids to withstand mutilation (Ellis 1913); the cryptic body form and coloration of such fishes as *Farlowella*; the heavy (and frequently spiny) body armor of many catfishes (especially in the families Doradidae and Loricariidae); and the protective dorsal and pectoral fin spines of most catfishes. In some catfishes these spines are very sharp and venomous, as in the Amazonian carataí, *Centromochlus heckelii*, and other auchenipterids. In other catfishes, the spines are stout and can be locked in erect position. *Centrochir crocodili* (Humboldt) of the Rio Magdalena is called "mata-caiman" because of the wounds it inflicts on crocodilians attempting to swallow it (Eigenmann 1922:47). It is said in Ghana that crocodiles are sometimes killed by trying to swallow *Auchenoglanis occidentalis*. The spines are not effective against piranhas, which bite out chunks of flesh or against various kinds of candirú. On the other hand, the electric eel of the Amazon (*Electrophorus electricus*) and the electric catfishes of Africa (*Malapterurus*) may be entirely exempt from predation. *Malapterurus*, only a few centimeters long, is capable of producing a jolting shock (personal observation).

PARTITIONING OF FOOD RESOURCES AND TROPHIC ADAPTATIONS

Another obvious aspect of biological interaction or accommodation is the partitioning of food resources. Many species have highly modified trophic structures, and some of the most peculiar types of feeding exhibit convergence between fishes of the Amazon and Congo basins. Were it not for partitioning of food resources and concomitant evolution of specialized feeding behavior, the astounding number of species probably would be unable to coexist. Partitioning appears to have proceeded to the point where utilization and cycling of energy and materials are very efficient. The upper limits presumably are determined by complicated factors, among them the variety of foods and their relative availability in space and time, and the capacity of fishes to survive (e.g., by fasting or facultative freedom) when such resources are unavailable and to reproduce when they become available. Food resources cannot be partitioned beyond the point where individual parcels of energy and materials are too small to support populations, even in habitats where catastrophic (i.e., nonbiological) causes of extinction tend to be minimal.

The main categories of food in big tropical rivers are relatively few: (1) other fishes; (2) insects, both aquatic and terrestrial, and aquatic insect larvae; (3) higher plants, including fruits and leaves fallen into the water and roots growing into the river from the banks, as well as some aquatic plants; and (4) mud or earth, including interstitial organisms, dead organic matter, and possibly bacteria. Each of these categories seems to have invoked various kinds of trophic, behavioral, and morphological adaptation.

PREDATORY FISHES

It would appear that nowhere else on earth have fishes evolved as many manners of preying on other fishes as in the Amazon. Excluding fin-eaters and scale-eaters, over 40 species of characoids are primarily or exclusively piscivorous. The more voracious species of *Serrasalmus* (known as *piranha chata* in Brazil) bite out chunks from larger fishes. Géry (1963:615-616) noted that for each species of *Serrasalmus* in the voracious subgenera *Pygocentrus* and *Taddyella,* there is a geographically corresponding species in the less-specialized subgenus *Serrasalmus.* He hypothesized a parallel evolution of sympatric species, in which the less aggressive *Serrasalmus* (known in Brazil as "pirambebas") benefited from association with *Pygocentrus* and *Taddyella.*

Several of the most archaic fishes of the Amazon and Congo basins are rapacious predators. Some species of *Polypterus* (e.g., *P. senegalus*) are insectivorous, but the two largest—*P. endlicheri* and *P. cogicus*—are piscivorous. The African osteoglossoid *Heterotis* is a filter-feeder, but the Amazonian *Osteoglossum* is piscivorous (feeding mainly on characins), as is *Arapaima gigas* (the largest osteoglossoid), which sometimes feeds on *Osteoglossum.* The African *Hepsetus,* apparently the most primitive living characoid (Roberts 1969), is a voracious piscivore, as are a number of phyletically isolated (and perhaps primitive) South American characoids.

Far from inhabiting situations that are geographically or ecologically isolated and open to relatively few organisms, these modern predaceous representatives of archaic fish groups are frequently dominant forms in shallow seas and especially in lowland rivers (e.g., the Congo), where the fish fauna is exceptionally rich.

SCALE-EATING AND FIN-EATING CHARACOIDS

South America has several scale-eating characoids (Roberts 1970). Four genera and about ten species in the Amazon have teeth obviously specialized for removing scales. They generally attack fishes larger than themselves, which presumably usually escape without mortal injuries. Although the Congo is rich in characoids, no scale-eaters have been reported there. The only other freshwater scale-eating fishes are cichlids from Lakes Nyasa, Tanganyika, and Victoria.

The Congo, on the other hand, has a remarkable group of fin-eating characoids, all in the family Ichthyboridae (Matthes 1961). *Gavialocharax* (from Cameroun), with its wonderfully elongated jaws, probably is a fin-eater, as are the species of *Belonophago* (from the Congo). Like scale-eaters, they attack fishes much larger than themselves. The fin-eater diet, rich in bone minerals, probably was prerequisite for the development of the exceedingly hard, plate-like dermal armor in *Belonophago* and *Phago.* In the Amazon, the only known fin-eaters are in the genus *Serrasalmus* (e.g., *S. elongatus*), and they tend to utilize other foods to a considerable degree.

FEEDING HABITS OF THE AMAZONIAN CATFISHES

Candirú is an Amerindian name for certain catfishes that attack other fishes and, occasionally, man. (For a delightful account of candirú attacks on man, see Gudger 1930). About 30 species have been described from the Amazon Basin, representing two unrelated families: Cetopsidae and Trichomycteridae.

The majority of candirú belong to the trichomycterid subfamilies, Stegophilinae, Tridentinae, and Vandelliinae. They are extremely slender and range from about 2 to perhaps 15 centimeters in length.

In Stegophilinae and Tridentinae, which are closely related, the mouth is wide and teeth in both jaws are very numerous and arranged in several rows. In Vandelliinae, the mouth is relatively narrow and the teeth are few and form only one or two rows in both jaws. According to Eigenmann (1918), Reinhardt in 1858 was the first to record that a species of candirú (*Stegophilus insidious* Reinhardt, from the Rio das Velhas, Rio São Francisco Basin) enters the gill chambers of other fishes. Eigenmann (1918) reported similar behavior for a species of Vandelliinae (*Branchioica bertonii* from the Rio Paraná). A number of Vandelliinae and Stegophilinae have been taken in the gill chambers of fishes caught on hook and line, and their stomachs were frequently gorged with blood. Except for their eyes and viscera, Vandelliinae are generally transparent in life. They are slender fishes, but are capable of considerable abdominal expansion to receive blood. There is no evidence that either Vandelliinae or Stegophilinae spend protracted periods in the gill chambers of other fish; perhaps members of both subfamilies gorge themselves fairly soon after entering and then swim out.

Jonathan Baskin, who is studying the family Trichomycteridae, called to my attention the scale-eating habit of the stegophiline *Apomatoceros alleni* Eigenman. The mouth of this species can be everted to form a discoid sucker about twice as wide as the head, and is provided with numerous bands of teeth. The evidence of scale-eating is provided by examination of an alizarin preparation and radiographs of two specimens (105 and 111 millimeters, cataloged, as no. 109804 in the fish collection of the Academy of Natural Sciences of Philadelphia), the alimentary canals of which are partially filled with scales about 3 millimeters long.

TROPHIC STRUCTURES OF AMAZONIAN FISHES

Production of phytoplankton is low in most Amazonian waters and zooplankton is often absent. The principal biotopes in which phytoplankton develops are the mouth-bays of clear water affluents, such as the Tapajós and the Xingú, and the shore lagoons or lakes of white water rivers. In some places, veritable plankton blooms occur and the shore lagoons are often favored fishing grounds. Fishes are scarce in the mouth-bays, however, and the main consumption of the phytoplankton produced there may occur in the white water rivers into which they flow (Sioli 1968).

A number of Amazonian fishes have trophic structures for utilization of plankton. Böhlke (1953) has described a minute (25-30 millimeters) herring-like characid from the upper Rio Negro, *Thrissobrycon pectinifer*, with "otter-board" maxillaries in a nearly toothless mouth and about 25 long gill rakers on the lower limb of the first gill arch; he inferred it to be an open-water, schooling planktophage. Amazonian clupeids and engraulids tend to be predators, but *Cetengraulis jurensis* Boulenger, with about 40 long, finely denticulate gill rakers on the lower limb of the first gill arch, is probably planktophagous. The Amazonian fishes with the most highly modified apparatus for straining minute organisms from the water are catfishes of the genus *Hypophthalmus* and the characin *Anodus elongatus* and one or two of its close relatives. In these presumably planktophagous catfishes and characins, the mouth is toothless and the gill silts are extremely long. The gill membranes are free from the isthmus. The gill openings and gill arches extend anteriorly almost to the symphysis of the lower jaw, so that virtually the entire floor of the oropharyngeal cavity is lined with gill rakers. The gill rakers on all of the gill arches are elongate and exceedingly numerous.

There are no less than three distinct species of *Hypophthalmus* in the Amazon, at least one of which attains a length of 60 centimeters. The trophic structures are highly distinctive. The opening of the mouth is large and its roof is smooth. Most of the gill rakers are borne on the elongate lower limbs of the gill arches. A 30-centimeter specimen of *H. edentatus* has about 240 gill rakers on the first gill arch. The length of the gill rakers tapers off at either end of the gill arch, but most of them are extremely long (about 15-17 mm). Rakers on succeeding arches are almost as numerous as those on preceding arches, and only slightly shorter. *Hypophthalmus* form large schools and undergo extensive migrations. They are one of the most important food fishes in the lower Tocantins and are among the fifteen or so commonest species in the fish market at Manaus.

The highly streamlined *Anodus elongatus* appears to have an even more perfect straining mechanism than *Hypophthalmus*. Both leading and trailing edges of its first four gill arches bear rakers, and the fifth arch bears rakers on its leading (free) edge. The first arch of a 20-centimeter specimen bears 80 + 110 rakers, most of which are 10 or 11 millimeters long. On the leading edge of each raker are two rows of about 100 or more tiny denticles approximately 0.2-0.3 millimeters long and 0.1 millimeter apart. The denticles of adjacent rakers mesh to form an exceedingly fine sieve. A related form, *Eigenmannina melanopogon*, from the upper Amazon has exceedingly numerous gill rakers, and is also presumably planktophagous.

In the Congo Basin, none of the catfishes or larger characins are planktophagous. The small characin *Clupeopetersius schoutedeni* Pellegrin in Lake Tumba is a pelagic planktophage (Matthes 1964:43, pl. 1b,d), as are some endemic Congo species of Pellonulinae.

PARALLELISM IN THE FEEDING HABITS OF MORMYROIDS AND GYMNOTOIDS

The nature of the electric faculties and their biological significance in gymnotoid and mormyroid fishes is now under intensive investigation. The best review is still by Lissmann (1958). (Important articles by M. V. L. Bennett (1971) on electric organs and electroreception were received too late for consideration in the present discussion.) Independent evolution of weakly electrogenic freshwater fishes in Africa and South America is a particularly striking example of parallelism because of its novelty and the pervasiveness of its effects. Some 18 genera and 35 species of gymnotoids have been recorded from the Amazon Basin; 10 genera and 93 species of mormyroids have been recorded from the Congo Basin. It is believed that all mormyroids and gymnotoids possess both electrogenic and electrosensory faculties, and this has been verified for at least one species in most genera. In both groups, virtually all aspects of the morphology and behavior have become specialized and integrated with the electric faculties. For instance, electrosignaling functions in territorial and aggressive behavior in gymnotoids and mormyroids, but its (presumably important) role in sexual behavior has not been described in either group.

The parallels in habitat selection, mode of locomotion, and feeding habits between gymnotoids and mormyroids are intimately bound up with their electric faculties and elucidation of the interrelations between electric behavior and feeding habits should contribute materially to understanding the evolutionary history and perhaps the origin of both groups. The strongest evidence that electric behavior has profoundly affected feeding habits lies in the repeated development in both groups of highly peculiar and remarkably similar trophic structures, e.g., diverse types of elongated tubular mouths with weak jaws and feeble dentition. These structures evidently permit efficient exploitation of rich bottom fauna of small worms and worm-like insect larvae (e.g., enchytraeids and chironomid larvae), which other fishes can use only marginally or not at all. The nocturnal feeding of some gymnotoids and mormyroids may coincide with the time that such prey is most susceptible to predation. Two further possibilities also merit consideration. The first is that mormyroids and gymnotoids are able to locate minute prey electrosensorily. The second is that the weak electric emanations of gymnotoids and mormyroids render these prey more susceptible to predation. Either of these last two possibilities (or both acting together) would, in my opinion, go a long way towards explaining the evolution of almost all of the more peculiar trophic modifications exhibited by mormyroids and gymnotoids. I would go even further, and suggest that the interrelation between electrical faculties and feeding played a decisive role in the initial divergence of the gymnotoids and mormyroids from nonelectrically specialized ancestors.

BOTTOM-FEEDING FISHES WITH GENERALIZED TROPHIC STRUCTURES

Roughly a third of Amazonian and Congo fishes are bottom feeders. Mormyroids, gymnotoids, and Chilodontidae and Hemiodontidae among the characoids, are highly selective in removing food items from substrate. Most bottom feeders, however, including members of the large South American characoid family Curimatidae, cyprinids of the genus *Labeo,* and many catfishes, ingest considerable amounts of substrate with their food. At first glance, this practice might seem to preclude fine selection of food resources, but I expect that this impression would disappear if we had more information about habitat selection and substrate preference of the bottom feeders.

Much of the present diversity of characoids is probably due to relatively late radiations and one of the main reasons that African characoids are less diverse than those in South America appears to be preemption by other groups of certain major potential food resources. Mormyroids and cyprinoids, to cite the two most important examples, have largely or entirely taken over bottom-feeding niches. In the Congo, the cypriniod genus *Labeo,* which parallels the family Prochilodontidae in certain respects, is represented by at least 22 species, most of them endemic.

FEEDING HABITS OF FISHES IN SMALL AMAZONIAN RAIN FOREST STREAMS

The stomach contents of 49 fish species from three rain forest streams near Manaus were analyzed by Knöppel (1970:343-346). His main conclusions were: (1) Terrestrial insects (especially ants), aquatic insect larvae (especially Ephemeroptera and Trichoptera), and vegetable remains were the major foods. (2) Most species exhibited considerable variability in the items ingested, and stomach contents of different families were relatively uniform. (3) Stomach contents of the same species collected at different times of the year (May, July, and November) were generally similar. (4) The fishes found their food in the whole living space, even those species that appear adapted to certain zones in the stream. (5) Distinct specialists in food ingestion were absent in the forest streams studied. (6) Neither the structure of the snout and denture, nor the morphological structure of the alimentary canal, nor even the intestinal ratio are useful indices to feeding habits.

The last three conclusions are not entirely in accord with statements in the main body of Knöppel's paper. The numerous, sharp conical teeth of *Hoplias* are clearly those of a piscivore, and Knöppel (1970:275) found that adult *Hoplias* ate only fishes. The fan-shaped teeth of *Poecilobrycon* and *Iguanodectes*, with numerous small cusps, are adapted to feeding on filamentous algae, and Knöppel found that considerable amounts of filamentous algae were ingested by *Iguanodectes* and *Poecilobrycon*. Concerning intestinal ratios, adult Curimatidae have extremely convoluted intestines, and this corresponds with their habit of ingesting large amounts of fine detritus, only a (small ?) portion of which is nutritional. In six specimens of *Curimatus spilurus* (?) from 26.0 to 42.3 millimeters in standard length, Knöppel (1970:276) found "sand (40 %) and detritus (54 %) in all stomachs"; in one stomach there was plant matter.

The generalization about the relative lack of narrow trophic specialists in such small Amazonian streams is partly valid, however. Most of the highly specialized predatory characoids, such as piranhas and scale-eaters, are entirely absent. The commonest piscivore is probably *Hoplias*, which swallows its prey whole. Many nonpredaceous characoids with highly specialized trophic structures do not occur in small streams. Hemiodontidae, Prochilodontidae, and most genera of Anostomidae are absent. On the other hand, few fishes are more "distinct specialists in food ingestion" than the leaf fish *Monocirrhus polyacanthus*, one of the species studied by Knöppel, which feeds exclusively on small live fish (see Liem 1970). Rhamphichthyid gymnotoids also have specialized means of ingesting their small prey.

Conclusions

In conclusion, it should be pointed out that many aspects of fish ecology in the Amazon and Congo basins have been subjected to little or no serious study. The role of electrogenic and electrosensory systems in the ecology of gymnotoid and mormyroid fishes is likely to receive further attention, but more general aspects of sensory perception and communication should not be neglected. We are far from fully understanding the varied color patterns displayed in fishes of the Amazon and Congo basins, and probably much further from an adequate appreciation of the roles played by the senses of taste and olfaction. It may be mentioned in passing that the function or functions of the so-called caudal glands of the glandulocaudine Characidae, suspected of producing sex pheromones, have yet to be elucidated, and that many Amazonian catfishes (including *Brachyplatystoma*) possess an exocrine gland in the axil of the pectoral fin that produces relatively large amounts of a milk- or cream-like substance of unknown function. Reproductive behavior and larval adaptations are unknown even in many of the commonest and largest species, and there is little precise information on fecundity, growth rates, longevity, and other factors needed to assess life-history strategies and population biology. Sophisticated studies of migrations, diel and seasonal rhythms, and even feeding habits are still lacking.

References

Bates, H. W. 1892. The naturalist on the River Amazon. lxxxix + 395 pp. Appleton. New York.

Beebe, W. 1945. Vertebrate fauna of a tropical dry season mud-hole. Zoologica 30(pt. 2):81-88, 2 pls.

Bell-Cross, G. 1965. Movement of fish across the Congo-Zambesi watershed in the Mwinilunga District of Northern Rhodesia. Proc. Central African Sci. Med. Congress 1963. Pergamon Press.

Bennett, M. V. L. 1971. *In* W. S. Hoar and D. J. Randall (Editors). Fish physiology. Vol. 5. Academic Press.

Böhlke, J. 1953. A minute new herring-like characid fish genus adapted for plankton feeding, from the Rio Negro. Stanford Ichth. Bull. 5(1):168-170.

Branner, J. C. 1884. The "pororoca," or bore, of the Amazon. Science 4(95):488-492.

Briggs, J. C. 1970. A faunal history of the North Atlantic Ocean. Syst. Zool. 19(1):19-34.

Carter, G. S. 1935. Respiratory adaptations of the fishes of the forest water, with descriptions of the accessory respiratory organs of *Electrophorus electricus* (Linn.) and *Plecos-*

tomus plecostomus (Linn.). J. Linn. Soc. (Zool.) 39(265): 219-233.

Carter, G. S., and L. C. Beadle. 1931. The fauna of the swamps of the Paraguayan Chaco in relation to its environment; II: Respiratory adaptations in the fishes. J. Linn. Soc. (Zool.) 37(252):327-367, pls. 19-23.

Collette, B. B. 1966. *Belonion*, a new genus of fresh-water needle-fishes from South America. Amer. Mus. Novit. 2274: 1-22.

Du Bois-Reymond, E. 1882. Uber die Fortpflanzung des Zitteraales (*Gymnotus electricus*). Archiv Anatomie und Physiologie (Leipzig), Physiol. Abth. 1882:76-80.

Eigenmann, C. H. 1918. The Pygidiidae: A family of South American catfishes. Mem. Carnegie Mus. 7(5):259-398, pls. 36-56.

————. 1922. The fishes of western South America, part 1. Mem. Carnegie Mus. 9(1):1-346, 35 pls., 1 map.

Ellis, M. M. 1913. The gymnotid eels of tropical America. Mem. Carnegie Mus. 6(3):109-204, pls. 15-23.

Freihofer, W. C., and E. H. Neil. 1967. Commensalism between midge larvae (Diptera: Chironomidae) and catfishes of the families Astroblepidae and Loricariidae. Copeia, 1967 (1):39-45.

Geisler, R. 1969. Untersuchungen über den Sauerstoffgehalt, den biochemischen Sauerstoffbedarf und den Sauerstoffverbrauch von Fischen in einem tropischen Schwarzwasser (Rio Negro, Amazonien, Brasilien). Arch. Hydrobiol. 6(3): 307-325.

Géry, J. 1960. New Cheirodontinae from French Guiana. Senckenbergiana Biol., 41(1 and 2):15-39, pl. 2.

————. 1963. Contributions à l'étude des poissons characoides, 27: Systématique et évolution de quelques piranhas (*Serrasalmus*). Vie et Milieu 14(3):597-617.

————. 1964. Une nouvelle famille de poissons dulcaquicolles africains: Les Grasseichthyidae. C. R. Acad. Sci. Paris, 259:4805-4807.

Gibbs, R. J. 1970. Mechanisms controlling world water chemistry. Science, 170:1088-1090.

Gosse, J. P. 1963. Le milieu aquatique et l'écologie des poissons dans la région de Yangambi. Ann. Mus. Roy. Afrique Centrale (Zool.) 116:113-270, pls. 1-10.

Greenwood, P. H., D. E. Rosen, S. H. Weitzman, and G. S. Myers. 1966. Phyletic studies of teleostean fishes, with a provisional classification of living forms. Bull. Amer. Mus. Nat. Hist. 131(4):339-455.

Gudger, E. W. 1930. The candirú, the only vertebrate parasite of man. xvii + 120 pp. Paul B. Hoeber, Inc. New York. [Reprinted with additions and corrections, from Amer. J. Surgery 8 (1 and 2).]

Knöppel, H. 1970. Food of central Amazonian fishes, contribution to the nutrient ecology of Amazonian rain-forest-streams. Amazoniana (Kiel) 2(3):257-352.

Lewis, W. M., Jr. 1970. Morphological adaptations of cyprinodontoids for inhabiting oxygen deficient waters. Copeia 1970(2):319-326.

Liem, K. F. 1970. Comparative functional anatomy of the Nandidae (Pisces: Teleostei). Fieldiana (Zool.) 56:1-166.

Lissmann, H. W. 1958. On the function and evolution of electric organs in fish. J. Exp. Biol. 35(1):156-191.

Marlier, G. 1967. Hydrobiology in the Amazon region. Atas do Simpósio sôbre a Biota Amazônica 3 (Limnologia):1-7.

Matthes, H. 1961. Feeding habit of some Central African freshwater fishes. Nature 192(4797):78-80.

————. 1963. A comparative study of the feeding mechanisms of some African Cyprinidae (Pisces, Cypriniformes). Bijdr. Dierk., 33:3-35, 12 pls.

————. 1964. Les poissons du lac Tumba et de la région d'Ikela. Ann. Mus. Roy. Afrique Centrale (Zool) 126:1-204, 2 maps, 6 pls.

McConnell, R. H. 1964. The fishes of the Rupununi savanna district of British Guiana, South America, Part 1: Ecological groupings of fish species and effects of the seasonal cycle on the fish. J. Linn. Soc. (Zool.) 45(304):103-144.

————. 1969. Speciation in tropical freshwater fishes. Biol. J. Linn. Soc. 1:51-75.

Moynihan, M. 1968. Social mimicry: Character convergence versus character displacement. Evolution, 22(2):315-331.

Myers, G. S. 1927. Descriptions of new South American fresh-water fishes collected by Dr. Carl Ternetz. Bull. Mus. Comp. Zool. 68(3):105-135.

————. 1944. Two extraordinary new blind nematognath fishes from the Rio Negro, representing a new subfamily of Pygidiidae, with a rearrangement of the genera of the family, and illustration of some previously described genera and species from Venezuela and Brazil. Proc. California Acad. Sci. 23(40):591-602, pls. 52-56.

————. 1949. Salt-tolerance of freshwater fish groups in relation to zoogeographical problems. Bijdr. Dierk. (Leiden) 28:315-322.

————. 1952. Sharks and sawfishes in the Amazon. Copeia 1952(4):268-269.

————. 1960a. The genera and ecological geography of the South American banjo catfishes, family Aspredinidae. Stanford Ichth. Bull. 7(4):132-139.

————. 1960b. The South American characid genera *Exodon, Gnathoplax*, and *Reoboexodon*, with notes on the ecology and taxonomy of characid fishes. Stanford Ichth. Bull. 7(4):206-211.

Myers, G. S., and T. R. Roberts. 1967. Note on the dentition of *Creagrudite maxillaris*, a characid fish from the Upper Orinoco-Upper Rio Negro system. Stanford Ichth. Bull. 8(4):248-249.

Oltman, R. E. 1968. Reconnaissance investigations of the discharge and water quality of the Amazon River. U. S. Geol. Survey Circ. 552:iii + 1-16.

Pfeiffer, W. 1963. Vergleichendende Untersuchungen über die Schreckreaktion und den Schreckstoff der Ostariophysen. Z. Vergl. Physiol. 47:111-147.

————. 1967. Schreckreaktion und Schreckstoffzellen bei Kneriidae und Phractolaemidae (Isospondyli). (Pisces). Naturwissenschaften 54(7):1-177.

Poll, M. 1957. Les genres des poissons d'eau douce de l'Afrique. 191 pp. La Direction de l'Agriculture, des Forêts et de l'Élevage. Brussels.

————. 1959. Recherches sur la faune ichthyologique de la région du Stanley Pool (zool.) 71:75-174, pls. 12-25, 1 map.

————. 1966. Genre et espèce nouveaux de Bagridae du fleuve Congo en région de Leopoldville. Rev. Zool. Bot. Afr., 74(3 and 4):425-428.

————. 1969. Contribution à la connaissance des *Parakneria*. Rev. Zool. Bot. Afr. 80(3-4):359-368.

Preston, F. W. 1962. The canonical distribution of commonness and rarity. Ecology 43 (pt. 1):185-215; 43(2):410-432.

Roberts, T. R. 1967. Tooth formation and replacement in characoid fishes. Stanford Ichth. Bull. 8(4):231-247.

————. 1968. *Rheoglanis dendrophorus* and *Zaireichthys zonatus*, bagrid catfishes from the lower rapids of the Congo River. Ichthyologica, the Aquarium Journal 39 (3-4):119-131.

————. 1969. Osteology and relationships of characoid fishes, particularly the genera *Hepsetus, Salminus, Hoplias, Ctenolucius*, and *Acestrorhynchus*. Proc. California Acad. Sci., 36(15):391-500.

———. 1970. Scale-eating American characoid fishes with special reference to *Probolodus heterostomus*. Proc. California Acad. Sci. 38(20):383-390.

Sioli, H. 1967. Studies in Amazonian waters. Atas do Simpósio sôbre a Biota Amazônica 3 (Limnologia):9-50.

———. 1968. Principal biotopes of primary production in the waters of Amazonia. *In* R. Misra and B. Gopal (Editors). Proc. Symp. Recent Adv. Trop. Ecol. 1968:591-600.

Weitzman, S. H. 1960. The systematic position of Piton's presumed characid fishes from the Eocene of central France. Stanford Ichth. Bull. 7(4):114-123.

Paleoclimates, Relief, and Species Multiplication in Equatorial Forests

P. E. VANZOLINI

Abundant evidence from many fields of research has recently begun to show that the Quaternary in both South America and Africa was a time of drastic climatic change. An alternation between drier and wetter episodes was reflected in cycles of spreading and retreat of the large forests and of the complementary open formations (for reviews of the literature, see Moreau 1966, Vanzolini 1970, Vuilleumier 1971). The importance of these cycles for species multiplication has been made evident for South American forest birds (Haffer 1969) and lizards (Vanzolini and Williams 1970).

Work on birds and lizards, although dealing with different materials and using completely different approaches, has arrived at the same speciation model. The basic pattern is a two-phase cycle, with one wetter and one drier episode. During the drier phase, open formations made impressive inroads in the Amazonian forest, which was reduced to isolated patches in areas where conditions remained relatively more favorable. From the viewpoint of the forest fauna, these areas functioned as refuges in which populations were isolated and consequently became differentiated.

There is no direct evidence as to the exact nature and distribution of the open plant formations involved, but the geomorphological evidence (laterites, stone lines, gravel beds, etc.; Vanzolini 1970) indicates that they were either xerophytic "caatingas" or very degraded "cerrados," or a patchy distribution of the two types (for descriptions and illustrations, see Hueck 1966). In cis-Andean tropical South America, they represent the most extreme contrasts to the hylea and thus constitute extremely effective barriers to movement of forest animals.

During the following wetter episode, the forest patches grew again until they coalesced to cover an area in the neighborhood of two million square kilometers. The formerly isolated animal populations followed the spread of the forest and when they met in the reforested areas, their interrelationships reflected the degree of genetic divergence reached during the period of isolation; some had reached full species status, some fit the usual concept of subspecies, and some showed a complex mosaic pattern of differentiation. This sequence was repeated at least three times during the last 100,000 years (or perhaps less), the two latest episodes dating approximately 11,000 (Damuth and Fairbridge 1970) and 2600 (Vanzolini and Ab'-Saber 1968) years before present.

It is a very important point of agreement between the analyses of Haffer and of Vanzolini and Williams that the areas of refuge indicated by the patterns of bird and lizard differentiation are strikingly similar and are supported by considerable geomorphological evidence. It is essential to our argument that the refuges are peripheral to the Amazon Basin and that they are determined by topographical features (Figure 1). With one or two possible exceptions (the easternmost refuge, east of the Rio Tocantins, and the southernmost one, in

P. E. VANZOLINI, Museu de Zoologia da Universidade de São Paulo, São Paulo, Brazil. ACKNOWLEDGMENTS: Aziz N. Ab'Saber and Ernest E. Williams have for years closely participated in the research that underlies this summary. I certainly owe many of my ideas to them, but it is now impossible to identify which ones. My work in Amazonia has been partly financed by the Fundação de Amparo à Pesquisa do Estado de São Paulo.

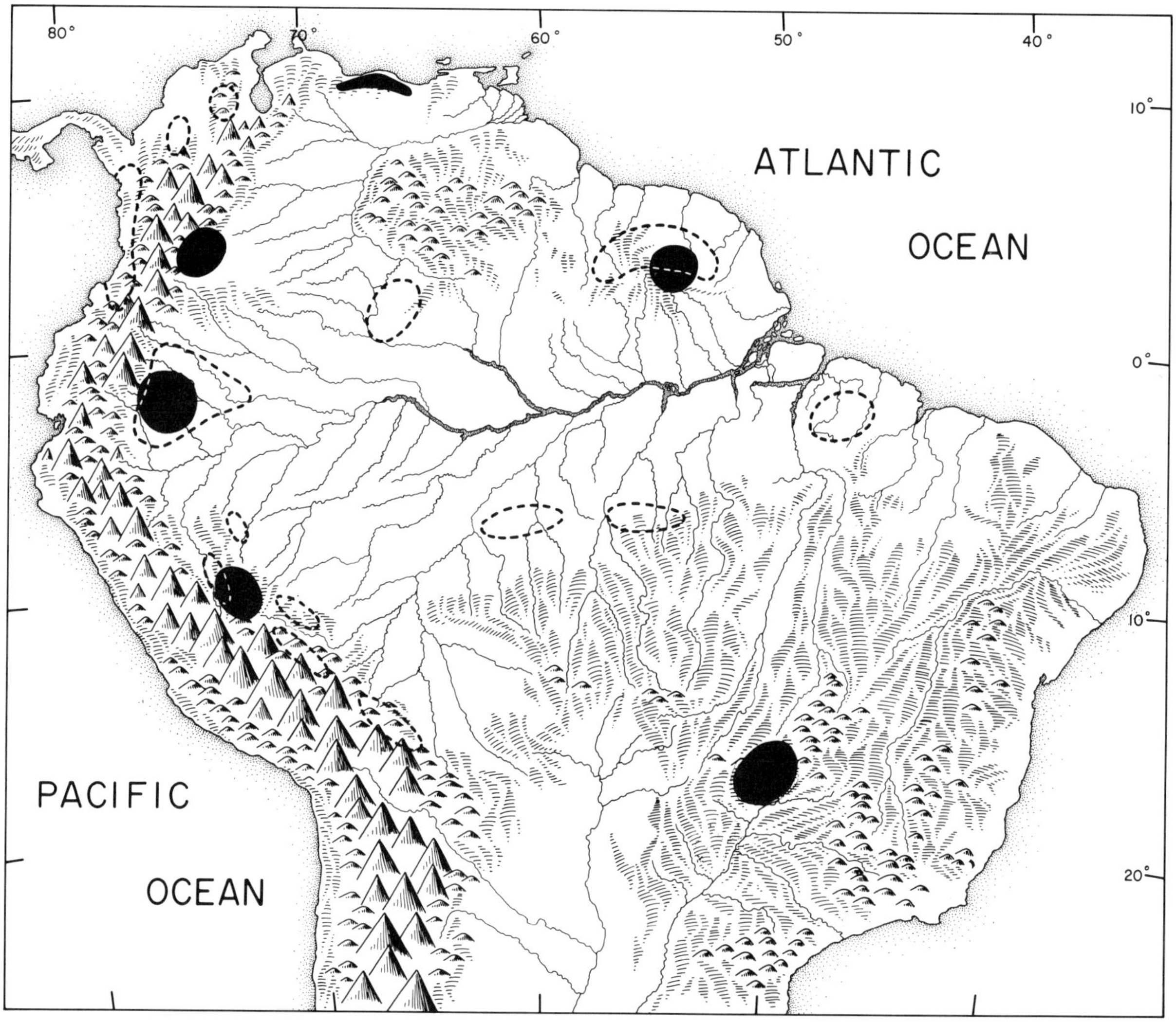

FIGURE 1. Location of refuge areas postulated by Vanzolini (solid black) and Haffer (dashed outline) during periods of shrinkage of the lowland forest in South America (areas after Vuilleumier 1971, fig. 4).

central Brazil), they are related to the relief: the Andes, the Tumuc-Humac range, the Cordillera de la Costa in Venezuela, the Brazilian highlands, etc. That they are not obligatorily montane is indicated by the varying relationships between similar present-day refuge areas in the dry caatingas of northeastern Brazil and the relief. They may be at the foot of a range, on its summit, or at intermediate heights (Ab'Saber 1968), since they are determined by a convergence of pedological, hydrological, and dynamic climatic factors.

The peripheral location of the Amazonian refuges and the rapidity of the climatic changes seem to explain the richness of the Amazonian forest fauna. Each new coalescence of expanding refuges would produce in the middle of the valley a complex mixing of the many populations that had differentiated during isolation. Such mixed populations should be expected to exhibit a high degree of variability and this is clearly shown by the data of Vanzolini and Williams (1970). The cycles being rapid, this variability would be available as raw material during the next cycle of isolation.

Although some of the relief associated with the establishment of refuges is geologically young (e.g., the Andes), it is old in terms of the time scale concerning us here. Thus, each retreat of the forest would be likely to involve the same areas, and this repetition would lead to a sort of resonance effect, reflecting the fact that each refuge would be inhabited by an already differentiated population.

The existence of multiple peripheral centers of differentiation and of rapid rates of evolution provides a mechanism to explain the great diversity of the Amazonian forest biota. That the same mechanism of species multiplication obtains in Africa would not be surprising. In fact Moreau (1966:12) suggests exactly this, although with reservations: "In birds speciation can be achieved not only in islands surrounded by water, but also in ecological islands, within no more than a fraction of the Pleistocene. Acceptance of this as a possibility is, of course, by no means the same as insisting that this has been the rule, but it is clearly a factor to be borne in mind." I believe that Haffer's work has placed this mechanism well beyond the range of a mere possibility. In fact, in terms of evolutionary theory, this is a rather orthodox position and well within the canons of the geographical speciation concept of the "synthetic theory" (Mayr 1942). That the phenomenon has remained so long unacknowledged is certainly due to the diehard dogma of the stability of the equatorial rain forest, which has stoutly resisted the accumulation of many kinds of contrary evidence (e.g., Schwabe 1969).

If we assume that the same mechanisms operated in the African lowland forest and in the Amazonian hylea, why is it that the latter is biologically more diversified than the former both in plants and in animals (Poynton 1967)? It cannot be a matter of area, since the Congo forest is only about 20 percent smaller than the hylea. Neither climate nor gross ecology affords any promising clues (Richards 1952). Moreau (1966:145) was inclined to ascribe the discrepancy to a supposed stabler climatic history in South America, but this we now know to be improbable.

I wish to offer for consideration the hypothesis that topographical features may have bearing on the problem. Judging from maps and from Moreau's extremely lucid account, I do not see around the Congo forest a ring of potential topographic refuges such as exists in South America. Moreau has been more concerned with historical changes in temperature than in humidity, and with the contrast between montane and lowland forest, but at one place he briefly (and with due caution) considers the effects of a dry episode on the Congo forest fauna (Moreau 1966:51). He postulates four potential refuge areas: (1) gallery forests, (2) areas on the Atlantic coast towards Gabon and the Gulf of Guinea, (3) areas on the highlands of the eastern edge of the basin, and (4) areas on the northern edge, provided the rainbelt was simply displaced northwards and not ruptured.

I consider it very doubtful that gallery forests, narrow and depleted as they are, can function as evolutionary core areas. I also think it doubtful that the rainbelt was simply displaced and not ruptured during dry episodes. I am inclined to believe that only two diametrically opposite refuge areas existed: the Atlantic coast and the eastern highlands. If so, climatic history in Africa would more closely resemble that in Eurasia at the time of the last retreat of the glaciers (Mayr 1942); it would represent the end of a bipolar longitudinal episode rather than a centrifugal-centripetal interplay of the kind that operated in the Amazonian hylea. Work on geographical differentiation in the area of the West African forest (Moreau 1969) has revealed clear evidence of one or more bipolar cycles. It is to be hoped that suitably oriented studies in the future on the Congo forest will provide a conclusive test of my hypothesis.

References

Ab'Saber, Aziz N. 1968. Províncias geológicas e domínios morfoclimáticos no Brasil. Centro Paulista de Estudos Geológicos, Geologia 3:85-123.

Damuth, J. E., and R. W. Fairbridge. 1970. Equatorial Atlantic deep-sea sand and ice-age aridity in tropical South America. Bull. Geol. Soc. Amer. 81:189-206.

Haffer, J. 1969. Speciation in Amazonian forest birds. Science 165:131-137.

Hueck, K. 1966. Die Wälder Südamerikas. Ökologie, Zusammensetzung und Wirtschaftliche Bedeutung. Gustav Fischer Verlag, Stuttgart.

Mayr, Ernst. 1942. Systematics and the origin of species. Columbia University Press. New York.

Moreau, R. E. 1966. The bird faunas of Africa and its islands. Academic Press. New York and London.

———. 1969. Climatic changes and the distribution of forest vertebrates in West Africa. J. Zool. (London) 158: 39-61.

Poynton, J. C. 1967. Santa Rosalia in Africa or why are there so many African birds? South Afr. J. Sci. 63:471-479.

Richards, Paul W. 1952. The tropical rain forest. Cambridge University Press. Cambridge.

Schwabe, G. H. 1969. Towards an ecological characterization of the South American continent. *In* Fittkau, E. J., and Others, Eds., Biogeography and Ecology in South America, Vol. 1. Monographiae Biologicae, vol. 18. W. Junk N.V. The Hague.

Vanzolini, P. E. 1970. Zoologia sistemática, geografia e a origem das espécies. Univ. S. Paulo, Inst. Geogr., Ser. Monografias e Teses 3.

Vanzolini, P. E., and A. N. Ab'Saber. 1968. Divergence rates in South American lizards of the genus *Liolaemus* (Sauria, Iguanidae). Papeis Avulsos Zool. S. Paulo 21:205-208.

Vanzolini, P. E., and E. E. Williams. 1970. South American anoles: the geographic differentiation and evolution of the *Anolis chrysolepis* species group (Sauria, Iguanidae). Arq. Zool. S. Paulo 19:1-240.

Vuilleumier, B. S. 1971. Pleistocene changes in the fauna and flora of South America. Science 173:771-780.

A Parallel Survey of Equatorial Amphibians and Reptiles in Africa and South America

R. F. LAURENT

After having been disdained for several decades, the theory of continental drift can no longer be denied and zoogeographers must keep it in mind when explaining the present-day distribution of animals. Continental drift seems to have surprisingly little bearing, however, on the relations between African and South American amphibians and reptiles. The bulk of the amphibian neotropical fauna consists of ceratophyrids, leptodactylids, and hylids, which are conspicuously lacking in Africa. The only exception is the South African *Heleophryne*, and although it is generally rated as a leptodactylid, it seems more likely to be a relict of an old stock of subcosmopolitan amphibians from which the more recent leptodactylids have not diverged widely. Conversely, the majority of the African frogs belong to the family Ranidae, which did not enter South America until the very end of the Cenozoic, and the Hyperoliidae, which is absent in the New World.

There are three families common to both continents: the Pipidae, the Microhylidae, and the Bufonidae. Although the first superfically suggests an African-Brazilian relationship, it is an archaic group that had a much wider range in the past. The Microhylidae is also a primitive family and its relict distributional pattern does not require postulation of a Gondwana origin. The toads are numerous all over the world, except in Madagascar and the Australian region. There is some evidence that they appeared in South America, but if so it is difficult to understand why they were not trapped there with the rest of the neotropical Cenozoic fauna. Whatever their origin, African toads show no special relationships to those of South America, with the possible but unexplained exception of *B. superciliaris*. To sum up, it appears that the similarities between African and neotropical amphibians result more from convergence than from a common ancestorship, although the latter is a factor among a few ancient and waning groups.

The tropical forest, which is our primary topic, is extensive on both continents. From the point of view of amphibian ecology, according to my Congolese experience, it can be divided into three main types of environment. One is an ecological extension of the savanna within the forest. There is often a narrow band along the main streams that is strongly illuminated and inhabited by savanna species or forms closely related to them. The same conditions are also common along roads, around villages, in cultivated areas, and in natural clearings. Since specimens are more easily collected in these habitats, these species have generally become known first, and have been considered as characteristic forest dwellers. Actually, they are only geographically so; ecologically they are savanna forms. For example, the common African toad, *Bufo regularis,* which has been considered a forest as well as a savanna form, lives only in savanna-like habitats. It is replaced by *B. maculatus* (a sibling species) in the secondary forest (and farmbush according to Schiøtz's expression) and *B. camerunensis* in the high primary forest. The farmbush association excellently described by Schiøtz (1967) in his book on West African tree frogs is also widespread in the Congo and its fauna is almost as well known as that of the clearings and human dwellings.

R. F. Laurent, Instituto Miguel Lillo, Tucuman, Argentina.

The high forest fauna is not so readily accessible, and therefore contains a much larger proportion of recently discovered, rare, or still unknown species. Schiøtz subdivided the high forest fauna into two groups: one associated with stagnant water, the other with running water. The canopy fauna constitutes a third category. It is not easy to explore, but I had an opportunity to make limited observations in the central Congo, during clearing of forest for a rubber plantation. Among the species collected were tree vipers (*Atheris squamigera*), an undescribed species of gecko (*Lygodactylus*), and females of two species of tree frogs (*Hyperolius phantasticus* and *H. brachiofasciatus*).

Let us now compare one by one the different groups of amphibians and reptiles in Africa and South America.

Amphibians

Coecilians:. The coecilians are far less common in Africa than in South America. Although well known in the Cameroon-Gabon area, they were considered absent from the Congo until one western species was collected in the Kivu. I tried unsuccessfully for several years to find another specimen, but two species have been collected in numbers by one of my assistants in the Mayombe (lower Congo) and we also know that one of these is relatively abundant near Kinshasa, which is on the opposite side of the Congo River.

Aquatic anurans. Purely aquatic anurans are present on both continents. The most exclusively aquatic genera belong to the same family, the Pipidae (*Pipa* and *Hemipipa* in South America, *Xenopus* and *Hymenochirus* in Africa), and are relicts of a formerly widespread group. They live in standing waters, as also do the South American Pseudidae, which have no relatives in Africa.

Semiaquatic forms occur in running water; the tadpoles are generally powerful and able to swim against the current and to adhere to stones by means of a buccal adaptation in the form of a sucker. They are confined to ancient hilly regions between Mount Cameroon and the lower Congo and to southeastern Brazil. In both continents, several groups are represented: the genera *Petropedetes* and *Conraua* (Ranidae) and the subfamily Astylosterninae (Hyperoliidae) in Africa, and the genus *Cyclorhamphus* and subfamily Elosiinae, which belong to the large family Leptodactylidae, in Brazil. Convergence has obviously occurred. The African genus *Conraua* is notable because it includes the biggest frog in the world, *Conraua goliath*, which also appears to be the only species that inhabits large rivers.

Terrestrial groups. The terrestrial fauna in both forests includes a considerable array of toads and frogs of different sizes. True toads of the genus *Bufo* are present both in Africa and in South America, but belong to different groups. The neotropical toads have cranial ridges and 22 chromosomes, while the African toads have no ridges and only 20 chromosomes, as far as is known. A possible exception is *Bufo superciliaris*, which may be closely related to the Ecuadorian *B. blombergi*, although the zoogeographic gap seems to be unbridgeable. On both continents, there are dwarfs (*B. gracilipes* and *B. dorbignyi*), middle-sized species (*B. maculatus* and *B. crucifer*), and giants (*B. superciliaris* and *B. marinus*); on both there are species that live in savanna-like forest clearings (*B. regularis* and *B. marinus*), in farmbush areas (*B. maculatus*), and in high forest (*B. superciliaris*, *B. tuberosus*, *B. camerunensis* in Africa, *B. typhonius*, and *B. guttatus* in America).

Small humicolous forms related to the toads are numerous in South America and generally brilliantly colored: *Atelopus*, *Brachycephalus*, *Dendrophryniscus*, *Melanophryniscus*, *Oreophrynella*, *Noblella*. In Africa, where they are rare and apparently restricted to the rich Cameroon area (genera *Didynamipus*, *Wolterstorffina*), their niche is partly occupied by the genus *Schoutedenella*, related to *Arthroleptis* and accordingly classified as a ranid by some authors and as a terrestrial hyperoliid by others. The subfamily Arthroleptinae includes three genera of terrestrial forest forms: *Cardioglossa*, which still has a tadpole stage; *Arthroleptis* and *Schoutedenella*, which have a direct development. While *Cardioglossa* may be an ecological equivalent of Dendrobatinae, the two other genera strikingly resemble the big neotropical genus *Eleutherodactylus*, not only in their terrestrial habitat, but also in their life history and general appearance. This is one of the most impressive cases of convergent evolution between Africa and South America.

Other convergences are equally interesting. In

Africa, a number of ranids are more or less terrestrial and more or less aquatic, at least in the breeding season. All of the larger frogs have in the past been classified in the genus *Rana* (which originally also included the *Salientia*) and still are by some lumpers. *Rana* (sensu stricto) is present in the lower Congo and the eastern part of the Congo forest. *Ptychadena* includes a number of species. *Hylarana,* with only three species, is partly arboreal. The genera *Dicroglossus* and *Aubria* are definitely more aquatic than terrestrial. In South America, comparable diversity exists in the genus *Leptodactylus.* The large species of the Pachypus group equate with *Dicroglossus occipitalis, Rana angolensis chapini,* and *R. desaegeri.* The middle-sized species of the Cavicola group correspond to the genus *Ptychadena* and the similarity between some species, particularly *Leptodactylus mystaceus* and *Ptychadena christyi* of the eastern Congo, is extraordinary. There are, however, two important differences: (1) the true frogs have webs, while *Leptodactylus* species do not; and (2) the latter lay eggs in spume, a habit also observed in the related genera *Pleurodema* and *Physalaemus,* but never in African frogs.

The small semiterrestrial frogs of the West African rain forest belong to the genera *Phrynobatrachus, Dimorphognathus,* and *Phrynodon.* In South America, comparable small species occur in the genera *Leptodactylus, Pleurodema,* and *Physalaemus.* There are some cases of superficial likeness, for instance, between *Phrynobatrachus plicatus* and *Physalaemus albonotatus.*

Burrowing forms are not very common in the forest, probably because the behavior is not a useful adaptation. Although numerous genera of microhylids, a family of burrowers, occur in South America, many lack this feature. In Africa, microhylids are absent from the forest, but several other amphibians burrow occasionally, among them some species of *Arthroleptis* (especially *A. variabilis*) and the females of *Trichobatrachus robustus.* The genus *Hemisus,* a good burrower and seemingly related to the Arthroleptinae, has two species in the rain forest: *H. olivaceus* in the eastern Congo and an undescribed species in the lower Congo. Characteristically, their metatarsal shovel is poorly developed in comparison with that of savanna species, a feature that reflects the considerably softer nature of forest soil.

Arboreal Groups. The principal adaptation to the rain forest environment is of course the arboreal one. In both continents, tree frogs are exceedingly numerous. Most live on plants growing within swamps or around them, including grass, herbs, bushes, branches, and creepers. Some climb up the trees to moderate heights (a few meters) while a few have even invaded the canopy. The richness and diversity of this fauna are greater in South America than in Africa. The South American genus *Hyla* is represented by a very large number of species, which equate ecologically, as well as morphologically, with almost the whole spectrum of the African Hyperoliinae: genera *Leptopelis* (which is also paralleled by the big neotropical frogs of the genus *Phrynohyas*) *Phlyctimantis, Cryptothylax, Opisthothylax, Afrixalus,* and *Hyperolius.*

The African genus *Acanthixalus* is noteworthy for its spiny skin and its peculiar habitat, which consists of watery holes in branches and tree trunks. The phragmotic hylids of the subfamily Trachycephalinae have similar ecology, but they have a further specialization that is unknown in Africa: they are able to plug the cavities in which they hide with their over-ossified head. Two other American adaptations not paralleled in Africa are found among the marsupial tree frogs (e.g., *Gastrotheca*), which incubate their eggs in a dorsal pouch, and the Phyllomedusinae, which have a prehensile first toe and move very slowly. However, the breeding habits of the Phyllomedusinae are comparable to those of the African *Chiromantis,* which is related to *Rhacophorus* from tropical Asia rather than to other African tree frogs. Some small hylids, among them *Hyla nana* and *H. minuta,* are very much like *Afrixalus laevis.* The light green Centroleninae are somewhat similar to some members of the genus *Hyperolius,* for instance *H. nasutus,* and the males of several high forest species (*H. chlorosteus* and others in West Africa, *H. phantasticus* and others in the Congo).

In several groups of green tree frogs, as well as *Chiromantis,* the bones are green and *Hyperolius frontalis* has green eggs. This phenomenon has also been observed in neotropical tree frogs (Centroleninae, *Hyla punctata, Phrynohyas venulosa*) by Barrio (1965), who showed that it results from a high rate of biliverdine in the blood. While tree frogs are more diverse in South America and exhibit several peculiarities unknown in Africa, the African genus *Hyperolius* is unique among all amphibians in its dichromatism, the females being brilliantly colored while the males are generally drab, possessing either a toned-down version of the female pattern or retaining the juvenile appearance.

Reptiles

The reptiles present a picture similar to that observed among the amphibia. Most of them originated after the separation of South America from Africa, so that there is little affinity between the two reptilian faunas. The few relationships that exist generally represent holarctic groups now extinct.

Sill (1968) has reviewed the zoogeography of Crocodilia and demonstrated quite convincingly that since the modern Crocodilia (i.e., the Eosuchia) originated during the early Cretaceous and expanded in late Cretaceous, their distribution has not been affected by the old southern interconnections.

Early in the Tertiary the Alligatorinae reached South America, where they underwent a secondary radiation (*Caiman, Paleosuchus, Melanosuchus*). The Crocodilinae, with only three species in Africa, entered northern South America along with the genus *Rana* during the major mammalian invasion. From the ecological point of view, the large *Melanosuchus niger* appears to be equivalent to the highly predaceous *Crocodylus niloticus*; the genus *Paleosuchus*, living in swift waters, is more or less comparable to *C. cataphractus*, while the genus *Caiman* is paralleled by the swamp-dwelling *Osteolaemus tetraspis* of Africa.

Let us now consider the turtles. Here also, there is no indication of direct affinities between the neotropical and the Ethiopian faunas. Although the *Pleurodira* and especially the family Pelomedusidae are represented on both continents, they appear in the fossil record only in late Cretaceous both in North and South America and in Europe and became widespread in the holarctic region during the Cenozoic. Therefore, their present distribution cannot be attributed to the splitting off of the southern landmass.

Ecologically, the turtle faunas of Amazonia and tropical lowland Africa are similar. Terrestrial tortoises like *Kinixys erosa* and *K. homeana* in Africa are replaced by a pair of neotropical species belonging to the genus *Geochelone* (Williams 1960). Aquatic species are, however, vastly more numerous in the South American forest, representing about 18 species belonging to three families (Pelomedusidae: 1 genus, 7 species; Chelydidae: 6 genera, 10 species; Kinosternidae: 1 genus, 1 species). In the African rain forest, there are only three species of the genus *Pelusios* (Pelomedusidae), one of the genus *Trionyx*, and one of *Cycloderma*, belonging to the softshelled family Trionychidae, which is lacking in South America. The ecological niches occupied by these species are not well known, but the existence of almost four times more species in the Amazonian region implies that specialization there is narrower and that possibly some niches are vacant in Africa.

The same ecological zones observed for the amphibians can be distinguished among lizards and snakes, most of which are normally terrestrial and not adapted to climbing or to burrowing. Many are of course arboreal, which is natural in a forest. Proportionally more reptiles than amphibians are burrowers. Why? Perhaps because frogs or toads burrow mainly to find shelter against heat and dryness, circumstances seldom encountered in forests; reptiles, being much more resistant to such conditions, might be driven underground by a predator pressure or attracted by specialized foods such as subterranean insects like white ants (*Typhlops*). Insectivorous forms in turn are the normal diet for other burrowing snakes, including *Miodon* and its allies, *Elapsoidea* and *Atractaspis*. Finally, there are also aquatic reptiles.

Lizards and snakes evolved after the continental split. Consequently, although we have conclusive evidence of direct relationships between West Africa and the neotropical forest lizards, they have nothing to do with the continental drift. Rather, they are the result of quite a different kind of drift, the flotsam and jetsam that has often carried many animals across the sea. The colonization of South America by *Hemidactylus mabouia* and *H. brooki* has been well documented by Kluge (1969). The African genus *Lygodactylus*, recently discovered in Brazil, surely came the same way, either as a drifter or an island-hopper. It is well known that geckos are good sailors and so are the skinks, which are fairly numerous in Africa but represented by only a few species of the genus *Mabuya* in South America. Such oceanic crossings are, of course, more likely for terrestrial and arboreal forms, and nowadays especially for anthropophilous species like *Hemidactylus brooki* and *H. mabouia,* although Kluge has provided evidence that their invasion predates a possible human intervention. Even burrowing species may be introduced that way; Vanzolini cites the example of *Amphisbaena ridleyi* from the archipelago of Fernando de Noronha. Although the migration of *Hemidactylus* has been quite recent, that of the skinks and some other geckos (*Briba* and *Bogertia*) must have been much earlier and if an amphisbaenian crossing of the Atlantic actually occurred, it must date still farther back in the Mesozoic.

The terrestrial lizards of the rain forest are almost always somewhat arboricolous, making it difficult to draw a clear-cut line between the two zones as far as they are concerned. In Africa, terrestrial lizards include the agamas (although most species often climb the trees), many lizards of the family Lacertidae (genera *Algyroides* and *Ichnotropis*), the genus *Gerrhosaurus,* the Nile monitor (which is also aquatic), and quite an array of skinks, mainly of the genus *Mabuya.* In South America, the agamas are ecologically replaced by the iguanas; the Lacertidae, Gerrhosauridae and monitors correspond respectively to the small, medium-sized, and large Tejidae; the skinks are supplanted by a few related species of skinks, the Anguidae, and some small Tejidae.

Among arboreal lizards, geckos are more numerous in South America, where the whole group of the Sphaerodactylinae (about 20 species) replaces the few African species of the genus *Cnemaspis.* The padded geckos, by contrast, are represented by a few forms in Brazil but are rather numerous in the Congolese forest. The African Agamidae can be equated with a host of neotropical Iguanidae, which can also be considered as ecological vicariants of the famous African chamaeleons, although the latter are so specialized that they lack a true neotropical counterpart. Three genera of the Iguanidae, *Corythophanes, Phenacosaurus,* and *Polychrus* are also slow in their gait, but only the last inhabits the western part of the Amazonian rain forest. Unexpectedly for such well-adapted arboreal reptiles. I have often collected the pigmy forest chamaeleon (*Rhampholeon spectrum boulengeri*) on the ground, lying on its side among dead leaves, which it closely resembles in appearance. Arboreal central African lacertids include *Centromastix echinata, Bedriagaia tropidopholis,* and the strange *Holaspis guentheri,* which can almost simulate a flying saucer. Similar adaptations do not appear to occur in Tejidae, which is understandable since the Iguanidae seem to have filled the arboreal environment much more completely than have the African agamas or chamaeleons.

Some terrestrial lizards have developed burrowing adaptations in reduction or loss of limbs, lengthening of the body, and reduction or loss of vision. This trend is polyphyletic. In Africa, it is exemplified by several families, but only the skinks have produced burrowers in the forest (for example *Feylinia currori, Melanoseps occidentalis*). In South America, the Anguidae and especially the Tejidae (the so-called Microtejidae) underwent the same evolution. Amphisbaenians occur on both continents, but are scarce in the African forests and rather common in the Brazilian hylea. Finally, some lizards have developed aquatic habits, notably the forest monitor (*Varanus niloticus*) in Africa and *Dracaena guyanensis* in South America.

Although snakes appear to have originated as primitively specialized burrowing lizards, their adaptive radiation involved reinvasion of other environments. Remnants of this ancestral phase, which in theory would have been hopelessly specialized and therefore doomed to extinction, appear to exist in the form of the Typhlopidae, which are a cosmotropical group, and the Leptotyphlopidae, which live in tropical America and Africa. They are perfect ecological vicariants but do not seem to derive from a common African-Brazilian origin, although their unknown antiquity cannot completely preclude such possibility.

Other burrowing snakes are less specialized than these primitive and aberrant families, and constitute secondary and polyphyletic adaptations. In spite of a high degree of convergence, which brought about striking similarities, the neotropical and African groups do not appear to be related. Among the African genera *Prosymna, Scaphiophis, Miodon, Aparallactus, Elapsoidea, Paranaja,* and *Atractaspis,* most of the venomous forms eat other snakes or lizards, although *Aparallactus* is a centipede eater. There is also a burrowing boa (*Cala-*

baria reinhardti) without parallel in South America. Burrowing forms are common among neotropical Colubridae: *Atractus* (very numerous species), *Elapomorphus*, and *Apostolepis*. The genus *Micrurus*, a neotropical elapid, corresponds in many respects to the African Elapsoidea, but is strikingly different in possessing red, yellow, and black rings, a pattern mimicked by several harmless South American snakes (e.g., *Simophis, Pseudoboa, Erythrolamprus*), but unknown in Africa.

Among the Colubridae, Elapidae, and Viperidae or Crotalidae, terrestrial species are a majority, but others are sometimes arboreal or aquatic. In the African equatorial forest, we have about a score of terrestrial genera in the Colubridae, among them *Mehelya* (a snake-eater) and *Dasypeltis* (an egg-eater); the genus *Naja* among the Elapidae; *Causus* and *Bitis* in the Viperidae; and the species *Python sebae* among the Boidae.

The neotropical forest harbors more than twice as many colubrid terrestrial genera, among them the mussurana (*Clelia*), a snake-eater like the African *Mehelya* but unrelated to it, and numerous species of the genus *Micrurus* among the Elapidae (they are half terrestrial, half burrowing). The bushmaster, a rattlesnake, and about a dozen of *Bothrops* species parallel the African vipers.

The arboricolous snakes in the African forest include six different groups comprising nine genera of Colubridae, three of Elapidae, and the tree viper *Atheris*. The latter is paralleled in South America by some species of *Bothrops* like *B. castelnaudi* or *B. bilineatus*. Of course, many other arboreal snakes also live in the Amazonian forest including about the same number of genera of the family Colubridae (but not related to the African groups) but containing many more species, no elapid, but two species of the genus *Boa* (formerly *Corallus*).

Finally, aquatic snakes strangely enough appear to be more common in Africa than in South America. In the latter area, only the huge anaconda (without parallel in Africa) and about seven species of a single colubrid genus *Helicops* occur, while in the African rain forest there are six genera of Colubridae (more than ten species) and two species of the aquatic cobra *Boulengerina*.

Although the ecology of many snakes is still so poorly known, especially in South America, that this picture may alter in the future, it seems evident that, as we noted in amphibians, some niches are empty in both realms. For instance, the large neotropical group of arboreal snail-eating snakes (Dipsadinae) has no African counterpart, although snails are quite abundant there. Conversely, the African egg-eater *Dasypeltis* seems to have no neotropical South American equivalent.

Summary

The amphibians and reptiles in the Amazonian and African forests show few close relationships because most groups evolved after the separation of the African-American continent. The existence of relationships between some of the most archaic families does not contradict this generalization because fossils from the Northern Hemisphere attest to their formerly contiguous distribution. On the other hand, examples of ecological convergence or parallel evolution between the two faunas are numerous, although some niches appear to remain empty on both continents. For instance, there are no marsupial frogs in Africa or egg-eating snakes in South America.

The large number of species in the tropical forest has often been cited as evidence for sympatric or ecological speciation. While isolating mechanisms are surely favored by natural selection, if only because they reduce wastage of gametes, they are unlikely to have played a significant role in initiating speciation. There is growing evidence to indicate that recurrent phases of contraction and expansion of the forest provided enough opportunities for geographic isolation to explain the multiplicity of the forest species on both continents (see Vanzolini and Williams 1970). Evidence from the Ethiopian region (Schiøtz 1967) and my own unpublished data on the Congo suggest that six regions must have been isolated in the African forest in the past (Figure 1): (1) a Guinea block from Sierra Leone to the Ivory Coast; (2) a Gold Coast block from the Ivory Coast to Ghana; (3) a Nigerian block from southern Dahomey to the Cross River; (4) a Cameroon block from southeastern Nigeria to the Congo estuary (the eastern limit is unknown, but may be the Congo watershed); (5) a northern and eastern Congo block, between the right bank of the Congo and Uganda and the Kivu mountains; and (6) a southwestern Congo block between the left bank of the Congo River and the gallery forests of the left affluents of the Kasai and Sankuru rivers.

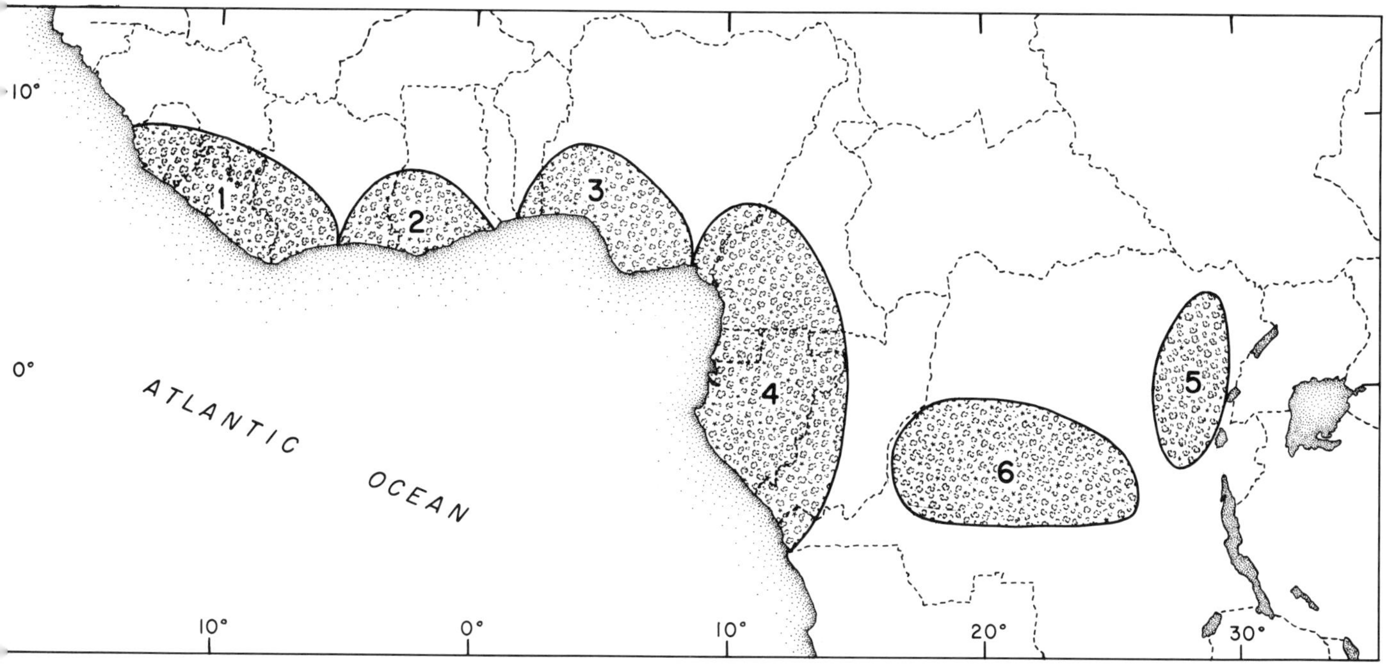

Figure 1. Postulated refuge areas during periods of contraction of the African lowland tropical forest: (1) Guinea block, (2) Gold Coast block, (3) Nigerian block, (4) Cameroon block, (5) northern and eastern Congo block, and (6) southwestern Congo block.

The first four areas have been quite convincingly described by Schiøtz (1967). The Congolese fauna is only well known in some eastern parts of the country. The northeastern endemisms include 16 species and 2 subspecies of frogs, 6 species and 11 subspecies of reptiles. The southwestern area (Area 6) is still poorly known; nevertheless it harbors many reptilian endemisms (12 species and 2 subspecies), while the amphibians include at least 8 species and 2 subspecies unknown elsewhere.

Similar conclusions have been drawn for South America by Haffer (1969) and Vanzolini and Williams (1970), who recognize four core areas: a Guiana block, a coastal Venezuelan block, an Andean block subdivided into three parts (Colombia, Peru, and Ecuador), and a Mato Grosso block, which is somewhat eccentric like my southwestern Congo refuge (Vanzolini, p. 256 herein).

References

Barrio, A. 1965. Cloricia fisiológica en batracios anuros. Physis 25:137-142.

Cei, J. M. 1968. Remarks on the geographical distribution and phyletic trends of South American toads. The Pearce-Sellards Series 13:1-21.

Haffer, J. 1969. Speciation in Amazonian forest birds. Science 165:131-137.

Kluge, A. G. 1969. The evolution and geographical origin of the New World *Hemidactylus mabouia-brooki* complex (Gekkonidae, Sauria). Misc. Publ. Mus. Zool., Univ. Michigan 138:1-78.

Laurent, R. F. 1955. A propos de l'origine des Esobe de la forêt équatoriale. Fol. Sci. Afr. Centralis, IRSAC, 1(4):1-13.

————. 1961. Étude d'une collection herpétologique du Mayombe, Première partie: Gymnophiones, Pipidae, Bufonidae et Astylosterninae. Rev. Zool. Bot. Afr. 63:262-276.

————. 1965. A contribution to the knowledge of the genus *Pelusios* (Wagler). Ann. Mus, Roy. Afr. Centr., sér. 8, Sci. Zool. 135:1-33, pls. 1-3.

————. 1971. Amphibia. *In* [Exploration du Parc National Albert (2)22:1-20, pl. 1-11.]

Lutz, B. 1952. Anfibios Anuros do Distrito Federal. Mem. Inst. Oswaldo Cruz 52:155-238.

————. 1966. Taxonomía de los anfibios anuros neotropicales. Physis 26:229-236.

Parker, H. W. 1934. A monograph of the frogs of the family Microhylidae. London.

Perret, J. L. 1966. Les amphibiens du Cameroun. Zool. Jahrb. Abt. System 93:289-460.

Peters, J. A., and B. R. Donoso Barros. 1970. Catalog of the Neotropical Squamata, Part II: Lizards and Amphisbaenians. Bull. U.S. Nat. Mus. 297:1-293.

Peters, J. A., and B. Orejas Miranda. 1970. Catalog of the

Neotropical Squamata, Part I: Snakes. Bull. U.S. Nat. Mus. 297:1-346.

Schiøtz, A. 1967. The tree frogs (Rhacophoridae) of West Africa. Spolia Zool. Mus. Haun. 25:1-346.

Sill, W. D. 1968. The zoogeography of the Crocodilia. Copeia 1968:76-88.

Vanzolini, P. E. 1967. Problems and programs in Amazonian zoology. Atas Simp. Biota Amaz. 5:85-95.

———. 1968. Geography of the South American Gekkonidae (Sauria). Arq. Zool. S. Paulo 17:85-111.

Vanzolini, P. E., and E. E. Williams. 1970. South American anoles: The geographic differentiation and evolution of the *Anolis chrysolepis* species group (Sauria, Iguanidae). Arq. Zool. S. Paulo 19:1-124.

Williams, E. E. 1960. Two species of tortoises in northern South America. Breviora 120:1-13.

Birds of the Congo and Amazon Forests: A Comparison

DEAN AMADON

Introduction

There are three main tracts in the world of what is variously known as lowland, tropical, humid, rain or evergreen forest, one in South and Central America, one in Africa; the third, more scattered, in portions of India, southeastern Asia and thence eastward over the islands to New Guinea and coastal eastern Australia. The present paper compares the avifaunas of portions of two of these areas; the Amazon forest of South America and the Congo forest of Africa. The term "Amazonia" is used for the former; there is no comparable term, apparently, for the forest of the Congo River Basin, but here, I shall use the word "Congo" in the restricted sense of applying only to that portion of the Congo drainage which is in rain forest.

The Congo River drainage as a whole contains vast areas of savanna, for example, in the southern Kasai district and along the lower Congo. On the other hand there are other tracts of forest to the north and east of the Congo Basin, in the Niger River drainage, in the Ivory Coast and elsewhere. These are rather closely related to those of the Congo botanically and faunally. The Nigerian forest is connected with that of the Congo (lower Guinea) by a narrow corridor between Mount Cameroon and the coast. Even this corridor presumably was not available during portions of the Pleistocene. The fauna, as might be expected, is somewhat different north and south of this corridor. Differences are even greater between the forests of Upper Guinea (Ivory Coast, Liberia, etc.) and those of Nigeria to the south, which are separated by a dry belt, the so-called Dahomey Gap, which extends right through to the coast. The discussion in this paper shall be confined to the Congo forest proper and exclude that of Nigeria and "Upper Guinea" (Chapin and some other authorities considered the Niger forest as part of "Upper Guinea.")

"East of the Cameroons the great Lower Guinea (Congo) forest extends east continuously to the foot of the mountains forming the eastern rim of the Congo basin, so that this block, which is extremely irregular on its southern edge, averages about 1300 miles from west to east by 500 from north to south" (Moreau 1966). This forest covers about 600,000 square miles. Miracle (p. 335 herein) refers to the entire Congo Basin as having an area of 1,425,000 square miles. Accepting these figures, a little less than half of the basin is in rain forest. For comparison, the Niger forest (now largely cut up) covered about 50,000 square miles and that of Upper Guinea 70,000, their combined extent being about one-fifth of the Congo forest.

Even within the Congo forest as here understood (which includes small areas in Gabon, Cameroons, Angola, the Sudan, and East Africa, as well as the Congo Republic proper) Chapin (1932:108, 215) distinguished a Mayombe or lower Congo segment. There the forest projects somewhat southward. The avifauna of this western portion is somewhat different, as regards subspecies and a few species, from that of the more eastern Congo forest; and to a slight extent it links the latter with those of Niger and Upper Guinea. However, the Congo forest is not subdivided in the following discussion (Figure 1).

DEAN AMADON, American Museum of Natural History, New York, New York 10024.

Turning to South America, Meggers (p. 311 herein) has given the area of Amazonia as "some six million square kilometers" (about 2,300,000 square miles). As Haffer (1969) has shown, there is wide regional variation in Amazonia in mean annual rainfall and, correlated with this, in the luxuriance of the forest and presumably in the diversity of the fauna. Moreover, large areas in the headwaters of some of the southern tributaries of the Amazon can scarcely be classified as lowland rain forest. In the north, the forests of the Amazon drainage sometimes grade insensibly into those of the Orinoco or other north coast rivers; in other areas there is a hilly hinterland between the two drainages, where the forest is less developed. In compiling the lists of birds given below for Amazonia, species listed as Guianan only were not included. But even if, for the reasons mentioned, we reduce the above estimate of 2,300,000 square miles to 1,750,000 of forest, it is still more than twice the area of the Congo forest (Figure 2).

A word is necessary as to other areas of lowland moist forest in tropical America. Tropical forest extends north along the eastern edges of the Andes to the north coast of the continent, then around and down the coasts of Colombia and Ecuador, as well as north through parts of Panama, Central America and Mexico. All these out-lying areas, though quite similar to Amazonia in flora and fauna and indeed an extension of it, are omitted from consideration here, just as are those of the Niger and Upper Guinea in Africa. The only other extensive moist lowland tropical forests in South America are those of southeastern Brazil and adjacent Paraguay and Argentina; they are less luxuriant than those of Amazonia and have a rather different fauna. No doubt there are similar variations from district to district in the Congo.

LITERATURE. Due largely to the labors of two men, the basic composition and relationships of the birds of the Congo forests are better known than those of any other major tract of rain forest in the world. The four volumes of the late J. P. Chapin *Birds of the Belgian Congo,* the last of which appeared in 1954, are the culmination of the work of a lifetime. The total work runs to more than 3055 pages, of which 391 pages in Volume 1 form a copiously illustrated analysis of the avifauna of the Congo Republic, its composition, ecology, and origin. I have also consulted an unpublished "List of Birds Known from the Belgian Congo" which Chapin compiled in 1950, but never published.

Chapin (1932:92-93) divided the Congo into eight avifaunal areas: (1) Cameroon-Mayombe forest (lower Congo forest), (2) upper Congo forest, (3) Ubangi-Uelle savanna, (4) southern Congo savanna, (5) Uganda-Unyoro savanna, (6) East Congo mountain forests, (7) Kivu highland grasslands, and (8) Rhodesian highland (Upper Katanga, Marungu). Areas 3 and 4, especially, are noted as having much gallery forest. As stated above, I am combining areas 1 and 2 for purposes of this paper. The birds themselves Chapin (1950) divided into 10 groups by habitat or ecology as follows: (1) coastal, (2) lowland rain forest, (3) clearings in forest, or young second growth, (4) gallery forest, or detached forested areas, (5) savanna or grass country, often with trees, (6) rocky hills or cliffs, (7) aquatic, mainly on shores of lakes or rivers, (8) mountain (subtropical) forest, (9) mountains, above 1525 meters but not always in forest, and (10) alpine (above 2750 meters).

To give an illustration of the complexity of such analyses, I have listed all of the areas and habitats, even though we are here concerned only with lowland forested ones.

More recently the late R. E. Moreau (1966) published his scholarly *The Bird Faunas of Africa and its Islands.* Treating as it does the entire continent of Africa, the avifauna of the Congo Basin is placed in proper perspective. This work has now been supplemented by Hall and Moreau's (1970) *Atlas of African Bird Species.*

For Amazonia less information is available. We have lists of the birds of South America and of Brazil. But unlike the Congo we do not have full analyses of how many of these species are peculiar to the Amazonian forest and how many to the various more or less drier types of habitats surrounding it. This information can, in many cases, be extracted from existing publications and specimens, but a full scale effort in that direction was out of the question. Instead, I have gone through de Schauensee's (1966) recent *Species of Birds of South America* and have decided, somewhat arbitrarily, which ones to regard as primarily Amazonian.

Several specialized studies also shed light on the Amazonian avifauna. One may mention Haffer's (1969) paper, which seeks to explain patterns within Amazonia on the basis of existing distribu-

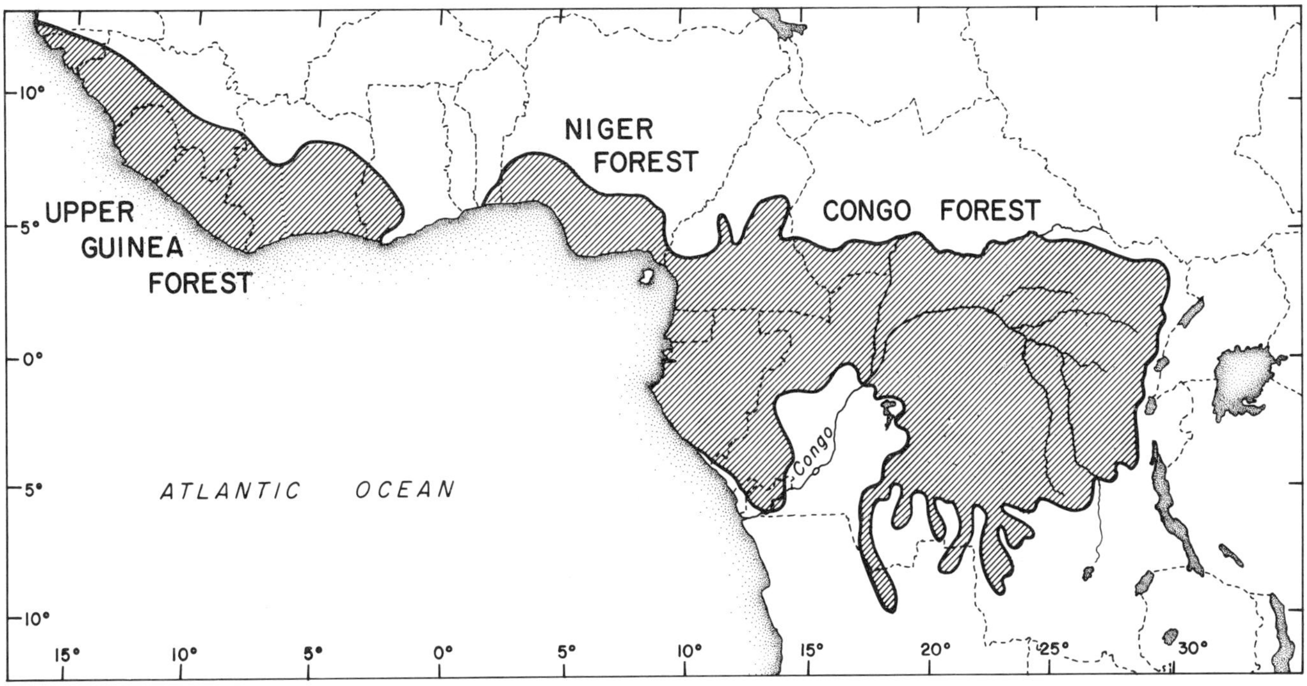

FIGURE 1. The African lowland tropical forest as defined for purposes of analysis of ornithological variation. Three primary subdivisions are recognized: (1) the Upper Guinea forest, (2) the Niger forest, and (3) the Congo forest.

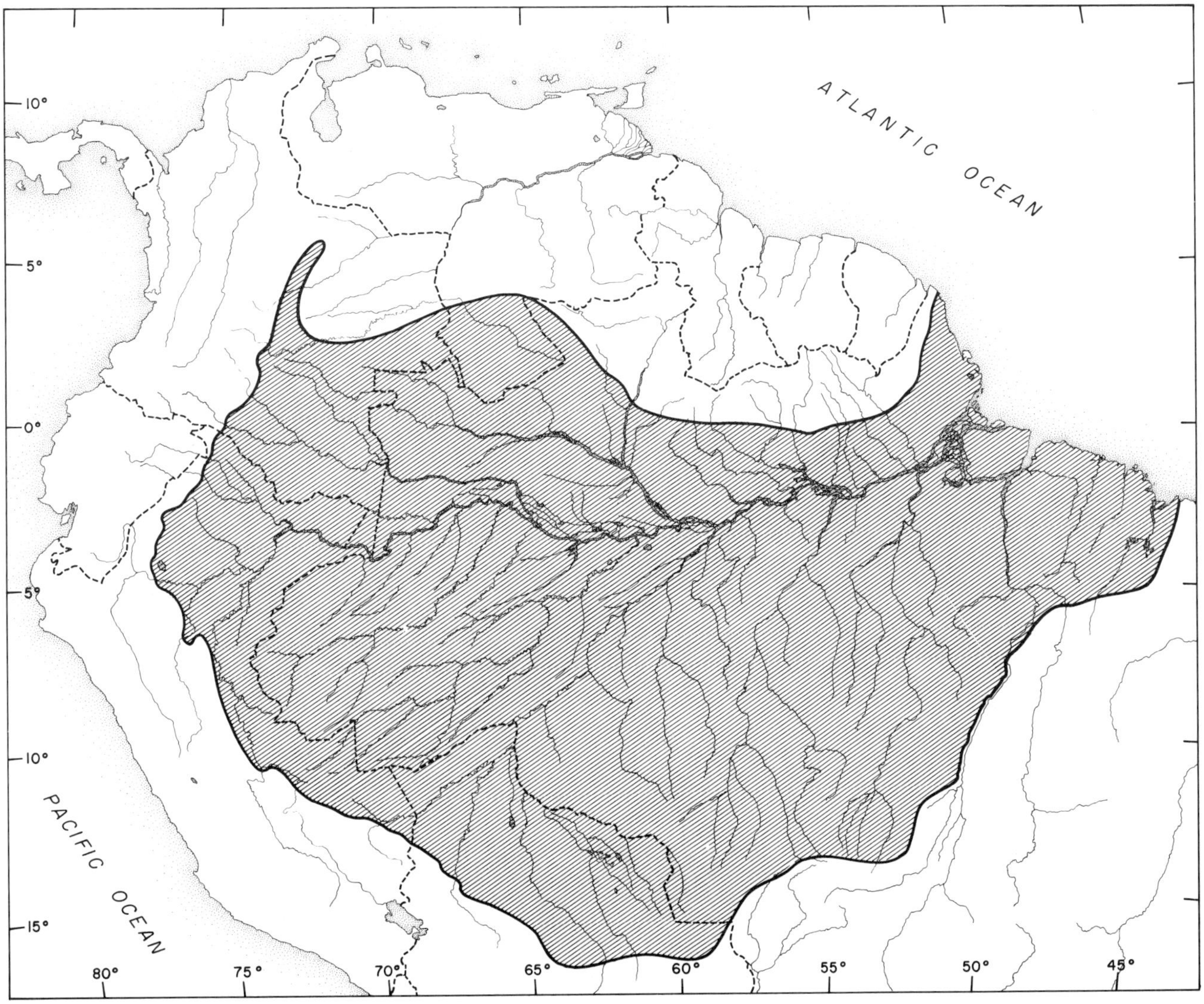

FIGURE 2. The Amazonian lowland tropical forest as defined for purposes of discussing ornithological variation.

tion of rainfall and Pleistocene climatic changes; various papers by Willis (1966, 1969), and Slud's (1960) monograph on the birds of a tract of rain forest in Costa Rica, which while outside Amazonia, is of comparative value. The late Maria Koepcke (pers. comm.) gave valuable information from Amazonian Peru, as did J. O'Neill.

What is a Forest Bird?

Chapin (1950) lists 20 species of herons (Ardeidae) from the Congo (excluding three of accidental or marginal occurrence). For 19 of the 20 he gave the habitat as "aquatic," that is, the shores of lakes or rivers or in marshes and swamps. For only one, the African tiger-heron, *Tigriornis leucolopha*, does Chapin give the habitat as "forest." Of this species, he (1932:423) wrote: "The African tiger-heron is truly a bird of the forest, seldom to be seen along river banks, even where heavily wooded, but keeping to the smaller streams entirely hidden within the depths of the forest, where the river herons evidently dare not venture . . . personally I saw just one [in years in the Congo forest] . . . which flew up along a brook . . . and in a moment was lost from view; while during all our hunting around Avakubi, neither Nekuma [a skilled African hunter] nor I ever secured a specimen."

Yet of the other 19 species of herons known from the Congo, no fewer than 13 (with 2 more doubtful) occur with greater or lesser frequency along the larger rivers or in marshes in the lowland forest. In comparing these with the tiger-heron one senses a fundamental difference. The tiger-heron is there because of the forest; where there is no forest it is absent. The others penetrate forested areas only locally and usually in small numbers *in spite of* the forest, or more explicitly where great rivers and marshes constitute clearings in the forest. Such species occur primarily in unforested areas, and in greater numbers.

Another problem exists when lowland forest birds ascend into montane forest, such as borders the Congo forest in the Kivu area in the east and the Cameroon highlands to the north, or in the case of Amazonia, to the west in the Andean foothills. As Moreau (1966) and others have pointed out this may vary from region to region even in a single species depending sometimes upon whether a related ecological analog, another species of the same genus, occurs in the montane area above to offer competition. That is to say, the failure of a lowland species to penetrate the mountains may reflect competitive exclusion, rather than an inability to live under the climatic conditions existing there. The fact, however, that a wide ranging lowland species also extends up into mountain forest should not exclude it from being considered a member of the lowland avifauna. On the other hand, a montane species that rarely or locally reaches the lowlands would not be regarded as a lowland species.

Some forest birds are restricted to large unbroken tracts of forest. Many others extend out into gallery forest along watercourses, or occur in smaller patches of forest. Anyone who has traveled in the vicinity of a rain forest has been impressed with the abruptness with which the forest often begins. One moment the traveler is in sunny grass or brush, the next in the forest. Forest does, however, creep out along watercourses where ground water supplements diminishing rainfall. As just noted, many forest birds, enter these tongues of forest. The fact that they do so does not mean that they are not to be included as components of the primary rain forest fauna. But when a species is found *only* in gallery or patchy forest and does not enter, or barely enters, the more extensive rain forest, it is to be excluded as not truly a forest species.

SPECIES OF FOREST EDGE OR CLEARING. Theoretically, rain forest species of birds (or other animals) might be regarded as only those that live in the interior of the forest, shunning edges, clearings, second growth, or the semi-openings and brush along streams. These are indeed the typical forest species and should receive special attention. But what of the birds that do in fact live in these more open situations? Consider a species, for example, that clings to the depths of the forest save in the nesting season, when the male tends to seek a singing perch at the edge of one of the slight openings caused by the fall of a giant tree. A parrot or hummingbird that feeds and lives in the sun-drenched canopy of the forest inhabits a very different world from one that walks about on the dimly lit forest floor far below. Ecologically the two are further apart than the canopy species is from a species that prefers trees in clearings (see Terborgh and Weske 1969). On the other hand, savanna or other open country birds that enter the forest only in clearings hacked by man are certainly not to be regarded as forest species.

One may suggest the same approach applied to other borderline situations: A species that is normally never found away from forest, even though it prefers slight clearings, riverine brush or the like, is a forest species; on the other hand, one that is found in open country as well is obviously in the forest clearing in spite of the forest, not because of it, and should be excluded. The case is parallel to the herons and other water birds mentioned above.

The question arises as to where birds of forest clearings that are not found elsewhere lived before man appeared on the scene. They must have evolved and lived in the scattered natural clearings resulting from windfalls, brush along streams, etc., but in smaller numbers than at present. A parallel may be mentioned from North America. The naturalist Audubon, who roamed eastern North America when it was still largely unbroken forest, saw in his lifetime very few individuals of the brush-inhabiting chestnut-sided warbler, *Dendroica pensylvanica*. Today, with most of the country cleared and the forest reduced to patches, this warbler is common.

MIGRANTS. Migrants that spend a portion of the year in the forest, or for that matter which pass through it, are to be analyzed along much the same lines suggested above for resident species. That is, if they are restricted essentially to the forest at that season, they must be regarded as components of the forest avifauna for the period of their stay (though of course listed separately from resident species). Although some North American migrants winter in Amazonia, many more stop in Central America or Panama. They are not a very important segment of the Amazonian fauna. The situation in the Congo forest appears to be similar.

TAXONOMIC TREATMENT OF SPECIES. In comparing the biotas of different areas, it is desirable to have a uniform taxonomy and level of information. So far as birds are concerned, this is not a serious problem at the species level. There is probably not a single undiscovered species of bird in the Congo lowlands, and half a dozen omissions would be an optimistic number for Amazonia. By and large, the taxonomies are well worked out, although Amazonia still has an appreciable number of dubious species or ones known only from one or two localities.

Another difference between Amazonia and the Congo that affects comparisons is partly, but only partly, a reflection of the larger area of Amazonia. This is the fact that in the Congo lowlands (when, as here, the forests of Nigeria and "Upper Guinea" are excluded) few species groups (superspecies) are represented by more than one geographically replacing (allopatric) species, whereas in Amazonia there are many examples. It would be better to count "superspecies" rather than their component species in making comparisons, but securing the relevant data for Amazonia would be a major research project in itself.

For these and other reasons, the following summary must be viewed as tentative, particularly for the Amazon. As mentioned above, for the most part I counted Amazonian species by scanning the ranges given in de Schauensee's (1966) list, wherein the ranges are perforce brief. The totals for Amazonia probably tend to be somewhat inflated as compared with those for the Congo, which are based on careful tabulations by Chapin (1950), Moreau (1966), and others.

Comparison of Avifaunas of the Congo and the Amazon Rain Forest

FAMILIES REPRESENTED IN BOTH AREAS

NON-PASSERES. Ardeidae (herons): Each area has a species of tiger-heron, probably closely related though placed in separate genera.

Threskiornithidae: One or two species of ibises have become forest dwellers on each continent.

Anatidae: Very few ducks are adapted to life along forest streams. The closely related comb ducks (*Sarkidiornis*), one in each continent, are among them (though not limited to forest); the related Muscovy (*Cairina*) of tropical America is another.

Accipitridae (kites, hawks, eagles): Moreau (1966) lists only six Congo species as primarily of lowland forest. The number in Amazonia would be much higher, about 20, with migrants adding one or two more. The choice as to whether to call certain hawks forest species is very arbitrary. The continents share two genera, *Accipiter* and *Spizaetus*; some others are related. No species is found in both areas.

Phasianidae: The Congo has two or three francolins (*Francolinus*), a true pheasant (*Afropavo*), and two guineafowl (Numidinae), all members of subfamilies not represented in Amazonia. In the latter, two or three species of American wood quail are the only members of the family.

Rallidae: Rails are aquatic marshland birds, but in various parts of the tropics there are terrestrial forest species. One or two in each area fall into this category.

Columbidae: The doves and pigeons are richly represented both in rain forest and drier country. Perhaps five species may be regarded as primarily of African lowland forest. In Amazonia the figure may be seven or eight. One genus, *Columba*, is shared, though there is some doubt as to whether the neotropical species are properly referred to it.

Psittacidae: Africa is poor in parrots with only two species in the lowland forest; South America is very rich with perhaps as many as 30 species to be considered as Amazonian.

Cuculidae: Chapin (1950) calls 10 of about 22 (excluding migrants) species of cuckoos occurring in the Congo forest-dwellers. The figures in South America are about the same. The two groups are not very closely related, but some are similar in habits.

Strigidae (owls): Another family well represented in both areas. About nine lowland forest species in the Congo; perhaps somewhat fewer (6) in Amazonia.

Caprimulgidae (nightjars): Chapin (1950) evidently regarded only two of the Congo nightjars as primarily of forest. The Amazon figure is greater, perhaps eight. The genus *Caprimulgus* is shared.

Apodidae: Swifts do all of their feeding above the forest, though descending to roost and nest in the trees. Some, however, are restricted to forest areas: perhaps three in the Congo, six in Amazonia.

Trogonidae: The trogons are an oft-cited example of a pantropical family. More diversified in America, there are about six species in Amazonia, only one in the Congo.

Alcedinidae: The opposite is true of the kingfishers. True forest types seem absent in Amazonia with one possible exception. Four in the Congo might be so classified.

Capitonidae (barbets): Congo 12, Amazon 4.

Picidae (woodpeckers): South America (and tropical Asia) are rich in woodpeckers, Africa poor. There are seven species in the Congo, about 20 in Amazonia.

PASSERES: PERCHING BIRDS. This vast order contains about three-fifths of the approximately 9000 species of birds. The order itself consists of two great divisions: the Mesomyodi (Clamatores) or "clamoring" birds, and the Oscines or true songbirds. It is in the first of these groups that one of the greatest differences between the neotropical region and all others, including the Ethiopian, is to be found. In the Congo there are only three species of mesomyodians, in Amazonia about 275!

Turdidae: The thrushes are one of the few families of perching birds fairly well represented in both river systems. The six Amazonian species of thrushes all belong to the cosmopolitan genus *Turdus*. The 11 Congo species belong to a closely related genus, *Geokichla*, whereas the genus *Turdus* itself occurs only in the mountains of the eastern Congo.

Sylviidae: There is some question as to whether the few sylviid warblers in South America actually belong to this family. At any rate there are two in Amazonia, about 11 in the Congo.

FAMILIES REPRESENTED ONLY IN THE CONGO (19)

Families restricted to the Ethiopian region (2) are marked with an asterisk.

NON-PASSERES. *Musophagidae (turacos) 4 species; Meropidae (bee-eaters) 2 species; *Phoeniculidae (tree-hoopoes) 1 species; Bucerotidae (hornbills) 7 species; Indicatoridae (honey-guides) 4 species.

SUB-OSCINE PASSERES. Eurylaimidae (broadbills) 2 species; Pittidae (pittas) 1 species.

OSCINE PASSERES, SONGBIRDS. Campephagidae (cuckoo-shrikes) 2 species; Pycnonotidae (bulbuls) 20 species; Timaliidae (babblers) 8 species; Muscicapidae (Old World flycatchers) 20 species; Dicruridae (drongos) 2 species; Laniidae (shrikes) 6 species; Prionopidae (wood-shrikes) 1 species; Paridae (titmice) 2 species; Oriolidae (orioles) 2 species; Sturnidae (starlings) 4 species; Nectariniidae (sunbirds) 10 species; Ploceidae (weaver-finches) 20 species (13 Ploceinae, 7 Estrildinae)

FAMILIES REPRESENTED ONLY IN AMAZONIA (31)

Those limited to the neotropical region (17) are marked with an asterisk.

NON-PASSERES. *Tinamidae (tinamous) 13 species; Cathartidae (New World vultures) 2 species; Falconidae (falcons) 5 species; *Cracidae (curassows) 9 species; *Psophiidae (trumpeters) 3 species; *Eurypygidae (sun-bitterns) 1 species; *Nyctibiidae (potoos) 2 species; Trochilidae (hummingbirds) 22 species; *Momotidae (motmots) 2 species; *Galbulidae (jacamars) 11 species; *Bucconidae (puffbirds) 17 species; *Ramphastidae (toucans) 16 species.

SUB-OSCINE PASSERES. *Dendrocolaptidae (woodcreepers) 23 species; *Furnariidae (ovenbirds) 32 species; *Formicariidae (antbirds) 94 species; *Conopophagidae (ant-pipits) 5 species; *Rhinocryptidae (tapaculos) 1 species; *Cotingidae (cotingas) 34 species; *Pipridae (manakins) 23 species; Tyrannidae (New World flycatchers) 60 species; Oxyruncidae (sharpbills) 1 species.

OSCINE PASSERES. Corvidae (jays) 2 species; Troglodytidae (wrens) 8 species; Mimidae (mockingbirds) 1 species; Vireonidae (vireos) 8 species; Icteridae (blackbirds) 9 species; Parulidae (wood warblers) 2 species; Coerebidae (honey-creepers) 9 species; *Tersinidae (swallow-tanagers) 1 species; Thraupidae (tanagers) 35 species; Fringillidae (finches) 12 species.

The preceding lists indicate that the avifaunas of the Congo and Amazon are tremendously different as regards taxonomic composition; also that the Amazon is more diversified. Eighteen families of birds occur in both areas, with a total of 94 species in the Congo, 134 in Amazonia. Of these 18 families, no fewer than seven deserve only slight consideration, either because they are essentially aquatic (or aerial as with the swifts) or because of taxonomic misgivings as to whether the forms on the two continents do in fact belong to the same family. These families are the Threskiornithidae, Anatidae, Rallidae, Heliornithidae, Apodidae, Alcedinidae, and Sylviidae. Of the eleven remaining families, only half a dozen have as many as five species in each area (never the same species, usually not the same genera). The thrushes are the only Passerine family in this group. Most of the others belong to older, widespread, more or less cosmopolitan families such as owls, hawks, and cuckoos.

Nineteen families (with a total of 118 species) found in the Congo are not represented in Amazonia. Only two of them, however, are endemic to Africa (Ethiopian region), the rest reaching or being shared with the Oriental and sometimes the palearctic, nearctic (Paridae), or Australian regions.

The number of families found in Amazonia but not the Congo is 31, of which 17 are limited to the neotropical region (including in some cases the Central American portion of Neotropica). These 31 families have an estimated 458 species in Amazonia.

The grand totals as estimated are: Congo—15 orders, 37 families, 212 species. Amazonia—16 orders, 49 families, 592 species.

Origins and History of the Two Avifaunas

Here there is opportunity to touch but briefly upon this subject. In passing, one may refer to Simpson's (1969) brilliant analysis of the origin of the mammals of South America; a group with a vastly greater fossil record. It is clear that the two avifaunas have been separated for a long time. They do not share any species and only a handful of genera. None of these are Passeriformes (perching birds), which because of their smaller size and more recent appearance would be less apt to get from one continent to the other.

Of the shared families, most are cosmopolitan so one need only assume that they got from the Old World to the New (or vice versa) via the long northern route. There are to be sure, three or four essentially tropical families—the barbets, parrots, trogons—shared by the Congo and Amazonia that might be thought to present a problem insofar as dispersal is concerned. This is not the case. We are concerned with events that transpired well back in the Tertiary, at a time when large portions of the Northern Hemisphere are known to have been tropical or subtropical. Fossils of tropical families, among them trogons, are known from farther north. In the second place, most "tropical" families of birds, if of appreciable size, contain some species that are quite tolerant of cold. To mention an instance or two: one parrot, the Kea (*Nestor notabilis*) nests high in the New Zealand Alps in winter; one kingfisher and one hummingbird reach Alaska; one wren crossed the Siberia-Alaska link. Thus over long spans of time, some more or less tropical families could get from hemisphere to hemisphere even under existing climatic conditions.

One group, the sub-Oscine or mesomyodian Passeres, requires special consideration. About 300 species and half a dozen families exist in Amazonia, while in the Congo there are only three species; two broadbills and a pitta. (To put this in a little more perspective, of the 9000 or so species of birds about 5000 belong in the single order Passeriformes and of the latter about 2000 species are Mesomyodi, 3000 Oscine. Of the 2000, approximately 100 are Old World; the rest are neotropical except for 15 or 20 in adjacent areas, chiefly southern nearctica).

In general the Oscine Passeres seem to be more "advanced" and successful than the sub-Oscines. Such dominant and widespread groups as the finches, sunbirds, and thrushes all belong to the former group. Oscines have a more complex syrinx, they are "songbirds." Aside from this, and in the absence of any significant fossil evidence, there is admittedly little basis for the general assumption that the sub-Oscines are a more primitive group that is being displaced by the Oscines. Yet their distribution in the Old World certainly seems to favor this view. Not only is the total number of sub-Oscines small, but most of them occur in island "refugia," e.g., Madagascar, New Zealand, the East Indies. On the continents they are rare and none inhabit temperate Asia and Europe. Following this line of deduction, some authorities (e.g., Mayr and Amadon 1951, Amadon 1957), have concluded, as stated above, that the sub-Oscines are being displaced by Oscines.

Neotropica and of course Amazonia are the great exceptions. Why? One recalls first that South America itself was an island for vast periods; and even now is connected with the rest of the world (faunistically speaking) only by the narrow, attenuated Central American corridor. Secondly, the Oscine Passeres presumably originated in the Old World tropics (just as perhaps the sub-Oscines had before them) and hence have been slow in reaching distant, isolated South America. Entire large families such as the Old World flycatchers have still not reached North, much less South America.

The assumption that the Oscines may eventually displace most of the sub-Oscines in South America too has been rather strenuously opposed by Slud (1960:137ff) on the basis that neotropical antbirds, ovenbirds, and others are active successful species. One cannot, however, rely on such evidence. The Nile crocodile or the red kangaroo, each in his chosen home, is active and successful, but there is no question that they belong to older and generally declining groups. For that matter, the sub-Oscines would occupy a higher perch on the avian phylogenetic tree than any of the non-Passerine orders.

It is possible, of course, that the South American sub-Oscines are now so well adapted that additional Oscine groups could not become established, much less displace them. This possibility is feasible in that the South American sub-Oscines have little affinity to those of the Old World. Olson (1971) has, however, found recent evidence that the older anatomists, who concluded that the Old World broadbills (Eurylaimidae) and the neotropical cotingas (Cotingidae) have a special relationship, were correct. All in all, I see no reason to abandon the classical view that the sub-Oscines are an early and in general less well-adapted group of Passeriformes that has persisted in South America longer than elsewhere because it was sheltered from the main wave of Oscine evolution.

In summary, then, the avifaunas of the Congo and Amazonia are very unlike, and this is without much doubt the result of the long, tenuous (even for flying birds) terrestrial connection between the two areas via the northern continents. The same conclusion holds if we compare the neotropical and Ethiopian avifaunas as a whole. But by including open country and freshwater birds, we do get more shared families, genera, and even an occasional species. It may be added that a handful of water birds seem to have hopped directly across the South Atlantic. Finally, one should point out that there are two or three families common to Africa and South America, members of which enter the forests under consideration in only one of these continents. There are two or three jays (Corvidae) in Amazonia, whereas none of the African corvids inhabits rain forest. The same is true of the falcons (Falconidae).

Comparison of Species Richness of the Two Forests

The estimate of 212 species of true lowland forest birds for the Congo (Lower Guinea) given above is evidently somewhat too low. Moreau (1966:161) stated that the forest bird fauna of Lower Guinea totals 266 species, that of Upper Guinea, 182, many of them of course duplications. I cannot quite reconcile this with his table 5, an analysis by habitat of the entire Ethiopian avifauna, in which he lists only 188 species for "primary lowland forest" but an additional 62 species for "secondary forest and edges." Presumably these latter prefer "edges," but are not found away from forest and so are to be included with the forest birds. Forbes-Watson (1970), if I understand him correctly, lists 41 families and 254 species as the total lowland, humid forest avifauna of Africa, including Upper and Lower Guinea. In any event we are concerned only with rough figures, and that for Amazonia is even more of a "guesstimate." So let us accept 250 species for the Congo (remembering that the family-by-family figures given earlier need to be revised upwards slightly) and 550 for Amazonia. My estimate of 212 species was based on Chapin's (1950) list, restricted to the political limits of the Congo Republic (Belgian Congo), and thus may exclude a few local species from other parts of Lower Guinea.

The total number of species in Amazonia is thus evidently somewhat more than twice that of the Congo. But Amazonia is more than twice as large and many of the sedentary forest birds are local and have been separated by the Amazon or its branches or by other barriers. Nevertheless, Amazonia is faunally very rich.

Is, however, the Congo forest significantly poorer in bird species at any particular place? Moreau (1966:161) thought that it was, and attributed this poverty to the greater stresses the Congo lowlands have undergone as recently as the early Pleistocene, when parts of the area were evidently arid. The flora is less varied than that of Amazonia, perhaps for the same reasons, and this might further inhibit proliferation by birds already decimated by being crowded into smaller refugia during these dry periods of the Pleistocene. This explanation needs reconsideration, however, in the light of growing evidence of similar climatic cycles in Amazonia (see Vanzolini, herein).

Moreau (1966:161-162) concluded that the total of 266 forest species for Lower Guinea was "remarkably unimpressive when put against the total of 269 resident species, which however includes a few belonging to clearings, recorded by Slud (1960) on no more than two square miles of forest in Costa Rica."

But although Slud's study was of a "tropical wet forest locality," the site as analyzed by him actually contained several other habitats, and this is reflected in some birds included in his list of 269 species. There are, for example, seven species of herons; it is doubtful from Slud's summary of their habitats whether even one of them is to be regarded as primarily a forest species. Actually, in Slud's (1960:116-132) ecological classification, which divides the birds among "forest," "second growth," "watercourses" and two other habitats, only 106 species appear to be primarily forest birds. Since Costa Rica is so much nearer the North Temperate Zone than either Amazonia or the Congo, it has more migrants, some of which winter in the forest. But it probably has no more than half as many resident primary lowland forest birds as Amazonia and that is more or less what one would expect on the basis of the great difference in the extent of Central and South American forests.

Detailed efforts to record the number of species of resident birds in any one area of lowland tropical forest, even by a person with much previous field experience, is a project for months or years, not days. It is not surprising therefore that few such lists exist. The late Maria Koepcke (pers. comm.) provided the following information for Amazonia. For a period exceeding two years, she and her husband attempted to list every species of bird on a tract of only one and a half square kilometers in eastern Amazonian Peru. The area is hilly and about 350 meters above sea level, not high enough to have any montane (subtropical) species, though the fauna as well as the terrain may be a little more diversified than in flat forest. There is a small clearing about the buildings, and a little stream nearby that attracts occasional nonforest birds (e.g., a migrant osprey, *Pandion*). On these 2.5 square kilometers, the Koepckes recorded 320 species of birds. Of these, however, only 210 were regarded as "primary forest species," a figure that might increase to 225 if certain rare or elusive birds are added. The number of wintering migrants in the forest is negligible, both as to species and individuals: three tyrannid flycatchers, *Contopus virens, Empidonax trailli*, and *Tyrannus tyrannus*, and one grosbeak, *Pheucticus ludovicianus*.

John O'Neill of Louisiana State University has visited another locality in eastern Peru, called Balta, every year except two between 1963-1971, each time for a period of at least three months. He has compiled the impressive total of 408 species of birds within one mile of Balta, and expects more! He wrote me:

> I think the only explanation for the high list at Balta is time. . . . This past spring (1971) I added 13 species. I suspect there are nearly 100 species [more] that will eventually be found there. Many of the Balta birds are recorded only on the basis of a single specimen.

O'Neill has generously shown me his Balta list of 408 species. The presence of open and aquatic habitat is implied by the fact that among perhaps a dozen northern migrants, one half are shorebirds. Still, without attempting further analysis, one may conclude that this one locality in Amazonia has 300 or more species of forest birds.

Turning to Africa, Forbes-Watson (1970) estimated that there may be as many as 205 species of forest birds at Mt. Nimba, Liberia. This seems remarkably high as compared with Moreau's (1966: 161) overall figures for Upper and Lower Guinea; however, it may include a few subtropical species. Moreau (1966:288) was not aware of a good count for any single locality in Lower Guinea. It seems quite safe to assume, however, that the figure might reach 200 species or slightly more. If one accepts these figures, the avian diversity of the Congo forest at any one locality is, as usually assumed, well below that of Amazonia. We are comparing two areas in which the breeding birds are resident (though O'Neill remarks that some of them wander more widely during the nonbreeding season than is realized) and in which migrants are negligible in numbers.

By contrast, Temperate Zone forests, even those with a rich varied flora, such as the southern Appalachians in eastern North America, would, I think attain 100 breeding species only by including those of open fields, clearings, and wetlands on a much more liberal basis than was done above for the tropical forest. In the tropics themselves, the bird life of the savannas or open woodland in areas such as East Africa has a considerable quota of large conspicuous species and often seems richer than that of the forest. Perhaps it is more diversified in some respects. I suspect, however, that local lists of breeding species would fall well below those of the forest if restricted to a comparable extent; that is if lakes, marshlands, grasslands, and open forest were not all included in a single "nonforest" category.

Assuming that tropical humid forest is richer in avian species diversity than is any other habitat, one can here only say that much has been written recently on the reasons for this. In general, terrestrial plantlife thrives best in a humid tropical climate; animals are dependent upon plants, so one would expect and does find the most animals there too. Further, other things being equal, the longer the time span the greater the amount of speciation and subdivision of the niches available (see also Orians 1969).

Though the tropical forest has experienced severe climatic shifts, it has probably had less disturbance than any other major terrestrial habitat. In fact most other habitats still experience violent annual changes at the present time. The Temperate Zone may be "arctic" in midwinter and "tropical" in summer; it is "temperate" only when one takes an average. Even in the tropics, the contrast between wet and dry seasons in the savanna may be enormous.

I agree with Mayr (1969) that there is no reason to assume or propose that speciation in the tropics is any more rapid or is different from speciation elsewhere. In general, however, the habitat is both richer and more stable than other terrestrial ones, so there is less turnover and extinction.

Birds and the Forest Ecosystem

If the tropical forest is the richest of terrestrial habitats, it should support the greatest number of individual birds (or at least total biomass), as well as species. Recalling the vast aggregations of birds to be found elsewhere, one is tempted to assert at once that such cannot be the case. For a reliable comparison, however, I think that we should exclude all birds securing their food from aquatic plants or animals, and not merely in the ocean but in fresh water as well. Also, perhaps, the animal life as a whole, not merely the birds, should be considered. For example, in forest in which fruit is heavily utilized by mammals (e.g., monkeys and bats), less is available for birds. I assume that the total animal biomass (including of course invertebrates) supported by humid tropical forest is greater than in other terrestrial habitats. But the role of birds in this complex may vary from region to region and perhaps is relatively minor everywhere.

Some visitors to the "jungle," such as the celebrated Bates (1895), have thought it to epitomize Darwin's struggle for existence and survival of the fittest, life rampant and pitted in a constant struggle. Others like von Steenis (1969) conclude that life is so profligate in the rain forest that all sorts of mutants, "hopeful monsters," and morphological peculiarities that would be weeded out elsewhere are able to survive and flourish. I think that Bates was nearer the truth. Ashton (1969) for example, has shown that the numerous, thinly dispersed species of trees of the rain forest cannot be explained by opportunistic or chance factors. Some of these trees are several decades old before they reach reproductive (flowering) condition. During all this long ontogeny, as well as their long later life, they species may, however implausible it at first seems, are subject to competition and deprivation. Each have its own niche, a niche that perhaps only one in ten thousand seedlings finds.

As regards birds, it is undoubtedly true that the dense vegetation of the "jungle" provides better cover from predators than is found elsewhere. Such "ornaments" as the brilliant colors and long trailing plumes of pheasants, birds of paradise, and the like, could rarely evolve in other contexts. It would be a mistake, however, to assume that such features are not adaptive and subject to natural selection. They are not adaptations for securing food, avoiding enemies, or coping with the environment; rather they are concerned with reproduction: securing a mate, repelling other males of the same species, and serving as species' recognition marks. It will, I am sure, be found, that superficially "extravagant" features of tropical flowers are equally adaptive. Indeed, one would have thought that Darwin, Lubbock, and others had long ago demonstrated the elaborate possibilities along such lines as regards orchids and by implication other plants.

Thus we confront the tropical forest: an infinitely complex asemblage of interrelated organisms. The birds, I am afraid, despite their number and variety, may be a rather superficial part of this intense turnover of organic materials. They eat a few insects, sample some fruit and nectar, but aside from a few plants dependent upon them for fertilization or dispersal, the jungle might soon adjust to their absence. But one cannot be sure, for the tropical rain forest is, paradoxically, the most vigorous and yet one of the most fragile of terrestrial habitats. Destroy the forest, as is being done at such a calamitous rate, and the soil is soon washed away by the torrential downpours. These forests, aside from their esthetic and scientific values, should be cherished as biological storehouses, the source of new plants, new drugs, new boons for mankind. Exploit them ruthlessly as at present, and the next generation will reap a whirlwind.

References

Amadon, D. 1957. Remarks on the classification of perching birds, order Passeriformes. Proc. Zool. Soc., Calcutta (Mookerjee Mem. Vol.) 1957:259-268.

Ashton, P. S. 1969. Speciation among tropical forest trees: Some deductions in the light of recent evidence. Biol. J. Linn. Soc. 1:155-196.

Bates, H. W. 1895. The naturalist on the River Amazons. 7th Edition. London.

Chapin, J. P. 1932. Birds of the Belgian Congo. Bull. Am. Mus. Nat. Hist. 65. [See also volumes 75(1939), 75A(1953), 75B(1954).]

————. 1950. List of Congo birds by habitat. American Museum of Natural History, New York. [In manuscript.]

de Schauensee, R. M. 1966. The species of birds of South America. Livingston Publishing Company. Narberth, Penn.

Forbes-Watson, A. D. 1970. The avifauna of the African lowland forest at its eastern and western extremities (Kakamega, Kenya and Mt. Nimba, Liberia). Abstracts XV Congressus International Ornithologicus. August-September 1970. The Hague.

Haffer, Jurgen. 1969. Speciation in Amazonian forest birds. Science 165:131-137.

Hall, B. P., and R. E. Moreau. 1970. An atlas of speciation in African Passerine birds. British Museum (Nat. Hist.). London.

Mayr, E. 1969. Bird speciation in the tropics. Biol. J. Linn. Soc. 1:1-17.

Mayr, E., and D. Amadon. 1951. A classification of Recent birds. Amer. Mus. Novit. 1496:1-42.

Moreau, R. E. 1966. The bird faunas of Africa and its islands. 424 pp. Academic Press. New York and London.

Olson, S. E. 1971. Taxonomic comments on the Eurylaimidae. Ibis.

Orians, G. H. 1969. The number of bird species in some tropical forests. Ecology 50:783-801.

Simpson, G. G. 1969. South American mammals: Biogeography and ecology in South America, 2:879-909.

Slud, P. 1960. The birds of Finca "La Selva," Costa Rica. Bull. Amer. Mus. Nat. Hist. 121 (art. 2).

Terborgh, J., and J. S. Weske. 1969. Colonization of secondary habitats by Peruvian birds. Ecology 50(5):765-782.

Von Steenis, C. G. G. J. 1969. Plant speciation in Malesia, with special reference to the theory of non-adaptive saltatory evolution. Biol. J. Linn. Soc. 1:97-133.

Willis, E. 1966. The role of migrant birds at swarms of army ants. The Living Bird 5:187-231.

————. 1969. On the behavior of five species of *Rhegmatorhina*, ant-following antbirds of the Amazon Basin. Wilson Bull. 81:363-395.

The Comparative Ecology of Rain Forest Mammals in Africa and Tropical America: Some Introductory Remarks

FRANCOIS BOURLIÈRE

Studies on the ecology and behavior of tropical rain forest mammals have barely begun. During the last ten years or so, a few pilot research projects have been carried out on the habitat preferences, social life, population structure, and dynamics of primates, bats, terrestrial rodents, ungulates, and pangolins. Some long-term studies on the same taxonomic groups are now under way, mainly in Africa. Practically nothing, however, is known of the ecology and behavior of such other groups as Insectivora, arboreal Rodentia and Carnivora, Xenarthra, and Hyracoïdea. It would therefore be premature, as well as unwise, to attempt to review the situation at the present time and to make extensive comparisons between the African and American tropics. All that can be done is to pin-point a few tentative conclusions, or working assumptions, from these preliminary studies, and to relate them to current knowledge of the functioning of the tropical rain forest ecosystem as a whole.

Faunal Richness and Diversity

Ornithologists have long pointed out the relative poverty of the lowland rain forest avifauna of Africa compared with that of tropical America. For instance, the forest bird fauna of Upper Guinea (i.e., the West African forest bloc) numbers 182 species, and that of the Congo forest 266 species, according to Moreau (1966). These totals are not impressive when compared with the 269 resident species recorded by Slud (1960) in a 2 square mile area of forest in Costa Rica.

The situation for mammals is somewhat different. The numbers of families, genera, and species inhabiting Africa (excluding the northwest corner of the continent and Madagascar) and the neotropical region as a whole are very similar (Table 1). If the parts of Africa that have an annual rainfall of less than 5 inches (120 mm) are excluded, the number of families and species per 100,000 square miles are again comparable: Africa, 0.63 and 9.39; neotropica, 0.70 and 11.3 (Keast 1969). The slightly higher figures for Latin America as a whole (including the Patagonian subregion) are referable to a greater richness in bats and rodents.

TABLE 1. Comparison of Mammal Faunas at the Continental Scale (from Keast 1969)

Area	*Families*	*Genera*	*Species*
Africa	51	240	756
Neotropica	50	278	810

This advantage partly persists, at least for bats, when comparison is restricted to countries of comparable size that contain large areas of tropical rain

FRANÇOIS BOURLIÈRE, Départment de Physiologie, Faculté de Médecine de Paris-Ouest, Paris 6, France. ACKNOWLEDGMENTS: I am most grateful to W. Aellen (Geneva), J. F. Eisenberg (Washington), C. O. Handley (Washington), C. M. Hladik (Makokou), X. Misonne (Brussels) and J. M. Thiollay (Lamto) for communication of unpublished information. I am particularly indebted to M. J. Hadley (Paris) for his constructive comments and criticisms, and his help in polishing the first draft of this paper.

TABLE 2. Comparison of Bat Faunas in Areas of Comparable Size in Africa and Latin America (from W. Aellen, in litt.)

Area	*Genera*	*Species*
Cameroon	28	76
Colombia	48	104
West Africa	31	97
Congo (Zaïre)	35	115
Brazil	56	98
Mexico*	55	154

* Mexican fauna includes both Neotropical and Nearctic faunistic elements.

forest (Table 2). The bewildering richness of the chiropteran fauna in the forested areas of the neotropics cannot be disputed. In some cases the number of species (and probably individuals) of bats may equal or, in some cases, even exceed that of all other mammals combined. For example, Hershkovitz (1969) recorded 196 species of native nonmarine mammals for Panama. Of these, 100 were bats and only 48 were rodents.

On the other hand, the faunal richness and diversity of Rodentia do not appear to differ greatly between tropical America and tropical Africa. Table 3 gives figures for two areas of comparable size, the Congo (Zaïre) and Brazil. The totals for the two areas are very similar. However, tropical America is much poorer in arboreal rodents than Africa, "flying squirrels" being totally absent in the New World.

TABLE 3. Comparison of Rodent Faunas in Areas of Comparable Size in Africa and Latin America (from X. Misonne, in litt.)

Area	*Families*	*Genera*	*Species*
Congo (Zaïre)	11	44	92
Brazil	9	40	95

The forest loving primates are also as well represented in Africa as in Latin America (Table 4), despite the fact that tropical rain forests cover about 32 percent of the neotropical region, compared to only 9 percent of tropical Africa. Not a single New World monkey has become adapted to the savanna environment, and the nocturnal Primates are only represented by a single genus and species (*Aotus trivirgatus*), whereas five genera and eight species of prosimians are to be found in the tropical rain forests of continental Africa.

As for ungulates, Africa undoubtedly leads the way. This preeminence is equally apparent in the forest and in the grassland biome, as shown in Table 5. Insectivora are also very well represented in the African lowland rain forests, whereas they are completely lacking within the same biome in South America. On the other hand, Marsupialia are totally absent from Africa.

TABLE 4. Number of Families, Genera, and Species of Primates in Tropical Africa and America

	Forest			Savanna		
Area	*Families*	*Genera*	*Species*	*Families*	*Genera*	*Species*
Africa	5	14	44	1	2	5
America	3	16	42			

The faunal lists of the few localities that have been intensively studied by mammalogists do not support the contention that the mammal fauna of the African lowland rain forests is poorer than that of ecologically similar habitats in tropical America. The 118 species of mammals so far found in the Makokou area (northeastern Gabon) by the French National Research Council (CNRS) team (Charles-Dominique 1971) compare favorably with the 73 species observed around Kartabo (British Guiana) by Beebe (1925) and the 82 species recorded in Barro Colorado (Panama) by Handley (pers. comm.). It should be remembered, however, that northern Gabon has been one of the few refuge areas where evergreen forests were able to persist during the climatic vicissitudes of the late Pleistocene (Moreau 1966), whereas apparently neither the Guyana lowlands nor the Canal Zone area escaped the climatic and ecological changes that took place at that time (Haffer 1969, Simpson Vuilleumier 1971).

To summarize, there is no great difference in faunal richness and diversity between terrestrial and arboreal mammals inhabiting the lowland rain forests of tropical Africa and America. The reason why there are more species of bats, as well as of birds, in the American humid tropics is not clear.

TABLE 5. Number of Families, Genera, and Species of Ungulates Found in the Lowland Rain Forest and Tropical Grassland Biomes in Africa and Latin America (from Keast 1959)

	AFRICA (Madagascar excluded)						LATIN AMERICA					
	Rain Forest Biome			Grassland Biome			Rain Forest Biome			Grassland Biome		
Family	*Families*	*Genera*	*Species*	*Families*	*Genera*	*Species*	*Families*	*Genera*	*Species*	*Families*	*Genera*	*Species*
Proboscidea	1	1	1	1	1	1	–	–	–	–	–	–
Equoidea	–	–	–	1	1	5	–	–	–	–	–	–
Tapiroidea	–	–	–	–	–	–	1	1	3	–	–	–
Rhinocerotoidea	–	–	–	1	2	2	–	–	–	–	–	–
Suoidea	1	2	2	1	1	1	1	1	2	1	1	1
Hippopotamidae	1	2	2	1	1	1	–	–	–	–	–	–
Traguloidea	1	1	1	–	–	–	–	–	–	–	–	–
Cervoidea	–	–	–	–	–	–	1	2	4	1	4	5
Giraffoidea	1	1	1	1	1	1	–	–	–	–	–	–
Bovoidea	1	7	20	1	24	57	–	–	–	–	–	–
Total	6	14	27	7	31	68	3	4	9	2	5	6

Convergence between Groups Living in Similar Niches

While comparing the contemporary mammalian faunas of the "southern continents," Keast (1969) emphasized that the major adaptive categories are common to South America, Africa, and Australia, despite their very different faunas and histories. Although levels of endemism and dominant groups differ in each case, "ecological equivalents or counterparts" are readily recognizable. This comparison has been carried a step further by Dubost (1968b) for the lowland rain forest mammals of West Africa and the Guianas, and by Eisenberg and McKay (1971). In his paper, Dubost pointed out that a remarkable morphological and behavioral convergence exists between the Cervidae and caviomorph rodents of South America and the Tragulidae and Cephalophinae of West Africa (Figure 1). This convergence is not limited to the size, body build, and horn (or antler) reduction, but also applies to habitat selection, diet, social structure, escape reactions, gait and communication. Obviously there are a number of differences between these taxonomically distinct but ecologically convergent groups. The small caviomorphs, in particular, are not ruminants as are their African (and Asian) ungulate counterparts. In addition *Dasyprocta* and *Myoprocta* share with many other rodents the characteristic of food hoarding, a trait unknown in any of the ungulates. However *Cuniculus paca* is not a food hoarder and the elaborate development of its coecum suggests that the animal can harbor bacterial symbionts, which facilitate the effective use of cellulose in the diet (Eisenberg and McKay 1971). Other cases of convergence are even more obvious, between Xenarthra and Pholidota for instance (giant armadillo–terrestrial pangolin, Figure 1).

The convergence in size and locomotor adaptation of such taxonomically different groups as ungulates and rodents is particularly striking. It can probably be best explained by the necessity to achieve a compromise between mobility in thick undergrowth, which favors a smaller size, and the ability to thrive upon a reduced diet, which is relatively enhanced by an increase in body size. That locomotion in the forest undergrowth is facilitated by a reduction in body size is suggested by the morphology of the forest races of those species that live both in closed and open environments, such as African buffaloes and elephants. As for the advantage of a larger body mass, it is well established that the larger a mammal, the smaller its daily food intake per unit of body weight, and the better its ability to withstand periods of starvation. Small mammals have a larger surface area and higher metabolic rate per unit body weight than large mammals, and therefore need more food to satisfy their metabolic requirements. All the other mechanical adaptations to locomotion in the dense undergrowth (arched back, head pointing downwards, reduction in horn or antler size, etc.) also contribute to a reduction of energy expenditures—a definite advantage for animals permanently living on a forest floor where food is seldom abundant.

The Stratification of Mammal Populations

Everywhere in the humid tropics, the mature forest is highly stratified. Trees generally form three layers: the upper and discontinuous storey, comprising scattered, very tall emergent trees; the canopy, which forms an almost continuous evergreen carpet 30 to 40 meters tall; and the understorey stratum that becomes dense only where there is a break in the canopy. Below this, there is a ground layer of scattered herbaceous plants and seedling trees; grass is entirely absent. The forest floor is covered with a thin litter of leaves mixed with dead twigs and branches, and rotting trunks of fallen trees. The situation is very different in the natural clearings opened by the falling of the tallest trees, or along the rivers. Here, the second growth is very thick and the amount of living plant material that can be used as food by terrestrial mammals much greater. Although the overall structure of lowland rain forest is very much the same in the New and the Old Worlds, there are nevertheless some important differences: for example, epiphytes and palms are much more numerous in the neotropics than in Africa.

The distribution of mammal populations within the three-dimensional habitat of the lowland rain forest is closely correlated with this stratification of the plant biomass and the uneven production of potential food by the different vegetation layers. The more extensive occupation of the "arboreal zone" in the tropical rain forest, as compared with temperate woodlands, has long been recognized. Thus, Davis (1962) reported that 45 percent of the nonflying mammals of the North Bornean rain forest are arboreal, compared to only 15 percent in the temperate woodlands of Virginia. Similar proportions to that in North Borneo are most likely in Africa and Latin America. The food resources of the canopy (buds, leaves, flowers, and fruits) are exploited by representatives of many orders: frugivorous and nectarivorous bats are particularly abundant in the top storey. Primates rank second, being almost exclusively arboreal, but numerous species of rodents also live permanently in the tree crowns and have developed many morphological and behavioral adaptations to arboreal locomotion and life. Other canopy browsers are found among Xenarthra (American sloths) and Hyracoïdea (African *Dendrohyrax*). Some insectivorous and carnivorous mammals also live and feed largely in the tree crowns. Such is the case for the "anteaters" (*Tamandua* and *Cyclopes* in America, *Manis tricuspis* and *M. longicaudata* in Africa, Pagès 1970) and a number of felids, procyonids, mustelids, and viverrids (*Felis aurata* and *Nandinia binotata* in Africa; *Potos flavus* and *Tayra barbara* in America).

Following Harrison (1962), the "middle-zone mammals" can be subdivided into two trophic groups: insectivorous bats and scansorial species of mixed feeding habits ranging up and down the trunks and the vines. Many primates, rodents, edentates, and carnivores fall into the latter category.

The ground mammals must also be split into three trophic groups: large and medium-sized ungulates browsing upon the scanty foliage available, searching for fallen fruit, seeking subterranean roots, corms and bulbs and occasionally grazing in clearings and on river banks; terrestrial rodents and shrews of varied diet searching the forest floor; and the few strictly terrestrial carnivores living upon the vertebrates roaming the undergrowth. No true fossorial mammals are to be found in the lowland rain forests; this absence is most probably due to the high water content of the soil, the rapid destruction of the litter and the scarcity of large soil invertebrates, such as earthworms.

The relatively low number of terrestrial species that characterizes the lowland rain forests of the tropics does not necessarily mean that most of the mammal biomass is found above ground level and that the forest undergrowth is almost empty, mammalogically speaking. In this respect, earlier calculations (Bourlière 1963) made on the basis of Collins' (1959) data are somewhat misleading. The Tano Nimri forest reserve in Ghana did not con-

FIGURE 1. Morphological convergences among African (left) and neotropical (right) rain forest mammals. From top to bottom and left to right: pigmy hippopotamus and capybara; African chevrotain and paca; royal antelope and agouti; yellow-back duiker and brocket deer (*Mazama gouazoubira*); terrestrial pangolin and giant armadillo. Each pair of animals is drawn to the same scale.

tain any of the large forest ungulates (elephant, buffalo, bongo, okapi, forest hog) found in undisturbed forests elsewhere in Africa. If these species had been present, the estimates would have been very different and the standing crop biomass of arboreal and terrestrial mammals probably similar. This is indeed the situation recorded by Eisenberg and McKay (1971) at Barro Colorado, though this island is thought to support an abnormally low level of larger forest ungulates. The standing crop biomasses of arboreal and terrestrial mammals, estimated from Eisenberg and McKay (1971, fig. 3), approximate to 950 and 825 kilograms per square kilometer, respectively. This quasi-equivalence of standing crop biomasses, despite the greater number of arboreal species, is obviously due to the smaller size of the latter. For a mammal spending its whole life in the trees, a large size is a definite handicap, precluding locomotion on smaller branches and twigs. In some cases the biomass of ground-living mammals in tropical forests may be considerable. Thus, in the mountain forests of Mount Kenya, Holloway (1962) has found very high densities of elephants (0.6-1.2/km^2) and buffaloes (1.3-10.0/km^2). Ignoring the other large mammals of the area (forest hog, bongo, duikers, bushbuck) these densities correspond to standing crop biomasses ranging from 2500 to 8500 kilograms per square kilometer. This situation is, however, probably artificial, such concentrations being attributable to immigration from surrounding areas progressively deforested and used for cultivation.

Although not a single comprehensive census of the total mammal fauna of a tropical forest is available, it has often been maintained that, whereas species diversity is always high, population densities on the contrary are low. Such a statement is obviously supported by the feelings of the casual visitor entering a mature rain forest for the first time: mammals are scarce and difficult to see, particularly since many of them are nocturnal in their habits. Usual trapping methods do not generally bring good returns, and it is not an easy task to set traps in the canopy even on the largest branches. Entirely new methods of observation and capture have to be devised before the population density of most forest mammals can be estimated confidently. Meanwhile, the generalization needs to be countered: in some cases at least, a given species of rain forest mammal is undeniably more numerous than other closely related forms. In the undisturbed mature forest of northeastern Gabon, *Crocidura poensis* and *Hipposideros caffer* are much more often caught and observed than the other species of shrews and insectivorous bats (Brosset 1966b). The abundance of *H. caffer* is apparently due to the lack of competition in its feeding niche. This species is the only bat in the area that can catch insects in the mature forest undergrowth.

The Ecological Niches of Rain Forest Mammals

Reviewing the available evidence for ecological segregation as a cause of increased species diversity among tropical birds, Lack (1971:268) comes to the following conclusion: "The number of bird species in an area depends essentially on the extent to which they can divide the available habitats and food resources. In particular, the greater number of species in the tropics than at high latitudes is associated with a greater diversity of habitats, a much greater number of vegetational layers, a greater number of foods of different types available throughout the year, and hence a greater number of food specialists which can be separated by feeding."

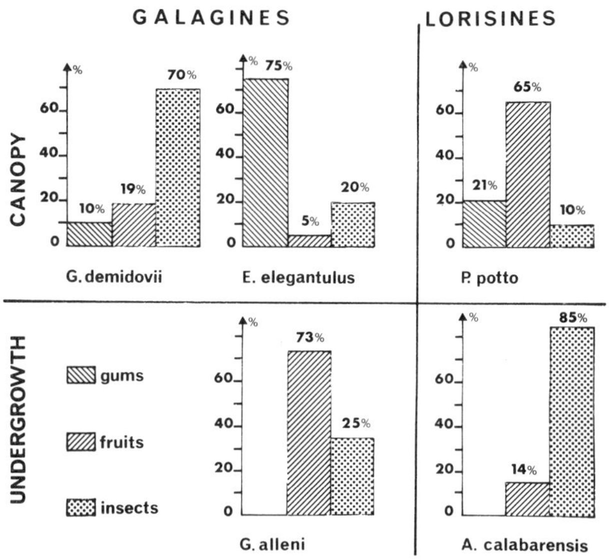

FIGURE 2. Food specialization among five sympatric species of prosimians in northeastern Gabon. Percentages of the fresh weight of major food items in the stomach contents (and coecum contents for vegetable gums). [After Charles-Dominique 1971.]

The feeding habits of the tropical rain forest mammals are still very poorly known, but the few data at hand tend to support Lack's viewpoint. Charles-Dominique (1971) has intensively studied the ecology of the five sympatric prosimians living in the forests of northeastern Gabon. His major findings are summarized in Figure 2. In the forest canopy, one species is predominantly insectivorous, another mainly folivorous, and the third eats large amounts of plant gums. In the undergrowth, one prosimian is predominantly frugivorous and the other insectivorous. Not only does each species differ in the percentage of the major food items consumed within its preferred forest-layer, but the hunting techniques are also different. The Lorisinae (*Perodicticus potto* and *Arctocebus calabarensis*) detect their prey by scent, and those selected are very often slow-moving and "repulsive" (stinging caterpillars, ants, centipedes, etc.). The Galaginae (*Galago alleni, G. dimidovii, Euoticus elegantulus*) feed mainly on Orthoptera, Lepidoptera, and Coleoptera, many of which are taken in flight by a rapid forward movement during which the bush baby grasps its support with hind limbs only. These food preferences are particularly marked during the dry season. Any direct competition with the sympatric diurnal monkeys is avoided, prosimians being active only at dusk and during the night. The preferences of the various species for different forest strata depend mainly on the type of support selected and consequently on their locomôtory adaptation.

A somewhat similar situation prevails among the monkeys of Panama, although their dietary specializations are less marked (Hladik and Hladik 1969). The bulk of the diet of all five species consists of fruits, but the howler, *Alouatta palliata,* eats much more foliage than the spider monkey, *Ateles geoffroyi,* whereas the capuchin, *Cebus capucinus,* is more opportunistic and supplements its vegetarian diet with large numbers of insects and even small vertebrates. The rufous-naped tamarin, *Leontocebus geoffroyi,* is even more insectivorous than the capuchin, but prefers young second growth and interface habitats. The night monkey, *Aotus trivirgatus,* has a diet rather similar to that of *Cebus,* but competition is prevented by its nocturnal habits.

Competitive exclusion also operates among the four species of "anteaters" living in northeastern Gabon (Pagès 1970). Differences in time of activity, food specialization, and preferences for different forest strata all contribute in limiting competition between these consumers of social insects. The nocturnal aardvark, *Orycteropus afer,* is a very active burrower; it can excavate to a depth of 3 meters in order to open termite nests. This animal very often destroys a termite colony in a single visit, and therefore needs to patrol a very extensive home range. The giant pangolin, *Manis gigantea,* is also nocturnal and confined to the ground, but it is not as efficient a burrower as the aardvark, being seldom able to penetrate the ground to a depth greater than 80 centimeters. *Macrotermes mülleri* is its staple food in northeastern Gabon, but the termite mounds are seldom destroyed by its visits; its home range is consequently smaller than that of *Orycteropus afer.* The longtailed pangolin, *Manis longicaudata,* is much smaller, strictly arboreal and diurnal: it feeds mainly upon arboreal ants (*Crematogaster* spp.). The nocturnal tree pangolin *Manis tricuspis* is more opportunistic, living in trees as well as on the ground and exploiting both ground-level and arboreal termite nests.

The ecological niches in tropical lowland rain forests are not always so clearly defined and mixed troops are often found, particularly among primates. In South America, capuchin monkeys often associate with squirrel monkeys (Baldwin and Baldwin 1971), but such associations are much more frequent in Africa. Their structure and functioning have been studied in northeastern Gabon by Gautier and Gautier-Hion (1969). Some species are more prompted to form mixed troops than others. For example, whereas *Cercopithecus cephus* has been found in association with various other Cercopithecines and Colobines in 86.6 percent of observations, *C. neglectus* invariably lives in small monospecific groups. The duration of these associations is also quite variable. In some cases troops mingle temporarily when a common food supply is superabundant at a given place (fruiting trees) and separate immediately afterwards. In most other cases, the monospecific troops forage separately during the day and join together late in the afternoon to spend the night either in adjacent trees or even intermingle completely until the following morning. Such mixed troops may be stable from one day to the next, and even for weeks and months. The species participating in these mixed troops apparently retain their specific food preferences, at least during periods of food shortage, but the polyspecific grouping affords them an extra advantage: A large mixed troop (safety-in-numbers) provides a better means of detection of potential predators, particularly at night.

Population Dispersion

Most of the tropical rain forest mammals, whatever the vegetation stratum they preferentially inhabit, have a vegetarian diet made up of fruits, buds, and fresh leaves. Herbaceous vegetation is always scantily taken. Even carnivores may seasonally subsist on fruits, the coati for example (Kaufmann 1962, Smythe 1970b). Edible plant material is consumed either where it is produced, i.e. mostly in the tree crowns, or on the ground, where the shower of fruits and seeds falling from the canopy makes up the bulk of the diet of the terrestrial herbivores.

There are few data on the annual production of leaves, fruits, and seeds in the various types of tropical rain forests. Some figures on total litter production are available but these are not generally broken down into components, such as leaves, bud scales, flowers, fruits, branches, and bark. Moreover, measurements of litter fall provide, by themselves, little indication of the amount of plant material consumed in the canopy by folivorous and frugivorous invertebrates and vertebrates. Two of the tentative conclusions of Bray and Gorham (1964) should be noted here. Firstly, that litter production increases as latitude decreases, averaging 11 tons per hectare in equatorial forests, as opposed to 5.5 and 3.5 tons per hectare, respectively, in warm and cool temperate forests; secondly, that litter destruction is much more rapid in the humid tropics than further north or south.

Some estimates of the actual production of fruits by a limited number of tropical tree species have been made. For instance, Hladik and Hladik (1969) estimated an annual production (fresh weight) of 5 tons per hectare for *Spondias mombin*, 0.5 to 2.0 tons per hectare for *Ficus insipida*, 0.3 tons per hectare for *Miconia argentea* and 0.02 tons per hectare for *Astrocaryum standleyanium* in Panama. But such fruit trees are very seldom found in dense stands and are usually widely scattered throughout the forest. It is therefore not possible to extrapolate from these estimates of specific production to the production of the forest as a whole. Smythe's (1970a) results, based upon material taken from lines of "fruit traps" set in the undergrowth of Barro Colorado's forest, cannot be considered as production figures either; they represent the amount of fruits and seeds available to the terrestrial herbivores. The monthly average of falling fruits, expressed in grams (fresh weight) per square meters per day, varied from a low of 0.061 in December to a high of 1.93 in June. The water content of the fruits ranged from 57.0 to 64.2 percent. This corresponds to a "fruit fall" of about 2.3 tons per hectare per year (fresh weight) and to a fruit component of the litter of about 0.9 tons per hectare per year (dry weight) if an average water content of 60 percent is assumed. Klinge and Rodrigues (1968) estimated a mean annual fruit fall of 0.4 tons per hectare (dry weight) in Amazonian forest near Manaus. These figures are of the same order of magnitude as that of 0.87 tons per hectare per year (dry weight) for the mean nonleaf, nonwood, litter production of seven deciduous forests in Belgium (Duvigneaud et al. 1971). It seems likely, therefore, that the amount of plant material usable for food by mammals in tropical rain forests approximates that available in temperate deciduous forests. This finding, if proved valid, would be in good accordance with the results of the few studies made to date on primary production in mature tropical forests (Müller and Nielsen 1965, Kira 1969). Unlike gross production, net production differs little between tropical and temperate woodlands, much of the gross production being dissipated through high rates of respiration. According to Kira (1969), this is mainly because the ratio of net production to gross production varies considerably according to the age composition of a forest. The small net production per gross production ratio recorded in tropical rain forests is probably an attribute of mature climax woodlands, rather than a peculiarity of tropical forest ecosystems per se.

In the rain forest, plant food is neither abundant nor does it remain available over a long time period due to the rapid decomposition of fallen fruits and leaves. Mammals have had, therefore, to adapt their population distributions to these scant resources. The various primary consumers have to "choose" between two alternative solutions. Some adopt a more or less uniform distribution of individuals (or small groups), each of them defending its permanent home range. Others pursue a continuing nomadism over a wider area in search of fruiting trees, whose temporarily superabundant food resources may be shared peaceably by different individuals or troops. These two solutions are not necessarily exclusive, a given sex- or age-category being able to change its spatial distribution according to the seasons and availability of food.

Most *nonarboreal* rain forest mammals are "solitary" and are characterized by a dispersed form of social organization, as described by Dubost (1968b) and Eisenberg and McKay (1971) in the case of ungulates and caviomorph rodents. The basic social unit is either the mother and her young or a temporary courting pair. In many cases, territorial behavior has been observed, the territory of the adult male often overlapping those of several females. There are nevertheless a number of notable exceptions to this generalization, such as the Suoidea (Tayassuidae and Suidae), which live in small "sounders," and the forest races of the African elephant and buffalo, which may form large, wide-ranging troops. However, it should be noted that the social groups in the two latter cases are always much smaller than those of populations of the same species living in more open environments.

Among *arboreal* rain forest mammals, two major trends are also discernible: a more or less uniform distribution of sedentary individuals, family groups or small troops, scattered all over the habitat range of the species, and a more "fluid" type of social organization where the different sex- and age-categories continuously wander over a large range and coalesce or separate according to prevalent food distribution and availability.

The "solitary" way of life prevails among most arboreal rodents (except flying squirrels of the genus *Idiurus*) and among the tree-top browsers (sloths in the neotropics and tree-hyraxes in Africa). Sedentary family groups are also found in a few diurnal primates: the white-handed gibbon, *Hylobates lar*, in the Old World (Carpenter 1940, Ellefson 1968) and *Callicebus moloch* in the New World (Mason 1968) are well-known examples. Among nocturnal African prosimians, the basic social unit appears to be the maternal group, comprising the mother and her young, which lives the whole year within a limited home range. Adult males have larger territories, which overlap the home ranges of several adult females; they join them in turn when they are in oestrus. Subadult males remain peripheral and vagrant (Charles-Dominique, 1972). Some sedentary monkeys live in larger social groups, more or less regularly distributed within their forest habitat. Territorial behavior, not always overtly displayed, is a characteristic of these "closed" troops. Thus, most members of the genus *Cercopithecus* live in "one-male groups" containing several adult females with their offspring but only one sexually and socially mature male (Bourlière et al. 1969, 1970; Gautier 1969; Struhsaker 1969; Aldrich-Blake 1970). A similar social structure has been recorded in the black and white colobus, *Colobus guereza* (Marler 1969). Such a social structure has been considered as adaptive to limited food resources, since procreation remains assured while the exploitation of food by adult males relative to females is reduced to a minimum (Crook 1970). Other forest monkeys, such as howlers (Carpenter 1934), spider monkeys (Klein, in Crook, 1970) and mangabeys (Chalmers 1968), live in "multimale groups." *Cebus capucinus* may form either one-male or multimale groups (Oppenheimer 1968). In all the above cases, typical group size is usually small, between 12 and 20 members, and home ranges reduced, 0.03 to 0.06 square miles. Larger multimale troops have also been observed among African talapoins (Gautier-Hion 1970, 1971) and neotropical squirrel monkeys (Baldwin and Baldwin 1971); both species are small-sized and partly insectivorous.

In contrast to these examples are the "open" social structures of a number of large rain forest primates, which roam over much greater home ranges than the above species. Thus, the drill, *Mandrillus leucophaeus,* occurs in the forests of southern Cameroon either in small one-male groups or in large troops totaling as many as 179 animals. The small "one-male units" may aggregate into large groups at certain times and subsequently separate (Gartlan 1970). Chimpanzee social organization is even looser. Groups are constantly changing membership, splitting apart, meeting others and joining them, congregating and dispersing. There are no stable groups other than the maternal family, that is the mother and her younger offspring. There is increasing evidence of lasting links between the mother and her sons and daughters, as well as between siblings themselves (Reynolds and Reynolds

1965, van Lawick-Goodall 1968). Each member of such population demes knows the other members individually and is antagonistic to the approach of strangers. However, there is no topographically limited territory. Such "fluid" types of social organization and opportunistic population distribution ensure efficient utilization of the resources of large areas of rain forests where food is patchily distributed, both in space and in time.

All tropical rain forest mammals whose diet is exclusively insectivorous (murine opossums, shrews, pangolins, arboreal anteaters) are "solitary" or live in small family groups. Most forest carnivores also live singly or in pairs.

On the other hand, the two major types of population distribution found in predominantly vegetarian mammals, "dispersed" and "clumped," are also to be found among flying mammals. A number of rain forest bats (whether insectivorous, nectarivorous, or frugivorous) roost, hunt, or feed singly or in pairs. Some have very bizarre hiding places during the day; for example, pairs of the small *Kerivoula harrisoni* roost under the large silk nests of the colonial spiders of the genus *Agelena*. Many more species, both of Macro- and Micro-chiroptera live in social groups of variable size. The "Falcon Cave" in northeastern Gabon harbors about 500,000 *Hipposideros caffer,* whose hunting range covers about 500 square kilometers and which consume about three tons of insects per night (Brosset 1966a, 1969).

Population Recruitment and Turnover

It has long been supposed that there were no definite breeding seasons in the uniform and warm climate of the humid tropical woodlands. In this "cradle of terrestrial life," secondary productivity was thought to be high, population turnover rapid and the rate of evolution consequently accelerated. This may be the case for plants and invertebrates, but the available data for mammals do not support these hypotheses.

On the basis of evidence at hand, it is apparent that a number, if not a majority, of tropical rain forest mammals breed seasonally rather than throughout the year. This is particularly obvious among primates. Figure 3 summarizes the situation for the Lowe's guenon, *Cercopithecus campbelli lowei,* in the Ivory Coast. This species gives birth at the end of the short rains and the beginning of the main dry season, but never at other periods of the

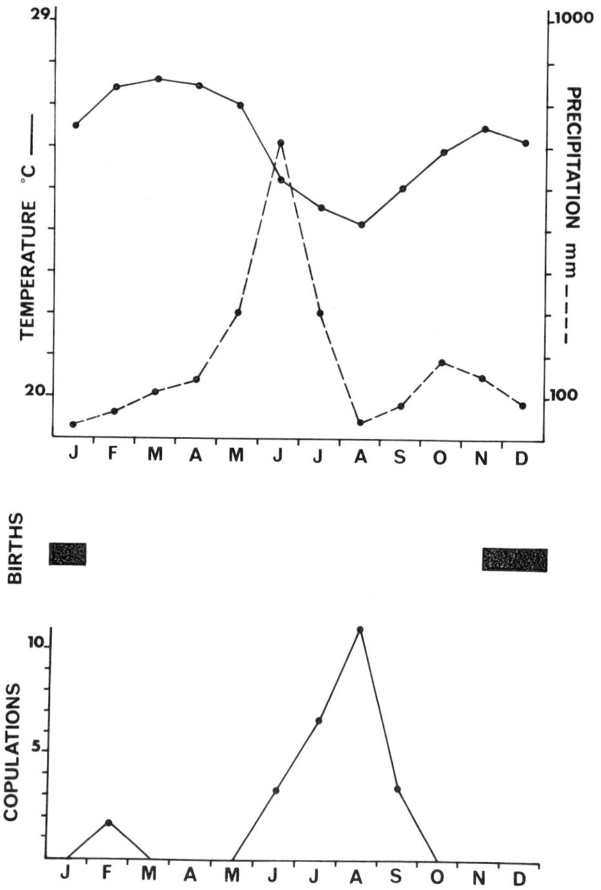

FIGURE 3. Seasonality in reproductive behavior among Lowe's guenons (*Cercopithecus campbelli lowei*) at Adiopodoumé, Ivory Coast. In the upper graph the solid line represents seasonal changes of the average monthly temperatures in °C. The dotted line shows the average monthly rainfall. The lower graph represents the number of copulations per 100 hours of observation each month. Note that most copulations occur when the monthly temperatures are the lowest.

year. The frequency of copulation also varies from month to month; most of the observed matings occurred between June and September, at the coolest period of the year, just after the long rains. A relationship between the conception time and declining temperature among primates has also been suggested by Lancaster and Lee (1965). A well-defined breeding periodicity has been observed for most of the representatives of genus *Cercopithecus* in northeastern Gabon, as well as for the talapoin (Gautier-Hion 1971). Seasonal breeding is also the rule among squirrel monkeys, the males manifesting a parallel cyclicity in their spermatogenetic cycle and their behavior (DuMond 1968). Seasonal peaks of reproduction have been recorded among the 16 species of Muridae living in northeastern Gabon, the highest percentages of pregnant females coinciding with periods of highest rainfall (Dubost 1968a). This correlation is apparent, but less obvious, among the 16 forest species of Muridae and 6 species of Sciuridae studied by Rahm (1970) in the Kivu province, eastern Zaïre. Delany (1971) has also found some correlation between breeding and rainfall among the forest rodents of Uganda. Obviously the environmental synchronizers of the breeding cycle vary from one species to another. Rainfall by itself is probably of little importance for a mammal; it acts indirectly through its influence on primary and secondary productivity; seasonal variations in abundance, nature, and availability of food being quite marked in this seemingly uniform environment (Figure 4). The reasons for these interspecific differences (and interpopulation differences within the same species) will become apparent only after more comprehensive ecological and physiological studies have been carried out.

The reputedly high reproductivity of tropical forest mammals has still to be documented. Among ungulates and primates, single births are the rule, and intervals between successive births may last up to several years in apes. Most forest rodents have smaller litter sizes than those of their savanna relatives (Dubost 1968a, b, Rahm 1970, Fleming 1971, Tesh 1970, Delany 1971), a situation very similar to that recorded in birds (Lack and Moreau 1965). Even rain forest insects may have a smaller reproductive potential than their relatives of temperate latitudes. This has been recently demonstrated in Gabonese strains of *Drosophila melanogaster* by David (1971).

Yet, until more studies of the average numbers of pregnancies per female are made, it cannot be concluded that tropical rain forest mammals are "less productive" than their savanna relatives. Thus, *Oryzomys caliginosus* of the Pacific lowlands of Colombia has a low litter-size (3), but can produce as many as five litters per year; its reproductive potential is, therefore, similar to that of its North American counterpart *O. palustris* (Arata and Thomas 1968). The small size, but possibly high frequency, of litters of most rain forest rodents might well be an adaptation to a sustained yet small production of their food.

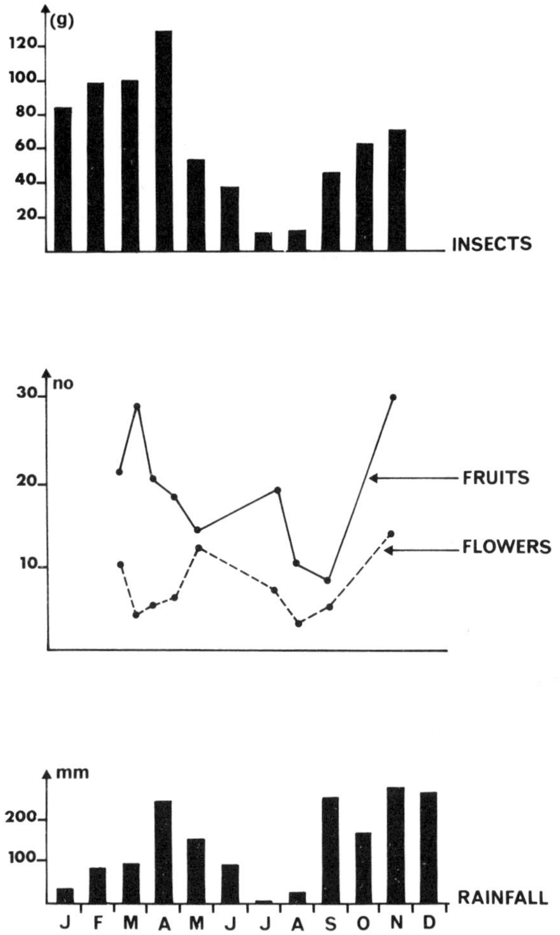

FIGURE 4. Seasonal variations in food availability and rainfall in the Makokou area, northeastern Gabon. *Top:* Monthly averages of insects taken by night in the same UV trap are used as an index of flying insect abundance. *Center:* The numbers of flowering and fruiting trees recorded each month along a 3 km line transect gives *(bottom)* an index of flower and fruit production. [After Charles-Dominique 1971.]

Another way for homeotherms to cope with a sustained yet scanty food production is to lower their metabolic rate, and in doing so to slow down their life processes. This is not possible in temperate latitudes where mammals and birds must constantly produce heat to keep constant their body temperature. But in hot-humid environments with ambient temperatures in the upper part of the thermoneutral range, a reduction in metabolic rate is possible and could even be a thermoregulatory mechanism.

No systematic study of the basal metabolic rate (BMR) of tropical warm-blooded vertebrates has yet been made. However a decrease of the BMR has been reported among Europeans living permanently in the humid tropics, and preliminary observations made on Ivory Coast shrews show that these animals can be maintained at a constant weight and in good health on smaller daily rations than their northern relatives (Hunkeler and Hunkeler 1970).

Data on mortality rates and population turnover of rain forest mammals are still lacking; long term studies, under field conditions, have to be made on life expectancies of representatives of major groups. Yet, some prosimians and flying foxes, as well as a number of tropical forest Passerines, are remarkably long-lived for their size (Snow 1962; Fogden 1972; Thiollay, CNRS team at Lamto, Ivory Coast, pers. comm.).

Ecosystem Interactions

Most of the characteristics discussed in the previous pages may be considered adaptive to the peculiar living conditions of vertebrates in the tropical rain forest ecosystem, where food is relatively scarce and patchily distributed in space and time. But the role played by mammals (and indeed birds) within this highly complex system is not confined to their position as primary or secondary consumers. There is rather a complex system of interactions and feedback relationships between mammals and their food sources. Without their mammal components, rain forests would not be self-perpetuating. Nectarivorous marsupials and bats contribute to pollinization of a number of species, but many more are dependent upon frugivorous vertebrates for the dissemination of reproductive structures. Primates, frugivorous bats, and ground herbivores in particular contribute in maintaining the floristic complexity of tropical forests and the remarkably scattered dispersion of many plant species. This is done in two different ways. First, through endozoochory, that is the ingestion and transportation of seeds without loss (even, in some cases, with increase) of their germinative potential (Hladik and Hladik 1969, van der Pijl 1969). Second, through "predation" itself. The ingestion by animals of the seeds and seedlings of a particular tree species will tend to lead to a reduction of the adults of that species and to an increased distance between newly mature trees and their parents. This tends to provide more space for other tree species (Janzen 1970).

Conclusion

Briefly, the ecology of lowland rain forest mammals in Africa and America does not appear to differ greatly, despite different evolutionary histories and taxonomic status. On both sides of the Atlantic, similar environmental pressures have "shaped" the mammalian members of these rich biotic communities in a similar way, leading to striking convergences in form, function, and behavior. However, much work remains to be done before the roles played by mammals within tropical forest ecosystems can be fully understood, or indeed even broadly appreciated.

References

Aldrich-Blake, F. P. G. 1970. Problems of social structure in forest monkeys. Pp. 79-101, *in* J. H. Crook (Editor). Social behaviour in birds and mammals. Academic Press. London.

Arata, A. A., and M. E. Thomas. 1968. Population studies of small neotropical mammals. IBP Small Mammal Newsletter (Warsaw) 2:140-141.

Baldwin, J. D., and J. L. Baldwin. 1971. Squirrel monkeys (*Saimiri*) in natural habitats in Panama, Colombia, Brazil and Peru. Primates 12:45-61.

Beebe, W. 1925. Studies of a tropical jungle. One quarter of a square mile of jungle at Kartabo, British Guiana. Zoologica 6:1-193.

Bourlière, F. 1963. Observations on the ecology of some large African mammals. Pp. 43-54, *in* F. C. Howell and F. Bourlière (Editors). African ecology and human evolution. Aldine. Chicago.

Bourlière, F., M. Bertrand, and C. Hunkeler. 1969. L'écologie de la Mone de Lowe (*Cercopithecus campbelli lowei*) en Côte d'Ivoire. La Terre et la Vie 23:135-163.

Bourlière, F., C. Hunkeler, and M. Bertrand. 1970. Ecology and behavior of Lowe's guenon (*Cercopithecus campbelli lowei*) in the Ivory Coast. Pp. 297-350, *in* J. R. Napier and P. H. Napier (Editors). Old World monkeys. Academic Press. New York.

Bray, J. R., and E. Gorham. 1964. Litter production in forests of the world. Vol. 2:101-157, *in* J. B. Cragg (Editor). Advances in ecological research. Academic Press. London and New York.

Brosset, A. 1966a. Les chiroptères du Haut-Ivindo (Gabon). Biologia Gabonica 2:47-86.

———. 1966b. Recherches sur la composition qualitative et quantitative des populations de vertébrés dans la forêt primaire du Gabon. Biologia Gabonica 2:163-177.

———. 1969. Recherches sur la biologie des chiroptères troglophiles dans le nord-est du Gabon. Biologia Gabonica 5:93-116.

Carpenter, C. R. 1934. A field study of the behavior and social relations of howling monkeys. Comp. Psychol. Monogr. 10(48):1-168.

————. 1940. A field study in Siam of the behavior and social relations of the gibbon, *Hylobates lar*. Comp. Psychol. Monogr. 16(5):1-212.

Chalmers, N. R. 1968. Group composition, ecology, and daily activities of free living mangabeys in Uganda. Folia Primat. 8:247-262.

Charles-Dominique, P. 1971. Eco-ethologie des prosimiens du Gabon. Biologia Gabonica 7:121-228.

————. 1972. Ecologie et vie sociale de *Galago demidovii*. Z. Tierpsychol. Beiheft 9:7-41.

Collins, W. B. 1959. The perpetual forest. Lippincott. Philadelphia.

Crook, J. H. 1970. The socio-ecology of primates. Pp. 103-166, *in* J. H. Crook (Editor). Social behavior in birds and mammals. Academic Press. London.

David, J. 1971. Particularités biométriques et faible potentiel reproducteur des souches de *Drosophila melanogaster* provenant du Gabon. Comptes-Rendus Acad. Sci. 272: 2191-2194.

Davis, D. D. 1962. Mammals of the lowland rain-forest of North Borneo. Bull. Nat. Mus. St. Singapore 31:1-129.

Delany, M. J. 1971. The biology of small rodents in Mayanja forest, Uganda. J. Zool., London 165:85-129.

Dubost, G. 1968a. Aperçu sur le rythme annuel de reproduction des muridés du nord-est du Gabon. Biologia Gabonica 4:227-239.

————. 1968b. Les niches écologiques des forêts tropicales sud-américaines et africaines, sources de convergences remarquables entre Rongeurs et Artiodactyles. La Terre et la Vie 22:3-28.

DuMond, F. V. 1968. The squirrel monkey in a seminatural environment. Pp. 87-145, *in* L. A. Rosenblum and R. W. Cooper (Editors). The squirrel monkey. Academic Press. New York.

Duvigneaud, P., S. Denaeyer, P. Ambroes, and J. Timperman. 1971. Recherches sur l'écosystème forêt. Mem. Inst. R. Sci. Nat. Belgique 164:1-101.

Eisenberg, J. F., and G. M. McKay. 1971. Comparison of ungulate adaptations in the New World and Old World tropical forests with special reference to Ceylon and the rain forest of Central America. Paper presented at the Symposium on the Behavior of Ungulates and its Relation to Management (Calgary, 2-5 November 1971).

Ellefson, J. O. 1968. Territorial behavior in the common whitehanded gibbon *Hylobates lar* Linn. Pp. 180-199, *in* P. C. Jay (Editor). Primates: Studies in adaptation and variability. Holt. New York.

Fleming, T. H. 1971. Population ecology of three species of neotropical rodents. Mus. Zool., Univ. Mich., Misc. Publ. 43:1-77.

Fogden, M. P. L. 1972. The seasonality and population dynamics of equatorial forest birds in Sarawak. Ibis 114:307-342.

Gartlan, J. S. 1970. Preliminary notes on the ecology and behavior of the drill *Mandrillus leucophaeus* Ritgen, 1824. Pp. 445-480, *in* J. R. Napier and P. H. Napier (Editors). Old World monkeys. Academic Press. New York.

Gautier-Hion, A. 1970. L'organisation sociale d'une bande de talapoins dans le nord-est du Gabon. Folia Primat. 12:116-141.

————. 1971. L'écologie du Talapoin du Gabon. La Terre et la Vie 25:427-490.

Gautier, J. P. 1969. Emissions sonores d'espacement et de ralliement par deux cercopithèques arboricoles. Biologia Gabonica 5:117-145.

Gautier, J. P., and A. Gautier-Hion. 1969. Les associations polyspécifiques chez les Cercopithecidae du Gabon. La Terre et la Vie 23:164-201.

Haffer, J. 1969. Speciation in Amazonian forest birds. Science 165:131-137.

Harrison, J. L. 1962. The distribution of feeding habits among animals in a tropical rain forest. J. Anim. Ecol. 31:53-64.

Hershkovitz, P. 1969. The evolution of mammals on southern continents, VI: The recent mammals of the neotropical region: a zoogeographic and ecological review. Quart. Rev. Biol. 44:1-70.

Hladik, A., and C. M. Hladik. 1969. Rapports trophiques entre vegetation et primates dans la forêt de Barro Colorado (Panama). La Terre et la Vie 23:25-117.

Holloway, C. W. 1962. The effect of big game forest management. Unpublished Thesis, Commonwealth Forestry Institute, Oxford.

Hunkeler, C., and P. Hunkeler. 1970. Besoins energétiques de quelques crocidures (insectivores) de Côte d'Ivoire. La Terre et la Vie 24:449-456.

Janzen, D. H. 1970. Herbivores and the number of tree species in tropical forests. Amer. Nat. 104:501-528.

Kaufmann, J. H. 1962. The ecology and social behavior of the coati, *Nasua narica*, on Barro Colorado Island, Panama. Univ. Calif. Publ. Zool. 60:95-222.

Keast, A. 1969. The evolution of mammals on southern continents, VII: Comparisons of the contemporary mammalian faunas of the southern continents. Quart. Rev. Biol. 44:121-167.

Kira, T. 1969. Primary productivity of tropical rain forest. Malayan Forester 32:375-384.

Klinge, H., and W. A. Rodrigues. 1968. Litter production in an area of Amazonian terra firme forest, I: Litter-fall, organic carbon and total nitrogen contents of litter. Amazoniana 1:287:302.

Lack, D. 1971. Ecological isolation in birds. Blackwell. Oxford and Edinburgh.

Lack, D., and R. E. Moreau. 1965. Clutch size in tropical passerine birds of forest and savanna. Oiseau 35 (special issue): 76-89.

Lancaster, J. B., and R. B. Lee. 1965. The annual reproductive cycle in monkeys and apes. Pp. 486-513, *in* I. DeVore (Editor). Primate behavior. Holt. New York.

van Lawick-Goodall, J. 1968. The behaviour of free-living chimpanzees in the Gombe Stream Reserve. Anim. Behav. Monogr. 1:161-311.

Marler, P. 1969. *Colobus guereza*: Territoriality and group composition. Science 163:93-95.

Mason, W. A. 1968. Use of space by *Callicebus* groups. Pp. 200-216, *in* P. C. Jay (Editor). Primates: Studies in adaptation and variability. Holt. New York.

Moreau, R. E. 1966. The bird faunas of Africa and its islands. Academic Press. New York and London.

Müller, D., and J. Nielsen. 1965. Production brute, pertes par respiration et production nette dans la forêt ombrophile tropicale. Forstl. Forsögsv. Danmark 29:69-160.

Oppenheimer, J. A. 1968. Behavior and ecology of the white-faced monkey *Cebus capucinus*. 181 pp. PhD dissertation, Univ. of Illinois, Urbana.

Pagès, E. 1970. Sur l'écologie et les adaptations de l'oryctérope et des pangolins sympatriques du Gabon. Biologia Gabonica 6:27-92.

van der Pijl, L. 1969. Principles of dispersal in higher plants. Springer-Verlag. Berlin.

Rahm, U. 1970. Note sur la reproduction des sciuridés et muridés dans la forêt équatoriale au Congo. Rev. Suisse Zool. 77:635-646.

Reynolds, V., and F. Reynolds. 1965. Chimpanzees of the Budongo forest. Pp. 368-424, *in* I. DeVore (Editor). Primate behavior. Holt. New York.

Simpson Vuilleumier, B. S. 1971. Pleistocene changes in the fauna and flora of South America. Science 173:771-780.

Slud, P. 1960. The birds of Finca "La Selva," Costa Rica: A tropical wet forest locality. Bull. Am. Mus. Nat. Hist. 121: 51-148.

Smythe, N. 1970a. Relationships between fruiting seasons and seed dispersal methods in a neotropical forest. Amer. Nat. 104:25-35.

————. 1970b. The adaptive value of the social organization of the coati (*Nasua narica*). J. Mammal. 51:818-820.

Snow, D. W. 1962. A field study of the black and white manakin *Manacus manacus* in Trinidad. Zoologica 47: 65-104.

Struhsaker, T. T. 1969. Correlates of ecology and social organization among African cercopithecines. Folio Primat. 11:80-118.

Tesh, R. B. 1970. Notes on the reproduction, growth, and development of Echimyid rodents in Panama. J. Mammal. 51:199-202.

Some Considerations of Biological Adaptation by Aboriginal Man to the Tropical Rain Forest

FRANK W. LOWENSTEIN

Introduction

Aboriginal man is usually defined as the original inhabitant of an ecosystem. In tropical west central Africa, the Pygmy and the Bushman are to be considered as the aboriginals although they have been pushed out of West Africa and also to a large extent Central Africa over the last 1500-2000 years by Bantu-speaking people coming from North and East Africa. In South America, the Indian, who came to the American continent in prehistoric times (20,000-40,000 years ago), was the only master of the vast rain forest covering the Amazon plain until about 500 years ago. With the arrival of the Portuguese colonizers, he retreated into the largely inaccessible upstream areas of both northern and southern tributaries of the Amazon River, where he remained in protective isolation until recently (50-100 years ago). Originally the aborigines of the tropical rain forest both in Africa and South America were nomadic or seminomadic hunters and food gatherers. They have all but disappeared from West Africa but are still to be found in the western part of the tropical forest of Central Africa. All are Pygmies and their number is estimated at several hundred thousand. The Bushman has completely disappeared from the forest and remnants live mainly in the Kalahari Desert. Strictly speaking, then, the present inhabitants of the west and central African rain forest are secondary invaders who brought with them early agriculture and some iron tools. They have been there between 1500 and 2000 years, perhaps sufficient time to adapt. In contrast, the Amazon Indian of today is the original inhabitant, though defeated and decimated. He is basically a seminomadic hunter-fisherman-food gatherer in spite of some agricultural crops which he has learned to plant (e.g., manioc, sweet potato).

If one wants to compare the aboriginals in the African and South American rain forests, then the only representative African group would be the Pygmies. Much less, however, is known about them than about the Amazon Indians, who have by no means been studied exhaustively. For this reason much of the data used here from Africa is based on studies on the present inhabitants although they are not strictly comparable to the South American Indian.

Based on the information discussed in the following chapters this hypothesis is presented: Aboriginal man in the tropical rain forest was well adapted to his environment prior to contact with invading stronger "civilized" peoples.

Climatic Adaptation

It is well known that weight and height (stature) show a strongly positive correlation the world over. As Roberts (1953) has shown, there is a negative correlation between weight and mean annual temperature; he found a correlation coefficient of minus .60 between mean weight and annual tem-

FRANK W. LOWENSTEIN, National Center of Health Statistics, U.S. Public Health Service, Rockville, Maryland 20850.

perature for 116 male samples distributed over the globe. He even developed a prediction formula for mean body weight based on height and temperature, according to which mean weight decreases by 1 kilogram for every 5° F increase in temperature, but increases by 1 kilogram for every increase in stature of 1.4 centimeters. Whether lean body mass or overall weight is associated with temperature is not known as we lack investigations on body composition of indigenous groups. One can agree with Roberts that there may be a mean weight level for a given climate, about which actual weights vary according to socioeconomic and nutritional standards. One must further agree with Roberts' conclusion that universal "norms" of weight based on European or North American standards should not be applied indiscriminately in nutrition and growth studies of other racial groups in different climates. This leads us to the important question of differences in body composition, about which very little is known. The skeleton of the American (Baker and Angel 1965) and South African Negro (Walker 1955) shows a greater bone mass than in the Caucasians, but this needs confirmation from tropical Africa. Studies by Edozien (1965b) in Ibadan have shown that the west Nigerian villager (belonging to the Yoruba tribe) has a mean extracellular fluid space per unit of body weight nearly 25 percent greater than adult European and Nigerian university staff members of the same age. This is due to (1) less body fat resulting in a higher proportion of water to total body weight; (2) probably a greater muscle mass (most villagers excreted high amounts of creatinine in their urines indicating high muscular activity); and (3) a greater amount of plasma proteins with a lower albumin and a much higher globulin level. These findings may be regarded as having adaptive value in the humid heat. Comparable data on body composition of the Amazon Indian are lacking completely. More work has been done on basal metabolic rate in Brazil, but usually with "civilized" Brazilians in different parts of the country. According to Galvão (1948, 1950), the heat produced per unit of active tissue mass is less in hot climates and the heat production is proportional to the metabolically active weight, independent of the heat loss through the body surface (invalidating Rubner-Richet's law for the tropics). A relative reduction of mass with respect to surface area is of value in a hot climate; a good correlation exists between the temperature of a part and its mass/surface ratio (Schreider 1957). On the other hand, recent work by Consolazio and Shapiro (1961) in North America and by Durnin et al. (1966) in Britain has shown that energy requirements of men subjected to severe heat or solar radiation are increased due to energy needed for the production and evaporation of sweat. If these findings should be confirmed for the tropics, a revision of internationally accepted standards for energy requirements would become necessary; present standards are based on a reduction of 5 percent in calories for each increase in mean temperature of 10° C.

BODY TEMPERATURE, PULSE RATE, SWEAT RATE

Regarding body temperature and pulse rates during physical exertion, Baker (1966) found that Shipibo Indians in the Peruvian Amazon showed no differences from mestizos in rectal temperature after a two hour walk at 5 kilometers per hour, but their pulse rates were significantly lower at the beginning of the test. Sweat rates were significantly greater in the Indians. Ladell (1952) found no differences in sweat rates of Nigerians in Oshodi and no differences in temperatures as compared to fully heat-trained Europeans. In Johannesburg, Wyndham et al. (1952) found lower rectal temperatures, heart rates, and significantly lower sweat rates (⅓) in African workers from Tanzania as compared to Europeans under similar conditions of heat stress (saturated air temperatures between 82° and 96° F). These results demonstrate the need for a standardized, uniform test to measure the effects of heat stress in different groups in comparable conditions.

Sweat rates are especially important as the ready onset of sweating, together with cutaneous vasodilatation, constitutes the main mechanism of adaptation to a hot, humid climate (Ladell 1964). Because of the high humidity, however, the cooling made possible by evaporation is limited. The endocrines, in particular aldosterone and the pituitary antidiuretic hormone (PAH), not only maintain the circulating blood volume, but also protect the extracellular space and the cells from dehydration. The fully hydrated normal individual has 2 to 3 liters of "free circulating water" that can be lost without impairment in functions. One factor that may account for differences in sweat rates is salt intake. The lower the salt (i.e., the sodium) intake, the lower the sweat rate and also the urinary excretion. Considerable differences in salt intake exist in the tropics and some Indian tribes as well as some tribes in West Africa use ashes of plants rich in potassium and practically devoid of sodium in place of salt (sodium chloride). Their food contains a sufficient amount of sodium, usually from meat and milk, to satisfy their minimum requirements. Nothing is known of the occurrence of any salt depletion syndrome among those people. This relatively low sodium intake ensures normal function of the adrenal cortex, whereas the habitually high sodium intake of the "civilized" people in the tropics may, if excessive, suppress aldosterone secretion and thus delay or even reverse adaptation. The loss of other nutrients in sweat needs study in Africans and Indians. High losses of potassium as well as of nitrogen and amino acids, and calcium have been described in unacclimatized Caucasians (Consolazio, Matoush et al. 1963, Consolazio, Nelson et al. 1963); however, with acclimatization these losses tend to diminish.

WATER INTAKE

Observations that Africans in the Congo and in Senegal drink only about one-third or less water than acclimatized Europeans under the same climatic conditions are of interest. It is known that the water content of the average African's food is about twice that of the average European's and this explains, at least in part, the need for less water in the African. Mazer (1970) showed that in Dakar water intake in nine African students was correlated with protein intake, apparently because the specific dynamic action of protein increases with higher temperatures. Mazer hypothesizes that the low protein intake observed in many hot regions can be considered an empirical adaptation of food habits. In view of the fact, however, that aboriginal man both in Africa and South America was (and still is) a hunter who consumes large amounts of animal foods, Mazer's hypothesis seems doubtful.

SKIN COLOR

The black skin color of the African presents a disadvantage from the thermal point of view, at least in the hot sun; in situations where energy must be expended to maintain body temperature, however, as at dawn and dusk in otherwise hot climates, the black color may maximize absorption of solar radiation (Hamilton and Heppner 1966). The thicker horny layer of the skin found in Negroes and the greater amount of pigment (melanin) account for their lesser sensibility to solar radiation (Thomsen 1955). In this case the black color is to be considered as protective against the ultraviolet rays. Dermatologists see no skin cancer in black people. However, the pigment may, possibly, interfere with the formation of vitamin D_2, a biosynthesis taking place under skin exposed to ultraviolet rays. Whether this may contribute to an increase in the incidence of rickets, a vitamin D deficiency disease in urban African slums, is an open question.

CARDIOVASCULAR SYSTEM

Burch et al. (1959) found the cardiac output increased by 43 percent average in 10 resting adult Negroes in the tropical summer of New Orleans. Whether similar changes exist in Africans or in Indians in the rain forest is not known. Pulse rate and oxygen consumption also increased, but the blood pressure decreased in the majority.

BLOOD PRESSURE

Blood pressure is generally lower in many "primitive" groups in the humid tropics than in "civilized" groups of comparable age and sex in urban areas and cooler climates (Lowenstein 1961). The low salt intake already mentioned may be a factor in this picture. In Table I, mean blood pressures in comparable age-and-sex groups of one African and three American Indian groups are presented. The lowest pressures are found in the Kalapalo and Kamaiura of the Upper Xingu, who are the most "primitive" of the three Amerindian tribes and use as salt the ashes of *Eichhornia*, a common water plant in tropical rivers (De Lima 1950). Analyses by Sick (1949) have shown that this plant ash contains 87.6 percent potassium chloride and no sodium whatsoever. The second lowest pressures were found by the writer (Lowenstein 1961) in the Karaja, a "primitive" tribe on the Araguaia with much longer contact with civilization than the Upper Xingu tribes. The Munduruku on the River Cururu, acculturated and christianized for 40 years, had the highest blood pressures of the three Amerindian groups. Their pressures showed the rising trend with age prevalent in all "civilized" groups. They have adopted sodium chloride as a table salt. The African Pygmies in the Congo forest

TABLE 1. Mean Blood Pressures in One African Pygmy and Three Amazon Indian Tribes by Age and Sex (Number in parentheses is sample size.)

	Pygmy		*Kalapalo-Kamaiura*		*Munduruku*		*Karaja*	
Age Groups	M	F	M	F	M	F	M	F
				Systolic Pressure				
15–19	119.8(10)	126.1(14)	94(18)	88(8)	104.7(3)	92.0(4)	109.5(7)	96.0(2)
20–29	124.4(20)	127.6(27)	95(20)	86(26)	117.2(10)	104.3(15)	109.8(9)	104.1(15)
30–39	129.5(27)	127.2(26)	87(13)	76(10)	108.9(13)	109.0(10)	107.3(11)	105.4(10)
40–49	120.7(21)	131.2(18)	91(5)	91(3)	115.8(8)	111.7(6)	100.9(7)	100.0(7)
50–59	128(13)	140.7(9)	–	–	132.8(5)	106.7(3)	109.0(6)	98.0(6)
60+	–	–	–	–	–	–	110.3(6)	108.0(3)
				Diastolic Pressure				
15–19	73.1(10)	71.7(14)	64(18)	58(8)	66.6(3)	59.2(4)	72.5(7)	69.0(2)
20–29	76.2(20)	76.5(28)	66(20)	59(26)	71.4(10)	64.8(15)	72.9(9)	70.7(15)
30–39	78.6(27)	75.2(26)	62(13)	54(10)	73.4(13)	72.8(10)	70.5(11)	72.2(10)
40–49	75.0(21)	80.7(18)	62(5)	57(3)	73.3(8)	70.7(6)	69.1(7)	69.4(7)
50–59	77.9(13)	84.9(9)	–	–	77.2(5)	70.0(3)	70.3(6)	67.7(6)
60+	–	–	–	–	–	–	69.7(6)	64.7(3)

(Mann et al. 1962) showed the highest mean blood pressures, with hypertensive levels at 1.6 percent (over 160/100), and in women the increase with age was significant. Since their way of life appears to be as "primitive" as that of the Amerindians, unknown factors such as urinary tract infections may be responsible for their relatively high pressure. Mean blood pressures in 150 rural Bakongo around Minduli and Kinkala (Acker et al. 1967) were in the same range as those of the Pygmies. Sixty-six Chavantes on the Rio dos Mortos in Brazil showed mean pressures of a similar level to the Karaja (Neel et al. 1964).

VISUAL ACUITY

Mattos (1958) studied the far visual acuity of 18 Indian groups in Brazil and found it to be superior to that of civilized Brazilians from central Brazil and São Paulo. No color blindness was observed.

Growth and Development

Figures on mean or median birthweights are available from various maternity hospitals in West and Central Africa. As one might expect, the lowest mean birth weights have been recorded among 40 Pygmy infants: 2635 grams (Mann et al. 1962). Newborn weights from nine other locations range from 2800 to 3205 grams. Where the data have been separated by sex, boys generally weigh between 100 and 200 grams more than girls at birth. Birthweights in rural areas are lower than in large cities, to judge from the data for Nigeria (Gans 1963, Edozien 1965a, Morley et al. 1968) and Gambia (McGregor et al. 1968). McGregor also reports seasonal variations in Gambia, birthweights during the rainy season being between 200 and 300 grams lower than during the dry season. The rainy season is the time of maximum physical activity, when everyone is out in the fields; it is also the season of highest incidence of malaria and other infections. In addition, the food supply is at its lowest ebb. All this must contribute to the seasonal variation in birthweight. An increase in birthweights over a 10 year period has been reported from parts of Africa (Dupin, Masse, and Correa 1962, McLaren 1959), probably reflecting the general improvement in maternal health and nutrition. No comparable information appears to exist on the American Indian.

FIRST SIX MONTHS

The first 6 months in the life of an African baby are the happiest. Breastmilk is usually available in sufficient quantity whenever desired, and the infant is in continuous close contact with its mother, well fed and secure. This sheltered existence and the presence of antibodies in the mother's milk offer protection against everpresent infections and infestations. Weight and height curves from seven localities in four countries of West and Central Africa indicate satisfactory gains up to the age of 6 months (Gans 1963, Edozien 1965a, Morley et al. 1968, Bascoulergue and Pierme 1961, Holemans 1960, McGregor et al. 1968, Gilles 1964). Birthweight was doubled at 6 months on the average, and after only 3 months in Yaounde, Cameroon (Bascoulergue and Pierme 1961) and Ituri, Congo (Tams 1959). The psychomotor development of the healthy, breastfed African infant in Kampala has been shown by Geber and Dean (1964) to be superior to that of the Caucasian infant. By applying Gesell tests to over 300 children, these authors were able to show mean development quotients of 130 to 140 for motility and manual dexterity and of 120 for adaptability, language, and social reactions (100 being the mean for the North American child). This precocious development, however, was not observed in African infants brought up in the Western way. Unfortunately, very little is known of the early development of tropical American Indian infants, so that no comparison is possible at this time. Observations on a few Karaja infants (Lowenstein 1958) have shown their weights at 3 and 6 months to be comparable to those of African village children in western Nigeria and Gambia.

SIX MONTHS TO FIVE YEARS

The African infant continues to grow as rapidly as the average Caucasian one if it has the good fortune to have parents who belong to the small elite (university teachers, high government functionaries, etc.). Weight curves of such elite children from Ibadan, Nigeria (Edozien 1965c), and Yaounde, Cameroon (Bascoulergue and Pierme 1961), are comparable to the 50th percentile of Caucasian children from Boston (Nelson 1965). The great majority of African children, however, show a decline in growth rate about the sixth month and this tendency becomes accentuated during the second year of life. At two years of age, the mean weight of African village children is between 2 and 3 kilograms less and their height is 5 to 6 centimeters lower than that of the optimal group. This situation continues so that at 5 years the weight deficit amounts to 3 to 4 kilograms and the height deficit reaches 9 to 11 centimeters as compared to more fortunate members of the same ethnic group. If one tries to analyze the causes for this serious growth retardation, one arrives at a fairly typical picture combining poor nutrition and a variety of ubiquitous tropical infections varying from malaria to measles.

The African infant usually begins to experience inadequate nutrition at 6 months, when the amount of breastmilk becomes insufficient for the increasing requirements and the traditional supplementary foods are of relatively low nutritional value (starchy paps from cassava, yam, or maize). Animal milk, the normal supplement in Europe and North America, is not available in the tropical forest. Other protein-rich animal foods, such as fish, eggs, and meat, are too expensive for the great majority of families; furthermore, they are considered taboo for young children in many traditional communities. Beans, the main source of vegetable protein, are usually available to the young child only in small amounts, if at all. This inadequate nutrition saps the resistance to the many infections to which the infant is exposed. If he survives malaria at 18 months, he may succumb to measles at 24 months. A vicious circle of malnutrition and infection develops, leading to a serious state of protein-calorie deficiency (Kwashiorkor) that ends fatally unless treated early (Lowenstein 1962). If the child recovers, he usually remains "retarded" in his growth and development for years.

The few observations by Lowenstein (1958) on young children of the Karaja Indians seem to indicate that their heights and weights at ages 2 to 5 years fall between those of the West African elite and village children. Mean heights and weights of the few children observed at a Munduruku village on the Cururu River (Lowenstein 1958) were significantly lower than those of the Karaja and comparable to African children from the Congo village of Feshi (Holemans 1960). Lowenstein (1958) noticed no serious mal- or undernutrition among the Munduruku children and would, thus, ascribe the differences to genetic factors. Heights of Cashinahua children of the Peruvian Amazon and the Trio and Wajana in Surinam fall between those of the two Brazilian tribes. Weights of the Cashinahua were superior to those of the Karaja (Figure 1).

SIX YEARS TO TWELVE YEARS

In West Africa, the gap in the growth curve between the optimal and village children persists in the 6 to 12 year age group (Edozien 1965a). Weight deficiencies are 5 to 6 kilograms in Nigeria (Edozien 1965a, Hauck and Tabrah 1963) and 3 to 4 kilograms in Ghana (Davey 1960–1962). Height deficits are 5 to 13 centimeters in Nigeria and 6 to 8 centimeters in Ghana. Cameroonian children (Pele 1967) continue to be taller and heavier than those from Ghana and Nigeria; however, the sample comes largely from the more privileged groups. The few data on Karaja children again show an intermediate position between those of optimal and village African ones (Lowenstein 1958); the Munduruku lag behind, although differences in weight decrease with increasing age. The Cashinahua, Trio, and Wajana heights and weights are intermediate (Johnston et al. 1971).

FIFTEEN TO EIGHTEEN YEARS

With the growth spurt at puberty, differences in weight and height between the few African rain forest groups for which data are available tend to diminish significantly, implying a sufficient improvement in nutrition at adolescence to enable normal functioning of the endocrine system. This phenomenon has also been observed at puberty among rural Amazonian children in Brazil (Lowenstein 1967). At age 18, Karaja boys are 10 centimeters taller and 14 kilograms heavier (Lowenstein 1958) than Ibo boys from east Nigeria (Hauck and Tabrah 1963); they are 17 centimeters taller and 12.7 kilograms heavier than Munduruku boys (Lowenstein 1958). Munduruku girls, by contrast, are still 20 centimeters shorter and 12 kilograms lighter than Ibo girls from eastern Nigeria. Heights of the Cashinahua, Trio, and Wajana more or less parallel those of the Munduruku; Cashinahua male weights, however, are superior to those of the Karaja (Johnston et al. 1971, Glanville and Geerdink 1970; see Figure 1). Unfortunately, no comparable data are available from African Pygmies, but one would expect them to be similar to the Munduruku in stature and weight.

ADULTS

Data on adult mean heights and weights are available from six African and four Amazonian groups; weights have also been recorded from several other African populations. All but two African (Holemans 1960, Hauck and Tabrah 1963) and two Indian groups (Neel et al. 1964, Lowenstein 1958) are different from those from which data have been recorded on children. Among the African groups, the Ghana Ashanti men are the tallest, followed by Yoruba men in western Nigeria; the Bakongo from Congo Kinshasa are the shortest (168.9 versus 157.2 cm). Among Amazonian Indians, the Karaja and Chavante men are the tallest, being comparable in height to the Ashanti and Yoruba, but weighing 5 to 9 kilograms more. The Munduruku are the shortest (155 cm) of all the groups and among the lowest in weight (52 kg). Differences between women are less pronounced than between men. Nevertheless, Ashanti and Yoruba women are 8 centimeters taller and 8 kilograms heavier than Munduruku women, with the Cashinahua, Trio, and Wajana intermediate. Differences between the sexes are marked, the men of the same ethnic group usually being 7 to 14 centimeters taller and 7 to 12 kilograms heavier than the women. An exception is the Bakongo, where differences in weight between the sexes are only 2 to 4 kilograms.

Weight data on adults at different ages are available from five ethnic groups. Among the Ewe in eastern Ghana (Davey 1960-1962), 60 to 69 year old men were 4.5 kilograms heavier than 30 to 39 year old ones. This trend was reversed among the Ashanti, where the older men weighed 6 kilograms less than the younger ones. In the Amazon, the older Karaja males weighed 7 kilograms less than younger ones, whereas no difference was observed among the Munduruku and Cashinahua (Lowenstein 1958). Women in the two West African groups showed little or no change in weight with age (1.0-1.5 kg), whereas the oldest Karaja women weighed 7.3 kilograms less than the younger ones. No difference existed among Munduruku and Cashinahua women.

Obesity is unknown among hunter-gatherers and poor farmers. It is a relatively recent phenomenon in urban areas of West and Central Africa among those few groups that have become sufficiently prosperous to keep eating as much or more with age as they did when they were young and more active.

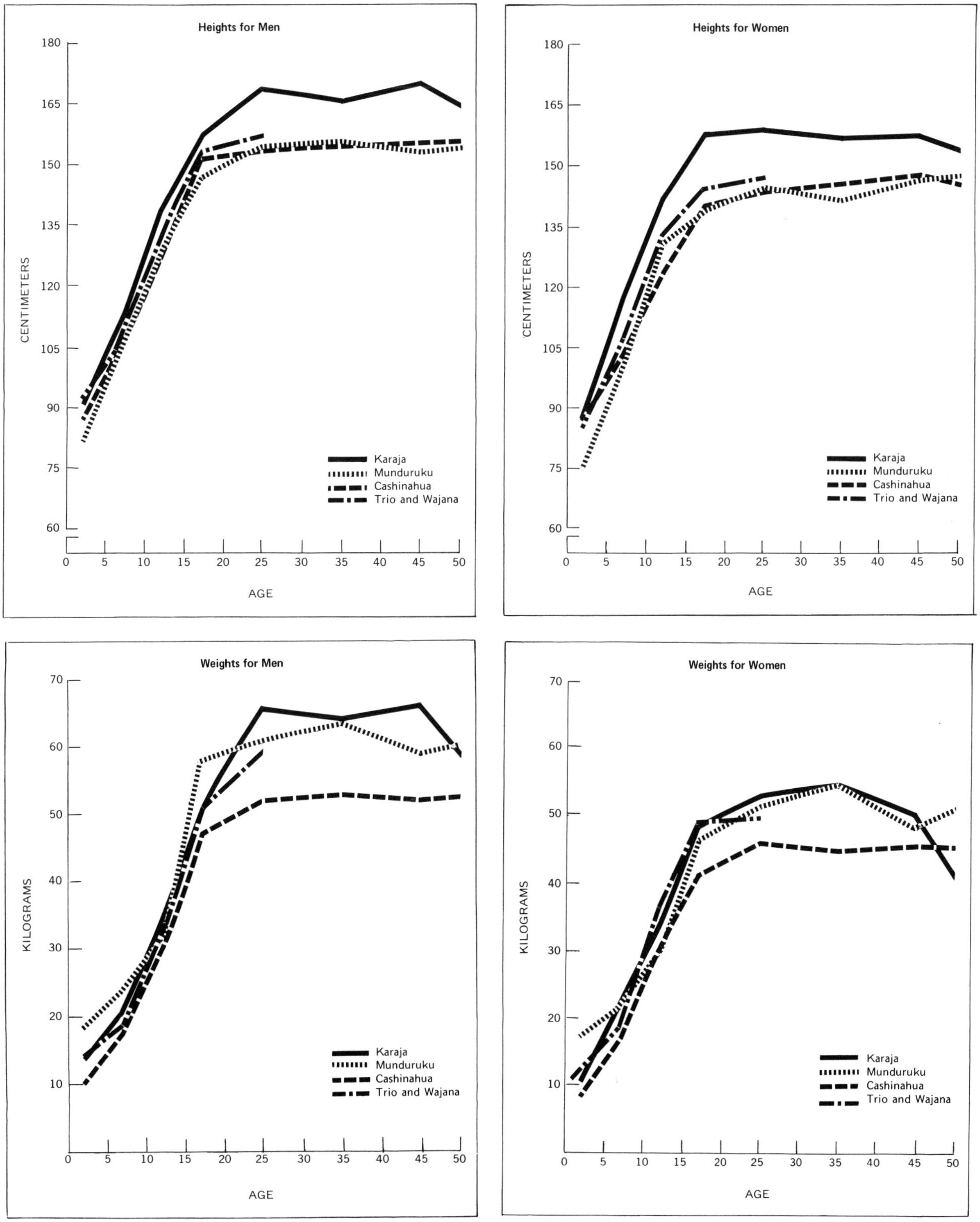

FIGURE 1. Growth curves in four Amazon Indian tribes.

Food and Food Intake

The diet of the African rain forest aborigines consisted (and still consists) mainly of "bush meat," fish, leaves, berries, roots, and indigenous grains (guinea corn and millets), sometimes with the addition of wild rice and chickpeas (Oke 1967). Hot chilipepper and salt from seawater or mines in the Sahara were used as condiments. Sheep and goats were introduced from the Middle East, as was cultivated rice. Portuguese slave traders brought sweet potatoes, maize, and manioc from South America at the end of the 15th century. Many other tropical fruits and vegetables were introduced from the Middle East or South America, among them bananas, citrus, mangos, okra, eggplant, and papayas. Among the newly introduced foods, manioc or cassava (*Manihot utilissima*) deserves special attention. The tuber contains hydrocyanic acid, a strong poison, and a tedious process of soaking, peeling, grating, pressing, and drying is required to make it safe to eat. Furthermore, its nutritional value is low because of its minimal protein content (1 percent or less). The green leaves, however, constitute a valuable food, containing 7 percent protein and high amounts of carotene and ascorbic acid. According to Terra (1964), considerable amounts of cassava leaves are consumed daily in various parts of West and Central Africa (up to 500 grams in Stanleyville). Many other kinds of leaves with protein content between 4 and 7 percent are rarely eaten.

The Amazonian Indian has at his disposal a multitude of edible wild plants and fruits, as well as game and a great variety of fish. From what we know of his food habits, he makes good use of these resources. Many groups clear a small patch of forest by the slash-and-burn technique for planting manioc, sweet potatoes, maize, papaya, and since European contact, bananas and sugar cane. The Karaja, who live on Ilha Bananal in the Araguaia River, are excellent fishermen and consume large quantities of fish and turtle. The Munduruku prefer game (Lowenstein 1958) and those living at the mission plant large gardens and earn part of their food by working for the missionaries. The Chavante plant beans and squash in addition to maize and obtain two harvests a year (Neel et al. 1964). Meat is their prestige food, but is not eaten every day. Recently, they have begun to eat fish, but only if meat is unavailable. The banana has become the staple among the Surara and Pakidai on the Demini and Araca Rivers, tributaries of the Rio Negro (Becher 1957). In addition, they cultivate the pupunha palm, the fruit of which is rich in carotene.

Dietary surveys in seven West and Central African countries provide data on food consumption; only forest areas are included in Table 2. With three exceptions (Casamance, Ghana in July, and East Cameroon), mean caloric intake is adequate to high. In Congo Brazzaville, calorie consumption per head decreases with an increase in family size from 2700 to 3400 calories in families of 1 to 4, to 2000 calories in families of more than 9 members (Acker 1968). Total protein intake is generally adequate except in some villages in Nigeria, Congo Brazzaville, and Congo Kinshasa (Zäire). The consumption of animal protein, however, is low or deficient in most areas. Consequently, the biological value of the protein in most diets is relatively low, with the sulfur-amino-acids and tryptophan the most limiting amino acids (Martineaud 1967). If the percentage of fat calories falls below 15 percent, it must be considered deficient; this is the case in five areas out of eight. The intake of calcium is low except in the Congo Kinshasa, whereas iron is sufficient or high except in the Ivory Coast and East Cameroon. Abundant supplies of red palm oil, and green and yellow vegetables provide ample vitamin A, except in the Ivory Coast and East Cameroon. Vitamin C intake is satisfactory only in the Cameroon and Congo Brazzaville. B vitamins are generally marginal or deficient because of low consumption of meat and whole grains. There are no quantitative data on food consumption by Pygmies or Amazonian Indians.

Health and Nutritional Status

Vital statistics provide some insight into the health and nutritional status of a population or community, provided they are reasonably accurate. Between 1954 and 1963, birth rates in five areas in West Africa were about 2.5 times those of many industrialized countries, with Ghana being lowest and eastern Nigeria highest (Edozien 1965a, Davey 1960-1962, Nicol 1958). Crude death rates from three countries were three times higher than in in-

TABLE 2. Food Consumption in Forest Villages of Seven Countries in West and Central Africa*

Provenience		*Daily Calories*		Other Nutients								*Daily Percent*
and Date Country	*No. of Villages*	*Intake (No.)*	*Requirement (%)*	*Protein (grams)*	*Calcium*	*Iron*	*Vit.* A	*Vit.* B_1	*Vit.* B_2	*Niacin*	*Vit.* C	*Calories from Fat*
1962												
Senegal	1	2058	86	72(9)	±	±	++		—	—	±	28
Casamance	2	1630	77	41(5)	±	±	++		—	—	±	21
1955–1956												
Ivory Coast	1	2054		56(2)		±	±		—	±	±	8
1960												
Ghana, July	?	1325	68									
Ghana, January	?	1636	84									
1955–1965												
Nigeria	10	1851	90	35(6)	±	++	++	±	—	—	±	
		to										
		2487	110	57.5(28)	±	++	++	±	—	—	±	
1958												
East Cameroon	?	1580	65–75	38	±	±	±	±	—	±	++	5–13
1956–1958												
Congo		1952	90	33(10)	±	±	++	±	±	±	++	8
Brazzaville	35	to										
		3630	170	58(41)	±	±	++	±	±	±	++	27
1955												
Congo Kinshasa	5	1689		17	+							1.5
and Kwango	?	to										
		2712		49.8	+							9

* Data derived from Bascoulergue and Bergot (1959), Cros (1967), Davey (1960-1962), Gilles (1964), Holemans (1960), and ICNND (1967). ++ = high; + = sufficient; ± = low; — = deficient; ( ) = percentage of animal protein.

dustrialized countries. The percent distribution shows, however, that 74 percent of all deaths in Ghana in 1960 were in the youngest age group (0 to 5 years); 54.3 percent of deaths in the Ivory Coast were within this group. The highest number died during the first two years. Correspondingly high figures are available from four Nigerian forest villages (Morley et al. 1968, Gilles 1964, Nicol 1958) and one village each in Gambia (McGregor et al. 1968), Senegal (Dupin, Masse, and Correa 1962), and Congo Kinshasa (Holemans 1960), where cumulative death rates in the same age group per 1000 live births range from 247 in the Congo to 572 in Imesi, western Nigeria. No comparable figures are available from Amazonia, but some indication can be gained from the cumulative death rates for ages 0 to 15 of 308 per 1000 among the Chavantes (Neel at al. 1964) and 475 per 1000 among the Karaja (Lowenstein 1958).

If he survives to age 5, an African or Amazonian child will probably continue living at least to middle age unless killed by an accident or, in the case of women, by complications of childbirth. The hazards of pregnancy are indirectly revealed by information on weight gain by pregnant women in three African countries. In Nigeria (Edozien 1965a) and Congo Kinshasa (Holemans 1960), the same women were observed from the beginning to the end of their pregnancy, while in Ghana (Davey 1960-1962) pregnant and nonpregnant women of the same age and tribe living in the same locality were compared. The mean weight gain in the Congo was 5 to 6 kilograms and in the western Nigerian village it was 7.1 kilograms; the western Nigerian reference group (Ibadan elite), by contrast, showed a mean gain of 12.3 kilograms, which is comparable to that in Europe and North America. The figures from the Ghana forest area were only 1.5 kilograms in women below the age of 20. Only in women who had their first baby between ages 20 and 24 was the weight gain comparable to that of Bakongo women (5 kg). No data exist on Amazonian Indian women. The fact that little or no gain during pregnancy has also been reported from tropical areas in southeast Asia (Venkatachalam, Shanker, and Gopalan 1960), however, suggests that this is a widespread phenomenon deserving attention. Although not yet entirely understood, this low weight gain certainly implies a caloric deficit. Numerous pregnancies in close succession force a woman to utilize her already precarious reserves of nutrients to nourish the fetus, leading to her premature ageing and death.

A number of signs are usually associated with nutritional deficiency diseases, but they are not necessarily diagnostic and should be interpreted in conjunction with biochemical assays and dietary data. Those most frequently observed in surveys in Ghana (Davey 1960-1962), Nigeria (ICNND 1967), and Congo Kinshasa (Holemans 1960) refer to the skin, tongue, lips, gums, and hair. Dryness and follicular hyperkeratosis of the skin occurred in between 4 and 48 percent of people examined in Ghana and Nigeria, and may indicate deficiency of vitamin A and/or fatty acids. Abnormalities in the surface of the tongue (smooth) and lips (dry and cracked, particularly at the angles of the mouth) were most frequent in Nigeria and probably reflect deficiency of vitamins of the B complex. Angular stomatitis was also found in 50 percent of 4498 children examined in Senegal, with a maximum frequency at age 4. It is similarly common among children in Togo and Cameroon (Martineaud 1967). Swollen, red gums, most frequent in Nigeria, may be due to vitamin C deficiency. Hair changes, usually consisting in dyspigmentation and loss of curliness, have been observed among Nigerian and Bakongo children between 1 and 3. These are early symptoms of protein deficiency. Endema, another sign of protein deficiency, was also recorded in the same age group in three Nigerian forest villages.

Indications of rickets, such as bossing and/or deformities of the extremities (bow legs, knock knees), are present in the Nigerian forest (ICNND 1967), in Senegal, and the Ivory Coast. In the latter country, 53.3 percent of the children examined exhibited deformities and many of the women had rachitic pelvis. Two other signs are of uncertain significance. One is bilateral swelling of the parotis gland, present in a fair percentage of older children in Ivory Coast (ORANA 1956). In the absence of an infection such as mumps, this may indicate a dietary imbalance combining too much carbohydrate with too little protein. The other is enlargement of the liver, frequent especially in children in Nigeria and Congo Kinshasa, which is probably due to the combined effects of chronic malaria and protein deficiency. Finally, enlargement of the thyroid gland, resulting from iodine deficiency, has been reported from Nigeria.

Surveys among two Amazonian tribes (Lowenstein 1958) have shown enlarged thyroid, with higher prevalence among the Karaja and among females (32 percent of the women as against 18 percent of the men). Similar sex differences existed in other signs. Dry skin with follicular hyperkeratosis occurred among 17 percent of Munduruku women and 10.4 percent of the men; fissured and smooth tongue was more common among the Karaja, where it was present in 25 percent of the women and 17 percent of the men. Some swelling and redness of gums existed in both tribes. A 1954 survey among the Chavantes (Saddock and Braga 1955), by contrast, revealed no signs of possible nutritional deficiency except for endemic goiter in one village. Similar results were reported in 1962 (Neel et al. 1964) and 1965 (Neel et al. 1967). These observers all remarked on the splendid physique and health, particularly of the younger men. No goitèr has been found among the Upper Xingu Indians in Brazil or the Trio and Wajana in Surinam, who are said to be "fit, well nourished and in good health."

HEMOGLOBIN AND BIOCHEMICAL DATA

Determination of red cell hemoglobin is useful as a screening device for the prevalence of anemia and relatively easy to ascertain, so that it is usually done during surveys. Differences in methods of determination and presentation of the data, however, make comparison of the data from four West and Central African countries difficult. Lowest mean hemoglobins were found in the forest areas of Ghana and Congo Kinshasa. Those from the Ivory Coast and Nigeria are higher but still significantly lower than the mean values for the Nigerian reference group in Ibadan. Values for the latter group are comparable to healthy Caucasians of the same age and sex. Differences between the sexes are marked in most places, with females having lower values except when very young. Pregnant women usually have lower values than nonpregnant ones. In Nigeria, significantly lower values have been obtained during the hot, dry season. One of the main causes of anemia the world over is iron deficiency. Investigations from Nigeria (Edozien 1965a) and Cameroon (Bascoulergue and Bergot 1959), however, have shown high normal values for serum iron and normal values for total iron binding capacity of the blood serum in all age groups, including Cameroon women at different stages of pregnancy. Serum iron was lower during the first and second trimester than in the last trimester of pregnancy. Serum copper values were normal, but higher in Cameroon women, especially during pregnancy. These findings, as well as the results from dietary surveys, suggest that iron deficiency is not a major cause of the anemia prevalent in many forest areas of Africa.

Mean hemoglobins among the Chavantes (Neel et al. 1964) are similar to those just described from West and Central Africa. Again, significantly lower values were found in females of all ages than in males. A hematological study of 69 Upper Xingu Indians between the ages of 5 and 55 years showed a generally normal distribution of values for hemoglobin, hematocrit, red and white blood cells, and serum iron (da Silva 1966).

SERUM PROTEINS

Serum proteins have been studied in four African countries. Lowest values for total proteins were found among the Bakongo in Congo Kinshasa (Holemans 1960) and highest values in western Nigeria (Edozien 1965a), with differences between 15 and 20 percent. Albumin values were higher in the Congo Kinshasa than in either Nigeria or Congo Brazzaville. Furthermore, they were highest in young Bakongo children and decreased with age; in the Nigerian Yoruba, they remained more or less stable. Different globulin fractions were analyzed in only two countries, Nigeria and Congo Brazzaville. Values for α_1 and β_2 globulins were similar, while those for α_2 globulin were twice as high in Nigerians as in Congolese. Gamma globulins showed equivalent values in the two groups and increased with age. They were about 30 to 50 percent higher in adult villagers, however, than in the Ibadan reference group. Among Pygmies, gamma globulin values are even higher than albumin values in many cases. Gamma globulin is known to contain several immune globulins involved in defense against malaria and possibly other tropical infections. The high levels found in Africans and other rural tropical groups can thus largely be explained on the basis of continuous exposure to a legion of infections and may be considered as having adaptive value. This conclusion is supported by Neel et al. (1964), who found gamma globulin levels among the Chavante as high or higher than those in Africans of comparable age groups; the other protein fractions showed similar values.

CALCIUM PHOSPHORUS LEVELS

Serum calcium levels in Nigerian (Edozien 1965a) and Congolese (Bascoulergue and Perme 1961) children were normal but lower than those of the Ibadan reference group. Serum phosphorus levels were the same for village and reference children. No equivalent data exist on Amazonian Indians. Plasma levels of vitamins A and C were generally satisfactory in Nigerians of all ages, whereas urinary excretion levels for vitamins B_1, B_2, and niacin were deficient or low in the majority (ICNND 1967). Plasma vitamin B_{12} levels were almost twice as high in Nigerians as in Europeans resident in Ibadan, whereas folic acid levels were lower in about 40 percent. Edozien (1965a) considers the high vitamin B_{12} levels as compensatory for the low folate levels.

Information on mean total cholesterol levels is available from Nigeria (ICNND 1967), Congo Brazzaville (Acker et al. 1967), and Congo Kinshasa (Holemans 1960) in Africa, the Chavantes in Brazil (Neel et al. 1964) and Amerindians in Surinam (Luyken and Luyken-Koning 1960). The lowest values, reported from the Pygmies (Mann et al. 1962), are 100.9 milligram percent. The only lower values were found among newborn Bakongo (89 mg percent). At comparable ages, the four ethnic groups all showed values between 110 and 160 milligrams; women showed an increase with age, whereas men did not. This differs from findings among industrialized Caucasians, where men also show a significant increase with age.

Infections and Infestations

In the tropical rain forest, man is exposed to a

complex ecosystem characterized by a great variety of species with few individuals per species; this includes infectious and parasitic organisms with many potential vectors, intermediate hosts, and alternative hosts. Numerous kinds of sexual infections, many organisms producing low intensities of infection, and low worm burdens exist, along with a large variety of asexual infections that lead to partial or permanent immunity. Finally, there are many species of arthropods and reptiles, few of them venomous, and in general with a low incidence of bites. Dunn (1968) has tabulated 20 species of parasitic helminths and protozoa in Pygmies of the African rain forest, showing the following situation:

Parasitic Group	*No. of Species*	*Infectious Agent*	*Percent of Population Affected*
Helminths	11	Hookworm eggs	71–92
		Onchocerca volvulus	5–100
Intestinal protozoa	6	*Entamoeba histolytica*	28–55
Blood protozoa	3	*Plasmodium malariae*	20–31

Data on Amazon Indians are less complete; however, Lowenstein's (1958) studies on the Karaja showed four species of intestinal helminths and five species of intestinal protozoa. Hookworm eggs occurred in 76.8 percent and *Entamoeba histolytica* in 7.2 percent of the sample. Comparable data on the Chavantes (Neel et al. 1968) and the Upper Xingu Indians (da Silva 1966) reveal a similar variety of intestinal parasites, with hookworm egg frequencies of 96.7 percent and 81 percent, respectively, and *E. histolytica* cyst frequencies of 48.3 and 61 percent. The helminth infections appeared to be light. Only three Chavante women had malaria parasites in their blood of a total of 76 examined. This contrasts with the situation among Upper Xingu Indians, where da Silva (1966) found 55 percent with enlarged spleens, and among two Surinam tribes (Schaad 1960), which exhibited nearly 100 percent frequency of enlarged spleens, presumably due to chronic malaria infection. A high percentage of Chavantes had specific antibodies in their blood against the following infectious agents: pertussis 45.5 percent, salmonella (4 strains) 45 to 85 percent of 63 tested, poliomyelitis I, II, III 93 percent, and measles 83.3 percent of 37 tested. The antibodies for measles and pertussis are very likely introduced, while salmonella, poliomyelitis, and certain arboviruses are endemic. In spite of these findings, the Chavantes do not appear disease ridden in comparison to other Indian or African groups. Isolated primitive man was in better equilibrium with his environment than is civilized man (Neel and Salzano 1967). Maternal gamma globulin available for placental transfer and the prolonged period of breastfeeding may permit a relatively smooth transition from passive to active immunity against many pathogens. Bacterial and viral stool antibodies found in breastfed infants, and derived apparently from breastmilk, may play an important role in enhancing this immunity.

The heavy impact of newly introduced diseases such as measles, tuberculosis, and influenza on Amazonian Indians has been documented by several investigators. Measles epidemics were witnessed by Sioli (1955) among the Munduruku in 1942 and by Ribeiro (1956) among the Urubu in 1953 and several Upper Xingu tribes in 1954. Lack of immunity against this new pathogen and helplessness during the acute phase of the infection are responsible for the very high death toll. With practically everybody stricken and immobilized, the group is unable to take care of the most urgent needs, such as procurement and preparation of food. When medical help was made available in 1954, the death rate declined from 27 to 7 percent on the Upper Xingu. Similar results have been described from initial introductions of influenza and/or tuberculosis (Biocca 1963). Tribes like the Bororo have become nearly extinct because of the deleterious effect of tuberculosis, and the percentage of positive tuberculin reactions is increasing in several tribes on the Upper Xingu (Nutels 1967) and in Surinam (Schaad 1960), with a fairly high incidence of clinical disease.

In his study of "convívio e contaminação," Ribeiro (1956) developed a classification of Brazilian tribes based on demographic data over the past 100 to 200 years and employed it for predicting the survival of these tribes. All are known to have diminished markedly during this period and their continued existence will depend mainly on two factors: (1) a birth rate sufficiently larger than the death rate to maintain the present numbers, if not augment them; and (2) a survival rate into mature adulthood sufficient to ensure a reproduction rate high enough to guarantee survival. Most tribes suffer from a childhood mortality rate that is as high or even higher than that among many African tribes. In contrast to Africa, however, the birth rate of many Indians does not compensate for this high mortality so that their number is constantly decreasing. Even groups like the Karaja, who have had more than 100 years contact with civilization, keep declining. What is known of the time before contact with "civilization" indicates that Amazon tribes had reached a population equilibrium. Not only has this equilibrium been disrupted, but in addition it appears that something more serious has happened to most of the tribes, which is expressed in loss of the will to survive as a group.

A brief review of the major infections and infestations among Africans in the rain forest reveals that malaria is by far the most important infectious disease, both in terms of numbers of people affected and of fatalities, especially among young children (Lowenstein 1970). Prevalence ranges between 25 and 98 percent. If a child survives an attack, he develops a certain immunity and although subject to reinfection, he is less and less severely affected as he grows up. Second to malaria is bilharziasis, usually the urinary type caused by *Schistosoma hematobium* (prevalence between 10 percent Akufo, Nigeria, and 76 percent Pokoase, Ghana). Several filaria infections follow, the most important being onchocerciasis, causing "river blindness." Investigations from West Africa have shown that onchocerciasis is usually less serious in the forest belt than on the savanna and causes few eye lesions. Trachoma, a viral eye infection, is also less frequent and less severe in the forest. Tuberculosis and tetanus are widespread in the West and Central African forest. Finally, intestinal parasites have been found in many surveys, with ascaris and hookworm present in 60 to 80 percent of children examined between ages 1 and 14. In short, while the African forest dweller is exposed to as many or more infections and infestations as the Amazon Indian, he seems not only to have "learned" over many centuries of exposure to live with them, but to increase in numbers in spite of a high mortality rate in young children. One should keep in mind, however, that most of these Africans are not the original inhabitants of the forest areas, so that the situations are not strictly comparable.

Some Special Problems of Adaptation

MALARIA AND THE SICKLE CELL GENE IN AFRICA

Sickle cell anemia is a very serious genetic disease in Black Africans everywhere. Those affected are homozygotic (SS), and they rarely survive to mature adulthood. The carriers, who are heterozygotic (SH), do not develop the disease and may live to old age; they show a relative immunity against malarial infection. Livingstone (1958) studied the distribution of the SH gene in West Africa and found it to be more frequent in the south (i.e., the forest zone), with the distribution of the malaria parasite following a similar gradient. In the rain forest area more than 15 percent of the people usually have the sickle cell trait and malaria is either hyperendemic or holoendemic (75-100% of the people show malaria parasites in their bloodsmear). In this environment, the carrier of the SH gene seems to have a selective advantage over the normal noncarrier. Although sicklers are almost as easily infected as nonsicklers, they show lower parasitic rates in the younger age groups and suffer less from cerebral malaria and blackwater fever. Since these are complications of falciparum malaria frequently leading to death, the sicklers have a lower mortality from falciparum malaria (Allison 1954, Raper 1956). In addition, sickling females may have lighter falciparum infections than normal females and thus may have a higher net reproduction rate; in this way, malaria maintains high frequencies of the sickle cell gene. There is evidence from archeology and ethnology that the Akan and other Kwa-speaking tribes from the east and northeast carried the sickle cell gene with them when they invaded the West African forest. Remnants of the aboriginal rain forest inhabitants in Portuguese Guinea and eastern Liberia have a low sickling frequency, as do the "true" Pygmies, the Babinga of the Congo. Finally, Livingstone (1958) has shown how agriculture in the forest created suitable conditions for the propagation of *Anopheles gambiae,* the carrier of malaria, and thus for the holoendemicity of malaria with a year-round cycle of infection.

TROPICAL NEUROPATHY

While the sickle cell gene, at least in the heterozygote carrier, seems to offer adaptive advantages to the forest dweller, another condition may constitute an example of lack of adaptation. For some years, Monekosso and Wilson (1966) have drawn attention to a very interesting neurological syndrome present in western Nigeria and leading to optic atrophy, nerve deafness, and sensory spinal ataxia. Nutritional amblyopia, particularly in young adults living on poor diets, had been described in the 1930s (Moore 1937) and was thought to be due to a deficiency of vitamin B_2 and/or a toxic factor in food. Recently, Osuntukun, Monekosso, and Wilson (1969) have shown the frequency of cassava consumption to be associated with higher plasma thiocyanate levels and low mean vitamin B_{12} levels in the serum in two Nigerian villages where a significantly higher percent of neurological abnormalities occurs (e.g., visual difficulties, paraesthesiae, impaired hearing, vibratory insensitivity). Sulfur amino acid levels in plasma were lower in patients suffering from tropical ataxic neuropathy. This may be due to the use of sulfur for the detoxification of the cyanide content of the cassava, which is high in the processed cassava consumed in western Nigeria. Whereas the Brazilian Indian had learned to rid the tuber of most of its cyanic acid through a tedious process, the African did not. The second important fact is the correlation between the quantity of cassava eaten and the development of neuropathy. A diet high in cassava is poor in essential nutrients, particularly protein and B vitamins, unless supplemented with animal protein (such as fish in the Amazon). The problem of the tropical neuropathy is by no means solved and offers a fruitful field for further investigations.

IODINE DEFICIENCY WITHOUT GOITER

A group of Yanomamö Indians on the Manaviche River, a tributary of the Orinoco in southern Venezuela, were shown to be in good health and nutritional state when examined by Rivière et al. (1968). Of 56 adults examined 45 had a thyroid of normal size, i.e. not palpable; in 10 the thyroid was palpable, but considered still of normal size and only one man had a small diffuse goiter. Radioactive iodine uptake studies of the thyroid showed a high mean uptake (74.1% of dose in 24 hours) in these Indians. Their mean urinary iodine excretion was very low (19.1 mmg/g creatinine). These two values are as low as comparable values found by Rivière et al. (1968) in a goitrous Andean region of Venezuela. They indicate that these goiter-free Indians are deficient in iodine. Similar findings have been reported by Roche (1959) among Indians from the Alto Ventuari regions in Venezuelan Amazonia and among Africans in the Congo Kinshasa by Delange, Thilly, and Ermans (1968). It appears that Indians as well as Africans fail to develop goiter in spite of iodine deficiency having been present through many generations. It is obvious that other factors are needed to develop goiter such as goitrogenic substances in food or water, some of which are known. These findings speak against the widely accepted theory that endemic goiter is an adaptation to iodine deficiency.

Summary and Conclusions

CLIMATIC ADAPTATION. Differences in stature and weight between inhabitants of the humid tropics and those of cooler climates seem to be associated with greater extra-cellular fluid space due to less body fat, a probably greater relative muscle mass and a greater amount of plasma protein. Although heat produced per unit of active tissue may be less in the tropics, energy requirements may be increased in the heat due to energy needed for the production and evaporation of sweat. A standardized, uniform test is needed to measure the effects of heat stress in different groups. Differences in sweat rates may be explained in part by differences in salt intake. Some groups in the tropics are still using plant ashes rich in potassium and practically void of sodium in place of salt. Their low sodium intake may actually be advantageous in insuring normal function of the adrenal cortex. The loss of other essential nutrients in sweat needs study in Africans and American Indians.

The fact that Congolese and Senegalese drink only one-third or less water than acclimatized Europeans under the same conditions may be due to the water content of African food, which is twice that of European food.

Blood pressures are generally lower in many "primitive" groups in the humid tropics than in comparable age and sex groups in "civilized" urban groups. Low sodium intake may be a factor in this picture. A blood pressure gradient correlates with the level of acculturation in Amazon Indian groups. Mean blood pressures in Africans tend to be higher than those of Amazon Indians, but are still lower than in Caucasians.

GROWTH AND DEVELOPMENT. Mean birth weights from 10 different locations in Africa ranged from 2635 grams for Pygmies to 3205 grams for Cameroonians. Boys usually weighed 100-200 grams more than girls. Variations were observed with season and rural versus urban areas. An increase in mean birth weights has been reported from two areas over the last two decades. No comparable data from Amazon Indians are available. Weight and height curves from seven localities in four West and Central African countries indicate satisfactory gains up to 6 months of age with a doubling of the birth weight between 3-6 months. The psychomotor development of the African infant is superior to that of the Caucasian infant. The scanty data available suggest that growth curves of Amazon Indian infants are comparable to those of Africans. While children of upper-class educated Africans continue to grow satisfactorily after 6 months, the great majority of African children show a slowing down in their growth, which becomes more accentuated during the second year of life. This retardation continues practically until puberty. At age 5, the mean weight deficit reaches 3-4 kilograms and the height deficit reaches 9-11 centimeters as compared to the elite group. Poor nutrition and a variety of infections are major causative factors. Weights and heights of Amazon Indian children are intermediate between those of the elite and village children in Africa. With the growth spurt of puberty, differences between the African groups decrease markedly. Differences between Africans and Amazon Indians in this age group are determined largely by genetic differences.

A comparison of six African and four Amazon Indian groups shows adult men to be of similar height (range 155-168.9 cm), but regarding weight the Karaja and Chavante Indians were 5-9 kilos heavier than the Africans of the same height. Differences between women were less pronounced. Differences between the two sexes of the same group were marked in all except the Bakongo. Differences in weight between younger and older men of the same group were considerable in some, but not consistent. No obesity was found in any of the groups.

FOOD AND FOOD INTAKE. Whereas many of the present African foods have been introduced from South America (e.g., cassava) no reverse traffic is known to have occurred in spite of a heavy slave trade from West Africa to the east coast of South America. Food consumption data from dietary surveys in seven West and Central African countries show a wide range of intakes with mean calories ranging from 1325 to 3630 per person per day depending on season and size of family. Protein intakes on the whole were marginal to satisfactory, but the percentage of animal protein was generally very low. Fat intake ranges from 5 to 28 percent of total calories. While intakes of vitamin A and C were generally satisfactory or high, those of the B-vitamins were low or deficient. Of the minerals studied iron intake was generally satisfactory or high, whereas calcium intake was low.

HEALTH AND NUTRITIONAL STATUS. Birth rates from five areas in West Africa between 1954 and 1963 were about 2.5 times higher than 1960 birth rates in the United States. Crude death rates from three countries in West Africa were three times those in the United States. More than 50 percent of all deaths in Africa, however, occurred in young children aged 0-5 years. From the few data available on two Amazon Indian tribes the situation seems to be similar.

Data on weight gains during pregnancy in three African countries show that the mean weight gain of the elite group in Ibadan was comparable to that of European or North American women while women in the villages gained about half as much or less than the elite group. This seems to be similar in tropical areas of Southeast Asia and, perhaps, South and Central America, indicating a caloric deficit in pregnant women. With a succession of numerous pregnancies this deficiency may eventually lead to depletion of reserves in the body, consequent premature aging, and death.

Physical signs found during surveys in three African countries indicated possible deficiencies of protein, vitamins A, F, C, and B complex and iodine. Similar findings were reported in two Amazon Indian tribes. In contrast other Amazon Indian tribes were practically free of any deficiency signs except for iodine. Although mean hemoglobins in four West and Central African countries were generally lower than those of the Ibadan reference group for all ages, the serum iron and total iron-binding capacity of the blood serum in two groups of women showed high normal values. It thus appears that iron deficiency is not a major cause of the anemia present in many forest areas of Africa. Hemoglobin values of some Amazon Indian groups were similar to those of the Africans. Total serum proteins from four African countries gave comparable values to those of Caucasians; albumin values, however, were generally lower and globulin values higher due to a relative increase in gamma globulin caused, most likely, by an almost continuous exposure to many infections. Similar findings were reported in the Chavante Indians. Mean total cholesterol levels reported from three African countries, the Chavante and Upper Xingu Indians were all low when compared to levels in urbanized Caucasians in Europe or North America. These low values may be an indication of a relatively low caloric and fat intake.

INFECTIONS AND INFESTATIONS. Man in the tropical rain forest is exposed to a great variety of infections and infestations of relatively low density. Whereas some infections such as salmonella are endemic, others, such as measles, have been recently introduced by the colonizers and have caused extremely serious epidemics that have decimated many Indian tribes, some almost to extinction. The equilibrium in which the Amazon Indian lived with his environment obviously has been seriously disturbed by the contact with "civilization" and it appears that in most cases he has been hurt beyond repair. The same may be true for the aboriginal in the African rain forest. However, the present inhabitant, who is already the second occupant, although plagued by as many (or more) infections and infestations as the Amazon Indian, has adapted quite well to his environment and more than compensates for a high loss of life during infancy and young childhood with a very high birth rate in contrast to the Amazon Indian.

SOME SPECIAL PROBLEMS OF ADAPTATION. In his specific fight against the greatest scourge, malaria, the African is helped by the sickle cell gene at least in heterozygote carriers. The latter show lower parasitic rates in the younger age groups and suffer less from deadly complications of malaria. On the other hand, a chronic neuropathy in western Nigeria seems to indicate a lack of adaptation to a high cassava consumption. This lack of adaptation may be due to failure to remove the toxic cyanic acid during the preparation of the tuber for consumption. The Amazon Indian, in contrast, apparently has learned over perhaps thousands of years through careful processing to rid the cassava tuber of its cyanic acid.

Iodine deficiency is present in several Amazon Indian groups and Africans in Zaïre. It appears, however, that some of these groups have tolerated iodine deficiency without developing goiter, which is considered by many an adaptation to iodine deficiency.

CONCLUSION. The most severe afflictions reported from tropical forest populations in recent decades are of two principal types: (1) introduced diseases to which the inhabitants had no natural resistance, and (2) nutritional deficiencies resulting from changes in diet attributable either to increased population density or higher dependence on foods with low protein, mineral, and vitamin content. When these categories are eliminated, the medical data indicate that the indigenous populations had achieved a successful adaptation to the physiological stresses of their environment.

References

Acker, P., A. M. Leclerc, J. Nicoli, C. Fourcade, and P. Trapet. 1967. Metabolisme lipidique des Congolais normal. Médicine Tropicale 27(4):408-416.

Acker, P., A. M. Leclerc, and P. Ramel. 1968. Quelques aliments traditionels du Congo (Congo-Brazzaville) envisagés sous l'angle de leur apport lipidique. Ann. Nutr. Alim. 22:17-24.

Allison, A. C. 1954. Protection afforded by sickle cell trait against subtertian malarial infection. Brit. Med. J. 1:290-294.

Baker, P. T. 1966. Ecological and physiological adaptation in indigenous South Americans. *In* The biology of human adaptability, Ch. 9. Clarendon Press. Oxford.

Baker, P. T., and J. L. Angel. 1965. Old age changes in bone density: Sex and race factors in the United States. Hum. Biol. 37(2):104-121.

Baruzzi, R. G. 1970. Contribution to the study of the toxoplasmosis epidemiology. Seriologic survey among the Indians of the upper Xingu River, central Brazil. Rev. Inst. Med. Trop., São Paulo, 12(2):93-104.

Bascoulergue, P., and J. L. Bergot. 1959. Alimentation rurale au Moyen-Congo. Service Commun de Lutte Contre les Grandes Endémies, Section Nutrition.

Bascoulergue, P., and M. L. Pierme. 1961. Poids de naissance et courbe de croissance ponderale des enfants noirs de Yaounde. I.R.C.A.M. Yaounde.

Becher, H. 1957. A importância da banana entre os índios Surára e Pakidái. Rev. de Antrop. 5(2):192-194.

Biocca, Ettore. 1963. A penetração branca e a difusão da tuberculose entre os índios do rio Negro. Rev. Mus. Paulista 14:203-212.

Burch, G. E., N. de Pasquale, A. Hyman, and A. C. de Graff.

1959. Influence of tropical weather on cardiac work and power of right and left ventricles of man resting in hospital. Arch. Internal Med. 104:553-560.
Cavelier, C., et al. 1970. Étude des teneurs du sérum en protides totaux, fer et cuivre chez la femme Camerounaise (Yaounde) en labor et au cours de la grossesse normale. Centre O.R.S.T.O.M. de Yaounde.
Collis, W. R. F., J. Dema, and A. Omolulu. 1962. On the ecology of child nutrition and health in Nigerian villages, II. Trop. and Geog. Med. 14:201-229.
Consolazio, C. F., L. O. Matoush, R. A. Nelson, R. S. Harding, and J. E. Canham. 1963. Excretion of sodium, potassium, magnesium and iron in the human sweat. J. Nutr. 79:407-415.
Consolazio, C. F., R. Nelson, L. O. Matoush, R. S. Harding, and J. E. Canham. 1963. Nitrogen excretion in sweat and its relation to nitrogen balance requirements. J. Nutr. 79:399-400.
Consolazio, C. F., and R. Shapiro. 1961. Energy requirements of man in extreme heat. J. Nutr. 73(2):126-134.
Cros, J. 1967. Enquête sondage sur la consommation des lipides dans quatre villages du Sénégal. Bul. Soc. Med. Afr. Noire Lgue. 12(2):153-176.
da Silva, M. P. 1966. Contribuição para o estudo do sangue periférico e da medula ossea em índios do Alto-Xingú. Thesis presented to the Escola Paulista de Medicina, São Paulo.
Davey, P. J. L. 1960-1962. Ghana national nutrition survey, 1960-1962. Unpublished.
Delange, F., C. Thilly, and A. M. Ermans. 1968. Iodine deficiency, a permissive condition in the development of endemic goiter. J. Clin. Endocrinol. and Metabol. 28:114-116.
De Lima, P. E. 1950. Niveis tensionais dos índios Kalapalo e Kamaiurá. Rev. Brasil. de Med. 7(12):787-788.
Dunn, F. L. 1968. Epidemiological factors: Health and disease in hunter-gatherers. Pages 221-228, *in* R. B. Lee and I. DeVore (Editors). Man the hunter. Aldine. Chicago.
Dupin, H., L. Masse, and P. Correa. 1962. Contribution à l'étude des poids de naissance à la maternité africaine de Dakar. Evolution au cours des années, variations saisonnières. Courier 12(4):1-30.
Durnin, J. V. G. A., M. F. Haisman, D. W. A. Tetteris, and L. Zurich. 1966. The effect of hot environments on the energy metabolism of man performing standardized physical work. Army Personnel Research Establ. Survey, Research Memo N-3. England.
Edozien, J. C. 1965a. A biochemical evaluation of the state of nutrition in Nigeria. J. West Afr. Sci. Assn. 10(1):22-38.
————. 1965b. The role of chemical pathology in rural health investigations. Ghana Med. J. 4:1-3.
————. 1965c. Establishment of a biochemical norm for the evaluation of nutritional status in West Africa. J. West Afr. Sci. Assn. 10(1):1-21.
Galvão, E. P. 1948. Human heat production in relation to body weight and body surface, I-II. J. Applied Physiol. 1:385-401.
————. 1950. Human heat production in relation to body weight and body surface, III-IV. J. Applied Physiol. 3:21-28.
Gans, B. 1963. Some socio-economic and cultural factors in West African pediatrics. Arch. Dis. in Childhood 38 (197):1-12.
Geber, M., and F. F. A. Dean. 1964. Le développement psychomoteur et somatique des jeunes enfants africains en Ouganda. Courier 14(4):425-437.
Gilles, H. H. 1964. Akufo, an environmental study of a Nigerian village community. Ibadan University Press. Ibadan.
Glanville, E. V., and R. A. Geerdink. 1970. Skinfold thickness, body measurements and age changes in Trio and Wajana Indians of Surinam. Amer. J. Phys. Anthrop. 32: 455-461.
Hamilton, W. J. and F. Heppner. 1966. Radiant solar energy and the function of black homeotherm pigmentation: An hypothesis. Science 155:196-197.
Hauck, H. M., and F. L. Tabrah. 1963. Heights and weights of Ibo of various ages. West Afr. Med. J. 27(April):64-74.
Holemans, K. 1960. Contribution à la protection maternelle et infantile en milieu rural du Kwango. Mem. N.S. X, Ac. Roy. des Sciences d'Outre Mer. Bruxelles.
ICNND. 1967. Republic of Nigeria nutrition survey, February-April 1965. A report of the Nutrition Section, Office of International Research, National Institutes of Health, U.S.P.H.S., Bethesda, Maryland.
Johnston, F. E., P. S. Gindhart, R. L. Jantz, K. M. Kensinger, and G. F. Walker. 1971. The anthropometric determination of body composition among the Peruvian Cashinahua. Amer. J. Phys. Anthrop. 34:409-416.
Ladell, W. S. S. 1952. The physiology of life and work in high ambient temperatures, desert research. Proc. of the Internat. Symposium held in Jerusalem May 7-14, pp. 187-198.
————. 1964. Terrestrial animals in humid heat: Man. *In* Dill, D. B., E. F. Adolf, and C. S. Wilber (Editors). Handbook of physiology, Section 4:625-659. American Physiological Society, Washington, D.C.
Livingstone, F. F. 1958. Anthropological implications of sickle cell gene distribution in West Africa. Amer. Anthrop. 60(3):533-562.
Lowenstein, F. W. 1958. Data on the Karajá and Mundurucú Indians. Unpublished.
————. 1961. Blood pressure in relation to age and sex in the tropics and subtropics. Lancet February: 389-392.
————. 1962. The vicious-circle mechanism in production of protein caloric malnutrition. Symposia of the Swedish Nutrition Foundation I. Bastad.
————. 1967. Report on nutrition surveys in eleven Brasilian Amazon communities between 1955 and 1959. Atas do Simpósio sôbre a Biota Amazônica 6:177-184.
————. 1970. Nutrition and infection in Africa. Nutr. Abstr. and Rev. 40(2):373-393.
Luyken, R., and F. W. M. Luyken-Koning. 1960. Studies on the physiology of nutrition in Surinam. Trop. and Geogr. Med. 12:313-314.
Mann, G. V., O. A. Roels, D. L. Price, and J. M. Merrill. 1962. Cardiovascular disease in African pygmies, a survey of the health status, serum lipids and diet of pygmies in the Congo. J. Chronic Diseases 15:341-371.
Martineaud, M. 1967. Malnutrition en zone tropicale. Afr. Med. (Dakar) 6:409-418.
Masseyeff, R., M. L. Pierme, and B. Bergeret. 1960. Une enquête sur l'alimentation dans la région de Batouri (Est-Cameroun). Recherches et Études Camerounaises 1:6-70.
Mattos, R. B. 1958. Acuidade visual para longe e frequência de discromatópsia em índios brasileiros. Thesis presented to the Escola Paulista de Medicina, São Paulo.
Mazer, A. T. 1970. Ration protidique et adaptation en climat tropical. Cahiers de Nutrition et de Diététique 5(1):57-63.
McGregor, I. A., A. Q. Rahman, B. Thompson, W. Z. Billewicz, and A. M. Thomsen. 1968. The growth of young children in a Gambian village. Trans. Roy. Soc. Trop. Med. and Hyg. 62(3):341-352.
McLaren, D. S. 1959. Records of birth weight and prematurity

in the Wasukuma of Lake Province, Tanganyika. Trans. Roy. Soc. Trop. Med. and Hyg. 53(2):173-178.

Monekosso, G. L., and J. Wilson. 1966. Plasma thiocyanate and vitamin B12 in Nigerian patients with degenerative neurological disease. Lancet 1:1062-1064.

Moore, D. G. F. 1937. Retrobulbar neuritis cum avitaminosis followed by postpartial optic atrophy—now shown to be of pellagrinous nature. West Afr. Med. J. 9:35.

Morley, D. C., M. Woodland, W. J. Martin, and I. Allen. 1968. Heights and weights of West African children from birth to age of five. West Afr. Med. J. 32(February):8-13.

Neel, J. V., W. M. Mikkelsen, D. L. Rucknagel, E. D. Weinstein, R. A. Goyer, and S. H. Abadie. 1968. Further studies of the Xavante Indians VIII. Some observations on blood, urine and stool specimens. Amer. J. Trop. Med. and Hyg. 17(3):474-485.

Neel, J. V., and F. M. Salzano. 1967. Further studies on the Chavante Indians. Amer. J. Hum. Genetics 19:554-574.

Neel, J. V., F. M. Salzano, P. C. Junqueira, F. Keiter, and D. Maybury-Lewis. 1964. Studies on the Xavante Indians of the Brazilian Matto Grosso. Amer. J. Hum. Genetics 16(1):52-140.

Nelson, W. E. (Editor). 1964. Textbook of pediatrics. 8th edition. W. B. Saunders Co. Philadelphia.

Nicol, B. 1953. Tribal nutrition and health in Nigeria. J. Clin. Nutr. 1:364-371.

————. 1958. Annual Report, Nutrition Unit, Fed. Med. Dept., Nigeria.

Nutels, N., M. Ayres, and F. M. Salzano. 1967. Tuberculin reactions, X-ray and bacteriological studies in the Cayapo Indians of Brazil. Tubercle (London) 48:195-200.

Oke, O. L. 1967. The present state of nutrition in Nigeria. World Rev. Nutr. and Dietetics 8:25-61.

ORANA. 1956. Enquête Nutrition-Niveau de Vie, Subdivision de Bongouanou 1955-1956. O.R.A.N.A. and C.S.R.S.O.M. Côte d'Ivoire.

Osuntukun, B. O., G. M. Monekosso, and J. Wilson. 1969. Relationship of a degenerative tropical neuropathy to diet. Report of a field survey. Brit. Med. J. 11(March): 547-550.

Pele, J. 1967. Essai pour une table de croissance des enfants Camerounais. Centre O.R.S.T.O.M. de Yaounde.

Raper, A. B. 1956. Sickling in relation to morbidity from malaria and other diseases. Brit. Med. J. 1:965-966.

Ribeiro, Darcy. 1956. Convívio e contaminação. Sociologia 18(1):3-50.

Rivière, R. 1972. [Study of Iodine Deficiency in the Venezuelan Andes.] (In press.)

Rivière, R., D. Comar, M. Colonomos, J. Deseune, and M. Roche. 1968. Iodine deficiency without goiter in isolated Yanomamö Indians: Preliminary notes. *In* Symposium of biomedical challenges presented by the American Indian. PAHO/WHO. Washington.

Roberts, D. F. 1953. Body weight, race and climate. Amer. J. Phys. Anthrop. 11:553-558.

Roche, M. 1959. Elevated thyroidal I^{131} uptake in the absence of goiter in isolated Venezuelan Indians. J. Clin. Endocr. and Metab. 19(11):1440-1445.

Saddock de Freitas, A., and N. Braga de Oliveira. 1955. Estudo sôbre o estado nutritivo dos Xavantes. Rev. Brasil. de Med. 12(7):482-486.

Schaad, J. D. G. 1960. Epidemiological observations in Bush Negroes and Amerindians in Surinam. Trop. and Geogr. Med. 12:38-46.

Schreider, E. 1957. Ecological rules and human body-heat regulation. Nature 5(19):915-916.

Sick, H. C. 1949. Sôbre a extração do sal de cinzas vegetais pelos índios do Brasil Central. Rev. Mus. Paulista 3:381-390.

Sioli, H. 1955. Eine Masern-epidemie bei den Munduruku Indianern im Amazonas-gebiet. Act. Trop. 12(1):38-52.

Tabrah, F. L. and H. M. Hauck. 1963. Some aspects of health and nutritional status, Awo Omamma, Nigeria. J. Amer. Diet. Assn. 43(4):321-326.

Tans, C. 1959. La croissance ponderale du nourisson pygmée (Bambuit-Ituri). Ann. Soc. Belge Med. Trop. 39:851-861.

Terra, G. J. A. 1964. The significance of leaf vegetables, especially of cassava in tropical nutrition. Tropic. and Geogr. Med. 2:97-108.

Thomson, M. L. 1955. Relative deficiency of pigment and horny layer thickness in protecting the skin of Europeans and Africans against solar ultra violet radiation. J. Physiol. 127:236-246.

Venkatachalam, P. S., K. Shanker, and C. Gopalan. 1960. Changes in body weight and body composition during pregnancy. Ind. J. Med. Res. 48:511-517.

Vincent, M. 1957. Quelques faits résultant de l'examen systématique des moyennes ponderales-relevées à travers le Congo chez les nouveau-nées et les nourissons. Ann. Soc. Belge Med. Trop. 37(6):973-980.

Walker, A. R. P. 1955. Composition and density of thoracic vertebral bodies from South African Bantu adults habituated to very high iron intake. S. Afr. J. Lab. Clin. Med. 1:254-262.

Wyndham, C. H., M. Bouwer, G. Devine, and A. E. Paterson. 1952. Physiological responses of African laborers at various saturated air temperatures, wind velocities and rates of energy expenditures. J. Appl. Physiol. 5:290-298.

Some Problems of Cultural Adaptation in Amazonia, with Emphasis on the Pre-European Period

BETTY J. MEGGERS

Amazonia can be defined in several ways, but for the purpose of the present analysis it will consist of that portion of South America east of the Andes that lies below 1500 meters in elevation, where rain falls on 130 or more days per year, where relative humidity normally exceeds 80 percent, and where annual average temperature variation does not exceed 3° C. These characteristics prevail over most of the Amazon drainage, with the exception of the headwaters of the longer tributaries, and extend over the Guianas to the mouth of the Orinoco to cover an area of some 6 million square kilometers. Vegetation consists of tropical rain forest broken by small enclaves of savanna where the soil is too porous to retain moisture during dry months.

A relatively homogeneous geological history and an equatorial location make Amazonia a remarkably uniform environment from the standpoint of human exploitation in spite of its vast size. Only two habitats possess sufficiently different subsistence potential to affect cultural adaptation significantly. The well-drained terra firme, which occupies 98 percent of the region, is characterized by ancient and severely leached soils. The remaining two percent is occupied by the várzea or flood plain of the Amazon, where fertility is annually renewed by silt carried down from the Andean highlands.

The terra firme consists of all land not subject to seasonal inundation. The northeastern and southeastern portions are occupied by the Guayana and Brazilian shields, which originated during the Pre-Cambrian. In the west, a large wedge between the Rio Japurá and the Rio Madeira is composed of Tertiary sediments, accumulated when this area was occupied by an immense freshwater lake. The rivers draining the terra firme are predominantly clear or black water, moderately to highly acid, free of sediment, and so devoid of dissolved minerals that they approach distilled water in their purity. This purity is a vivid expression of the low fertility of terra firme soils, which are the end products of hundreds of millennia of leaching and erosion under warm temperatures and high rainfall.

The vegetation has made such a successful adaptation to the rigorous climatic and edaphic conditions of the terra firme that temperate observers generally find it incredible that its luxuriance does not imply highly propitious conditions for intensive agriculture. A large body of data exists, however, to demonstrate the mechanisms by which the tropical forest conserves nutrients, preserves the soil from erosion, and minimizes the destructive potential of predators and disease. By maintaining soil temperature a degree or two below that at which the rate of humus destruction exceeds the rate of accumulation, the forest improves the water and nutrient-holding capacity of the soil; by virtue of its rapid rate of growth, it retrieves nutrients released from the litter so efficiently that there is little or no loss by leaching. In contrast to temperate regions, where a major portion of the nutrients are stored in the soil, the storage role in the tropics has been taken over by the leaves, branches, stems, and roots of the plants. In addition to serving as the storehouse for nutrients, the vegetation protects the ground from erosion by reducing the impact of

BETTY J. MEGGERS, Department of Anthropology, National Museum of Natural History, Smithsonian Institution, Washington, D.C. 20560.

rainfall; it also absorbs large quantities of water so that a significant proportion is withheld from the rivers and excessive flooding is thereby averted.

Successful adaptation by the vegetation to these climatic and edaphic conditions has been achieved, however, at the sacrifice of qualities important to the primary consumers. Among the most significant defects from the standpoint of plant-eaters, including man, are: (1) low nutrient value per unit of bulk, (2) low protein and mineral content, and (3) dispersed distribution of individuals of the same species. The staple cultigens, manioc (*Manihot esculenta*) and sweet potatoes (*Ipomoea batatas*), share the first two defects, ruling out the possibility of dependence on agriculture as the primary source of subsistence. Slash-and-burn or shifting cultivation, so often unjustly deprecated by temperate observers, is well adapted to lowland tropical environmental conditions. Leaving trunks and branches where they fall helps protect the soil from sunlight and erosion, while intercropping minimizes the loss of nutrients by distributing plants with different requirements throughout the field. In spite of these adaptive features, nutrient loss proceeds so rapidly that most fields are abandoned after the third harvest. Damage to the soil is not irreversible, however, and with time the forest is able to reconstitute itself.

The low nutrient content of most tropical cultigens makes it necessary to depend heavily on wild foods for a balanced diet. Since the only aboriginally domesticated animal was the dog, which was not eaten, the primary source of protein was the fauna. Because the majority of terrestrial vertebrates are small and solitary, productivity of hunting per man hour expended is usually low and the supply of game within an accessible radius of the village is generally depleted after a few years of predation. Fish are not abundant in clear water rivers and are even less numerous in black water ones, so that they offer a supplemental rather than an alternative protein source in most parts of the terra firme. Birds, turtles, insects, reptiles, and other types of fauna were consumed by many groups. Fruits and nuts are seasonally available and some (particularly the Brazil nut) are highly nutritious.

In summary, food resources on the terra firme are varied and adequate under proper utilization, but tend to be dispersed and readily depleted. Successful adaptation to these conditions consists in the development of an annual subsistence cycle that not only provides the essential nutrients and assures an adequate quantity of food, but simultaneously prevents overexploitation and resultant irreversible damage to the habitat. Fortunately, the greatest abundance of wild and cultivated plant foods occurs at different times of the year, making periods of hunger rare.

The várzea, a maze of channels, ponds, lakes, and islands occupying only about 60,000 square kilometers, offers an extraordinary contrast to conditions on the terra firme. On this tiny sliver, narrowing from 50 or more kilometers on the lower Amazon to less than 25 kilometers above the Rio Negro, the seasons are marked by the rise and fall of the water rather than changes in the pattern of rainfall. Tributaries originating in the Andean highlands carry a heavy burden of soluble minerals and silt, which is deposited annually on the Amazon flood plain, replacing nutrients lost through leaching or removed by harvesting of cultivated plants. White water channels and lakes support abundant and nutritious aquatic flora and fauna, including the *Victoria regia*, wild rice and other grasses, turtles, manatees, caimans, water birds, and a great variety of fish. As the river level falls, many of these resources are concentrated in shrinking lakes where they are obtainable in large quantities with relatively little effort. Furthermore, maintenance of soil fertility not only allows indefinite utilization of the same garden plots without decline in yield, but also permits the raising of maize (*Zea mays*), which is more nutritious than manioc but does poorly on terra firme soils. The regime of the Amazon is also ideal for agricultural exploitation because the water rises gradually over about 6 months, allowing ample time for harvest of crops before inundation, and then drops rapidly, freeing the land for early planting and making possible two maize harvests per year. As a consequence, the várzea subsistence potential during normal years compares favorably both in concentration and long-range productivity with that of temperate regions.

Unfortunately, this productivity is unreliable. If the rains begin too soon or last too long, the result is an abnormally high crest, severe crop loss, and depletion of many of the wild food resources. Even in normal years, the peak of agricultural activity coincides with the period of greatest abundance of wild resources, so that hunting, fishing, gathering, and farming activities must be practiced simultaneously and sufficiently intensively to produce a surplus for consumption during the leaner months when inundation permits dispersal of the fish and destroys any crops left unharvested. Whereas the major adaptive problem on the terra firme is prevention of irreversible deterioration of the habitat by man, the challenge of the várzea is to minimize the detrimental effects on the human population of seasonal and sometimes unpredictable reductions in the normal level of subsistence support.

Culture as an Adaptive Mechanism

Although most textbooks on anthropology assert that the acquisition of culture freed man from environmental constraints, the opposite view is more compatible with ecological evidence. Here, *Homo sapiens* will be treated as an animal and culture will be considered a unique and superior form of behavioral adaptation, whose extraordinary flexibility has permitted mankind to occupy a wider variety of habitats than any other kind of mammal. The acquisition of culture has not only permitted our species to overcome hostile aspects of the environment by providing insulation against heat and cold, protection against predators, and improved means of securing food; it has also made possible major modifications, notably in the quantity and quality of the local food supply. Culture does not exempt man from the pressures of natural selection; rather, under normal circumstances it provides a more rapid and efficient behavioral mechanism for responding to these pressures. Nowhere is this function more obvious than in Amazonia.

The first human immigrants into the tropical lowlands were small bands of hunters and gatherers. The time of their arrival is still a matter of speculation, but evidence from the adjacent Andean area indicates that it could have been 10,000 or more years ago. Archeological remains from the preceramic period have not been reported as yet from Amazonia and no groups that survived to historic times depended solely on wild foods. By analogy with the situation in other areas, however, we can infer that as the population increased, bands were obliged to restrict their wandering within a recognized territory and to evolve an annual subsistence schedule that took advantage of sequentially available local resources. Differences in the kinds of wild foods exploited by more recent Amazonian agriculturalists can most readily be explained as survivals of this process of adaptation.

As long as man remained dependent on wild foods, his impact on the ecosystem was not significantly different from that of other species of large fauna. This relationship changed with the introduction of cultivated plants. The time and place of origin of the principal lowland staples, manioc and sweet potatoes, are still unknown, but indirect evidence indicates that bitter manioc had become a basic food in some lowland regions by 1000 B.C. and suggests that maize had been introduced to the várzea by about the same time. Crops require planting, weeding, and tending to produce well, and agriculture is consequently associated with lowered community mobility. The fact that settled life enhances the probability of survival for ill, aged, or very young individuals physically unable to travel or to obtain their own food, and also permits more adequate provision for defense against natural and human enemies, makes it beneficial to the group. On the other hand, in an environment like that of Amazonia prolonged residence in one spot intensifies exploitation of local resources with consequent risk of irreversible damage to their long-range productivity. The cultural advance from food collecting to food production thus required the development of new regulatory mechanisms to create a balance in which the maximum benefits of settled life were achieved at the cost of minimal permanent damage to the habitat.

Several of the most characteristic features of Amazonian culture, such as infanticide, sorcery, and warfare, are adaptive responses to this requirement. Although many people view such practices as "barbaric," most biologists have no difficulty in recognizing them as more sophisticated versions of the behavioral adaptations to selective pressures exhibited by noncultural animals. Close examination of existing ethnographic data on Amazonian peoples reveals systems of checks and balances so intricate and so strongly reinforced that it seems safe to conclude that the aboriginal culture pattern of the terra firme represents an equilibrium adaptation to the special characteristics of the rain forest environment. The same appears to be true for the aboriginal populations of the várzea, although the rapid extinction of the indigenous inhabitants following European contact limits available information to fragmentary comments in early chronicles.

The Tropical Forest Cultural Pattern

The principal features of the tropical forest pattern of culture can be summarized as follows: The primary economic and social unit is the extended family, composed of a man, his wife or wives, preadolescent children, married sons (if residence is patrilocal) or daughters (if it is matrilocal), and their children. This kin group, which may number 50 or more individuals, occupies a communal house, which lacks interior partitions and windows and is typically provided with two doors. A village may consist of a single house or of several houses arranged in a circle, and is generally moved to a new location about every 5 years.

Social organization is based on kinship ties and sexual division of labor. Each household has a chief, usually the oldest active male, but he lacks power to issue orders or to enforce obedience. A village composed of several houses has a village chief and often a council made up of the heads of households, who discuss matters of community concern and arrive at decisions that are binding on all members. Each person has well-defined rights and obligations according to his or her age, sex, and relationship to other members of the community. Since no individual commands all of the basic skills, mutual aid is essential to survival and rules of conduct are seldom violated. On rare occasions when someone becomes a threat to the community, ostracism is the customary penalty. The only specialized occupation is shamanism and the shaman's principal duty is the curing of illness, generally by practicing or counteracting sorcery.

Intergroup relations are typically hostile and range from avoidance to incessant warfare. All males become warriors after puberty and killing of an enemy is often a prerequisite for full adult status. The most common motive for aggression is revenge of death by sorcery; food, material goods, and land are never sought. Although the primary aim is to kill the enemy, women or children may be captured and adopted into the community, acquiring thereby equal status with members by birth and the concomitant rights and obligations. Trading between villages is minimal and usually undertaken more as an excuse for social interaction than because of economic necessity. There are no formal markets and exchange of products between households within the village is rare.

The variability of expression of this cultural pattern and some of its adaptive aspects can be illustrated by examining selected features of two tribes occupying widely separated parts of the terra firme: the Jívaro who live along the base of the Andes and the Kayapó who occupy the southeastern sector of the region (Figure 1).

THE JÍVARO

The Jívaro inhabit an area of some 65,000 square kilometers in the eastern lowlands of Ecuador, where there is no dry season and rainfall averages between 2000 and 2600 millimeters annually. The population is about 20,000 and females outnumber males more than 2 to 1. In spite of a common language and culture, there is no permanent social or political cohesion between villages; on the contrary, blood revenge and warfare are more intense between Jívaro communities than between the Jívaro and neighboring groups.

Each village consists of a single house occupied by a patrilineal extended family with between 15 and 46 members. It is abandoned when hunting becomes unproductive, fields are exhausted, or the head of the household dies, so that a settlement has an average permanency of about 6 years. Polygyny is the prevailing form of marriage and women have high status because of their primary role in subsistence and their special relationship with the spirit world. The highest permanent authority is the male head of a household, although several households form temporary alliances under an elected chief during warfare. Although each village is an independent economic and political unit, it maintains social relations with neighboring households. Members of this community exchange visits, celebrate ceremonial occasions together, intermarry, and unite for attack on enemies.

Jívaro subsistence depends heavily on agriculture. The staple, sweet manioc, is utilized in the form of a fermented beverage. Fermentation is induced by chewing the boiled tubers and is a time-consuming daily female task. Manioc foliage,

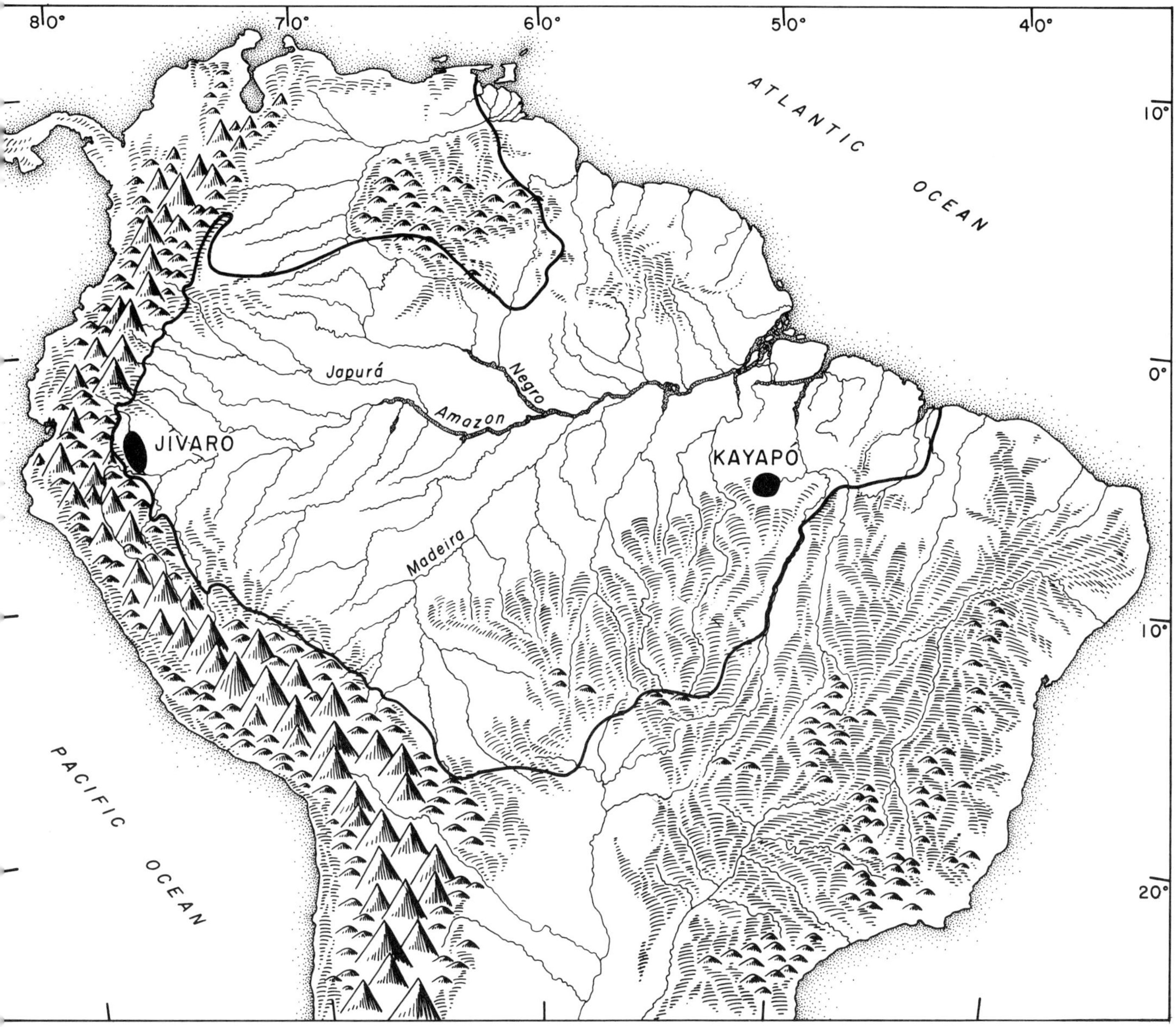

FIGURE 1. Boundary of the lowland tropical forest of Amazonia (heavy line) and location of two aboriginal groups, the Jívaro and the Kayapó. [Relief after Guerra 1959, fig. 7.]

which has a high protein content, is boiled and eaten. Other important cultigens are chonta or peach palm (*Guilielma utilis*), sweet potatoes, squash, papaya, and peanuts. Fields are cleared by the men; planting, weeding, harvesting, and garden magic are the responsibility of women.

Hunting is a daily male activity, but the skill of the hunters and the relative abundance of game makes it possible to supply the household with a couple of hours of work. Deer and tapir are avoided for supernatural reasons, but all other kinds of game are consumed. Fish are also eaten, but wild plants are unimportant except for palmito, utilized principally while traveling. Several kinds of insects are relished, along with turtle eggs (Figure 2).

Sexual division of labor prevails in manufacturing activities. Men perform all tasks involving wood, including house, bed, and loom construction, canoe-making, carving of paddles, lances and other implements, gathering firewood, and making fire.

They also weave baskets, spin and weave cotton, and make many types of ornaments. Women are associated with the earth and consequently not only do all of the agricultural work, but make pottery and dye cotton. Most of the daily household tasks also fall to them, such as cooking, preparing drinks, caring for children and dogs, carrying burdens, and fetching water. The only part-time occupational specialist is the shaman, who performs cures, identifies the guilty person in the case of a death, and causes death or illness. All adult deaths are viewed as unnatural events produced by sorcery.

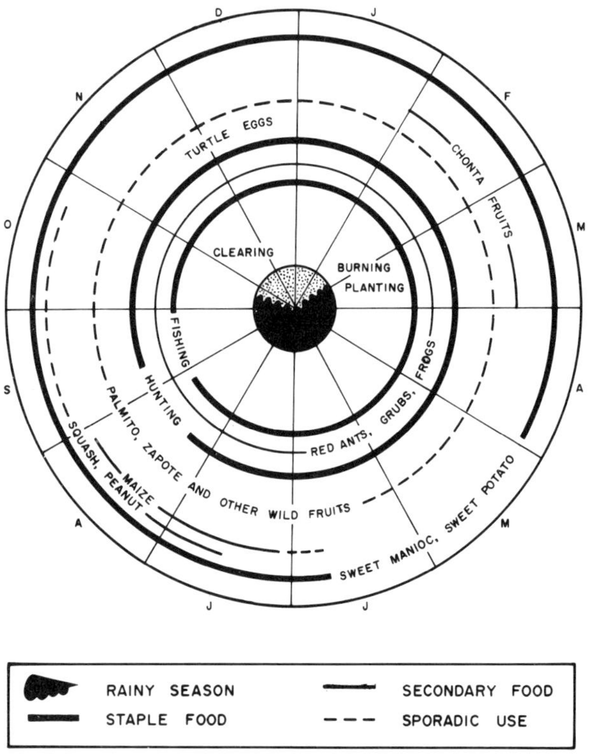

FIGURE 2. Annual subsistence cycle of the Jívaro. Game, fish, cultivated sweet manioc, and sweet potatoes are the staple foods; other cultigens and various wild products are exploited seasonally or throughout the year.

Intergroup hostility is the overriding theme of Jívaro society. Aggression may be provoked by several kinds of incidents, among them abduction of a woman, murder of a member of the household while traveling through alien territory, or a series of deaths indicating sorcery. Adult male members of the immediate family are required to take blood revenge and a single death restores the balance. Warfare, on the other hand, aims at annihilation of the enemy village, including death of all inhabitants, burning of the house and contents, and uprooting of the gardens. After the heads of adults have been taken as trophies, the victors hastily retreat to their own territory where they are safe from retaliation by unfriendly spirits. Since killing is mandatory and only males participate in raids, an unbalanced sex ratio is inevitable. Jívaro society has taken advantage of this situation by assigning to females the preparation of the fermented beverage that is the primary dietary constituent, but which is far too time consuming a process to be feasible without polygyny. Because game is relatively abundant, an excess of females does not significantly increase the burden of the male subsistence contribution.

Sorcery and warfare are mechanisms of high adaptive value in the terra firme environment because a larger population can be maintained indefinitely if it is thinly distributed and relatively mobile than if it is concentrated and permanently sedentary. Fear of sorcery is an effective means of promoting dispersal because safety lies in being surrounded with the smallest number of individuals consistent with physical survival. Sorcery also helps to stem population growth since blood revenge and warfare intensify as density increases and diminish as it declines. The strength of the sanctions brought to bear on Jívaro males testifies to the adaptive importance of these functions. Success in battle is the principal means of achieving status, and a man who evades his responsibility to take blood revenge or to participate in a raid not only risks retaliation by the spirits against himself, but places his whole family, his crops, and his hunting dogs in jeopardy.

Another means of limiting population increase is periodic suspension of sexual relations between husband and wife. While such taboos are common among precivilized groups, those observed by Jívaro men are unusually prolonged. Continence is mandatory for 3 to 6 months after the taking of a head and from before the birth of a child until weaning, which takes place between the ages of 2 and 3. Since each man has two or more wives, and since adultery by females is punishable by death, such taboos can significantly affect the birth rate.

Throughout the terra firme, villages are relocated when hunting in the vicinity becomes unproductive or the surrounding land is exhausted. The risk of environmental degradation apparently is so great, however, that cultural practices have developed to supplement these natural incentives. Most widespread is the requirement that a house be abandoned upon the death of an occupant. Among the Jívaro, this is obligatory on the death of the male head, which the hazards of warfare make a more likely event than the death of a female.

THE KAYAPÓ

The Kayapó, who occupy the region between the Rio Araguaia and the middle Rio Xingú in the Brazilian state of Pará, evolved an adaptation that differs in many details from that of the Jívaro, although the general cultural pattern is basically the same. In this part of Amazonia, annual rainfall of 1500 to 2000 millimeters is concentrated between September and May, creating a dry season of several months duration. Population figures are unreliable, but 5000 Kayapó were estimated to exist in 1897.

As among the Jívaro, the village is the largest political entity, and the linguistic and cultural similarities do not prevent active hostility between Kayapó villages. The household unit remains the extended family, but here villages are composed of one or two concentric rings of communal houses and may reach populations of several hundred. In the center of the plaza or at one side is a larger structure, which is the men's house.

Kayapó social organization exhibits several features in keeping with their larger population concentrations. There is a village chief, who is selected for his experience and judgment, but who exercises no coercive power. Community matters are decided by the adult males, who meet nightly in the men's house. Each household consists of a woman, her married and unmarried daughters, and their children. After the age of 10, boys spend most of their time in the men's house, where food is brought to them by a female relative and where they learn the arts and crafts, the skills of warfare, and the tribal customs. At puberty, a boy enters the warrior class and embarks on the most prestigious and exciting period of his life. Since this status is lost with the birth of his first child, he attempts to postpone fatherhood as long as possible. Girls are equally reluctant to assume the responsibilities of family life and employ oral contraceptives as well as mechanical means of abortion to avoid successful pregnancy. Monogamy is universal, but adultery is common.

Kayapó subsistence exhibits much greater seasonal variation than that of the Jívaro. During the rainy season, the main emphasis is on garden produce, principally sweet potatoes. These are supplemented by maize, Brazil nuts, piqui fruits, game, and fish (Figure 3). Clearing of the fields is done by men, while planting and weeding are performed by family groups. Harvesting and food

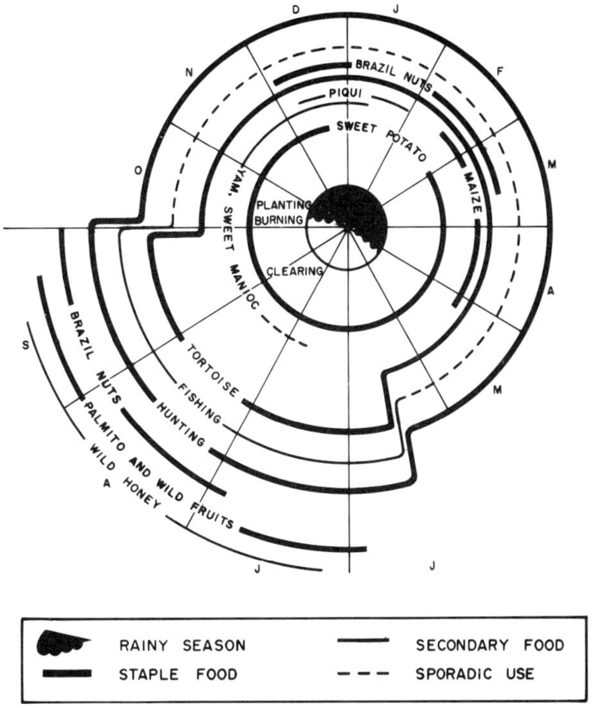

FIGURE 3. Annual subsistence cycle of the Kayapó. Greater emphasis is placed on wild than on domesticated foods, the sweet potato being the primary cultigen. During the dry season, the community splits into family bands, which wander through the forest hunting, fishing, and gathering.

preparation are the responsibility of women. The traditional dish is a kind of bread made from manioc, sweet potato, or maize dough. During the dry season, the community splits into extended family bands, which wander in distant parts of the forest subsisting on wild foods. Only persons too ill or infirm to travel remain in the village in the care of a relative.

Hunting is the favorite male activity and is often done in the company of relatives. Small animals are shared within the family, but large game, such as deer, tapir, and peccary, is delivered to the chief for distribution. Fishing is important during the dry season. Land tortoises, which abound in the region, are frequently kept alive until needed. Brazil nuts, which are rich sources of protein, are exploited during the dry season and are also collected and stored for use during the period of sedentary village life.

Sexual division of labor is more pronounced in the arts and crafts than in subsistence activities. Baskets, twine, weapons, and most ornaments are made by men, while women spin cotton and make ornaments of seeds and cotton. Many of the usual Amazonian products, including hammocks, cotton cloth, canoes, and pottery, are not made or used. Household tasks such as food preparation, child care, firewood collecting, and water carrying are assigned to women. The only specialist is the shaman, who is primarily a curer. Most illness is believed to be caused by intrusion of a foreign object into the body by sorcery and the shaman attempts to remove this with the aid of supernatural assistants. Payment is made whether or not the patient recovers and deaths are not avenged.

A Kayapó man achieves full status only after having killed an enemy, a designation that applies to anyone not a member of the village. As among the Jívaro, hostility is more intense between adjacent and related communities than toward members of other tribes. Kayapó rules do not require each man to make his own kill, however, and all members of a war party can share the glory of victory by striking a blow at an enemy even after he is dead. The village is burned and women and children taken prisoner are adopted into the community.

ADAPTIVE ASPECTS OF CULTURAL PRACTICES

The Kayapó and Jívaro ways of life differ markedly in several respects, one of the most striking being settlement pattern. The Kayapó are able to achieve population concentrations twenty times larger than those of the Jívaro and to maintain the village indefinitely in the same location, but only at the cost of a periodic temporary increase in community mobility. During the dry season, groups of families wander in the forest, moving camp every few days, and subsisting on wild foods. This annual dispersal not only relieves the pressure on flora and fauna in the vicinity of the village but also permits more uniform exploitation of the resources of the entire sustaining area. Heavy dependence on wild foods by a population of several hundred, however, requires a relatively large territory. Since wandering takes place about 5 days distance from the village, Kayapó villages must be spaced no less than two weeks travel apart. Maintenance of adequate separation is assured by the hostility that prevails between adjacent communities and which erupts periodically into warfare.

The Jívaro are able to maintain a sedentary existence not only throughout the year but for a period of several years by limiting their population concentrations to a single extended family. The adaptive superiority of this solution under the climatic conditions of western Amazonia, where rainfall is high and there is no dry season, is obvious. The higher intensity of warfare and the prevalence of sorcery and blood revenge suggest that population increase is a greater potential hazard here than among the Kayapó. Both tribes employ infanticide under certain circumstances and prohibit sexual relations between parents from the inception of pregnancy until the child is about 2 years old. Although the rationale is supernatural, the function of these practices is also clearly regulation of population size.

While the settlement pattern, subsistence emphasis, social organization, and material culture of the Jívaro and Kayapó differ considerably in detail, both complexes appear to be successful adaptations to the terra firme environment. Both insure utilization of a wide range of forest resources and regulate their exploitation so that long-range deterioration of the habitat is prevented and a stable food supply incorporating all of the necessary nutrients is assured. In other subregions of the terra firme with different combinations of climatic, edaphic, and biotic variables, other cultural variations of the generalized tropical forest pattern emerged through the operation of natural selection. The unifying theme and significant aspect of all these adaptations is maintenance of a level of population density compatible with the long-range exploitation of the subsistence resources of the terra firme environment.

Aboriginal Adaptation to the Várzea

The first Europeans to explore the Amazon reported its margins densely populated. Villages occupied the high banks, where they were secure from flooding, while subsistence activities concentrated on the floodplain where hunting and fishing assured a large supply of turtles, manatees, and fish, and careful scheduling of planting produced bountiful harvests. Surpluses of grain, fish, and meat were preserved and stored for consumption during months of high water.

The basic social unit remained the extended family household, but the village chief had greater authority than anywhere on the terra firme. Above him was a multi-village or "province" chief, who was strictly obeyed by his subjects. It seems likely that division of labor existed in arts and crafts and probably also in subsistence activities, because of the necessity of exploiting simultaneously the varied resources of the várzea, all of which were most productive during the period of low water. Religion was also more highly elaborated and supernatural beings were represented by idols kept in shrines, where they received the prayers and offerings of the populace and were attended by priests. Warfare here too was a major activity, but taking captives rather than killing was the primary motive. Captives were not incorporated into the kinship structure by adoption, but entered the society as a distinct servile class.

The cultural elaboration and population density achieved on the várzea has led some observers to the conclusion that the Amazon floodplain possessed a cultural potential comparable to that of the Old World centers of early civilization, and that a similar level of development would have been achieved here as in Mesopotamia or Egypt if the process had not been terminated by European contact. Because of its equatorial location, its high rainfall, and its geological antiquity, however, the Amazon floodplain is not comparable to those of the Tigris-Euphrates, the Nile, or the Indus. The vastness of the river makes it immune even today to control measures that would eliminate periodic unpredictable declines in productivity resulting from abnormally high or low water. This situation sets the carrying capacity at the level sustainable during lean years rather than normal years.

Furthermore, although the hostile interaction between floodplain farmers and less sedentary interior groups often cited as a stimulus to cultural advance existed in Amazonia, the function of warfare appears to have been different. Whereas in the Near East, the food surpluses of the floodplain farmers made them vulnerable to raids from inland groups unable to make adequate provision against periods of winter scarcity, the situation in Amazonia was reversed. Here, it was the floodplain farmers who suffered years of famine, while the terra firme residents with their broader subsistence base and dependence on root crops with year-around productivity were far better insulated against food shortage. Available evidence thus suggests that cultural development on the várzea had reached a climax by the time of European contact and would not have evolved to a higher level of complexity even had it remained undisturbed.

The Post-Contact Period

Shortly after A.D. 1500, the millennia-long isolation of Amazonia was disrupted by a new wave of human immigrants, coming this time from Europe and Africa. Unlike the first invaders, who filtered in over land and were constantly subject to the adaptive pressures of the local environment, the newcomers arrived by sea possessed of an alien culture that had evolved under completely different ecological conditions. This transplanted culture was founded on intensive food production, large population concentration, highly diversified division of labor, private land ownership, and permanency of settlement. Exploitation of the natural resources of the tropical forest was motivated by economic gain, which was controlled by the demands of an extracontinental market rather than by the biological needs of the local population. Accessibility to transportation facilities dictated a pattern of settlement that hugged the river banks and left the vast hinterland vacant or occupied by those Indians who managed to avoid slave raids and foreign disease. Dietary preferences brought from overseas and prejudices against certain local foods favored the introduction of cattle and other domestic animals poorly suited to the region.

As a consequence of these nonadaptive implantations, the mestizo population that emerged during the colonial period was both culturally and physically impoverished by comparison with the precontact inhabitants. Substitution of a monetary economy for the aboriginal kinship-based distribution system, and of craft specialization for sexual division of labor, made many essential items available only by purchase. To acquire clothing, sugar, salt, knives, and other necessities, a man had to obtain rubber, hides, Brazil nuts, or some other commercially desirable product. The unfavorable price structure left little or no time for subsistence activities and the result was an inadequate diet that aggravated the effects of endemic disease to lower the state of health of the population. Short-term profit took precedence over long-term productivity and rising prices resulting from increasing scarcity accelerated the rate of depredation. Culture, which originated as an improved form of behavioral adaptation, has clearly been too successful. Like Frankenstein, natural selection has created a monster that it can no longer control.

Post-contact Amazonia presents a marked contrast to post-Columbian Africa. In the Congo, the introduction of New World cultigens, including staples such as maize and manioc, increased agricultural productivity and stimulated population growth. Amazonia received in exchange a remarkable number of virulent diseases, which decimated the aboriginal inhabitants and facilitated the intrusion of an alien and incompatible culture. Ironically, the survival of the ecosystem into the twentieth century is probably due to the introduction of smallpox, malaria, yellow fever, and other effective natural controls, since they minimized the intensity of settlement and commercial exploitation until neutralized by medical advances after World War II. On the other hand, this situation has fostered the erroneous impression that eradication of disease is all that stands in the way of intensive exploitation of the "natural wealth" of the region, diverting attention from the intrinsic environmental deficiencies of Amazonia as a habitat for man.

References

Carvajal, Gaspar de. 1934. The discovery of the Amazon, according to the accounts of Friar Gaspar de Carvajal and other documents. Compiled by José Toribio Medina, edited by H. C. Heaton. American Geographical Society Special Publication 17. New York.

Frikel, Protásio. 1968. Os Xikrín: Equipamento e técnicas de subsistência. Museu Paraense Emílio Goeldi, Publs. Avulsas 7. Belém.

Guerra, Antonio Teixeira, Editor. 1959. Geografia do Brasil: Grande região norte. Biblioteca Geográfica Brasileira 1(15). Conselho Nacional de Geografia, Rio de Janeiro.

Karsten, Rafael. 1935. The head-hunters of western Amazonas: The life and culture of the Jibaro Indians of eastern Ecuador and Peru. Commentationes Humanarum Litterarum 8(1). Societas Scientarum Fennica, Helsingfors.

Meggers, Betty J. 1971. Amazonia: Man and culture in a counterfeit paradise. Aldine-Atherton. Chicago and New York.

Richards, Paul W. 1952. The tropical rain forest. Cambridge University Press. Cambridge.

Recent Human Activities in the Brazilian Amazon Region and Their Ecological Effects

HARALD SIOLI

Like all other organisms, man, as long as he has lived on earth, has interacted with his environment and will continue to do so. In an ecological sense, "life" can be defined in the same way for man as for all other creatures: "Life" is not only the process of certain physicochemical reactions in the living matter, but it is also the constant interaction between the living organism, which has its internal laws (i.e., physiology, ethology, psychology, etc.), and the environment, which has internal laws according to which it is structured and operates. Between these two systems—organism and environment—there is a tension field that has to be overcome actively and passively by the organism for it to continue its life. That tension field is the stage on which the play of life is conducted by partners with equal rights, namely organism and environment, which influence, alter, and shape each other. Feedbacks are the rule so that any alteration produced in one partner reacts against the other. The result of that interplay is a functional unity on earth, for which we use the new, rather vague word: ecosystem.

The intensiveness of the interactions between organism and environment can be of very different degrees, ranging from almost imperceptible modifications inflicted on one or both partners to a ruthless struggle that may end with a total breakdown and death of one of them, a consequence that will then affect the survivor. Between these two extremes extends a whole scale of degrees of alteration. Their classification becomes a problem if we want to relate them to eventual advantages or disadvantages for the partner who causes them.

This interaction applies to man and his environment. The partner given to man by nature, is, sensu largo, the landscape, or, in modern scientific terms, the geosynergy or biogeocenosis, which he encountered when he made his appearance on the stage. From the beginning, man has dealt with his landscape-environment, has lived from it, has more or less altered it, and sometimes has even died with it. The natural landscapes he first encountered, as well as the altered landscapes produced by his interaction, are all more or less "favorable" or "hostile" for his existence. Thus, if we want to do more than describe the changes man has wrought in certain landscapes, and to evaluate them in regard to human life, we must try to construct a value-scale related to man's material as well as spiritual requirements.

A philosophical or aesthetic value would be too subjective, as well as too transitory temporally and geographically. Individual emotional values—ranging from the enthusiasm for nature to the demand that the "equilibrium of nature" not be disturbed—are equally arbitrary. A natural equilibrium, in the sense of an absolutely ideal "steady state," does not exist. Equilibrium in nature is neither stable nor labile, but indifferent. It resembles the behavior of a ball on the surface of an even table; again

HARALD SIOLI, Max-Planck-Institut für Limnologie, Plön, West Germany.

and again the ball gets pushed from diverse directions, causing it to roll a bit before coming to rest at a new point on the table until the next impulse sends it on again. If too violent a stroke rolls it over the edge of the table, the whole play has come to an end.

Seen from the whole, it does not matter on what part of the table—the world stage—the ball rests. In the life of man, however, the different points are not of equal value, for they represent different portions of his environment, which, as we have seen, may be more or less "favorable" or "hostile." These values must be expressed in any scale we try to develop, and as a guiding line for our attempt, the "quality" of a landscape-as-living-space has been proposed (Sioli 1969b:309). Quality is to be understood more or less as a product (not a sum!) of the variety of the existing life (number of species of plants and animals), of the biomass (standing crop) and number of individual plants and animals, of the productivity (primary, secondary, etc., production) in organic matter and its energy-content, of the number of people who can live on the specific landscape, of their cultural level and content, and of the richness of their personal experiences. These factors cannot be expressed, at least at present, in quantitative terms, but perhaps they can provide a starting point and some tentative directives for evaluation of landscapes as habitats—as ecologists we would say: of biotopes—for mankind.

With these prefatory remarks, let us see what man has done with the landscape of Amazonia, what has been the result of his past activities in different regions, what are the present tendencies, and what are the outlooks and possibilities for the future.

Man's presence in the Amazonian lowlands dates back only a few millennia. Compared with other continents, man here is a newcomer, who probably entered the humid warm hylea in several waves, coming mainly from the northwestern direction and generally following the river courses. He evidently came as a hunter and fisherman and collector of wild plant foods, but most groups ultimately adopted agriculture to some degree. Aboriginal man existed in Amazonia in relatively small numbers, living only in tribal communities, and never building real cities. He collected, hunted, fished, and planted only to satisfy his own needs. Thus, he did not significantly interfere with the structure and dynamics of the natural ecosystems. The waters remained full of fishes, turtles, manatees, and caymans, and in the jungle there was plenty of game, as Carvajal, Rojas, Acuña (1941), and other early discoverers reported. Also, the small plantations, forming isolated spots in the storied forest cover and abandoned after a few years, did not damage the continuity of the great forest. Settlements were mostly located near the banks of the rivers, the preferred places being uninundated margins of the Amazon itself or its great shore lagoons (the várzea lakes), or along the courses of major affluents, such as, Rio Tapajós and Rio Xingú.

Some settlements either remained for considerable periods at the same place or were often reoccupied, to judge from the accumulation of by-products of their "metabolism": ashes, charcoal, bones, etc., along with broken pottery, transforming and overlaying the original yellow-brown forest soil, or sometimes the white sandy soil of "campina" vegetation, with "terra preta" (black earth), which exists in layers of up to 1 meter or more in thickness. In contrast to the original acid "brown loam," which is poor in nutrients for plant growth, or the even more sterile bleached sands, the terras pretas are more or less neutral and enriched in calcium, phosphate, etc. As a result of their renowned fertility, they are often utilized by the neo-Brazilian settlers for their plantations.

The terras pretas, however, are only small spots of local importance; relative to Amazonia as a whole, they represent nothing at all. The pre-Columbian Indians did not change the biogeocenosis, the landscape of their country, which did not offer favorable conditions for development of high cultures and greater population centers. No easily accessible metal ores are found within the hylea and, in most of Amazonia (that portion covered by the Pliocene-Pleistocene "Barreiras" sediments), even rocks and stones are lacking. Timber and palmleaves were the raw material for the construction of shelters, which vanished soon after the departure of the builders. They left no ruins of abandoned cities or other monuments to testify to a life-form that had achieved relative independence from the original conditions of the surroundings. On the contrary, the life-form or cultural pattern of the aboriginal peoples of Amazonia seems to have been shaped thoroughly by the environment, even in its spiritual aspects, which filled both the celestial and underwater kingdoms with replicas of the jungle animals in the form of ghosts.

After the discovery of Amazonia and during the first centuries of Portuguese colonization, i.e., until the second half of last century, little changed in the general situation. The terra firme, which occupied most of the area, remained practically untouched and exploitation activities concentrated more on the waters and on the várzea of the Amazon, as the flood-plain of the great river is called. Let us start with the waters, which have suffered less from the effects of human interaction.

Since pre-European times, the rivers have been the main and until very recent years almost the only traffic routes. In spite of this, all watercourses still flow in their original beds and retain their natural shores. No industry has established itself in the Amazonian lowland, and consequently there is no problem of pollution with industrial wastes. Since no agricultural fields in Amazonia are treated according to modern methods with mineral fertilizers, run-off does not introduce such fertilizer salts into the waters, starting their eutrophication. Pollution by domestic (urban) sewage fortunately does not yet affect the waters to any important degree. In all rivers (with a few exceptions such as the waterfront of Manaus) one still can drink the water without danger of intestinal infections. The watermass of the Amazon system is so enormous, the final discharge into the ocean being an annual average of above 200,000 cubic meters per second, and the human population by comparison is still so sparse, that the self-purification capacity of the waters is more than sufficient to digest completely the end products of human metabolism in a short distance. It seems to me desirable that this situation, which is almost unique in the world, should be maintained, but it is endangered by the fact that many industries are looking for places where clean water is cheaply available and the expense of purification of the discharged waters can be avoided.

A few species of aquatic fauna have been heavily reduced in number by the neo-Brazilian settlers, who, unlike the natives, exploit the resources for exportation in addition to satisfaction of their own needs. Thus, the manatee (*Trichechus inunguis*) has practically vanished not only from the lower Amazon and its shore lakes but also from most other places where formerly it was plentiful. The great river turtle (*Podocnemis expansa*) has virtually disappeared from the same region, along with the smaller "tracajá" (*Podocnemis dumeriliana*). These animals were caught just when they emerged onto the sandy river beaches to lay their eggs, and then sold on the markets of Manaus and smaller cities; the eggs have also been collected by the millions. The great "pirarucú" fish (*Arapaima gigas*), which furnished when dried the main supply of durable protein food in the Amazonian interior, has become so decimated during the past 2-3 decades that its meat is now scarce and too expensive for general consumption. The fish is speared just after moving from the river into the várzea lakes, where it spawns, rears, and protects its young. The cayman or "jacaré" (*Caiman niger*), still abundant throughout Amazonian waters during the first half of this century, has now almost completely been transformed into belts, purses, billfolds, etc., and it is now rare luck to see one specimen even on a long trip through the rivers, lakes, and igapós. For years, the tanneries of Manaus alone have received 5 million or more jacaré skins annually. The disappearance of the jacarés from certain water bodies, however, has not, as one might expect, increased the number of fish; on the contrary, the fish seem to have decreased along with the jacarés. The ecological reason for this curious phenomenon has been analyzed by Fittkau (1970).

Passing now from the aquatic to the terrestrial biotopes, we must first speak about the ecological differences between terra firme and várzea. Terra firme designates terrain with sufficient elevation to escape inundation. It occupies by far the largest area and consists in central Amazonia of deposits laid down in the enormous Pliocene-Pleistocene inland lake that extended from the foot of the Andes to the Atlantic. These deposits, called "series of the Barreiras," are up to 300 meters thick. The soil was derived principally from the Guayana shield in the north and the central Brazilian upland in the south, both ancient regions composed of granitic and gneissic rocks and some sandstones. Both are bedrocks poor in inorganic nutrients for plant growth. The resulting weathering products were intensely leached by equatorial rains and then transported into the lake, where the water acted again on them as they settled on the lake bottom. After the disappearance of the lake, these soils and underlaying material were reexposed to the leaching effects of the heavy rains of the Amazonian climate. The result is extremely poor and acid soils, with almost no reserves of inorganic nutrients for plant growth.

In spite of these edaphic conditions, the terra firme is, with only few local exceptions, generally covered by high, closed forest, which gives the region an aspect of exuberant vitality falsely attributed to extremely fertile soils. The apparent fertility of that landscape, however, represents nutrients accumulated during centuries or millennia in the *living* matter, and their constant and uninterrupted circulation through generations of vegetation with a minimum of loss. The soil serves more as a mechanical substratum for the support of the tall trees and as a reservoir for water than as a supplier of nutrients. This statement is based on recent research on the waters and the soils of Amazonia. Only the western part of the Amazonian lowland, adjacent to the Andean range, has received younger sediments eroded from the Andes and consequently developed richer soils.

The ecological situation on the várzea is different from that of the central and lower Amazonian terra firme. When the glacial period ended, the rise in ocean level drowned the wide river valleys of lower Amazonia and the rivers started to fill with alluvium brought from their headwater zones, forming new flood plains related to the altered river level. The largest area was along the Amazon itself, because of its enormous discharge of turbid white water. The várzea shall be emphasized here since the sedimentation zones of the affluents are relatively unimportant. The Amazon's sediment load does *not* come from the ancient Guayana shield or central Brazil (as did the material composing the terra firme), but by far the greatest amount is derived directly or indirectly from the Andes with their complex lithology and young weathering crust, including abundant volcanic material. Therefore, the sediment particles carried by the Amazon (the same is true for such as the Rio Madeira) are less leached than the material of the terra firme and contain much larger reserves of plant nutrients. Every year during the flood period, the Amazon deposits a new layer of fresh, more or less neutral, and minerally rich sediments on the surface of the flooded stretches of its várzea, thus annually renewing the fertility of those areas. Along the lower Amazon, the várzea is covered by forest only near the banks. The deepest depressions are filled by shallow várzea lakes and the shore-dam-forest is substituted by natural floodable grasslands or the várzea savannas. On the várzea of the upper Amazon, these vast campos are missing and the forest extends into the water of the várzea lakes.

Because of these differences, the effects of agricultural utilization on terra firme and the várzea are different, even when the same methods are employed, namely, cutting and burning the original forest and subsequently planting some crop for a shorter or a longer period. Agriculture was and still is the main occupation of the European colonizers and the subsequent neo-Brazilian settlers—interrupted only temporarily by the "rubber-boom," or locally by gold and diamond rushes. Only very recently has some industry been established in a few areas. Let us start, therefore, with the practices of agriculture in Amazonia and their effects on the ecology.

Naturally, ecological thinking has not been the basis for the expansion of agriculture or for the introduction of European agricultural principles into Amazonia. Some 25 years ago, however, the then Director of Instituto Agronômico do Norte at Belém, Felisberto C. de Camargo, began with his own insight into the ecology of the region and the results of studies by his collaborators to develop general ideas for a future, lasting, and productive utilization of the enormous area. Until after World War II, the "mentality of extractivism" dominated without restrictions the economic thinking in Amazonia. Although from the beginning of the Portuguese colonization and settlement, diverse attempts had been made to establish agricultural zones, until the 1880s aboriginal practices predominated. Wherever it seemed appropriate, small manioc plantations were made and then abandoned after one or two harvests, and a new "roça" or field was prepared by cutting and burning another small piece of jungle, used for the next two to three years, and then abandoned, etc.

This very extensive system of shifting cultivation was, however, well suited for the land and did no harm as long as the human population remained extremely low and the roças constituted only small and widely separated openings in the generally closed forest cover, which healed rapidly after their abandonment. The buffering capacity of the forest ecosystem was not exceeded, and after some 30 to 40

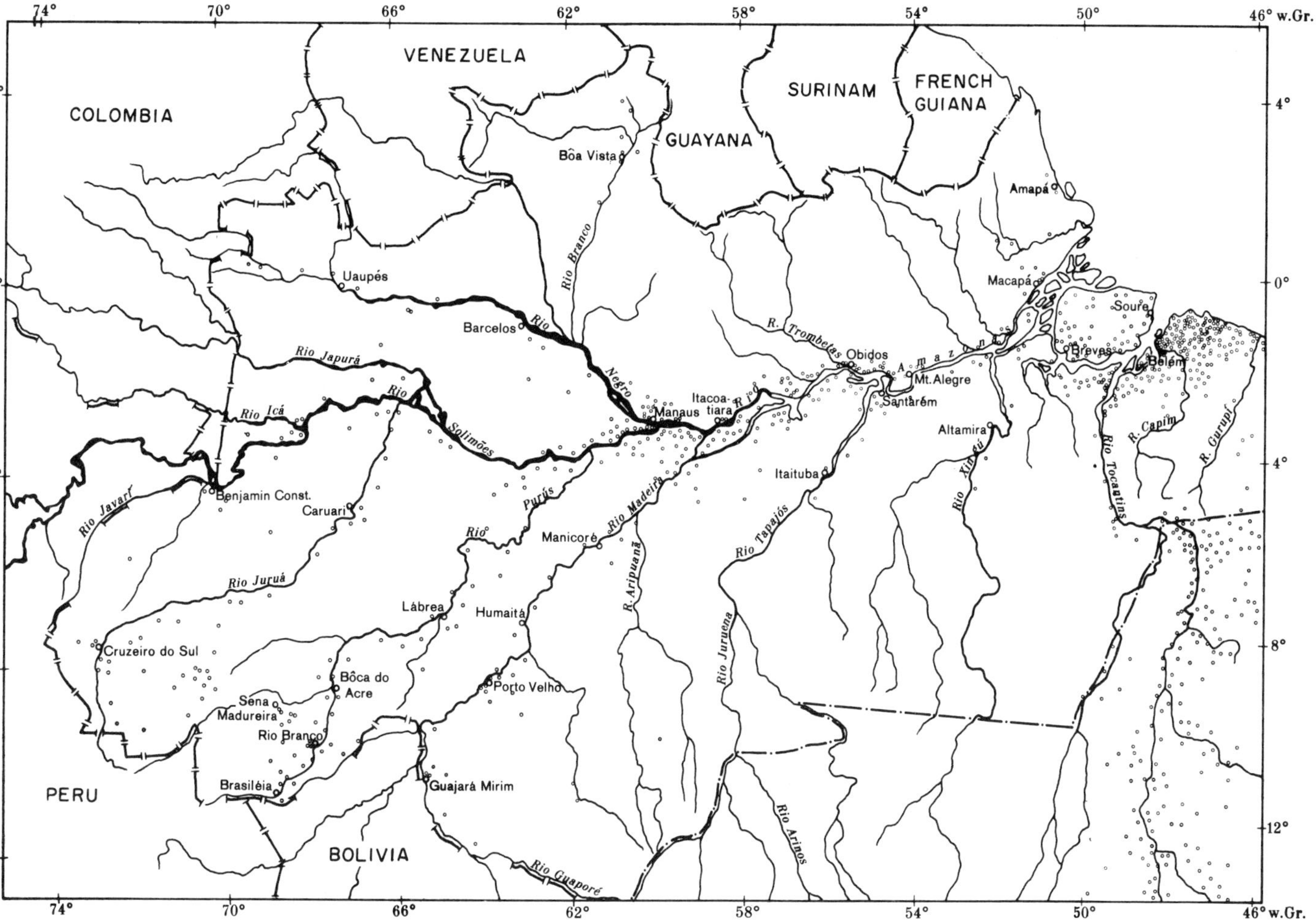

FIGURE 1. Distribution of population in Brazilian Amazonia. One dot equals 2500 inhabitants. [After Sioli 1969 and Atlas Nacional do Brasil 1966.]

years only a botanist could distinguish former roças from the virgin forest by the species composition of the tree vegetation, the general aspect being the same. Even today, nothing can be said against maintenance, for the present, of the traditional system in regions where population density is still extremely low and the crop is only used to supply the local demands. Fortunately, the population density in most parts of Amazonia is in fact one of the lowest on earth: the State of Pará (excluding the capital of Belém) has only about 0.9 inhabitants per square kilometer. The State of Amazonas, omitting the capital Manaus (~200,000 inhabitants), has less than 0.3 inhabitants per square kilometer, the spaces between the river courses being practically empty (Figure 1).

Only once during the colonial period was agriculture attempted on a larger scale on the lower Amazonian terra firme. Between the end of the 17th and the middle of the 18th centuries, the Jesuit missionaries concentrated some 10,000 Indians around Vila Franca on the left bank of lower Rio Tapajós, where they taught them agricultural practices. It is doubtful whether this project would have had lasting success even if it had not been prematurely terminated by the famous edict of the Portuguese Foreign Minister, Marquês de Pombal, which expelled the Jesuits from Brazil in 1756. Today, the region contains many large and small areas in which the forest has been replaced by unproductive, meager savannas and there are many indications that they are of human origin. We now know how poor terra firme soils are in nutrient reserves necessary for plant growth, and in the Tapajós zone that deficiency is aggravated by a relatively dry climate, so that a series of four weeks without a drop of rain is a common annual phenomenon. Once the forest has been cleared on larger stretches, that "drought" period, together with increasingly sandy soil resulting from selective erosion, may be sufficient to disturb the water table to such a degree that young forest trees cannot develop and the terrain is consequently taken over by savanna vegetation similar to that of the Cerrado of Central Brazil.

Other new subsistence practices have been introduced only on the natural várzea grasslands along the lower Amazon. One of them was cattle raising, not only for the use of the local settlers, but also for consumption in newly founded communities like Santarém and Ôbidos. Another was cocoa plantations of larger or smaller extent. These plantations were worked with slaves, and when slavery was abolished in Brazil in the 1880s these plantations were gradually abandoned. Only small "cacauais" can still be found, mainly in the region from just above Ôbidos to below Santarém, and their production is economically insignificant. The majority of them rapidly reverted to forest, so that former human activity has left no definitive impact on the várzea landscape. Extensive cattle-raising, however, has persisted and is still of great economic importance in spite of the difficulties and the heavy losses inflicted on the herds by the annual floods.

During the 1880s, a new period began in the exploitation of Amazonia. First of all, those years marked the beginning of the rubber-boom, of the "golden rubber-time," in which the "mentality of extractivism" celebrated its greatest triumphs. Wild rubber was gathered in the farthest reaches of the interior of the vast country, and the price for raw rubber rose to one pound sterling for a pound (454 g) of rubber, which made other human occupations and activities uninteresting and forgotten. Thousands and thousands of "seringueiros" or rubber-tappers, were imported, mostly from the relatively crowded states of arid northeastern Brazil that suffered from periodic droughts, and sent into the interior of Amazonia to work on enormous estates sometimes extending over several thousand square kilometers of virgin forest, in which the *Hevea* trees grew wild. Nobody thought of establishing artificial plantations of rubber trees, since the general idea was that the world would depend forever on the Amazonian rubber. Where the distance was too remote to warrant installation of a "center" for exploiting the natural rubber trees, expeditions attempted to get as much rubber as possible on one trip. That was not done by tapping the trees, so that they could recover quickly and remain productive for years, but by felling them to extract all the latex they contained. Thus, the forest was impoverished in native *Hevea* trees (as is occurring today in certain regions in regard to rose-wood trees and some good timber). The forest as a whole, however, was fortunately not destroyed.

When the first plantation rubber from Southeast Asia came onto the world market in 1912 and the rubber price dropped, the whole nightmare of expeditions and destruction of rubber trees far in the Amazonian interior came to an end. In that regard, the "golden age of rubber" was a transitional phase for Amazonia. The human tragedy was more enduring. Although the rubber tappers were able with time to adapt their life-style to the pattern of the forest-and-water ecosystem by adopting many former Indian practices, the two big cities, Belém and Manaus, which had grown and lived in luxury during that time, fell into decay.

At the beginning of the rubber boom Belém, the entrance to the Amazon, became a rapidly growing city, and the promise of economic wealth led the governors to view the apparent fertility of the forest country east of Belém as a potential permanent source of supply for that town. A colonization scheme was planned and administered by an official agency and a 300-kilometer long railroad was constructed from Belém to Bragança between 1883 and 1908 in order to facilitate access, to stimulate settlement by immigrants, and to provide rapid and reliable transportation of the products to the consumers. The areas of the "colônias" (i.e., of the settlements) were fixed by the government and in the course of time some ten thousand agricultural settlers—mostly Spaniards, Portuguese from the Azores, and Frenchmen—were brought into the so-called Zona Bragantina (Figure 2). In spite of all these efforts, however, the governmental colônias were not successful and the majority of the settlers wandered off again. Eugênia Gonçalves Egler (1961:533), has written: "The reasons for this fiasco are always seen, by the successive administrators, in

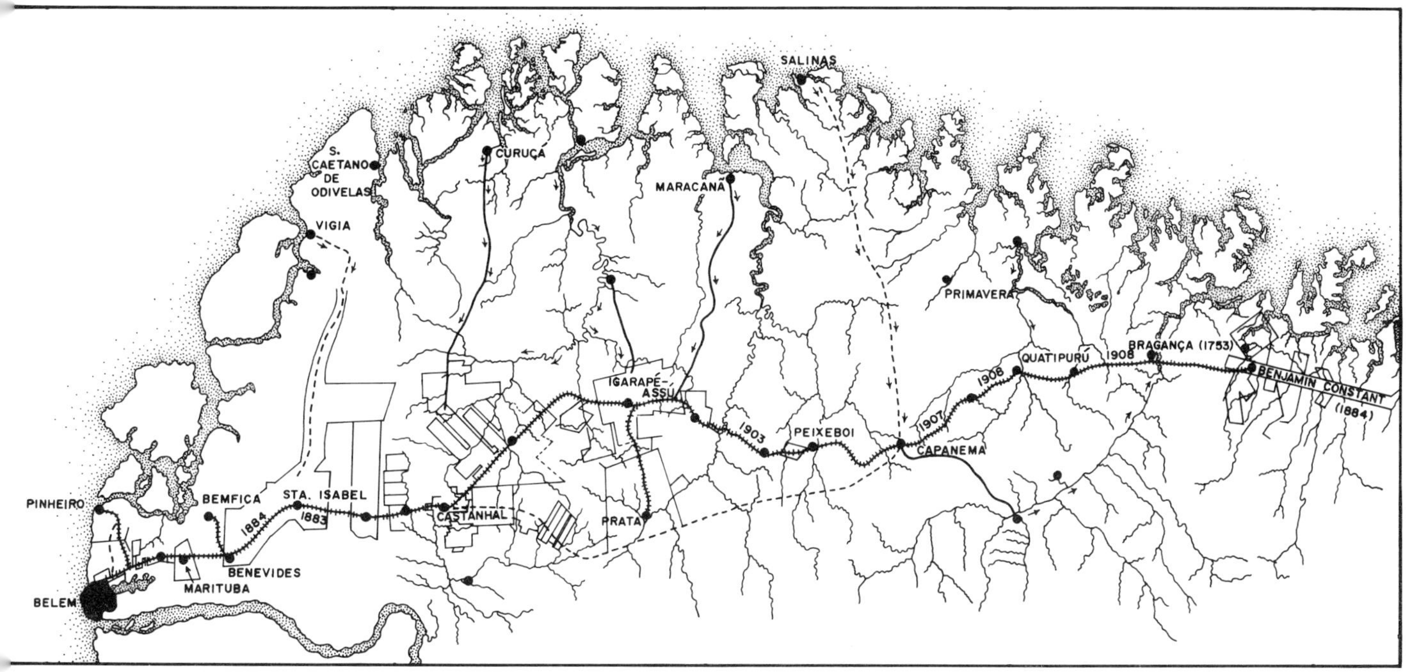

FIGURE 2. Agricultural colonization in the Zona Bragantina. [After Egler 1961.]

a lack of adequate administration of the 'colônias,' in insufficient official support and lack of propaganda in foreign countries for attracting greater numbers of immigrants. Local causes like low fertility of the soils, climatic difficulties or the appearance of pests in the plantations are never mentioned" (translation by H.S.).

Subsequently, even greater numbers of immigrants came on their own initiative, settling near the railroad, as did some 30,000 Cearenses after the great drought in the State of Ceará in 1915. They felled and burned the high forest in order to start plantations, and little by little they occupied the whole zone. But "the work of those people consisted in destroying gold for the production of silver," as a prudent Brazilian, Amaro Theodoro Damasceno, Jr., once expressed it, and Camargo (1948:128) who cites him, agrees: "Indeed, the felling of the forest was destroying gold, and producing manioc-meal, rice and other cereals, was producing silver. The physician Damasceno couldn't have expressed the situation better, but the worst is that the scandalous destruction of those forest riches still continues today" (translation by H.S.).

We must add that now, 22 years after Camargo's sad statement, annihilation of the high forest has not only expanded over practically the whole area of the Zona Bragantina, i.e., about 30,000 square kilometers, but has also extended far beyond its southern border, which had been more or less along the Rio Guamá, where a strip about 20 kilometers wide is being destroyed along the new Belém-Brasília highway.

The method of utilization applied here has been and continues to be the same as elsewhere in the interior of Amazonia; namely, the already mentioned "shifting cultivation," involving cutting and burning an area of jungle, planting generally manioc (in the Zona Bragantina preferably corn, rice, sugarcane, cotton, and tobacco), and abandoning the roça after two or at most (and exceptionally) three harvests. Capoeira, a meagre secondary growth that takes over the abandoned roça, is cut and burned after 8-10 years, but yields only one harvest. After that, further utilization of the same area is generally not attempted in the interior of Amazonia. In the Zona Bragantina, however, it may be repeated about every 10 years.

Since colonization started in that zone, new areas of forest were always treated in the same manner. The ancient method of shifting cultivation had been tolerated by the forest when practiced only in small isolated spots, but here the rapid population growth (8 inhabitants/km^2 by 1950, excluding Belém; Soares 1956:187) brought the roças into increasing proximity and caused complete destruction of the landscape over vast continuous expanses. The jungle receded farther and farther from the traffic routes of the region, first the railway and later on the roads, giving place to a monotonous sequence of capoeiras. Luxuriant high forest was transformed into extensive stretches of stunted scrub and only a few skeletons, now becoming rarer and rarer, of isolated jungle trees still testify to the former exuberant growth in the region. The nutrient storage of the former forest community, as well as the water-holding capacity of the soil have been upset, and local changes of the climate in form of longer droughts have been produced as a result of large bare areas. The final result of this effort at "development" was reached in a relatively short time. The introduction of fiber plants, especially Malva (*Pavonia malacophylla*) and Uacima (*Urena lobata*), brought a short-term recovery in the general decline. But in general these new crops—as well as the use of the capoeiras for charcoal production—serve mainly to complete the process that produced a "ghost-landscape," as Eugenia Egler has called it, in less than 50 years.

Meanwhile, similar "agricultural colonies" were established in other parts of the lower Amazon, for example around Santarém, Monte Alegre, and Alenquer (Figure 3) and expanded toward the interior. Generally, they had no better success than the Zona Bragantina, but fortunately they were of smaller size so that damage to the soil and vegetation has been less extensive.

With regard to other agricultural experiments on the lower Amazonian terra firme, two private attempts to establish agricultural settlements with Japanese peasant immigrants should be mentioned. The first, on the Rio Uaicurapá (south of Parintins) (Figure 3) was dissolved after a few years, since the Japanese settlers despaired at their lack of success and revolted. The other center, Tomé-assú at Rio Acará-pequeno (150 km south of Belém; Figure 3), was initially dedicated to development of a great cocoa plantation which, however, also failed. The colony began to decline, a process that was intensified and accelerated by a severe malaria epidemic that killed many Japanese settlers and caused others either to move to southern Brazil or to go back to Japan. Those who remained shifted to intensive production of vegetables for the Belém market.

The failures of these Japanese companies, however, were compensated and finally outweighed by the ultimate successes attained with new agricultural crops, namely pepper and jute. About 1930, a farsighted Japanese immigrant of the Tomé-assú group had taken black pepper seeds (*Piper nigrum*) with him to Amazonia. During World War II, when the Brazilian government transformed the former Japanese concession of Tomé-assú into a state colony and internment camp, the Japanese settlers started to expand the pepper plantations. By the end of the war, the product, which was a new one for Amazonia, was available to the Brazilian and soon also the world market. Pepper then became the basis of a completely new, intensive utilization of the Amazonian terra firme. Tomé-assú began to flourish, and pepper cultivation soon started to spread to the Zona Bragantina and the vicinity of Manaus.

Another Japanese immigrant had brought some seeds of jute (*Corchorus capsularis*) with him from India. After the decay of the colony at Rio Uaicurapá, he planted the jute near the small town of Parintins on the lower Amazon, but on the várzea rather than on the terra firme. The experiment was so successful that jute cultivation expanded rapidly along the várzea banks of the lower Solimões and the Amazon to below Santarém. Within less than 10 years, jute rose to second place among the local products of economic importance and changed the appearance of long stretches of the banks of the lower Amazon, replacing the original forest with rows of planted jute.

For centuries, settlers had established huts and small plantations of corn, beans, etc., on other parts of the várzea margin. But these alterations of the vegetation cover of the várzea, as well as the yearly burning of the natural grasslands in the dry season for quick production of fresh fodder for the cattle, did not and do not destroy the high productivity of the várzea as a whole. The great river itself is constantly altering the vast flood plain, eroding away one side, depositing fresh soils on the other side, and annually precipitating a new, fertile layer of silt and clay over all areas covered by the turbid white flood water. Thus, the fertility of that zone is periodically renewed.

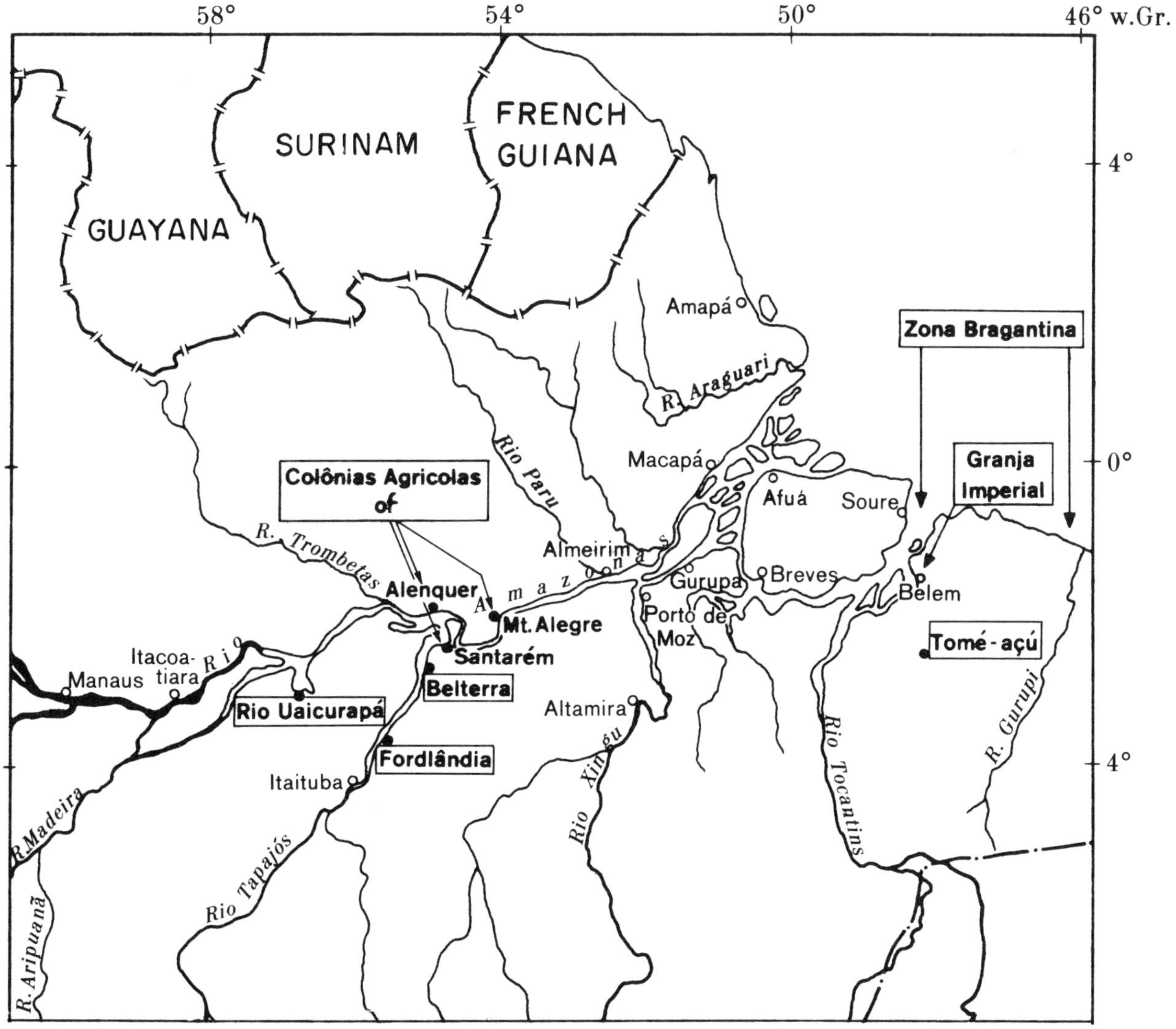

FIGURE 3. Agricultural colonies on the lower Amazon. [After Sioli 1969.]

In order to extend the area of fertile soil farther inland, Felisberto C. de Camargo developed a plan for increasing the amount of turbid, sediment-rich Amazon water that normally enters the várzea lakes either by overflowing the natural levees or through a few, narrow, natural water courses, called furos, which connect the river with the lakes. At the experimental station of Maycurú, a number of artificial channels have been dredged through the levee and now a strong current of Amazon water flows into the lakes, where it stagnates and deposits its sediment load, raising the level of the ground so that ultimately it can be used for plantations or pasture. The fissures that appear in the dry season give an idea of the absorption capacity of that fresh soil and may be taken as an indirect sign of its fertility.

What other effects this filling of parts of the várzea lakes by directed sedimentation may have in the future is still uncertain. The influx of a greater amount of water into the várzea lakes, which some times are of enormous extension (20 x 60 km and more), situated in a zone with extremely low gradient seems to cause a higher rise of the lake level on the downstream shores and to cause flooding of larger areas on that side. Whether the ultimate consequences of this great experiment will be more beneficial or more detrimental for the life of mankind on the várzea cannot yet be predicted. It remains true, however, that up to now human effects on the várzea are slight in comparison with the constant changes, destruction and rebuilding of the whole várzea terrain caused by the activity of the mighty Amazon itself. Only if man should carry out the fantastic project of the construction of a large dam across the lower Amazon, drowning an area of 300-400,000 square kilometers, would his interaction have a lasting and definite effect on the várzea, namely its disappearance. But I hope and I think we all hope and shall do our best to prevent this plan from ever becoming a reality.

There are many other possible human activities that do less harm to the Amazonian landscape (in the sense of the tentative value scale mentioned earlier) and to the productivity of the ecosystem. The basis for every intervention, however, must be an understanding of its ecology. The first ecologically based ideas for utilization of the Amazon region were developed and propagated by Felisberto Cardoso de Camargo some 20 years ago. In those years, the general poverty of most of the Amazonian terra firme soils was documented by many studies, mainly by Camargo and his scientific staff at Instituto Agronômico do Norte. The failures of agricultural efforts had been the first indicator, but soil analyses (Camargo 1948, 1958) and chemical analyses of natural waters (Sioli 1950, 1954, 1957b), both carried out at the Instituto, provided definite proof. The várzea, however, was built up by different matter and is constantly being renewed, as we have seen.

Based on these findings, Camargo made a fundamental distinction between these two major terrestrial biotopes of Amazonia with regard to their prospective utilization. For each of them he worked out and presented a different practical plan for agricultural exploitation. Short-lived agriculture was to be restricted to the várzea, while the terra firme was reserved for long-term plantations and mainly for forestry, which over longer periods of time can produce a profit from the extremely scarce reserves of inorganic nutrients in the soil and out of the rainwater. The scheme designed by Camargo (1948, 1958) (Figure 4) for the region of Belém (Rio Guamá) is valid in principle for all of the Amazonian lowlands. The only difference is that the lower Amazon with its campo-covered várzeas permits extensive animal (cattle) husbandry by creation of artificial reserve pastures on nearby terra firme terrain for use during the flood season.

This scheme is to be considered a preliminary result of the application of landscape ecological thinking on an agricultural utilization of the region, a utilization that does not aim at a short-term exploitation according to the philosophy "après nous le déluge," but at a conservation of the natural habitability of the Amazonian landscapes and their productivity. The scheme has especially great

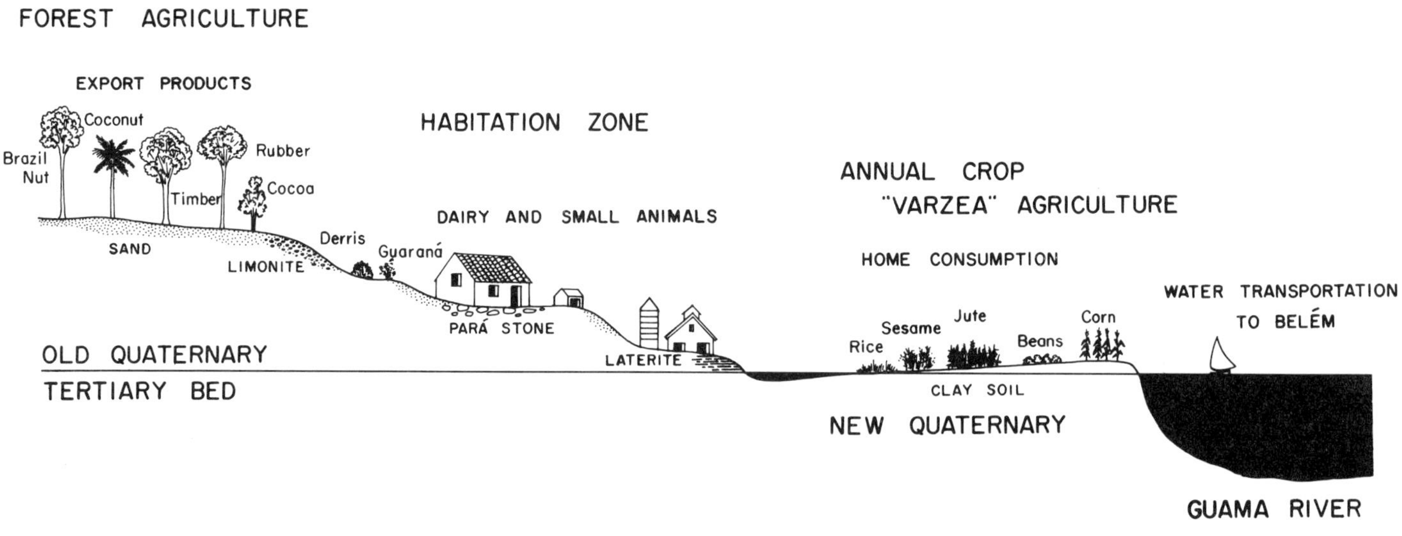

FIGURE 4. Scheme for differential permanent exploitation of terra firme and várzea environments recommended by Felisberto de Camargo. [After Camargo 1958.]

potential for cultivation of the várzeas in the estuary zone of the Amazon in regard to further development of rice culture (Lima 1956). Theoretically, there is the possibility, which Camargo has already envisaged, of surrounding the flat sediment islands by low dikes provided with floodgates. After removal of the várzea forest, vast areas would be available for wet rice; their productivity could satisfy half of the world's needs (Sioli 1966). The appearance of such a landscape would certainly be very similar to that of the Ganges Delta, which after some thousands of years still retains its fertility and productivity, and also a peculiar beauty.

The cultivation of the terra firme, however, is a different problem. Forestry, which would be the most advisable method of utilization for large extensions, is still in its infancy. The first large scale experiment was the famous rubber plantations initiated by the Ford Motor Company in 1926 at Fordlândia and Belterra on the lower Rio Tapajós (Figure 3). The first one, Fordlândia was founded for economic reasons: The price of rubber was controlled by the British and the Dutch, who at that time held a monopoly over plantation rubber, produced on their great estates in Southeast Asia and Indonesia. For the establishment of American-owned rubber plantations, Ford chose almost exactly the same spot from which 50 years before, in 1876, Sir Henry Wickham had secretly exported seeds of *Hevea brasiliensis,* brought them to Kew Gardens, and initiated the story of South Asiatic plantation rubber. Opposite the little village of Boim on Rio Tapajós, Ford secured from the Brazilian government a concession of 10,000 square kilometers, which was named "Fordlândia." In a short time, great areas of virgin forest were felled and burned, hygienic settlements were installed, and around 800,000 *Hevea* trees were planted. The project was envisaged as having a great future and planned for a final population of 100,000. Large amounts of capital were invested, including construction of a great saw mill (at that time the biggest in the whole of South America) making use of the forest trees. At the bank of the Rio Tapajós, a port was built suitable for ocean-going vessels, since the river is navigable during high waters for steamers of up to 10,000 tons.

In spite of these efforts, the ambitious enterprise failed for several reasons. First, the hilly terrain of Fordlândia, selected on the basis of historical rather than scientific criteria, was not particularly suitable for the collection of latex. Second, the usable timber was exhausted in a very short time, not only from the plantation itself, but also from the surrounding neighborhood from which trees could be brought by rafting, and within two years the sawmill had to be shut down. These two failures, however, did not cause abandonment of the Fordlândia project. The disaster began a few years later, around 1932, when the rubber trees were attacked by the "South American leaf disease." The fungus (*Dothidella ulei*) attacked the leaves, so that the trees were heavily damaged and finally died. This disease is unknown in the rubber plantations in the Orient. It occurs only in the American tropics in heavy epidemic form and only in plantations with a great number and high concentration of *Hevea* trees. It is insignificant among the wild rubber trees in the Amazonian jungle, which are a hundred or more meters distant from one another and thus isolated and protected against attack by the fungus spores by intervening trees of other species and genera.

In Fordlândia, however, the disease soon spread over the whole plantation, and could never be eradicated. But Ford did not give up; in 1934 he exchanged one quarter of the concession for another terrain, Belterra, situated 100 kilometers down river on a flat plateau about 150 meters above Tapajós level (165 m above sea level), where he started a new and even larger plantation, which finally reached 2,000,000 trees. The *Dothidella* disease appeared here also, however, and the plantation could only be saved by the trick of double grafting. Young trees were grown from seeds of native *H. brasiliensis,* adapted to the Amazonian soil conditions. After two years, they were cut shortly above the ground and scions of oriental clones of *H. brasiliensis,* selected for high rubber yield, were grafted on them. These Asiatic clones, however, were especially sensitive to *Dothidella,* so that after another two years the crowns had to be cut off and replaced by the crowns of another species of the genus *Hevea,* namely *H. benthamiana* or *H. spruceana* both resistant to *Dothidella.* Since the latex is extracted from the trunk, this did not inhibit production, and the yield of such trees, each composed of three individuals belonging to two different species, was comparable to that of pure *H. brasiliensis* individuals.

A rubber plantation requiring double grafting is not profitable in economic terms, however, and when the need for rubber declined after the end of World War II, Ford handed his plantations over to the Brazilian government, which consigned them to the Instituto Agronômico do Norte. Under the direction of Felisberto de Camargo, the first large-scale experimental plantations of Amazonian timber were initiated at Fordlândia, with somewhat promising results. In addition, large stretches of infected rubber trees were transformed into pasture and breeding of selected races and herds of zebu cattle and of water buffalos was introduced.

In Belterra, experiments on selection and crossing to produce *Hevea* varieties resistant to *Dothidella* continued, as well as the management of the rubber plantation. The latex yield of individual rubber trees was relatively good except for a much lower dry-rubber content; but the growth of the planted forest was noticeably slow. This may reflect the fact that latex contains a great amount of phosphate, which is drained from the trees by the periodic tappings, while the soil is extremely poor in this mineral. Camargo experimented with application of a complete (N-P-K) mineral fertilizer, and that year the rubber forest showed significant improvement. The experiment could not be repeated, however, because of its expense. The decline of the Ford Rubber Plantations on the Rio Tapajós could not be stopped. They remained a subsidized undertaking until they were separated from the Instituto Agronômico do Norte some ten years ago and since then have lost their significance as examples of utilization of the Amazonian terra firme by plantations.

It is interesting that up to now only two smaller scale efforts to exploit the terra firme have been successful in the sense of offering a new stability with economic value for mankind. Both are based on an intuitive or rational insight into the ecological realities and interrelations in the Amazonian landscape. One was a trial conceived and twice executed by Sakae Oti, a former member of the Japanese company that founded the Uaicurapá settlement. It was based on the ancient method of shifting cultivation, but it avoided the "shifting" and envisaged, instead of abandoned areas of capoeiras on impoverished soils, a new stable forest with a constant yield. To achieve this, a modest area (the second experiment comprised about 25 hectares) of virgin forest was cut and burned in the usual way for preparing a roça. The subsequent treatment was different, however, and consisted of simultaneous planting several crops: rice as densely as possible, manioc 1.5 meters apart, guaraná (*Paullinia sorbilis*) 6 meters apart, and Brazil nut trees (*Bertholetia excelsa*) 18 meters apart. The rice grew quickly, fertilized by the ash of the burned jungle, and produced a sufficient harvest within three months to pay the whole expense of preparation of the area. After 1½ years, the manioc was harvested, paying the expense of maintaining the plantation clean of invading capoeira vegetation. The guaraná started to bear fruits in about 6 years, and finally, after about 15 years, the Brazil nut trees began to produce. The result was a new, planted, and productive forest replacing the original jungle, covering and protecting the soil, and yielding annually a good return for the human settler. Another favorable aspect of that small-scale enterprise was its economy. No large capital had to be invested for a long period, all stages paying for themselves by the consecutive harvests within short periods. This system can and should be varied by trying different crops and forest trees. Its main and lasting advantage, however, is maintenance of a stable forest ecosystem.

The second economically sound practice has been elaborated in recent years by a German engineer, Ernesto Rettelbusch, who was a son of an ancient peasant family. The goal was not forestry, but stable agriculture on terra firme terrain, and it was achieved—after years of failure with the usual ancient system—near Marituba, 20 kilometers east of Belém, in the Zona Bragantina (Figure 3). Since the knowledge and the methods developed here may serve as general guidelines for future intensive cultivation of other parts of the Amazonian terra firme, some details are relevant.

The basis for lasting agricultural production is, as is well known, a very simple arithmetical problem. A good harvest depends on several environmental factors, the most important ones being: land, solar energy, warmth, water, and inorganic nutrients for plant growth. Of the latter, what is removed from the soils with the harvests (and by other losses) must be replaced either from reserves remaining in the soils or by artificial introduction. On the Amazonian terra firme, the following are present in sufficient amount: land for the plantations, solar energy and warmth, and usually water. What are lacking are the inorganic nutrients, especially phosphate and potassium, nitrogen perhaps to smaller degree, and sometimes also trace elements (e.g., cobalt). As a result, a reasonable harvest requires supplying artificially those substances necessary for plant growth.

The most simple solution to this problem would theoretically be the application of mineral fertilizers, which naturally must correspond in their composition to the local necessities. Such a procedure, however, represents the transfer to Amazonia of a technique developed in and suited to conditions in the northern temperate climatic zone without taking into consideration the peculiarities of that region, particularly the physical properties of the soils and the quantities of rainfall. The lower Amazonian terra firme soils are not only chemically poor, they also contain almost no colloids for fixation of added fertilizer salts. The mineral fertilizers consequently would be leached out by the frequent heavy rains, and therewith lost to the plants. For the crops to profit, it would be necessary to repeat the fertilization every one or two weeks, a procedure that would be irrational and uneconomical. The agrotechnical problem, therefore, was to find another method of providing a lasting and economical supply of fertilizer.

That problem has been solved by Engineer Ernesto Rettelbusch in principle and in practice by a combination of intensive animal husbandry with intensive agriculture. His "Granja Imperial" (covering 240 hectares, of which at present only about one-third is in use) is a model for further enterprise. On the terrain are some 20,000 pepper plants, up to 20,000 chickens, 500 pigs, and around 200 cattle. Instead of mineral fertilizer, most of the fodder for the animals is imported, mainly from southern Brazil. The animals consume the fodder, utilizing the proteins, fats, and carbohydrates, while most of the minerals, including trace elements and organic colloids are released in the form of manure, which for centuries has been the basis for maintaining the fertility of European fields. When this animal manure is applied to the plantation, it lasts much longer as a supplier of nutrients than mineral fertilizers. Furthermore, the meat, milk, eggs, and other products of the animals pay most of the expenses of the farm, while the sale of pepper provides the profits.

Naturally, the interlocking of animal husbandry and agriculture is much more complicated in its details than can be described here. Also, readjustments to accommodate to the changing local and international market situation will always be necessary, so that no "permanent recipe" for lucrative agriculture in Amazonia can be offered. Such readjustments are, however, only variations on a general theme and do not change the basic principle for a sound utilization of the lower Amazonian terra firme, adapted to local ecological conditions. It takes into account the chemical and physical properties of the soil, as well as the climate. Since it aims at the management of relatively small areas, it represents a renunciation of the old colonial system of increasingly extensive exploitation. Consequently, it reduces the danger of erosion and enhances the possibility of preserving portions of the natural landscape, in form of national parks or something similar, for future generations when an increase in the human population will cause stronger pressure on the land. Finally, it transforms parts of the "wilderness," exuberant with life, into a "cultural landscape," which is equally stable because of constant regulation by man. It also meets our requirements for maximum utilization by offering to a greater number of human beings, not only food but, because of the variety of its internal structure, an enriched quality of life.

Granja Imperial is a model that is being increasingly adopted by other farmers in the formerly devastated Zona Bragantina, with results that have changed the aspect of the countryside. May it be a symbol that the future of humanity does not lie in a destructive battle against nature, but rather in working together with nature toward a new harmonic unity.

References

Bluntschli, Hans. 1921. Die Amazonasniederung als harmonischer Organismus. Geographische Zeitschrift 27(3/4): 49-67.

Camargo, F. C. de. 1948. Terra e colonisação no antigo e novo Quaternário na Zona da Estrada de Ferro de Bragança, Estado do Pará-Brasil. Bol. Mus. Para. Emílio Goeldi 10:123-147.

———. 1958. Report on the Amazon region. Problems of humid tropical regions, 11-24. Humid Tropics Research. Unesco. Paris.

Carvajal, Gaspar de, Alonso de Rojas, and Cristobal de Acuña. 1941. Descobrimentos do Rio das Amazonas. Trans-

lated and annotated by C. de Melo-Leitão. Brasiliana 203, Ser. 2, São Paulo.
Egler, E. G. 1961. A Zona Bragantina no Estado do Pará. Rev. Bras. Geogr. 23:527-555.
Fittkau, E.-J. 1970. Role of caimans in the nutrient regime of mouth-lakes of Amazon affluents (An hypothesis). Biotropica 2:138-142.
Hilbert, Peter Paul. 1968. Archäologische Untersuchungen am mittleren Amazonas. Marburger Studien zur Völkerkunde 1.
Lima, R. R. 1956. A agricultura nas várzeas do estuário do Amazonas. Bol. Tec. Inst. Agron. 33:1-164. Belém.
Meggers, Betty J. 1971. Amazonia: Man and culture in a counterfeit paradise. Aldine-Atherton. Chicago and New York.
Oti, Sakae. 1947. A historia da juta na Amazônia. A Vanguarda 9.7. Belém.
Sioli, Harald. 1950. Das Wasser im Amazonasgebiet. Forschungen und Fortschritte 26:274-280.
————. 1954. Beiträge zur regionalen Limnologie des Amazonasgebietes, II: Der Rio Arapiuns. Arch. Hydrobiol. 49:448-518.
————. 1956. Über Natur und Mensch im brasilianischen Amazonasgebiet. Erdkunde 10:89-109.
————. 1957a. Sedimentation im Amazonasgebiet. Geol. Rdsch. 45:508-633.
————. 1957b. Beiträge zur regionalen Limnologie des Amazonasgebietes, IV: Limnologische Untersuchungen in der Region der Eisenbahnlinie Belém-Bragança ("Zona Bragantina") im Staate Pará, Brasilien. Arch. Hydrobiol. 53:161-222.
————. 1966. Soils in the estuary of the Amazon. Pp. 89-96 *in* Scientific problems of the humid tropical zone deltas and their implications. Humid Tropics Research, Proc. Dacca Symp., Unesco, Paris.
————. 1969a. Zur Ökologie des Amazonas-Gebietes. *In* Biogeography and ecology in South America 1:137-170. Monogr. Biol. 18. W. Junk. The Hague.
————. 1969b. Entwicklung und Aussichten der Landwirtschaft im brasilianischen Amazonasgebiet. Die Erde 100(2-4):307-326.
————. 1969c. Die Biosphäre und der Mensch: Probleme der Umwelt in der heutigen Weltzivilisation. Universitas 24(10):1081-1088.
Sioli, H., and H. Klinge. 1966. Anthropogene Vegetation im brasilianischen Amazonasgebiet. Anthropogene Vegetation. Bericht über das Internationale Symposium in Stolzenau, 1961 der Internationalen Vereinigung für Vegetationskunde, 357-367. Den Haag.
Soares, Lúcio de Castro. 1956. Excursion guidebook No. 8: Amazônia. 18th Internat. Geograph. Congr. Brasil, Internat. Geograph. Union. Rio de Janeiro.
Valverde, Orlando, and Catharina Vergolino Dias. 1967. A rodovia Belém-Brasília: Estudo de geografia regional. Bibliotéca Geográfica Brasileira 22, Sér. A. Rio de Janeiro.

The Congo Basin as a Habitat for Man

MARVIN P. MIRACLE

Few, if any, of the world's regions have undergone such dramatic changes in their food-producing resources in the last four centuries as the Congo Basin, but for reasons not well understood it still supports a far smaller population than similar habitats elsewhere in tropical Africa.

The first section of this paper briefly outlines the physical setting; the second section focuses on the present population and agricultural methods employed; the third section deals with the uneveness of population densities within the basin; and the final section, almost as long as the first three sections combined, is addressed to variations in population and the food producing capacity of the basin over time.

The Physical Setting

The Congo Basin, a 1,425,000 square mile saucer in the middle of the African continent, is covered mainly with humid rain forest or savanna. It is tropical throughout, except around part of its perimeter where the rim of this gigantic bowl at times rises abruptly to elevations of 1500 to 1800 meters—and even higher in one or two relatively short stretches of its eastern border (Figure 1). Toward its center, where it is bisected by the equator, rain falls every month of the year and the "dry season" is limited to two months or less. (For a fuller account of the physical environment see Miracle 1967).

On either side of the equator humid rain forest is found; but as one moves toward either pole rainfall diminishes in quantity and reliability, and is more concentrated in time. Dense rain forest gives way first to open forest, then to vegetation that is dominated by tall grass rather than trees, and finally to a short-grass savanna with occasional bushes and few trees (Figure 2). At the northern edge of the basin the dry season is three to four months long; at the southern edge, more than twice as far from the equator, it lasts from 6 to 8 months. Latitude is, of course, not the only determinant of vegetation. For example, on the relatively high eastern rim of the basin forest gives way to savannas in some areas because of altitude; some of the savanna belt adjacent to the present edge of the rain forest almost certainly has been transformed from forest to grassland because of the activities of man.

This is a favorable habitat in some ways but a hostile one in others. Little clothing or shelter are required, but on the other hand soils lose their fertility readily and rapidly under most cultivation systems that are economic at the present time. And, as in the tropics generally, the environment suits a large variety of pests and diseases that plague man, his crops, and his animals. One of the better known and most economically important in many areas of the Congo Basin is the tsetse fly which, because it is a carrier of trypanosomiasis, precludes livestock other than goats, some sheep, and dwarf cattle unsuited for draft purposes.

Present Population and Agricultural Methods

The Congo Basin is one of the least urbanized areas left in the tropical world—over 90 percent of its population is still rural and the towns and cities now found are a recent phenomena. At the beginning of this century, less than one percent of the population was in urban areas.

MARVIN P. MIRACLE, Department of Agricultural Economics, University of Wisconsin, Madison, Wisconsin 53706.

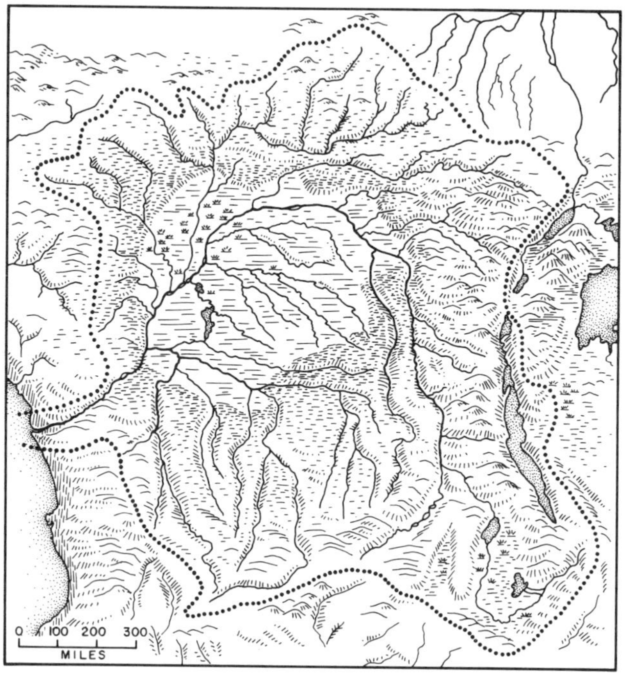

FIGURE 1. Relief of the Congo Basin and limits of the drainage area (dotted line).

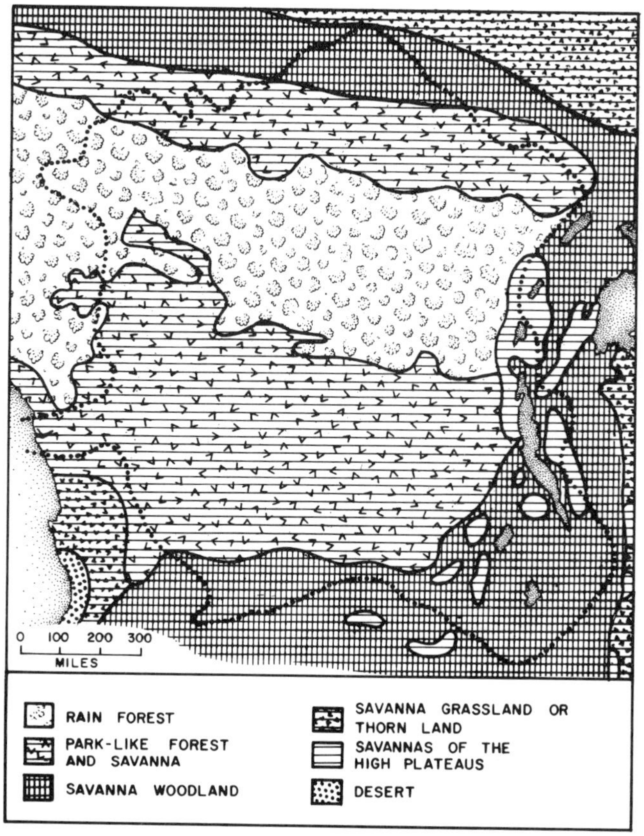

FIGURE 2. Major vegetation zones in the Congo Basin. [Adapted from Delevoy, n.d.]

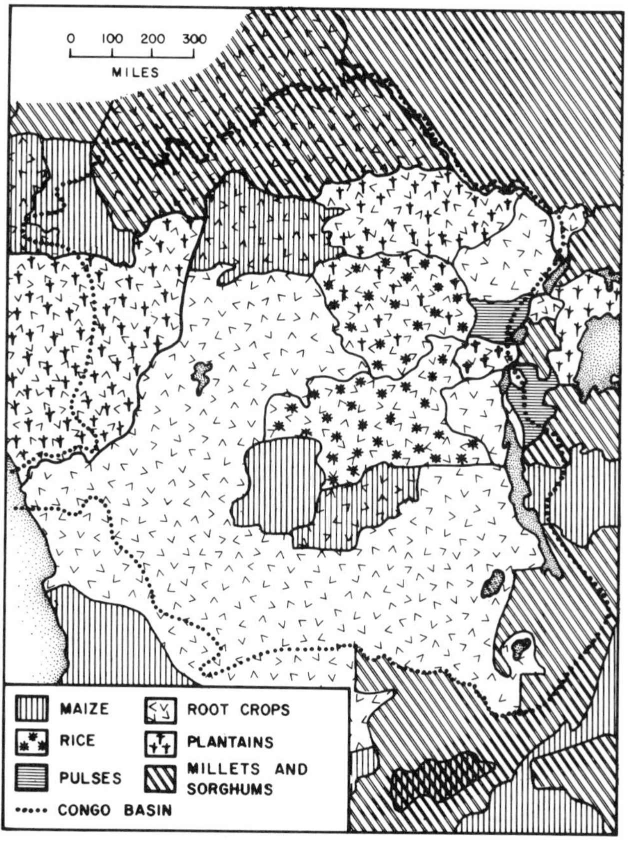

FIGURE 3. Distribution of principal food crops about 1950. [After Johnson 1958.]

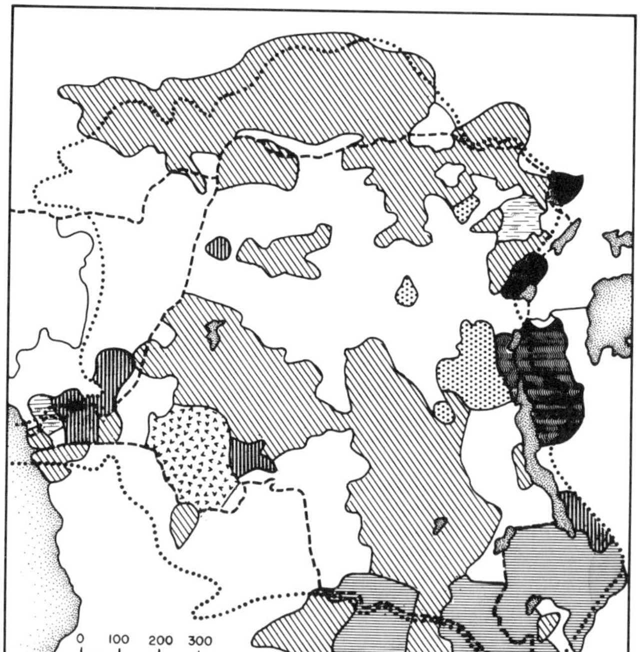

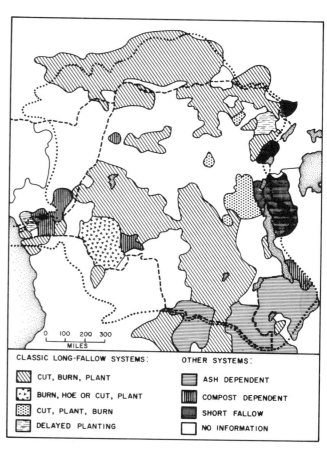

FIGURE 4. Distribution of different systems of shifting cultivation in the Congo Basin.

Population densities are generally low compared with similar environments in some other parts of Africa—typically less than 4 persons per square kilometer—although there are a few pockets (such as along the eastern rim of the basin) where much higher densities are found—over 100 persons per square kilometer.

Agriculture is the primary economic activity throughout most of the Congo Basin, although hunting and gathering continue to be found everywhere, and fishing is the main occupation of a handful of the basin's two hundred-odd tribes.

Although economic activity everywhere centers around production of food and drink, there is great variation in the number of products available and in how they are derived. The most direct method is merely harvesting useful products of the forest or grassland and consuming them raw on the spot. At the opposite extreme is complete reliance on domestic plants and animals, with foodstuffs being subjected to several stages of processing before they are consumed. Neither of these extremes is found in the rural economies of the Congo Basin. Even some Pygmies who rely largely on hunting, fishing, and gathering subject some of their foodstuffs to considerable processing. In the most complex rural economies some food is still obtained by hunting, fishing, and gathering, even though the level of agricultural production may be fairly high relative to needs.

CROPS

Most of the rural economies produce a wide range of commodities. The Azande of the northern Congo Basin were growing at least 52 different crops in 1945 when they were studied by Pierre de Schlippe (1956); the Banda had 60 cultivated plants in 1953; 30 crops were cultivated by the Mandja in 1911, the date of the only detailed study of their agriculture; 16 crops are listed for the Bas-Congo area, near the mouth of the Congo River for 1948; and the same number are reported for the Bemba in the southeastern corner of the basin for the 1930s when Audrey I. Richards (1939:407-408) made an extremely detailed survey of their agriculture (see also Tisserant 1953, Gaud 1911:211, Drachoussoff 1947).

The main activity in terms of labor and land input is production of staple foodstuffs—mainly manioc (or cassava), yams, taro, sweet potatoes, maize, rice, millets and sorghums, and plantains—which according to consumption surveys collectively account for from 72 to 95 percent of total caloric intake (Figure 3).

Unirrigated hoe culture in which staple crops are produced by one of several types of shifting cultivation prevails almost everywhere (Figure 4). In some cases shortage of land requires that fields be fallowed only a year or two before they must be cleared and put into production again. In other areas land may be fallowed twenty years or more. In many areas villages are shifted from time to time when the distance to rested uncleared land comes to be considered excessive by the cultivators.

In addition to main fields three other types of fields are commonly also found: (1) special distant fields established to take advantage of special microenvironments such as pockets of fertile soil or better moisture conditions near a stream or in a depression; (2) plots around old village sites where soils are relatively fertile because of accumulated human and animal wastes; and (3) plots near the homestead which benefit from kitchen refuse, other wastes, and which receive more intensive care.

ANIMAL HUSBANDRY

Relatively few tribes in the Congo Basin keep cattle, except in Rwanda-Burundi, in a zone of high altitude extending along the rim of the basin from Lake Kivu to Lake Albert, and in scattered areas of the west, northwest, and southwest. Goats, sheep, and chickens are found widely, although prior to European rule several tribes kept no goats or sheep. Swine are commonly raised in the southwest portion of the basin, in parts of the southeast, and in some of the north-central and eastern portions of the Congo watershed. Pigeon keeping is also found in much of the south-central portion of the basin.

Data now available from surveys by administrators and others on consumption patterns suggest that the major sources of nonvegetable protein for much of the basin are fish, near rivers and lakes, and game elsewhere, except where cattle are raised. In several areas insects—mainly caterpillars—appear to be the most important source of nonvegetable protein, and, although the data are too rough to make meaningful quantitative comparisons, insects may be third behind fish and game as a source of animal protein for the whole basin (Figure 5).

TRADE

Intertribal and interregional trade, conducted either through large and active marketplaces (some of which are attended by over 1000 people) or through caravans, has existed for as long as we have record; and many commodities move a hundred miles from producer to consumer. The main items of trade are foods and foodstuffs, hoes and knives, salt, cloth, and body ornaments, sold in open-air marketplaces, in stores, or by hawkers that go from village to village. Caravans were important before the turn of the century in the extreme north and southeastern corner of the basin, areas that then did not have marketplaces. Caravans, however, have been discontinued with the development of towns and cities.

ADEQUACY OF FOOD SUPPLIES

In a number of areas of the drier parts of the savanna zone a few weeks of food shortage in the preharvest period may still be experienced some years. At such times there is much more gathering of wild plants than usual. (In most areas there are a number of plants—the so-called famine foods—that are eaten only at such times.)

Apart from such episodes, hunger is rare. Produc-

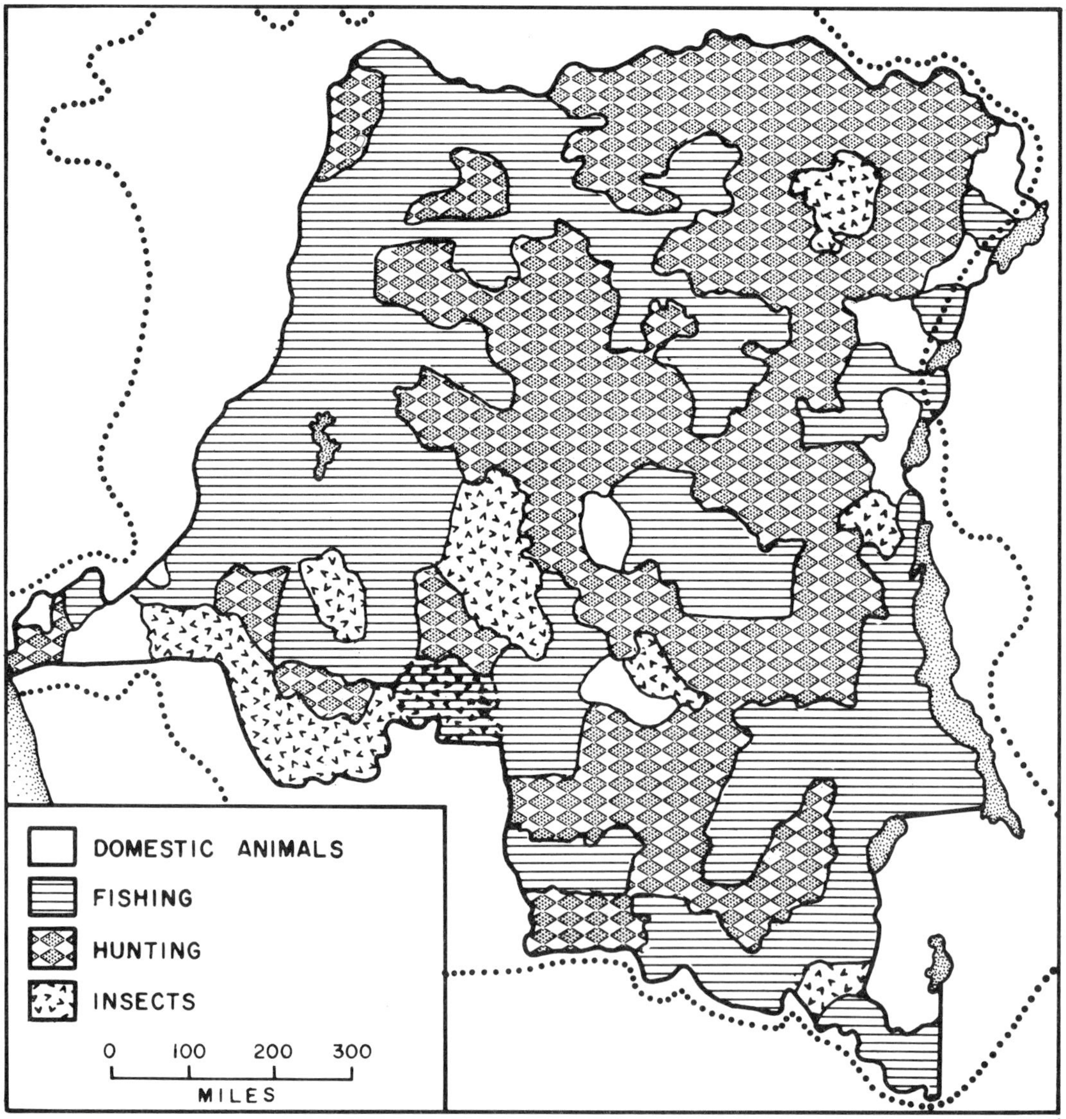

FIGURE 5. Primary sources of animal protein in the Belgian Congo. [Data from Gomez, Halut, and Collin 1961.]

tion techniques now employed may not produce as large a surplus as is desired, but there is no evidence to suggest that the population is generally poorly fed.

It is true that food balance sheets have been published which propose to show that per capita food intake is below standards recommended by some nutritionists, but the reliability of quantitative data on population alone—not to mention that on food production, waste, and storage losses—is so questionable that this sort of data cannot be used to demonstrate anything about hunger.

Food consumption surveys are somewhat more reliable. Researchers live with a sample of families for a few weeks—preferably spaced over time to cover all the seasons—and attempt to measure both amounts served at meals and amounts wasted or fed to animals. In addition, family members are questioned about food eaten away from home. Typically the results of these surveys have shown measured food consumption to be either above recommended levels or within the measurement and sampling margins of error.

Still another type of data is clinical evidence which, again, does not suggest that hunger is widely found. Where populations have been carefully examined by medical teams, nutritional deficiency syndromes typically have been found only in a very small proportion of the population or in age groups that have inferior status and therefore may receive less than the average allocation of food available. One example of the latter is children between two and four or five who may get *kwashiorkor,* a protein deficiency syndrome, because they are too young to fend well for themselves and do not yet have enough status in the family to get their share of prized protein rich foods such as meat and fish.

Variations in Population Density Within the Congo Basin

Except for parts of the eastern rim of the Congo Basin—which are high enough to have subtropical or temperate climates and often have population densities more than 10 times greater than the rest of the Congo Basin—it is not clear what the relationship between physical environment and population density is.

The rain forest is generally very sparsely populated (less than 4 persons per square kilometer). On the other hand, the humid (or tall-grass) savannas, adjacent to the rain forest both in the north and south, tend to have population densities considerable greater (five times larger in several areas) than either the rain forest or the drier savannas on the poleward sides of the humid savannas. Within the drier savanna belt, however, patterns of population densities do not seem to correspond to climatic variations. The southern belt of dry savanna (roughly the southern fifth of the Congo Basin), is sparsely populated throughout (less than 4 persons per square kilometer). Considering climate, however, we would expect considerably denser population in the western segment, an area within a zone with a pattern of rainfall that will allow the growing of two successive staple crops per year, than in the eastern segment, an area where, because of shortage of rainfall, only one staple crop per year can be produced (Bennett 1962:208-213).

Neither the sparse population of the rain forest, nor the differences in population densities within the savannas are well understood. Similar rain forest in Nigeria supports over 25 times as many people per square kilometer, and, in contrast, supports a larger population density than the adjacent humid savanna.

A possible explanation of the relatively denser population in the humid savanna to the south of the rain forest in the Congo Basin is the prevalence of alluvial soils along the Kasai River and its tributaries. But if soil is a major factor, why do we fail to find relatively dense populations on the alluvial soils in the rain forest along the Congo River? Also to be explained is the concentration of population in the humid savanna north of the rain forest, a belt with no large pockets of alluvial soils (Kimble 1960, 1:77).

VARIATIONS IN POPULATION OVER TIME

Trends in the population of the Congo Basin are of interest here insofar as they reflect changes in the character of the basin as a habitat for man. We have good reason to believe that the present human population is the largest the Congo Basin has ever known, and that population growth has been especially rapid since World War I. To what extent does this reflect changes in the environment that have increased its carrying capacity, and to what extent is it a reflection of other factors? And, can this habitat support still further increases in population without major changes in production techniques?

We have no reliable quantitative data on population trends in the Congo Basin. No serious attempt was made to accurately estimate the size of the population until the Congo Free State became a Belgian colony in 1908, and for a variety of reasons, including lack of trained manpower and inadequacies of the transportation network, there is still a large margin of error in official estimates. However, a variety of other evidence suggests that over the last 75 years population has increased rapidly.

Both intertribal warfare and the slave trade were responsible for sizeable annual population losses in the basin during the precolonial period, losses that ceased after the advent of colonial rule. But the coming of colonial rule had perhaps an even greater impact by provision of medical services which have drastically cut the death rate.

Changes in Food-producing Capacity

What we know about changes in the food-producing capacity of the Congo Basin extends back to the early seventeenth century. Part of the evidence—that on the introduction and spread of staple crops—suggests that the food-producing capacity of the basin may have been increased strikingly over the past century; but the data are too crude to make useful quantitative estimates of the magnitude of these increases or to determine the magnitude of such increases relative to population changes over the same period.

During the period between 1830 and 1960 alone, the introduction of 73 crops or domestic animals has been documented for one area of the basin or another—e.g., maize, manioc, Irish potatoes, plantain, breadfruit, oil palms, peanuts, avocado, sesame, sugar cane, cotton, onions, tomatoes, egg plant, oranges, lemons, mangoes, tobacco, horses cattle, and swine—and there almost certainly were a number of other introductions for which there is no written record (Miracle 1967:296-301).

Maize and manioc—perhaps the most important of these introductions—are now dietary staples in many areas and by themselves often account for over 60 percent of total calories ingested. Both appear to be introductions from the New World and are still growing in importance in the diets of some Congo Basin people even though they were first introduced to other areas of the basin as long ago as before the seventeenth century. There is good reason to think that these staples have progressively pushed out lower yielding crops as they have gradually made their way over the basin. Hence their spread probably indicates a substantial growth in the food-producing capacity of the Congo Basin over time.

Maize now clearly gives considerably higher yields in calories per acre than other cereals in much of the humid savannas of the Congo Basin, and manioc generally gives greater yields than any other starchy staple in the forest areas. Both crops may have been brought to the Congo Basin from the New World by the Portuguese before 1600.

Portuguese voyagers reached the mouth of the Congo River in 1482 and established amicable relations with the ruler of the Kingdom of Congo, which extended from the mouth of the Congo River to about the present city of Kinshasa. Within twenty years the Kingdom of Congo was sufficiently influenced by the Portuguese for the Pope to recognize it as a Christian state. In 1521 the son of the king of Congo, who as a boy had been sent to Portugal for education, was made the first bishop of Congo. Thus any superior crops that the Portuguese knew about at this time could easily have been introduced either by the Portuguese or by Africans sent to Portugal during this period. However, despite numerous assertions in the literature that the Portuguese did in fact introduce maize, manioc, and a host of other crops to the African continent from the New World, we have no direct evidence, such as statements by sailors or administrators, to confirm this, although it is admittedly a plausible hypothesis. (There is, however, a statement in the memoirs of Baron Grant, who lived on Mauritius as early as 1740 and spent 20 years there, that Governor de la Bourdonnais, who began his administration in 1735, introduced manioc to Mauritius from the Cape Verde Islands and Brazil; and there is other written evidence that it was introduced in 1738 to Reunion and was carried on a ship named *Le Griffion;* Miracle 1966:97.)

We have some evidence on the early spread of manioc through the Congo Basin. According to the oral tradition of the Kuba, a tribe about 600 miles inland from the mouth of the Congo River, a popular ruler, Shamba Bolongongo, was responsible for the introduction of manioc to the Kuba area sometime between 1600 and 1620. Unlike most events reported in oral traditions, the date of this crop introduction can be checked by the date of a natural phenomenon—an eclipse of the sun referred to in the oral history of the Kuba. The only eclipse visible in the area during the seventeenth and eighteenth centuries was on 30 March 1680, a date that is consistent with dates given for the introduction of manioc in the Kuba area.

A similar tradition among the Bena Kalundwe group of the Luba, located in western Katanga about 400 miles southeast of the Kuba, attributes the introduction of manioc to a famous ruler, Tambo Kanonge, about 1885. We also have it from written records of missionaries that it was not known to the Tabwa, a tribe about 300 miles east of the Bena Kalundwe on the western edge of Lake Tanganyika, until 1893, when it was introduced by the first missionaries in the area. The documents further record that it quickly replaced traditional staples of the Tabwa once it was introduced, a point also made in both of the oral traditions cited (Miracle 1967:245, 274-275).

While there appears to have been a striking increase in the food-producing capacity of the Congo Basin over time through introduction of new crops and livestock—or new varieties or breeds of those already known—there have also been smaller, but notable, reductions in the food-producing capacity of the environment, of which probably the most important has been a drastic reduction in the size of game populations with no corresponding increase in livestock numbers.

Game populations have been severely depleted over the past 50 years by both white and black hunters. Black hunters have become increasingly more efficient because they have been able to obtain more and better firearms. White hunters were rare until the turn of the century but since have shot game for sport, to provide meat for labor gangs (such as those building railroads or working in mines), and to obtain ivory and skins for the international market.

We have no reasonably trustworthy data on trends in livestock numbers—livestock statistics in Africa are even more unreliable than data on human populations—but there is no reason to think that livestock numbers have increased generally. There is evidence that among at least four tribes in the Congo Basin—and probably others as well—cattle were once kept but have not been raised since the end of the precolonial period (Miracle 1967:14). The record is not entirely clear, but the major reason for this seems to be that soldiers or administrators of the Congo Free State confiscated herds of these people and cattle keeping has never been successfully reintroduced.

Apart from changes in the total numbers of game and livestock, the loss of game has without doubt meant a loss in the production potential of the environment in many areas of the basin—in areas where cattle cannot be kept because of the tsetse fly and game is a major source of meat.

Before leaving changes in the food-producing capacity of the environment we must also take account of changes in techniques by which crops and livestock are produced. Thus far we have been considering the resources farmers and herders have had to work with, but spread of superior methods of transforming resources into desired products is also relevant in assessing the size of population a given environment can support.

Except for introduction of simple new tools—e.g., replacement of digging sticks with hoes and introduction of hand-operated grain grinding mills—there have been no striking changes in tools documented for most areas of the Congo Basin. For nearly all farmers adoption of such innovations as the plow, irrigation techniques, machines (other than grinding mills), chemical fertilizers, or pesticides is yet to come. There may have been important innovations in systems of shifting cultivation—e.g., in crop combinations and sequences employed—but we have little evidence on this. The most frequently noted development in this domain is not adoption of superior methods but rather an adjustment of old ones, which suggests that important new techniques have not been adopted. This development is a widely observed tendency to reduce the fallow portion of the shifting cultivation cycle, suggesting that population pressure is now excessive for present production techniques.

Under all of the more than 30 types of shifting cultivation found in the Congo Basin, cultivators clear new land every year or so and at the same time allow a corresponding amount of land to revert to natural vegetation. Farmers have rights to most land only while it is supporting a crop and most fields must be fallowed for relatively long periods—often longer than they are cultivated—to allow nature to restore part or all of the nutrients taken in the cropping portion of the cycle. If land is sufficiently abundant, the fallow period may be 20 years or more, as noted earlier.

If production techniques remain unchanged while population steadily increases, some point will ultimately be reached at which cultivators will no longer find rested land to clear; they will be forced to clear land that is not fully rested. Because it is not fully rested such land will either give lower yields per acre, or will be exhausted sooner, or both. In either case farmers will have to clear larger plots to compensate, or will have to clear partially rested land more frequently, with the result that land cleared in the next cycle will have had still less rest. Thus, unless new land is found (say by migration), new crops or livestock enterprises are introduced, new production techniques are developed, or nonagricultural activities (e.g., trade) are increased, the response of farmers to land shortage leads to cumulative deterioration of land, a trend which might eventually be stopped by insufficient food per capita, followed by an increase in the death rate until balance is restored.

In the long run population control by resort to such measures as sexual abstinence, abortion, or infanticide may sufficiently reduce population pressure to help restore the balance between population and resources. My impression is that such practices are not common among the 200-odd tribes of the Congo Basin. Indeed, examination of value systems of the region suggests that in most areas there is a strong desire to maximize the rate of population growth, as is reflected in barrenness being commonly a cause for divorce and by the practice of wives being inherited by their husbands' male heirs.

In the short run warfare, slave raiding, and migration are more likely to keep population within the limits imposed by resources and technology. In the Congo Basin these in fact seem to have been the main avenues for relieving population pressure in the precolonial period.

Migrations have a prominent place in the oral histories of many Congo Basin tribes, and all accounts suggest that slave raiding was common even before there was an overseas market for slaves. Warfare may have increased with the rising demand for slaves. Apart from warfare connected with the quest for captives, however, periods of prolonged peace appear to have been rare for most basin people, in part, perhaps, because there were few large kingdoms, and those that did rise were relatively short lived (Vansina 1966).

Colonial administrations not only stopped the slave trade and established peace, as has been already noted, but also fixed tribal boundaries, prohibiting migration as a means of relieving population pressure. Thus one of the main impacts of colonial rule was that many—probably most—tribes were forced to consider new measures for maintaining an equilibrium between population and resources, measures such as voluntary population control, adoption of new technology, and migration of able bodied males to other areas where they could be employed as migrant laborers or traders.

Conclusion

Perhaps the most striking feature of the Congo Basin as a habitat for man is the large influx of new crops and domesticated animals in the past three centuries—and particularly the past 90 years—and the impact of colonial rule in forcing new methods of keeping an equilibrium between population, resources, and technology.

Reports that cultivators in several parts of the basin have reduced the length of fallow periods in recent years would seem to suggest that the maximum population density that present agricultural techniques will support has already been reached in many areas. However, doubt about the accuracy of some of these reports is raised by the fact that in parts of the rain forest belt in Nigeria agriculture based on hoe culture and unirrigated shifting cultivation supports at least 25 times as many people per square kilometer as in the Congo rain forest.

Soil surveys are not detailed enough to rule out important differences in soil fertility, but it seems unlikely that the entire difference can be explained by soil. It may be that differences in productivity within shifting cultivation are much larger than heretofore commonly supposed. Careful examination of crop combinations and sequences in the Congo Basin, for example, shows no two tribes with exactly the same system, and there are major variations in techniques of tilling the soil, methods of using composts, wastes, and ashes to enrich fields, and in the thoroughness of weeding. Much of the explanation of the differences in the population density of these two forest zones may be in terms of just such factors. As is shown in the appendix, greater reliance on manioc among people who currently draw 40 percent or more of their calories from other starchy staples, such as yams, rice, or taro, could easily increase total calories available by 117 percent.

It is well possible, it seems to me, that without any striking breakthroughs in control of plant and animal diseases, or adoption of drastically different techniques, the rain forest at least—the core of the Congo Basin—can support a population that is several times larger than at present.

References

Bennett, M. K. 1962. An agroclimatic mapping of Africa. Food Research Institute Studies 3:208-213.

Delevoy, Gaston. n.d. Le Congo forestier. Encyclopédie du Congo Belge, vol. 2. Brussels, Editions Bievels.

de Schlippe, Pierre. 1956. Shifting cultivation in Africa. London.

Drachoussoff, V. 1947. Essai sur l'agriculture indigène au Bas-Congo. Bull. Agric. Congo Belge 37:798-806.

FAO. 1966. Agricultural development in Nigeria 1965-1980. Rome.

Gaud, Fernand. 1911. Les Mandja. Brussels.

Gomez, P. A., R. Halut, and A. Collin. 1961. Production des proteines animales au Congo. Bull. Agric. Congo Belge 52:689-817.

Johnson, B. F. 1958. The staple food economies of western tropical Africa. Stanford Univ. Press. Stanford, California.

Kimble, George H. T. 1960. Tropical Africa. Twentieth Century Fund. New York.

Miracle, Marvin P. 1966. Maize in tropical Africa. Univ. of Wisconsin Press. Madison, Wisconsin.

————. 1967. Agriculture in the Congo Basin: Tradition and change in African rural economies. Univ. of Wisconsin Press. Madison, Wisconsin.

Nigeria. 1965. Rural economic survey of Nigeria, Farm survey 1964-65.

Richards, Audrey I. 1939. Land, labour and diet in Northern Rhodesia: An economic study of the Bemba tribe. London.

Tisserant, C. 1953. L'Agriculture dans les savanes de l'Oubangui. Bull. Inst. Études Centrafricanes, n.s. 6:212-257.

Vansina, Jan M. 1966. Kingdoms of the savanna. Univ. of Wisconsin Press. Madison, Wisconsin.

Appendix: Increases in Production of Calories Within Shifting Cultivation by Greater Reliance on Manioc

One simple, but striking, means of increasing the output of systems of shifting cultivation in the African rain forest is to rely more heavily on manioc, an adjustment which by itself can easily double the caloric output per acre.

The following data are for the rain forest of eastern Nigeria and were collected in order to calculate the maximum food-producing capacity of areas suffering from food shortages associated with the Nigerian civil war. They illustrate a type of shift in cropping patterns that is equally feasible in the Congo Basin.

According to the latest agricultural census in Nigeria (1964-65) yields of manioc in the former eastern region were strikingly greater than for any of the other starchy staples. Manioc produced from 88 to 357 percent more calories per acre than any other starchy staple (Nigeria 1965:9). (Yields, in pounds per acre, obtained in this survey were manioc, 10,961; yams, 6,058; taro, 6,059; maize, 726; and rice, 1,199. In converting these yields into calories per acre we used the following conversion factors (calories per 100 grams): manioc, 109; yams, 90; taro, 86; maize, 360; and rice, 360.)

Crop	*Increase in calories per acre, manioc over other crops (percent)*
Yams	98
Taro	88
Maize	357
Rice	198
Weighted average*	132

* Using weights of yams 6, taro 2, maize 1, rice 1.

The differential is further increased if net yields are compared, i.e., if adjustments are made for the proportion of each harvest that must be saved for seed and the proportion represented by storage losses. Seed requirements for yams and taro are unusually high (commonly as much as one-third in West Africa); seed requirements for maize and rice are considerably less. Manioc, however, is propagated by cuttings and no seeds are required.

The seed saving of manioc over alternative crops is approximately the following:

Crop	*Percent of harvest saved for seed*
Yams	33
Taro	33
Rice	5
Maize	2.5
Weighted average	27.5

Storage losses are not as well known, but are positive and considerable for all other starchy staples, whereas with manioc storage can be avoided. Harvesting of a manioc field commonly is done gradually, beginning six to nine months after planting for varieties commonly grown in eastern Nigeria and continuing several months, depending on the timing of needs. Roots left in the ground continue to grow slowly and may be harvested as long as 24 months after the field was planted.

FAO researchers say that yams in western Nigeria have storage losses of 20 to 25 percent and that similar figures hold for maize and other cereals (FAO 1966:352). I did work on this for maize in 1958 in western Nigeria and obtained a smaller figure for maize, the only crop I investigated. Samples taken from a large number of farm granaries in the forest belt of western Nigeria indicated that maize storage losses were less than 5 percent in most cases, although when maize was stored in urban areas losses were high indeed, 20 percent or more sometimes being lost in a single month (Miracle 1967, ch. 8). (Maize is rarely stored longer in urban areas for this reason.)

The following are our estimates of storage losses at farm level in eastern Nigeria:

Crop	*Storage losses (percent)*
Yams	20
Taro	10
Maize	5
Rice	5
Weighted average	15

Still another gain in calories by increasing the reliance on manioc derives from the fact that in the rain forest manioc can be planted throughout the year, as mentioned earlier, thus land need never be idle, whereas the other starchy staples will grow only seasonally.

To summarize, total probable increases in calories, available by completely replacing all other starchy-staple acreage with manioc, are the following:

Item	*Increases in starchy-staple calories (percent)*
Increased time land is supporting a growing crop (12 months rather than 6)	100
Yield increases (including adjustment for manioc's greater tolerance of poor soils)	150
Elimination of seed	27.5
Elimination of storage losses	15.0
Total percentage	292.5

If nonmanioc starchy-staple calories are now 40 percent of aggregate calories produced by agriculture, this substitution of manioc for other starchy staples increases total available calories by 117 percent. If nonmanioc calories are more than 40 percent of aggregate calories from agriculture, the increase would be more than 117 percent.

Temperate Zone Influence on Tropical Forest Land Use: A Plea for Sanity

F. R. FOSBERG

Before the impact of Western culture, the carrying capacity of the lowland humid tropics for human populations as measured by "standing crop" or sheer numbers varied enormously from place to place, but was generally low. A few areas favored by rich volcanic ash soils, such as Bali, Mexico, and Guatemala, or by annually renewed floodplain soils, as the Nile, Indus, and Ganges valleys, became densely populated and evolved civilizations characterized by advanced social stratification and division of labor, complex technologies, and highly elaborated religions. These civilizations and populations have waxed and waned; some have disappeared entirely and been replaced by others; generally, however, these developments have been restricted to the same favorable regions. Such areas, unfortunately, form only a very small part of the tropical zone.

The vast preponderance of the tropics shows no evidence of ever having had more than an extremely low population density. The carrying capacity of the typical lowland rain forest is, as very well shown by Meggers (1971), very low indeed. She demonstrates that the Amazonian tribes have evolved cultural adaptations which, in pre-European times, restricted the sizes of the populations to the limited capacity of the rain forest environment to support them.

That the same sort of adaptation prevailed in the African equatorial forest is suggested by Turnbull's descriptions (1961, 1965) of the still-existing hunting-and-gathering culture of the Pygmies of the Ituri Forest in the Congo. The situation in Africa is complicated by long-standing pressures from outside peoples. The extent of the "derived savanna" zones along the margins of the rain forest (Keay 1959), generally regarded as forest converted to savanna by human activity, may be evidence of these measures. Where the agricultural practices originated that resulted in this change is not obvious, but they have been very destructive. The rain forest adjacent was, in pre-European times, and still is—to the extent that any rain forest is left—sparsely populated. Its carrying capacity, on a sustained basis, is very low. Mangenot (1955) has noted the appearance, from the air, of small circular areas of different vegetation in the Ivory Coast rain forest that may have been clearings made by small shifting human groups. He reports occasional potsherds, scattered *Elais* trees, and other indications of human presence in the rain forest, seen during ground explorations.

The low potential of the humid tropical forest is no surprise to those who have spent much time in it. However, there are some who are deceived by its incredible luxuriance and even urge plans for its "development" to feed the ever-increasing population of the world.

Several factors combine to bring about the low level of fertility of humid tropical soils. The most important is probably long-continued leaching of mineral nutrients by the high rainfall of typically humid tropical areas. Bases and even silica have been washed out of the upper layers of typical tropical soils and either deposited in the deep subsoil or carried away by the streams. What remains is a

F. R. Fosberg, Department of Botany, National Museum of Natural History, Smithsonian Institution, Washington, D.C. 20560.

mixture of sesquioxides of iron and aluminum, with essentially no plant nutrients. In areas where the parent materials of the soil are predominantly silica, the ultimate soil may be sand or a mixture of sesquioxides and sand, likewise lacking in plant nutrients. There are no clay minerals to provide the colloidal fraction of the soil.

A second major reason for the sterility of tropical soils is the rapid and complete decomposition and disappearance of organic matter that take place at the high temperatures prevailing in these regions. In a typical lowland tropical rain forest soil there is litter on the surface, but no significant dark humus-stained layer or A1 horizon. Organic colloids, as well as mineral colloids, are essentially lacking, and the soil has little capacity to absorb nutrient ions from percolating rainwater that has leached them from living leaves and from litter, excreta, or fertilizers. There is, further, no inherent supply of such ions to be released by further weathering of soil minerals.

The age of many lowland tropical soils enhances the effects of the above factors. There has been ample time for all complex silicates to be thoroughly and completely weathered, so that all nutrient elements have been released and lost.

How then, can such luxuriant vegetation as the tall lowland rain forest flourish on such sterile soils? The full answer to this question cannot yet be documented, but some pieces of the puzzle are in hand and others can be reconstructed, speculatively, but with some confidence.

We know something of the nature of the tightly integrated, almost closed, system of nutrient recycling that enables the forest to continue under present conditions. It has been shown (d'Hoore 1961, and below) in African rain forests that at any one time at least 70 percent of the total mineral nutrient supply in the system occurs within the living biomass of the forest. As soon as any dead plant parts are decomposed, the released nutrients are immediately absorbed by the very efficient surface root network of the trees, possibly aided by mycorrhizae (Went and Stark 1968).

The most difficult question remaining is how such a system ever became established in the first place. When completely destroyed over a significant area, tropical rain forest does not reestablish itself in anything like its original form and composition. It is likely, except in the very wettest areas, to be replaced by savanna. The presumption is that the tropical lowland rain forest is an extremely ancient plant formation. Supporting paleontological evidence is abundant, even back to the Eocene, as shown by such fossil assemblages as the London Clay flora. It is entirely likely that when new substrata became available, with high nutrient levels, high forest established itself, which has persisted during the extended period of leaching and soil degradation leading to the present equilibrium. So long as no catastrophic destruction takes place, the closely integrated nutrient recycling permits the forest to continue. The small amounts of nutrients normally lost may well be made up by air-borne volcanic ash and other dust, as well as nitrogen and sulfur oxides brought down by rain (d'Hoore 1961).

To summarize: The properties of the humid tropical ecosystem that are basic to an understanding of what is now happening are: (1) the rapid leaching out of the soluble bases and other nutrient elements from the soils; (2) the resulting low fertility of the soil proper and the lack of weatherable mineral compounds, e.g., silicates; (3) the concentration of the greater part of the available nutrient supply in the biomass of the tropical forest; (4) the frequent accumulation of nutrients in the lower subsoil zones; (5) the tendency of the sesquioxide mixture dominant in laterized soils to harden to ironstone on exposure to the sun and air; and (6) accelerated erosion that starts immediately when bare soil is exposed to the often torrential tropical rains.

This long introduction, superfluous for readers who have a clear understanding of the humid tropical ecosystem, leads to a consideration of the influence of modern man on this system. Of particular concern is the complex of concepts and attitudes of Temperate Zone man, especially those characteristic of Western European civilization.

Much of Temperate Zone agriculture depends on the continuous exploitation of the upper 20-30 centimeters of the soil mantle. This, in good Temperate Zone soils, has a loose friable consistency, a moderate clay content, and a fairly high humus content. Temperatures in such latitudes permit the accumulation of humus—rates of addition exceed or at least equal rates of decomposition. Populations of earthworms and other soil animals are high and their activities result in a "mull" type of humus, intimately mixed with the mineral components of the soil. Aeration tends to be good and the better soils have a circumneutral reaction, pH 6.5-7.5.

Such a combination of circumstances, with ample but not excessive rainfall, permits the retention in this layer, in available form, of a large reservoir of mineral nutrients, which the shallow-rooted herbaceous temperate crop-plants are admirably adapted to take advantage of. Overexploitation of this nutrient store is easily compensated for by addition of fertilizers and manures, and carefully planned and executed farming can, under ideal circumstances, be carried on indefinitely on the same land. This, of course, presupposes that accelerated erosion is very carefully guarded against.

Although such careful farming is by no means the rule, the penalties of careless farming, or even of deliberate overexploitation or "mining" of the soil's resources, are usually slow in coming. The same lands have been farmed for many hundreds of years in western Europe and Japan, and for almost 200 years in certain parts of the United States.

The coincidence of such an agricultural situation with the development of the aggressive Western European culture was not accidental. The forms of agriculture that developed in temperate areas were enormously successful and produced great surpluses of food and raw materials, so long as other factors such as wars and disease held the populations in check, and, even in some places where they increased rapidly.

Under the influence of the Judaeo-Christian-Moslem philosophy (and independently elsewhere), development of a commercial life-style followed the success of Temperate Zone agriculture, largely to exploit the fruits of the farmers' labors. The bearers of this Western culture have always had an overweening and usually very intolerant confidence in the superiority of their way of life and methods of doing things. It was, and, indeed, still is, practically inconceivable to those possessing this culture that any other pattern of living could be in any way as good as their own. And very early they set out to enlighten the rest of the world and to make it over in the image of Western civilization. Other peoples accepted this conversion or were replaced, or both.

It was not long before this wave reached the tropics. Even though the warm regions were definitely uncomfortable for Europeans and their North American offshoots, the determination that the culture they brought was best for the "natives" resulted in temporary European empires over almost the entire tropical zone. The mores and modus operandi developed in Western Europe, including Temperate Zone agricultural practices and philosophies, were forced on the tropical peoples. The highly visible material benefits of technology brought by "civilized" intruders made this cultural wave irresistible, even after the empires began to break up and the subjugated peoples to regain their political independence.

What this meant, and still means, for tropical agriculture is the substitution of Temperate Zone methods and machinery for the admittedly more primitive, but, in many cases, better adapted indigenous practices. Many existing tropically oriented agricultural methods were, indeed, destructive enough. Others had elements of permanency and inherent stability. Folk wisdom had often developed in harmony with the landscape and this tended to last as long as the populations did not become too large.

Outstanding among the "permanent agricultural systems" were those taking advantage of annual flooding and silt deposit on river flood plains and deltas, marsh taro culture, and wet-land rice cultivation, especially where night-soil is used. The first of these systems is, in essence, a recycling of soil material and fertility lost by erosion from lands around the upper reaches of the river systems. The Nile Valley, until the river was dammed, was an outstanding example, continuously supporting large human populations for 6000 years or more, and permitting the development of several cultural peaks. Of course, the abundance of food produced by this agriculture led to the production of surplus people, who had to be killed off by wars and disease if they were not to starve even in a highly productive system of agriculture.

In the last century or two war and disease, as population regulators, have come to be viewed with repugnance by most exponents of Western thought (at least when the wars are local in origin and not carried on by the dominant powers, themselves). Hence, there has been a general tendency to impose a "Pax Romana" on the weaker tropical peoples. At the same time advances in medicine, public health, and sanitation have been brought to the tropics.

The combined effects have been an alarming increase in numbers of people, almost invariably to higher levels than traditional tropical agricultural methods, and even transplanted temperate agriculture, can support. The rather low carrying-capacity of these tropical habitats has been far exceeded. The response has been transplantation to the tropics of yet another complex of temperate ideas and practices, one that, even in the Temperate Zone, has led to serious deterioration of the quality of the environment and of life. This is generally called "Development." "Development" is one of that group of concepts, words or incantations, including Motherhood, Virginity, and Progress that are axiomatically accepted as good, and seldom challenged. They may once have had a survival value, but that time has long since passed, though the traditional attitudes toward them persist.

"Development" usually implies industrialization, which is the large-scale conversion of raw materials to usable products, usually through assembly-line mass production technologies. Initially, these processes employ large numbers of laborers who are attracted from rural, low-productivity agricultural areas. Concentrations of people built up in industrial areas must be fed. Traditional agriculture cannot support such numbers. Even though their wages would suffice to buy the necessary food, it is often not available, given the low productivity of traditional agriculture and diminishing numbers of farmers left in the countryside.

The obvious answer, in Temperate Zone terms of course, is large-scale, intensive agricultural "Development:" Big machinery, extensive clearing of jungle, heavy fertilization, planting of pioneer-type, sun-requiring crops, clean cultivation, and pesticide application, in a typical Temperate Zone agricultural pattern.

This sort of agricultural development, and even of smaller-scale clean cultivation of herbaceous dryland crops, leads, in nonsandy areas, to the formation of extensive thick ironstone crusts or cuirasses (Aubert 1961), and in sandy areas to unproductive sand barrens. The processes of soil depletion that occur in bare-soil types of cultivation under tropical conditions are mostly irreversible, resulting in degraded landscapes and impoverished and undernourished populations.

The extensive type of cattle-raising that is the dominant form of modern land-use over much of the tropics causes destruction of the forest and exposure of the soil that is even more detrimental, as it covers enormously larger areas in proportion to the human populations benefitted. Overstocking is the rule, resulting in overgrazing and overtrampling. Whether this can be blamed on Temperate Zone man's influence in Africa is not altogether clear, as cattle were brought from South Asia, at least to East Africa, in pre-European times. They may well have spread from there to the Fulani and other nomadic West African peoples. In tropical America, cattle-raising is certainly a Spanish and Portuguese introduction. In both continents it is a very destructive cultural pattern. It is doubtful if anywhere in the humid tropics grazing on a large scale could have evolved as an indigenous cultural pattern, as traits that seriously degrade the habitat are nonadaptive and would be selected against. It is usually when they are transplanted to unsuitable habitats that obvious disequilibria appear and rapid deterioration takes place. It may be that even in India the practice of greatly overstocking with cattle is of recent origin, in terms of the length of cultural history in the subcontinent. Possibly this tendency came in with the Mogul invasion, though cattle obviously have a far longer history in India, judging by the distinctness of the local varieties.

The effects of overgrazing in accelerating erosion, changing the vegetation, and modifying water regimes are well known and obvious. A cattle-degraded landscape in the Magdalena Valley, Colombia, was once aptly described by a visitor as "tierra cansada" (tired land). It is most probable that the vast preponderance of savanna landscapes in reasonably humid regions of the tropics is the result of grazing by free-ranging cattle with the annual burning that always accompanies it. Some fires are doubtless set by lightning, but most are deliberately started by the herders to release nutrients and provide a stimulus to the growth of new grass more palatable to the cattle than the old dry accumulation from the previous growing season. Large areas of West African savannas are characterized by massive ironstone crusts, suggesting that the savannas have replaced humid forests. How much of this antedates the arrival of Europeans is not clear, but cultural traits such as cattle-grazing and cultivation of annual, pioneer-type crop plants may have originated in temperate Eurasia and been carried into Africa by migrating peoples before the spread of

Western European culture. The effects are, at any rate, very widespread and the continent has lost enormously in productive capacity and biotic diversity as a result of these temperate practices.

Despite the almost universal political independence of tropical peoples at the present time, the Temperate Zone cultural conquest seems irresistible and destined to be completed. This is unfortunate, and even more so are the distressingly slow diffusion into the tropics of the recent trends toward environmental enlightenment and the resistance to such trends that is encountered there. Tropical peoples suspect, understandably, that attempts to introduce ideas of conservation and environmental preservation are merely designed to deny them material benefits from rapid exploitation of their resources. Yet, it is distressing to see them repeating the same mistakes that have brought about serious degradation of temperate environments, perpetuating them, in fact, with the increased tempo characteristic of the tropics and augmented by modern technology. Even though there is no quick or easy way to convince the people, the fact remains that environmental losses anywhere are losses to the world as a whole and legitimate cause for concern even outside the tropics. Therefore, it seems not only appropriate but urgent that any possible ecologically preferable alternatives be suggested.

The several forms of more or less permanent agriculture mentioned above (marsh taro culture and wet-land rice culture) where nutrients are caught, accumulated, and recycled in undrained or poorly drained low or terraced areas, and the farming of floodplains where the soil fertility is annually renewed by flood-deposited silt are perhaps the soundest patterns available where the local topography permits. Flood-control engineering is generally antithetic to permanent floodplain agriculture and is another Temperate Zone practice usually unsuitable for tropical regions. The planner from a "developed" country usually finds it impossible to conceive of floods as anything but bad, so flood-control usually accompanies "development" whether or not it is ecologically desirable.

Plantation-type sugarcane culture, where practiced on flat land, can be managed on a reasonably permanent basis, but unless the wastes are returned to the soil it will require heavy fertilization. Soil loss by erosion can be controlled by leveling and diking the land. Even though sugarcane, when well grown, covers the soil rather densely, cane culture on sloping ground usually results in substantial soil loss. This can be seen by the amount of silt carried by streams issuing from cane-planted slopes, as on the Hamakua Coast of Hawaii.

By and large, the least destructive cultures on sloping land in the wet tropics are those based on tree crops. This is true only where clean cultivation is not practiced. Even where tree crowns form a continuous canopy, if bare ground is exposed beneath it soil loss by erosion is almost certain to occur. The success of many forms of tree agriculture derives from the ability of tree roots to reach the deeper soil layers where at least some of the nutrients are deposited after being leached out of the surface layers. A considerable amount of recycling of nutrients can take place even in a tree-based monoculture, such as *Hevea* rubber, coconuts, or shadeless coffee. More effective are combinations, such as the coconut-jakfruit-areca-mango tree-garden culture on the lower interfluves of western Ceylon. In general, where such tree or shrub crops as coffee, tea and cacao are shaded by an open to closed tree stratum, there is less soil deterioration, especially if such leguminous trees as *Erythrina* and *Inga* are used for shade. They have the virtue of adding nitrogen to the soil, as does *Casuarina*.

The humid tropics are almost entirely a naturally forested region. Experience has generally shown that the character of the natural vegetation is a good guide to the general type of cultivated vegetation suitable for an area. Of course this cannot be applied literally, as the natural vegetation is not very likely to produce what is needed. But a cultivated vegetation belonging to essentially the same structural formation is likely to succeed and cause a minimum of harm to the soil. Tree plantations, while not ordinarily identical in structure with tropical rain forest, are near enough to show some of the same environmental characteristics.

The above remarks on suitable agricultural methods for the humid tropics are only very general hints. Environmentally oriented agricultural research in the tropics is quite insufficient so far to allow us to be very sure about anything, except that Temperate Zone methods have proven very destructive when transplanted to the tropics.

In any consideration of what is happening in the tropics, we cannot escape the fact that the natural ecosystems there are being destroyed at an exponentially increasing rate. It is probably no exaggeration to say that 90 percent of the species of plants and animals in the entire world are found between the Tropic of Cancer and the Tropic of Capricorn. A great many of these have not even been discovered, and many more have never been described. Unquestionably, many species existing in the 20th century either have been or will become extinct before they are discovered and studied by man.

Whether one looks at these facts from a purely practical, an intellectual, or an esthetic point of view, the loss of even one species before it is thoroughly known is a tragedy. These organisms cannot be utilized until they are known. Their roles in the ecosystems in which they live are mostly unknown, and our understanding of tropical ecosystems must continue to be deficient if we cannot learn the functions of their living components. Also, although the health and stability of an ecosystem seem dependent on its internal diversity, we are reducing this diversity at an increasing rate. This bodes ill for the future, and not only in the tropics.

Now that we know that what we are doing is wrong, what are the chances that we will change our direction and slow down our destruction of the tropical environment?

Do other forms of life have any "right" to live and share the resources of the earth ecosystem with man? A great many people would concede that at least the "good" or "beneficial" animals and plants should coexist with humans. A few true nature-lovers would even insist that the "bad" species should not be completely eliminated. The number of people who take expensive trips to remote places to see wild animals, redwood trees, and tropical coral reef fish is rapidly increasing, but, of course, only as general prosperity and affluence increase.

What happens as crowding becomes more severe, resources become scarce, and competition builds up is another story. Where poverty is rife and food scarce, anything at all edible tends to be looked upon strictly as food, and used to satisfy immediate need. Keeping any wild animal or plant as breeding stock or as seed for next year's crop is certain to be regarded as foolishness. Someone else might get it. And the harder the pinch of poverty and hunger becomes, the greater the certainty that anything edible that shows itself, no matter how rare, how beautiful, or how interesting it is, will be eaten. All that stands in the way of immediate food production or economic activities will certainly be eliminated.

Such is only the functioning of the normal instinct for self-preservation and it is not to be denied as long as man insists on increasing the numbers of his own species. The time will arrive, granted continuing exponential increase in numbers, when stronger groups of humans themselves will prey upon weaker groups as other food becomes so scarce as to be generally unavailable. This occurs, in effect, even now. It may soon become literal as well as figurative fact.

Thus, in spite of the emergence of a certain amount of nature appreciation and other evidences of civilization among the more moderately populated countries, the greater part of the world is losing its natural biotic features at an appalling rate. Where the population is or has been close to the limit of productivity—actual rather than potential—the mores and habits of the people are such that it is useless to urge preservation of rare species or threatened habitats. It must become the task of authority, and authority can succeed only if, by whatever means, the people can be induced to support it, or at least, to tolerate it.

The prospect is discouraging, indeed. That tropical countries will ever learn from the mistakes of temperate ones is infinitely less likely than that such mistakes will be imitated and repeated on a much more disastrous scale.

References

Aubert, G. 1961. Influence des divers types de végétation sur les caractères et l'évolution des sols en régions équatoriales . . . UNESCO, Tropical soils and vegetation, 41-47, Paris (Abidjan Symposium, 1959).

d'Hoore, J. 1961. Influence de la mise en culture sur l'évolution des sols dans la zone de forêt dense de basse et moyenne altitude. Pp. 49-58, *in* UNESCO, Tropical soils and vegetation. Paris (Abidjan Symposium, 1959).

Keay, R. W. J. 1959. Vegetation map of Africa south of the Tropic of Cancer. Oxford University Press.

Mangenot, G. 1955. Étude sur les forêts des plaines et plateaux de la Côte-d'Ivoire. Études Eburnéenes 4:5-61.

Meggers, B. J. 1971. Amazonia: Man and culture in a counterfeit paradise. Aldine-Atherton. Chicago and New York.

Turnbull, C. 1961. The forest people. Simon and Schuster. New York.

———. 1965. Wayward servants: The two worlds of the African Pygmies. Natural History Press. Garden City, New York.

Went, F. W., and N. Stark. 1968. Mycorrhiza. BioScience 18:1035-1039.